Blood-Brain Barriers

Edited by
Rolf Dermietzel,
David C. Spray,
Maiken Nedergaard

Related Titles

R. A. Meyers

Encyclopedia of Molecular Cell Biology and Molecular Medicine

2nd Edition

2005. ISBN 3-527-30542-4

O. von Bohlen und Halbach, R. Dermietzel

Neurotransmitters and Neuromodulators

Handbook of Receptors and Biological Effects

2nd Edition

2006. ISBN 3-527-31307-9

G. Thiel (Ed.)

Transcription Factors in the Nervous System

Development, Brain Function, and Diseases

2006. ISBN 3-527-31285-4

M. Bähr (Ed.)

Neuroprotection

Models, Mechanisms and Therapies

2004. ISBN 3-527-30816-4

S. Frings, J. Bradley (Eds.)

Transduction Channels in Sensory Cells

2004. ISBN 3-527-30836-9

T. A. Woolsey, J. Hanaway, M. H. Gado

The Brain Atlas

A Visual Guide to the Human Central Nervous System

2002. ISBN 0-471-43058-7

S. H. Koslow, S. Subramaniam (Eds.)

Databasing the Brain: From Data to Knowledge (Neuroinformatics)

2005. ISBN 0-471-30921-4

C. U. M. Smith

Elements of Molecular Neurobiology

2002. ISBN 0-470-84353-5

R. Webster (Ed.)

Neurotransmitters, Drugs and Brain Function

2001. ISBN 0-471-97819-1

Blood-Brain Barriers

From Ontogeny to Artificial Interfaces

Volume 1

Edited by
Rolf Dermietzel, David C. Spray,
Maiken Nedergaard

WILEY-VCH Verlag GmbH & Co. KGaA

Editors:

Prof. Dr. Rolf Dermietzel
Department of Neuroanatomy
and Molecular Brain Research
Ruhr University Bochum
Universitätsstrasse 150
44780 Bochum
Germany

Prof. Dr. David Spray
Department of Neuroscience
Albert Einstein College of Medicine
1410 Pelham Parkway S
Bronx, NY 10464
USA

Prof. Dr. Maiken Nedergaard
School of Medicine and Dentistry
University of Rochester
601 Elmwood Avenue
Rochester, NY 14642
USA

■ All books published by Wiley-VCH are carefully produced. Nevertheless, authors, editors, and publisher do not warrant the information contained in these books, including this book, to be free of errors. Readers are advised to keep in mind that statements, data, illustrations, procedural details or other items may inadvertently be inaccurate.

Library of Congress Card No.: applied for

British Library Cataloguing-in-Publication Data:
A catalogue record for this book is available from the British Library

Bibliographic information published by Die Deutsche Bibliothek
Die Deutsche Bibliothek lists this publication in the Deutsche Nationalbibliografie; detailed bibliographic data is available in the internet at http://dnb.ddb.de

© 2006 WILEY-VCH Verlag GmbH & Co. KGaA, Weinheim, Germany

All rights reserved (including those of translation in other languages). No part of this book may be reproduced in any form – by photoprinting, microfilm, or any other means – nor transmitted or translated into a machine language without written permission from the publishers. Registered names, trademarks, etc. used in this book, even when not specifically marked as such, are not to be considered unprotected by law.

Typesetting K+V Fotosatz GmbH, Beerfelden
Printing betz-druck GmbH, Darmstadt
Binding Litges & Dopf Buchbinderei GmbH, Heppenheim

Printed in the Federal Republic of Germany
Printed on acid-free paper

ISBN-13: 978-3-527-31088-3
ISBN-10: 3-527-31088-6

This handbook is dedicated to Eva

Contents

Preface *XXIII*

List of Contributors *XXV*

VOLUME 1

Introduction
The Blood-Brain Barrier: An Integrated Concept *1*
Rolf Dermietzel, David C. Spray, and Maiken Nedergaard

Part I   Ontogeny of the Blood-Brain Barrier *9*

1   Development of the Blood-Brain Interface *11*
          Britta Engelhardt

1.1	Introduction	*11*
1.2	Pioneering Research on the Blood-Brain Barrier	*11*
1.3	The Mature Blood-Brain Interface	*13*
1.4	Development of the CNS Vasculature	*17*
1.5	Differentiation of the Blood-Brain Barrier	*20*
1.5.1	Permeability	*20*
1.5.2	Transport Systems and Markers	*21*
1.5.3	Extracellular Matrix	*23*
1.5.4	Putative Inductive Mechanisms	*23*
1.6	Maintenance of the Blood-Brain Barrier	*27*
1.7	Outlook	*28*
	References	*29*

2		**Brain Angiogenesis and Barriergenesis** *41* Jeong Ae Park, Yoon Kyung Choi, Sae-Won Kim, and Kyu-Won Kim
	2.1	Introduction *41*
	2.2	Brain Angiogenesis *42*
	2.2.1	Hypoxia-Regulated HIF-1 in the Development of the Brain *42*
	2.2.2	Hypoxia-Inducible Factor *43*
	2.2.3	Hypoxia-Induced VEGF *46*
	2.2.4	Other Neuroglia-Derived Angiogenic Factors *47*
	2.3	Oxygenation in the Brain: Brain Barriergenesis *49*
	2.3.1	Cellular and Molecular Responses Following Brain Oxygenation *49*
	2.3.2	Role of src-Suppressed C Kinase Substrate in the Induction of Barriergenesis *50*
	2.3.3	Barriergenic Factors in Perivascular Astrocytes and Pericytes Following Brain Oxygenation *51*
	2.4	Perspectives *53* *References* *55*
3		**Microvascular Influences on Progenitor Cell Mobilization and Fate in the Adult Brain** *61* Christina Lilliehook and Steven A. Goldman
	3.1	Introduction *61*
	3.2	Angiogenic Foci Persist in the Adult Brain *61*
	3.3	Neurotrophic Cytokines Can Be of Vascular Origin *62*
	3.4	Angiogenesis and Neurogenesis are Linked in the Adult Avian Brain *63*
	3.5	Angiogenesis-Neurogenesis Interactions in the Adult Mammalian Brain *65*
	3.6	Purinergic Signaling to Neural Progenitors Cells: the Gliovascular Unit as a Functional Entity *66*
	3.7	Nitric Oxide is a Local Modulator of Progenitor Cell Mobilization *67*
	3.8	Parenchymal Neural Progenitor Cells May Reside Among Microvascular Pericytes *68*
	3.9	The Role of the Vasculature in Post-Ischemic Mobilization of Progenitor Cells *69* *References* *70*

Part II The Cells of the Blood-Brain Interface *75*

4 The Endothelial Frontier *77*
Hartwig Wolburg

4.1 Introduction *77*
4.2 The Brain Capillary Endothelial Cell *78*
4.3 Endothelial Structures Regulating Transendothelial Permeability *83*
4.3.1 Tight Junctions *83*
4.3.1.1 Morphology of Tight Junctions *83*
4.3.1.2 Molecular Biology of Tight Junctions *84*
4.3.2 Caveolae *89*
4.3.3 Transporters in the Blood-Brain Barrier Endothelium *92*
4.4 Brief Consideration of the Neuroglio-Vascular Complex *94*
4.5 Conclusions *98*
  References *99*

5 Pericytes and Their Contribution to the Blood-Brain Barrier *109*
Markus Ramsauer

5.1 Introduction *109*
5.2 Pericyte Structure and Positioning *110*
5.3 Pericyte Markers *113*
5.4 Pericytes in Culture *113*
5.5 Contractility and Regulation of Blood Flow *115*
5.6 Macrophage Function *116*
5.7 Regulation of Homeostasis and Integrity *118*
5.8 Angiogenesis and Stability *120*
5.8.1 PDGF-B and Pericyte Recruitment *120*
5.8.2 TGF-β1 and Differentiation *121*
5.8.3 Ang-1 and Maturation *122*
5.9 Conclusion *123*
  References *123*

6 Brain Macrophages: Enigmas and Conundrums *129*
Frederic Mercier, Sebastien Mambie, and Glenn I. Hatton

6.1 Introduction *129*
6.2 Different Types and Locations of Brain Macrophages *130*
6.2.1 Macrophage Structure and Ultrastructure *133*
6.2.1.1 Perivascular Macrophages *133*
6.2.1.2 Meningeal Macrophages *138*
6.2.1.3 Dendritic Cells *138*
6.2.1.4 Ventricular Macrophages *139*
6.2.2 Immunotyping by Cell Surface Antigens *139*
6.2.3 Macrophages Contact Basal Laminae *140*

6.2.4	Network of Macrophages Through the Brain	*141*
6.3	Migration of Brain Macrophages	*142*
6.4	Fast Renewal of Brain Macrophages	*142*
6.5	Functions *143*	
6.5.1	Known Functions of Brain Macrophages	*143*
6.5.1.1	Phagocytosis *144*	
6.5.1.2	Immune Function *144*	
6.5.1.3	Production of Growth Factors, Cytokines, and Chemokines	*144*
6.5.1.4	Production and Degradation of the Extracellular Matrix	*145*
6.5.1.5	Repair After Injury *145*	
6.5.2	Potential Functions of Brain Macrophages	*146*
6.5.2.1	Interactions with Meningeal/Vascular Cells, Neurons, and Astrocytes *146*	
6.5.2.2	Do Macrophages Govern the Neural Stem Cell Niche in Adulthood? *147*	
6.5.2.3	Role of Macrophages in CNS Angiogenesis	*151*
6.5.2.4	Role of Macrophages in CNS Plasticity	*152*
6.6	Conclusion: Macrophages as Architects of the CNS Throughout Adulthood *153*	
	References 154	
7	**The Microglial Component** *167*	
	Ingo Bechmann, Angelika Rappert, Josef Priller, and Robert Nitsch	
7.1	Microglia: Intrinsic Immune Sensor Cells of the CNS	*167*
7.1.1	Development *167*	
7.1.2	Microglial Activation *168*	
7.1.3	Antigen Presentation/Cytotoxicity *169*	
7.2	Terminology: Subtypes and Their Location in Regard to Brain Vessels *170*	
7.2.1	Perivascular Macrophages *170*	
7.2.2	Juxtavascular and Other Microglia *170*	
7.3	Turnover of Brain Mononuclear Cells by Precursor Recruitment Across the BBB *173*	
7.3.1	Perivascular Cells *173*	
7.3.2	Microglia *174*	
7.3.3	Turnover Used by "Trojan Horses" *174*	
7.4	Microglial Impact on BBB Function *175*	
7.4.1	Concept of the BBB *175*	
7.4.2	Chemokines – an Overview *176*	
7.4.3	GAG/Duffy *176*	
7.4.4	Chemokine Expression in the CNS *177*	
7.4.5	CCL2 and CCR2 *177*	
7.4.6	CCL3 and CCL5 *178*	
7.4.7	CXCR3 and CXCL10 *178*	

7.4.8	Microglia-Endothelial Cell Dialogue	*179*
7.4.9	Microglial Effects on Tight Junctions	*179*
7.5	Concluding Remarks	*180*
	References 181	

8   The Bipolar Astrocyte: Polarized Features of Astrocytic Glia Underlying Physiology, with Particular Reference to the Blood-Brain Barrier *189*
N. Joan Abbott

8.1	Introduction	*189*
8.2	Formation of the Neural Tube	*189*
8.3	Origin of Neurons and Glia	*190*
8.4	Morphology of Glial Polarity in Adult CNS	*193*
8.5	Astrocyte Spacing and Boundary Layers	*195*
8.6	Origin and Molecular Basis of Cell Polarity	*196*
8.7	Functional Polarity of Astrocytes and Other Ependymoglial Derivatives	*197*
8.8	Secretory Functions of Astrocytes	*199*
8.9	Induction of BBB Properties in Brain Endothelium	*199*
8.10	Astrocyte-Endothelial Signaling	*202*
8.11	Conclusion	*202*
	References 203	

9   Responsive Astrocytic Endfeet: the Role of AQP4 in BBB Development and Functioning *209*
Grazia P. Nicchia, Beatrice Nico, Laura M. A. Camassa, Maria G. Mola, Domenico Ribatti, David C. Spray, Alejandra Bosco, Maria Svelto, and Antonio Frigeri

9.1	Introduction	*209*
9.2	Astrocyte Endfeet and BBB Maintenance	*210*
9.3	Astrocyte Endfeet and BBB Development	*212*
9.4	Astrocyte Endfeet and BBB Damage	*215*
9.5	The Role of Aquaporins in BBB Maintenance and Brain Edema	*216*
9.5.1	AQP Expression and Functional Roles	*216*
9.5.2	The Role of AQP4 in Brain Edema	*220*
9.5.2.1	DMD Animal Models	*220*
9.5.2.2	The α-Syntrophin Null Mice	*223*
9.5.2.3	The Effect of Lipopolysaccharide on AQP4 Expression	*224*
9.5.2.4	AQP4 in Astrocytomas	*224*
9.6	AQP4 Expression in Astrocyte-Endothelial Cocultures	*225*
	References 230	

Part III	Hormonal and Enzymatic Control of Brain Vessels 237
10	**The Role of Fibroblast Growth Factor 2 in the Establishment and Maintenance of the Blood-Brain Barrier** 239
	Bernhard Reuss
10.1	Introduction 239
10.2	Role of FGF-2 in the Regulation of BBB Formation 239
10.2.1	Expression of FGF-2 in Astrocytes and Endothelial Cells of the Rodent Brain 239
10.2.2	Induction of BBB Properties in Endothelial Cells by Soluble Factors 240
10.2.3	Indirect Astrocyte Mediated Effects Seem to Play a Role in FGF-2-Dependent Changes in Endothelial Cell Differentiation 241
10.2.4	Involvement of FGF-2 in the Regulation of BBB Properties in the Pathologically Altered Brain 242
10.3	Future Perspectives 243
	References 244
11	**Cytokines Interact with the Blood-Brain Barrier** 247
	Weihong Pan, Shulin Xiang, Hong Tu, and Abba J. Kastin
11.1	Introduction 247
11.2	Identification of the Phenomena 248
11.2.1	Cytokines That Cross the BBB by Specific Transport Systems 248
11.2.2	Cytokines That Permeate the BBB by Simple Diffusion 249
11.2.3	Cytokines That Have Known Effects on Endothelial Cells 250
11.3	Mechanisms of Cytokine Interactions with the BBB 253
11.3.1	Endocytosis of Cytokines by the Apical Surface of Endothelial Cells 253
11.3.2	Intracellular Trafficking Pathways 253
11.3.3	Signal Transduction in Endothelial Cells 254
11.3.4	Involvement of Other Cells Comprising the BBB 254
11.4	Regulation of the Interactions of Cytokines with the BBB 254
11.5	Stroke and Other Vasculopathy 255
11.6	Neurodegenerative Disorders 256
11.7	Summary 257
	References 258

12 Insulin and the Blood-Brain Barrier 265
William A. Banks and Wee Shiong Lim

12.1 Introduction 265
12.1.1 Early Studies 265
12.1.2 Debates Related to the Question of Permeability of the BBB to Insulin 269
12.1.3 Does Insulin Cross the BBB? The Middle Years 270
12.1.4 Insulin, the BBB, and Pathophysiology: The Past Decade 272
12.2 Pathophysiology of Insulin Transport 274
12.2.1 Insulin, Obesity, and Diabetes 274
12.2.2 Insulin Resistance and Inflammatory States 275
12.2.3 Insulin and Alzheimer's Disease 277
References 280

13 Glucocorticoid Hormones and Estrogens: Their Interaction with the Endothelial Cells of the Blood-Brain Barrier 287
Jean-Bernard Dietrich

13.1 Introduction 287
13.2 Glucocorticoids and the Endothelial Cells of the BBB 288
13.2.1 Mechanisms of Action of Glucocorticoids 288
13.2.2 Glucocorticoids and Inflammation 289
13.2.3 Regulation of Adhesion Molecules Expression by GC in Endothelial Cells 290
13.2.4 Effects of GC on Leukocyte-Endothelial Cell Interactions 293
13.2.5 Glucocorticoids, Cerebral Endothelium and Multiple Sclerosis 294
13.3 Estrogens and the Endothelial Cells of the BBB 297
13.3.1 Mechanisms of Action of Estrogens 297
13.3.2 Endothelial Cells as Targets of Estrogens 298
13.3.3 Adhesion Molecules are Regulated by Estrogens in Endothelial Cells 298
13.3.4 Estrogens and Experimental Autoimmune Encephalomyelitis 299
13.4 Conclusions and Perspectives 301
References 302

14 Metalloproteinases and the Brain Microvasculature 313
Dorothee Krause and Christina Lohmann

14.1 Introduction 313
14.2 Metalloproteinases in Brain Microvessels: Types and Functions 314
14.3 Cerebral Endothelial Cells and Metalloproteinases 318
14.4 Perivascular Cells and Metalloproteinases 321
14.4.1 Pericytes 321
14.4.2 Astrocytes and Microglia 322
14.5 Metalloproteinases and the Blood-Liquor Barrier 324

14.6	Metalloproteinases and Brain Diseases	*325*
14.6.1	Metalloproteinases and Cerebral Ischemia	*326*
14.6.2	Metalloproteinases and Brain Tumors	*327*
14.6.3	Metalloproteinases and Multiple Sclerosis	*327*
14.6.4	MMPs and Migraine	*328*
14.7	Conclusion	*328*
	References	*328*

Part IV   Culturing the Blood-Brain Barrier *335*

15   Modeling the Blood-Brain Barrier *337*
Roméo Cecchelli, Caroline Coisne, Lucie Dehouck, Florence Miller, Marie-Pierre Dehouck, Valérie Buée-Scherrer, and Bénédicte Dehouck

15.1	Introduction	*337*
15.1.1	In Vitro BBB Model Interests	*337*
15.1.2	The BBB: Brain Capillary Endothelial Cells and Brain Parenchyma Cells	*338*
15.2	Culturing Brain Capillary Endothelial Cells	*338*
15.2.1	Brain Capillary Endothelial Cell Isolation	*338*
15.2.1.1	Brain Capillary Endothelial Cell Isolation	*338*
15.2.1.2	Endothelial Cell Culture From Capillaries	*340*
15.2.1.3	Primary Endothelial Cells and Subculture of Brain Capillary Endothelial Cells	*341*
15.2.1.4	Immortalization	*342*
15.2.1.5	Purity	*344*
15.2.1.6	Species	*346*
15.2.2	Coculture	*347*
15.3	Characteristics Required for a Useful In Vitro BBB Model	*347*
15.3.1	Confluent Monolayer	*347*
15.3.2	Tight Junctions and Paracellular Permeability	*349*
15.3.3	Transcellular Transport, Receptor Mediated Transport	*350*
15.3.4	Expression of Endothelial Adhesion Molecules/Vascular Inflammatory Markers	*351*
15.4	Conclusion	*352*
	References	*352*

16   Induction of Blood-Brain Barrier Properties in Cultured Endothelial Cells *357*
Alla Zozulya, Christian Weidenfeller, and Hans-Joachim Galla

16.1	Introduction	*357*
16.2	In Vitro BBB Models	*359*
16.3	Hydrocortisone Reinforces the Barrier Properties of Primary Cultured Cerebral Endothelial Cells	*360*

16.3.1	In Vitro Model Based on Pig Brain Capillary Endothelial Cells (PBCEC) *360*	
16.3.2	In Vitro Model Based on Mouse Brain Capillary Endothelial Cells (MBCEC) *361*	
16.4	The Involvement of Serum Effects *362*	
16.5	Hydrocortisone Improves the Culture Substrate by Suppressing the Expression of Matrix Metalloproteinases In Vitro *363*	
16.5.1	ECIS Analysis of Improved Endothelial Cell-Cell and Cell-Substrate Contacts in HC-Supplemented Medium *363*	
16.5.2	Low Degradation of ECM in HC-Supplemented Medium Leads to Improved Cell-Substrate Contacts of Cerebral Endothelial Cells *365*	
16.6	The Role of Endogenously Derived ECM for the BBB Properties of Cerebral Endothelial Cells In Vitro *368*	
16.7	Conclusions *370*	
	References *371*	

17 Artificial Blood-Brain Barriers *375*
Luca Cucullo, Emily Oby, Kerri Hallene, Barbara Aumayr, Ed Rapp, and Damir Janigro

17.1	Introduction: The Blood-Brain Barrier *375*	
17.2	Requirements for a Good BBB Model *378*	
17.3	Immobilized Artificial Membranes *379*	
17.4	Cell Culture-Based in vitro BBB Models *379*	
17.4.1	Cell Lines *380*	
17.4.1.1	Immortalized Rat Brain Endothelial Cells *380*	
17.4.1.2	Other Cells From Non-Cerebral Sources *380*	
17.4.1.3	Brain Capillary Endothelial Cell Cultures *381*	
17.4.2	Monoculture-Based in vitro BBB Models *382*	
17.4.3	Coculture-Based in vitro BBB Models *385*	
17.5	Shear Stress and Cell Differentiation *386*	
17.6	Flow-Based in vitro BBB Systems *387*	
17.6.1	Dynamic in vitro BBB: Standard Model *387*	
17.6.2	Dynamic in vitro BBB: New Model *390*	
17.7	A Look Into The Future: Automated Flow Based in vitro BBBs *392*	
17.8	Conclusion *392*	
	References *394*	

18 In Silico Prediction Models for Blood-Brain Barrier Permeation *403*
Gerhard F. Ecker and Christian R. Noe

18.1	Introduction: The In Silico World *403*	
18.2	The Blood-Brain Barrier *404*	
18.3	Data Sets Available *405*	
18.4	Computational Models *410*	
18.5	Passive Diffusion *410*	

18.5.1	Regression Models 410
18.5.2	Classification Systems 419
18.6	Field-Based Methods 421
18.7	Active Transport 423
18.8	Conclusions and Future Directions 426
	References 426

VOLUME 2

Part V Drug Delivery to the Brain 429

19 The Blood-Brain Barrier:
Roles of the Multidrug Resistance Transporter P-Glycoprotein 131
Sandra Turcotte, Michel Demeule, Anthony Régina, Chantal Fournier, Julie Jodoin, Albert Moghrabi, and Richard Béliveau

19.1	Introduction 431
19.2	The Multidrug Transporter P-Glycoprotein 432
19.2.1	P-gp Isoforms 432
19.2.2	Structure 433
19.2.3	P-gp Substrates 435
19.3	Localization and Transport Activity of P-gp in the CNS 438
19.3.1	Normal Brain 438
19.3.2	Brain Diseases 440
19.3.2.1	Malignant Brain Tumors 440
19.3.2.2	Brain Metastases 441
19.3.3	Expression of Other ABC Transporters at the BBB 442
19.3.4	Subcellular Localization of P-gp 442
19.4	Polymorphisms of P-gp 444
19.4.1	MDR1 Polymorphisms at the BBB 444
19.4.2	MDR1 Polymorphism and Brain Pathologies 445
19.5	Role of P-gp at the BBB 446
19.5.1	Protection Against Xenobiotics 448
19.5.2	Secretion of Endogenous Brain Substrates and Endothelial Secretion 448
19.5.3	Caveolar Trafficking 449
19.6	Conclusions 450
	References 451

20 Targeting of Neuropharmaceuticals by Chemical Delivery Systems 463
Nicholas Bodor and Peter Buchwald

20.1	Introduction 463
20.2	The Blood-Brain Barrier 464

20.2.1	Structural Aspects *464*	
20.2.2	Enzymatic and Transporter-Related Aspects *466*	
20.3	Brain-Targeted Drug Delivery *467*	
20.3.1	Lipophilicity and Its Role in CNS Entry *468*	
20.3.2	Quantifying Brain-Targeting: Site-Targeting Index and Targeting Enhancement Factors *470*	
20.4	Chemical Delivery Systems *472*	
20.5	Brain-Targeting CDSs *473*	
20.5.1	Design Principles *473*	
20.5.2	Zidovudine-CDS *475*	
20.5.3	Ganciclovir-CDS *476*	
20.5.4	Benzylpenicillin-CDS *478*	
20.5.5	Estradiol-CDS *479*	
20.5.6	Cyclodextrin Complexes *484*	
20.6	Molecular Packaging *484*	
20.6.1	Leu-Enkephalin Analogs *485*	
20.6.2	TRH Analogs *486*	
20.6.3	Kyotorphin Analogs *487*	
20.6.4	Brain-Targeted Redox Analogs *489*	
	References 490	

21   **Drug Delivery to the Brain by Internalizing Receptors at the Blood-Brain Barrier** *501*
Pieter J. Gaillard, Corine C. Visser, and Albertus (Bert) G. de Boer

21.1	Introduction *501*
21.2	Blood-Brain Barrier Transport Opportunities *502*
21.3	Drug Delivery and Targeting Strategies to the Brain *504*
21.4	Receptor-Mediated Drug Delivery to the Brain *506*
21.5	Transferrin Receptor *506*
21.6	Insulin Receptor *508*
21.7	LRP1 and LRP2 Receptors *509*
21.8	Diphtheria Toxin Receptor *511*
21.9	Conclusions *512*
	References 514

Part VI   **Vascular Perfusion** *521*

22   **Blood-Brain Transfer and Metabolism of Oxygen** *523*
Albert Gjedde

22.1	Introduction *523*
22.2	Blood-Brain Transfer of Oxygen *525*
22.2.1	Capillary Model of Oxygen Transfer *525*
22.2.2	Compartment Model of Oxygen Transfer *528*
22.3	Oxygen in Brain Tissue *529*

22.3.1	Cytochrome Oxidation 529
22.3.2	Mitochondrial Oxygen Tension 531
22.3.2.1	Distributed Model of Tissue and Mitochondrial Oxygen 531
22.3.2.2	Compartment Models of Tissue and Mitochondrial Oxygen 533
22.4	Flow-Metabolism Coupling of Oxygen 535
22.5	Limits to Oxygen Supply 538
22.5.1	Distributed Model of Insufficient Oxygen Delivery 538
22.5.2	Compartment Model of Insufficient Oxygen Delivery 541
22.6	Experimental Results 542
22.6.1	Brain Tissue and Mitochondrial Oxygen Tensions 542
22.6.2	Flow-Metabolism Coupling 543
22.6.3	Ischemic Limits of Oxygen Diffusibility 546
	References 547

23 Functional Brain Imaging 551
Gerald A. Dienel

23.1	Molecular Imaging of Biological Processes in Living Brain 551
23.1.1	Introduction 551
23.1.2	Molecular Imaging 551
23.1.3	Influence of Blood-Brain Interface on Functional Imaging 552
23.2	Overview of Brain Imaging Methodologies 554
23.2.1	Computed Tomography 556
23.2.2	Magnetic Resonance Imaging 556
23.2.3	Functional MRI 557
23.2.4	Radionuclide Imaging 560
23.2.5	Optical Imaging 562
23.2.6	Thermal and Optico-Acoustic Imaging 564
23.2.7	Summary 565
23.3	Imaging Biological Processes in Living Brain: Watching and Measuring Brain Work 565
23.3.1	Functional Activity, Brain Work, and Metabolic Imaging 565
23.3.2	Quantitative Measurement of Regional Blood Flow and Metabolism in Living Brain 567
23.3.2.1	Assays at the Blood-Brain Interface: Global Methods 567
23.3.2.2	Highly Diffusible Tracers to Measure CBF 568
23.3.2.3	Metabolizable Glucose Analogs to Measure Hexokinase Activity and CMR_{glc} 569
23.3.2.4	Non-Metabolizable Analogs to Assay Transport and Tissue Concentration 572
23.3.2.5	Cellular Basis of Glucose Utilization 572
23.3.2.6	Acetate is an "Astrocyte Reporter Molecule" 573
23.3.2.7	Summary 573
23.4	Molecular Probes are Used for a Broad Spectrum of Imaging Assays in Living Brain 574

23.4.1	Potassium Uptake and CMR_{glc} During Functional Activation	*574*
23.4.2	Multimodal Assays in Serial Sections of Brain	*574*
23.4.3	Imaging Human Brain Tumors	*576*
23.4.4	Functional Imaging Studies of Sensory and Cognitive Activity Reveal Disproportionate Increases in CBF and CMR_{glc} Compared to CMR_{O_2} During Activation	*576*
23.4.5	Imaging Brain Maturation and Aging, Neurotransmitter Systems, and Effects of Drugs of Abuse	*580*
23.4.6	Imaging Electrolyte Transport Across the Blood-Brain Interface and Shifts in Calcium Homeostasis Under Pathophysiological Conditions	*584*
23.4.7	Summary	*587*
23.5	Optical Imaging of Functional Activity by Means of Extrinsic and Intrinsic Fluorescent Compounds	*587*
23.6	Tracking Dynamic Movement of Cellular Processes and Cell Types	*589*
23.7	Evaluation of Exogenous Genes, Cells, and Therapeutic Efficacy	*590*
23.8	Summary and Perspectives	*594*
	References	*595*

Part VII **Disease-Related Response** *601*

24 **Inflammatory Response of the Blood-Brain Interface** *603*
Pedro M. Faustmann and Claus G. Haase

24.1	Introduction	*603*
24.2	Diagnostic Features of Cerebrospinal Fluid	*603*
24.2.1	Cell Count and Cell Differention	*604*
24.2.2	Protein Level	*605*
24.2.3	Glucose Level	*605*
24.3	Acute Bacterial Meningitis	*605*
24.3.1	Bacteria and the Blood-Brain Barrier	*606*
24.3.2	Leukocyte Migration into the CNS	*607*
24.3.3	Cytokines	*611*
24.4	Inflammatory Response in Acute Trauma	*612*
24.5	Inflammatory Response in Alzheimer's Disease	*613*
	References	*614*

25 **Stroke and the Blood-Brain Interface** *619*
Marilyn J. Cipolla *619*

25.1	Introduction	*619*
25.2	Brain Edema Formation During Stroke	*620*
25.2.1	Cytotoxic Versus Vasogenic Edema	*620*
25.3	Role of Astrocytes in Mediating Edema During Ischemia	*621*
25.3.1	Aquaporins and Cerebral Edema During Ischemia	*624*

25.4	Cellular Regulation of Cerebrovascular Permeability	626
25.4.1	Role of Actin	626
25.4.2	Ischemia and Hypoxia Effects on EC Actin and Permeability	627
25.5	Reperfusion Injury	628
25.5.1	Nitric Oxide and Other Reactive Oxygen Species as Mediators of EC Permeability During Ischemia And Reperfusion	628
25.5.2	Thresholds of Injury	629
25.6	Transcellular Transport as a Mechanism of BBB Disruption During Ischemia	630
25.7	Mediators of EC Permeability During Ischemia	631
25.8	Hyperglycemic Stroke	633
25.8.1	Role of PKC Activity in Mediating Enhanced BBB Permeability During Hyperglycemic Stroke	633
25.9	Hemodynamic Changes During Ischemia and Reperfusion and its Role in Cerebral Edema	634
	References	635

26 Diabetes and the Consequences for the Blood-Brain Barrier 649
Arshag D. Mooradian

26.1	Introduction	649
26.2	Histological Changes in the Cerebral Microvessels	650
26.3	Functional Changes in the Blood-Brain Barrier	650
26.3.1	Diabetes-Related Changes in the BBB Transport Function	651
26.4	Potential Mechanisms of Changes in the BBB	655
26.4.1	Hemodynamic Changes	655
26.4.2	Biophysical and Biochemical Changes	656
26.4.3	Changes in Neurotransmitter Activity in Cerebral Capillaries	657
26.5	Potential Clinical Consequences of Changes in the BBB	658
26.6	Conclusions	661
	References	661

27 Human Parasitic Disease in the Context of the Blood-Brain Barrier – Effects, Interactions, and Transgressions 671
Mahalia S. Desruisseaux, Louis M. Weiss, Herbert B. Tanowitz, Adam Mott, and Danny A. Milner

27.1	Introduction (by D. Milner)	671
27.2	Malaria: The *Plasmodium berghei* Mouse Model and the Severe Falciparum Malaria in Man (by M.S. Desruisseaux and D. Milner)	683
27.2.1	Upregulation of Intercellular Adhesion Molecules	685
27.2.2	Secretion of TNF-α	686
27.2.3	Microglial Activation	687
27.2.4	Vascular Damage	688

27.3	Trypanosomiasis: African and American Parasites of Two Distinct Flavors (*by H. Tanowitz, M. S. Desruisseaux, and A. Mott*) *689*	
27.3.1	CNS Pathology in African Trypanosomiasis *690*	
27.3.2	Cytokines and Endothelial Cell Activation *690*	
27.3.3	Astrocytosis and Microglial Cell Activation *691*	
27.3.3.1	Nitric Oxide *691*	
27.3.4	American Trypanosomiasis (Chagas' Disease) *692*	
27.4	Toxoplasmosis: Transgression, Quiescence, and Destructive Infections (*by L. Weiss*) *694*	
27.5	Conclusion *696*	
	References *696*	
28	**The Blood Retinal Interface: Similarities and Contrasts with the Blood-Brain Interface** *701*	
	Tailoi Chan-Ling	
28.1	Introduction *701*	
28.2	The Inner and Outer BRB *702*	
28.3	The Choroidal Vasculature *702*	
28.4	Characteristics of Intraretinal Blood Vessels *703*	
28.5	Ensheathment and Induction of the Inner BRB by Astrocytes and Müller Glia *703*	
28.6	BRB Properties of Newly Formed Vessels *706*	
28.7	Pericytes and the BRB *707*	
28.8	Membrane Proteins of Tight Junctions *710*	
28.9	Localization of Occludin and Claudin-1 to Tight Junctions of Retinal Vascular Endothelial Cells *710*	
28.10	Expression of Occludin by RPE Cells and Lack of Occludin Expression by Choroidal Vessels *711*	
28.11	Inherent Weakness of the BRB and Existence of Resident MHC Class II$^+$ Cells Predisposes the Optic Nerve Head to Inflammatory Attack *712*	
28.11.1	Compromised BBB Where CNS Meets Peripheral Vascular Bed *713*	
28.12	Clinical and Experimental Determination of the Blood-Retinal Barrier *713*	
28.13	Conclusions *716*	
	References *717*	

Subject Index *725*

Preface

Writing on the blood-brain barrier (BBB) has become a multidisciplinary task. In this two-volume handbook we try to cover all the different aspects that need to be considered to picture a modern image of this fascinating physiological phenomenon. During the past decade our knowledge on the BBB has largely improved, and as mentioned in our introduction, the focus has shifted from an endothelio-centric point of view to a more generalized perspective that includes additional cellular adjuncts like astrocytes, pericytes, microglial cells, and even stem cells. Insofar we prefer the term blood-brain interfaces when describing the various constituents that comprise this more holistic architecture of the BBB. It is useless to say that such an endeavor can never be complete, in particular when the different facets involve a host of structural, molecular, and cell biological aspects ranging from ontogenesis to the diseased BBB. The different chapters provide basic knowledge and state-of-the-art information that can be used for further perusal. Especially, the reference lists collected by the authors are treasures of their own, since they set pathways for additional in-depth studies. All contributors are experts in their field with a profound background in BBB research; and the editors are thankful for their enthusiasm which helped them to concentrate on their subjects outside the daily rush of paperwork. From this perspective, we hope this handbook will serve as a rich source of information for a wide audience, including graduate students, advanced undergraduates, and professionals.

Our sincere thanks go to all contributors for contributing excellent and comprehensive chapters, to Mrs. Monika Birkelbach for secretarial assistance and to Dr. Andreas Sendtko and his colleagues at Wiley-VCH for continuous and encouraging support throughout the preparation of this handbook.

Bochum, New York, Rochester
January 2006

Rolf Dermietzel
David C. Spray
Maiken Nedergaard

Blood-Brain Interfaces: From Ontogeny to Artificial Barriers.
Edited by R. Dermietzel, D.C. Spray, M. Nedergaard
Copyright © 2006 WILEY-VCH Verlag GmbH & Co. KGaA, Weinheim
ISBN: 3-527-31088-6

List of Contributors

Joan Abbott
Wolfson Centre for Age Related
Diseases
King's College London
Hodgkin Building
Guy's Campus
London, SE1 1UL
UK

Barbara Aumayr
Department of Neurosurgery
Cleveland Clinic Foundation
9500 Euclid Avenue
Cleveland, OH 44106
USA

William A. Banks
Veterans Affairs Medical Center
915 N. Grand Boulevard
St. Louis, MO 63106
USA

Ingo Bechmann
Institute of Cell Biology and
Neurobiology
Charité University Hospital Berlin
Schumannstrasse 20/21
10098 Berlin
Germany

Richard Beliveau
Département de Chimie-Biochimie
Université du Québec à Montréal
C.P. 8888, Succursale Centre-ville
Montréal, H3C 3P8
Canada

Nicholas Bodor
Center for Drug Discovery
University of Florida
PO Box 100497
Gainesville, FL 32610-0497
USA

Alejandra Bosco
Department of Neuroscience
Albert Einstein College of Medicine
1410 Pelham Parkway S
Bronx, NY 10464
USA

Peter Buchwald
IVAX Research, Inc.
4400 Biscayne Boulevard
Miami, FL 33137
USA

Valérie Buée-Scherrer
Faculté des Sciences Jean Perrin
Université d'Artois
SP 18, rue Jean Souvraz
62307 Lens
France

Blood-Brain Interfaces: From Ontogeny to Artificial Barriers.
Edited by R. Dermietzel, D.C. Spray, M. Nedergaard
Copyright © 2006 WILEY-VCH Verlag GmbH & Co. KGaA, Weinheim
ISBN: 3-527-31088-6

Laura M.A. Camassa
Department of General and
Environmental Physiology
and Centre of Excellence in
Comparative Genomics (CEGBA)
University of Bari
Via Amendola 165/A
70126 Bari
Italy

Roméo Cecchelli
Faculté des Sciences Jean Perrin
Université d'Artois
SP 18, rue Jean Souvraz
62307 Lens
France

Tailoi Chan-Ling
Department of Anatomy
Institute for Biomedical Research
University of Sydney
Sydney, NSW 2006
Australia

Yoon Kyung Choi
Research Institute of Pharmaceutical
Sciences and College of Pharmacy
Seoul National University
Seoul, 151-742
Korea

Marilyn J. Cipolla
Department of Neurology
University of Vermont
89 Beaumont Avenue, Given C454
Burlington, VT 05405
USA

Caroline Coisne
Faculté des Sciences Jean Perrin
Université d'Artois
SP 18, rue Jean Souvraz
62307 Lens
France

Luca Cucullo
Department of Neurosurgery
Cleveland Clinic Foundation
9500 Euclid Avenue
Cleveland, OH 44106
USA

Albertus (Bert) G. de Boer
Blood-Brain Barrier Research Group
Division of Pharmacology
Leiden University
Einsteinweg 55
2333 CC, Leiden
The Netherlands

Bénédicte Dehouck
Faculté des Sciences Jean Perrin
Université d'Artois
SP 18, rue Jean Souvraz
62307 Lens
France

Lucie Dehouck
Faculté des Sciences Jean Perrin
Université d'Artois
SP 18, rue Jean Souvraz
62307 Lens
France

Marie-Pierre Dehouck
Faculté des Sciences Jean Perrin
Université d'Artois
SP 18, rue Jean Souvraz
62307 Lens
France

Michel Demeule
Département de Chimie-Biochimie
Université du Québec à Montréal
C.P. 8888, Succursale Centre-ville
Montréal, H3C 3P8
Canada

Rolf Dermietzel
Department of Neuroanatomy and
Molecular Brain Research
Ruhr University Bochum
Universitätsstrasse 150
44780 Bochum
Germany

Mahalia S. Desruisseaux
Montefiore Medical Center
Albert Einstein College of Medicine
1300 Morris Park Avenue
Bronx, NY 10461
USA

Gerald A. Dienel
University of Arkansas for Medical
Sciences
Shorey Building
4301 W. Markham Street
Little Rock, AR 72205
USA

Jean-Bernard Dietrich
Inserm U 575
Université Louis Pasteur
5, rue B. Pascal
67084 Strasbourg
France

Gerhard F. Ecker
Department of Medicinal Chemistry
University of Vienna
Althanstrasse 14
1090 Vienna
Austria

Britta Engelhardt
Theodor Kocher Institute
University of Bern
Freiestrasse 1
3012 Bern
Switzerland

Pedro M. Faustmann
Neuroanatomy/Mol. Brain Research
Ruhr University Bochum
Universitätsstrasse 150
44780 Bochum
Germany

Chantal Fournier
Département de Chimie-Biochimie
Université du Québec à Montréal
C.P. 8888, Succursale Centre-ville
Montréal, H3C 3P8
Canada

Antonio Frigeri
Department of General and
Environmental Physiology
University of Bari
Via Amendola 165/A
70126 Bari
Italy

Pieter J. Gaillard
to-BBB technologies BV
Bio Science Park Leiden
Gorlaeus Laboratories,
LACDR Facilities-FCOL
Einsteinweg 55
2333 CC, Leiden
The Netherlands

Hans-Joachim Galla
Department of Biochemistry
University of Münster
Wilhelm-Klemm-Strasse 2
48149 Münster
Germany

Albert Gjedde
Pathophysiology and Experimental
Tomography Center and
Center of Functionally Integrative
Neuroscience
Aarhus University Hospital
44 Norrebrogade
8000 Aarhus
Denmark

Steven Goldman
Department of Neurology
University of Rochester Medical
Center
601 Elmwood Rd., Box 645
Rochester, NY 14642
USA

Claus G. Haase
Department of Neurology and Clinical
Neurophysiology
Knappschafts-Hospital
Dorstener Strasse 151
45657 Recklinghausen
Germany

Kerri Hallene
Department of Neurosurgery
Cleveland Clinic Foundation
9500 Euclid Avenue
Cleveland, OH 44106
USA

Glenn I. Hatton
Department of Cell Biology and
Neuroscience
University of California, Riverside
1208 Spieth Hall
Riverside, CA 92521
USA

Damir Janigro
Department of Neurosurgery
Cleveland Clinic Foundation
9500 Euclid Avenue
Cleveland, OH 44106
USA

Julie Jodoin
Département de Chimie-Biochimie
Université du Québec à Montréal
C.P. 8888, Succursale Centre-ville
Montréal, H3C 3P8
Canada

Abba J. Kastin
Pennington Biomedical Research
Center
Blood-Brain Barrier Laboratory
6400 Perkins Road
Baton Rouge, LA 70808
USA

Kyu-Won Kim
Research Institute of Pharmaceutical
Sciences and College of Pharmacy
Seoul National University
Seoul, 151-742
Korea

Sae-Won Kim
Research Institute of Pharmaceutical
Sciences and College of Pharmacy
Seoul National University
Seoul, 151-742
Korea

Dorothee Krause
Institute of Neuroanatomy and
Molecular Brain Research
Ruhr University Bochum
Universitätsstrasse 150
44780 Bochum
Germany

Christina Lilliehook
Department of Neurology
University of Rochester Medical Center
601 Elmwood Road, Box 645
Rochester, NY 14642
USA

Wee Shiong Lim
Department of Geriatric Medicine
Tan Tock Seng Hospital
11 Jalan Tan Tock Seng
Singapore 308433
Republic of Singapore

Christina Lohmann
Discovery DMPK&BA
AstraZeneca R&D Lund
Scheelevägen 8
22187 Lund
Sweden

Sebastien Mambie
Pacific Biomedical Research Center
University of Hawaii
1993 East-West Road
Honolulu, HI 96822
USA

Frederic Mercier
Pacific Biomedical Research Center
University of Hawaii
1993 East-West Road
Honolulu, HI 96822
USA

Florence Miller
Faculté des Sciences Jean Perrin
Université d'Artois
SP 18, rue Jean Souvraz
62307 Lens
France

Danny A. Milner
Harvard Medical School
Department of Pathology
Brigham & Women's Hospital
75 Francis Street, Amory 3
Boston, MA 02115
USA

Albert Moghrabi
Centre de Cancérologie
Charles Bruneau
Université du Québec à Montréal
3175 Chemin Côte-Ste-Catherine
Montréal, H3T 1C5
Canada

Maria G. Mola
Department of General and
Environmental Physiology
and Centre of Excellence in
Comparative Genomics (CEGBA)
University of Bari
Via Amendola 165/A
70126 Bari
Italy

Arshag D. Mooradian
Division of Endocrinology
Saint Louis University
1402 South Grand Blvd
Saint Louis, MO 63104
USA

Adam Mott
Department of Immunology &
Infectious Diseases
Harvard School of Public Health
665 Huntington Avenue
Boston, MA 02115
USA

Maiken Nedergaard
School of Medicine and Dentistry
University of Rochester
601 Elmwood Avenue
Rochester, NY 14642
USA

Grazia P. Nicchia
Department of General and
Environmental Physiology
and Centre of Excellence in
Comparative Genomics (CEGBA)
University of Bari
Via Amendola 165/A
70126 Bari
Italy

Beatrice Nico
Department of Human Anatomy
and Histology
University of Bari
Via Amendola 165/A
70126 Bari
Italy

Robert Nitsch
Institute of Cell Biology and
Neurobiology
Charité University Hospital Berlin
Schumannstrasse 20/21
10098 Berlin
Germany

Christian R. Noe
Department of Medicinal Chemistry
University of Vienna
Althanstrasse 14
1090 Vienna
Austria

Emily Oby
Department of Neurosurgery
Cleveland Clinic Foundation
9500 Euclid Avenue
Cleveland, OH 44106
USA

Weihong Pan
Pennington Biomedical Research
Center
Blood-Brain Barrier Laboratory
6400 Perkins Road
Baton Rouge, LA 70808
USA

Jeong Ae Park
Research Institute of Pharmaceutical
Sciences and College of Pharmacy
Seoul National University
Seoul, 151-742
Korea

Josef Priller
Department of Psychiatry
Charité University Hospital Berlin
Schumannstrasse 20/21
10098 Berlin
Germany

Markus Ramsauer
Buchenrain 5
4106 Therwil
Switzerland

Ed Rapp
Cleveland Medical Devices Inc.
4415 Euclid Avenue
Cleveland, OH 44103
USA

Angelika Rappert
Institute of Cell Biology and
Neurobiology
Charité University Hospital Berlin
Schumannstrasse 20/21
10098 Berlin
Germany

Anthony Régina
Département de Chimie-Biochimie
Université du Québec à Montréal
C.P. 8888, Succursale Centre-ville
Montréal, H3C 3P8
Canada

Bernhard Reuss
Center for Anatomy
University of Göttingen
Kreuzbergring 36
37075 Göttingen
Germany

Domenico Ribatti
Department of Human Anatomy
and Histology
University of Bari
Via Amendola 165/A
70126 Bari
Italy

David C. Spray
Department of Neuroscience
Albert Einstein College of Medicine
1410 Pelham Parkway S
Bronx, NY 10464
USA

Maria Svelto
Department of General and
Environmental Physiology
and Centre of Excellence in
Comparative Genomics (CEGBA)
University of Bari
Via Amendola 165/A
70126 Bari
Italy

Herbert B. Tanowitz
Division of Infectious Diseases
Albert Einstein College of Medicine
1300 Morris Park Avenue
Bronx, NY 10461
USA

Hong Tu
Pennington Biomedical Research
Center
Blood-Brain Barrier Laboratory
6400 Perkins Road
Baton Rouge, LA 70808
USA

Sandra Turcotte
Département de Chimie-Biochimie
Université du Québec à Montréal
C.P. 8888, Succursale Centre-ville
Montréal, H3C 3P8
Canada

Corine C. Visser
Blood-Brain Barrier Research Group
Division of Pharmacology
Leiden University
Einsteinweg 55
2333 CC, Leiden
The Netherlands

Christian Weidenfeller
Department of Chemical and
Biological Engineering
University of Wisconsin-Madison
1415 Engineering Drive
Madison, WI 53706
USA

Louis M. Weiss
Division of Infectious Diseases
Albert Einstein College of Medicine
1300 Morris Park Avenue
Bronx, NY 10461
USA

Hartwig Wolburg
Institute of Pathology
University of Tübingen
Liebermeisterstrasse 8
72076 Tübingen
Germany

Shulin Xiang
Pennington Biomedical Research
Center
Blood-Brain Barrier Laboratory
6400 Perkins Road
Baton Rouge, LA 70808
USA

Alla Zozulya
Department of Pathology and
Laboratory Medicine
University of Wisconsin-Madison
1300 University Avenue
Madison, WI 53706
USA

Introduction
The Blood-Brain Barrier: An Integrated Concept

Rolf Dermietzel, David C. Spray, and Maiken Nedergaard

Writing about the blood-brain barrier (BBB) these days has become a multidisciplinary enterprise. While the classic view formulated by the pioneering work of Ehrlich (1885) and Goldman (1913) was based on evidence that blood-borne substances were excluded from the brain, this concept was systematically reformulated over the past century, cumulating in the compartment concept put forward by Davson (1967) in which he redefined the BBB as the "sum of all bidirectional exchange processes which occur at the morphological blood-brain interfaces". This definition has shifted the more or less endotheliocentric view to a more integrated concept that takes into account not only the bidirectionality of the exchange processes, but also the discovery that besides the endothelium additional components constitute integral parts of the barrier mechanisms. These additional components include the perivascular structural adjuncts such as pericytes, microglia and macrophages and the astroglial interface, interacting in an orchestrated fashion in order to achieve the "functional complex" that allows the transmission of metabolic and homeostatic information between the blood and the brain parenchyma and vice versa. In a former review on the BBB, one of the editors of this handbook described the function of this barrier as a dynamic *homeostat* (Dermietzel and Krause 1991) that regulates the interchange between the body and the brain border. By its nature, this homeostat is not simply constituted by the sum of its parts. The integration of all molecular and structural components and their concerted interplay at the critical locations gives rise to the "more" that we believe is the essence of synergism. Consequently, when conceptualizing this handbook we had to take into account the entire spectrum of disciplines that contribute to our modern understanding of the BBB. The picture we tried to paint cannot, of course, be complete and the reader may find missing a contribution that exactly covers his field of interest. In this case, we must direct the interested scholar to the list of references that are included in each of the articles and which may be well suited as a guideline for further deepening his knowledge.

The collection of articles is thus structured in a way that reflects our understanding of the BBB as a dynamic homeostat. The first part deals with the ontogenesis of brain vasculature and begins with a basic article written by *Britta Engelhardt* that not only describes the maturation of the BBB and its structural substrate, but also goes into details regarding molecular aspects of the developmental processes. What we have learned during recent years, especially through the work by the late *Werner Risau* and his coworkers, is that when the BBB is first established it is by no means a static structure, but rather a dynamic construct that requires permanent feeding by growth and differentiation factors in order to maintain and elaborate its complex machinery as an exquisite barrier. As we now know, the microvasculature is not uniformly tight throughout the brain, but is a composite building that contains segments of leakiness integrated into the mass of tight vessels. These leaky segments prevail at sites where a requirement for differential exchange between the blood site and the brain site (or vice versa) exists, i.e. neurohaemal regions which entail neurosecretory and neurosensory areas for controlling specific parameters of the body fluid. An understanding of how the brain manages to keep leaky segments maintained in an embedment of tight vessels will provide a key to the understanding of BBB development. The article by *Jeong Ae Park*, *Yoon Kyung Choi*, *Sae-Won Kim* and *Kyu-Won Kim* focuses on some of these factors with emphasis on hypoxia inducible factor (HIF), VEGF and other neuroglia-derived factors. It describes the main sources of these potent humoral factors and their effects during angiogenesis on the developing brain including aspects for barriergenesis.

An extension of the discussion on humoral regulation of brain angiogenesis is provided by *Christina Lilliehook* and *Steven Goldman*. Their focus is on a rather new and exciting field of BBB research which considers the brain microvasculature as a niche that conditions brain stem cells for further lineaging. The observation that hippocampal stem cells before they integrate into the granular cells of the dentate gyrus associate with cerebral capillaries has made this niche a favorable subject for unravelling the conditioning mechanisms taking place at the sites of contact. With the concept of a *vascular niche* a further aspect of the BBB emerges, namely that brain vessels not only provide targets for autocrine and paracrine regulation to maintain their own degree of differentiation, but constitute a putative source of fostering brain stem cells.

Part II is devoted to the different structural components that define the BBB. Without surprise, this part begins with an account on our recent understanding of the structural and molecular complement of the cerebral endothelium written by *Hartwig Wolburg*. The endothelium plays a crucial role in separating the blood from the two major fluid compartments of the brain: the interstitial cerebral compartment and the cerebrospinal compartment. As mentioned above, this separation entails a constitutive dynamic component which is responsive to the actual metabolic situation on both sites of the barrier. The chapter on the endothelium is not intended to give a complete survey covering all the various carrier and transport mechanisms taking place at the endothelial frontier. Such an encyclopedic detailed overview is outside the scope of this handbook; and we refer here to recent reviews and monographs that deal with these particular aspects of the BBB (Pardridge 1993; Paulson et al. 1999; Nag 2003). Rather, this chapter presents the major facts and our recent knowledge regarding the two major components of the barrier: its structural and metabolic substrates.

The subsequent articles in this Section concentrate on the cellular adjuncts that surround the vessel wall and should be highlighted together in our conceptual context. Pericytes, presented by *Markus Ramsauer*, brain macrophages by *Frederic Mercier, Sebastien Mambie* and *Glenn Hatton* and microglial cells by *Ingo Bechmann, Angelika Rappert, Joseph Priller* and *Robert Nitsch* are all cellular components which permanently or transiently settle in the perivascular space. Cerebral pericytes have long been neglected by the BBB community although already described in the last third of the 19th century by Rouget. In spite of the fact that brain and retinal capillaries possess the highest density of pericytes per endothelial cell (about 1:3), their functional properties and morphological phenotype are still difficult to describe. They seem to be involved in microvascular perfusion by reciprocal contraction as well as in immune responses, a property that has god-fathered them as a "second line of defense" by their ability to enfold macrophagic activity under challenging conditions. Recent discoveries feature pericytes as key players of BBB differentiation, together with astrocytes and regulators of endothelial growth as well as stabilizers of the vascular wall. Their potential to differentiate into other mesenchymal cell types makes them a versatile cellular pool with progenitor features for macrophages and even endothelial cells.

Macrophages, which are of mononuclear origin, reside in the whole CNS vasculature. Together with microglial cells they are responsible for the immune response of the brain and their mostly perivascular association makes them part of the BBB complex. Macrophages are potent producers of extracellular matrix molecules, growth factors and cytokines and are thought to play key roles in regulating glial and neuronal cell function. In contrast to microglial cells which have been detected by recent efforts (Nimmerjahn et al. 2005) to be primarily resident even when BBB injury commences, macrophages seem to be more migratory, patrolling the brain parenchyma and cleansing it from debris after injury or inflammatory attacks. A recent concept pointed out by the paper of *Frederic Mercier, Sebastien Mambie* and *Glenn I. Hatton* calls attention to the presumed function of macrophages as partners in the neurogenic niche, which is supposed to play a crucial role in neural precursor or stem cell priming. Microglial cells are by far the most abundant cellular entity (about 15% of brain cells) involved in the immune response of the CNS. They serve as sensors for pathologic events in the brain tissue and reside in the brain parenchyma and in perivascular and juxtavascular positions. Here, besides their paramount function as executors of the innate and adaptive immune responses of the brain tissue, they seem to function in concert with perivascular macrophages in regulating the tightness of the barrier through chemokine secretion. Both cellular entities, perivascular macrophages and microglial cells, perform fundamental functions in the regulation of the leakiness of the BBB under inflammatory conditions, an issue that is addressed in more detail in part VII of the handbook.

The following two articles by *Joan Abbott* and the group of *Antonio Frigeri* (*Grazia P. Nicchia, Batrice Nico, Laura M.A. Camassa, Maria G. Mola, Domenica Ribatti, David C. Spray, Alejandra Bosco, Maria Svelto* and *Antonio Frigeri*) concentrate on the astroglial surrounding of the blood-brain interface. Since the work by Goldstein and Betz (1986) and later by Janzer and Raff (1987), the astroglia and in particular the perivascular endfeet have been considered to be crucial elements for maintaining the BBB complex. Astrocytes not only participate in regulating transport through the endothelium and provide trophic support for the tightness of the endothelium, but they also provide the essential link for vascular-neuronal signaling. Because of their ideal strategical position and their molecular and structural polarization, the astroglia are destined to link the neuronal site with the vascular bed of the brain parenchyma and serve as a pathway for metabolic and ionic transfer. Most importantly, astrocytes seem to represent the most dynamic part of the BBB complex, since they are able to respond to neuronal activity and transmit signals to the blood front to regulate local perfusion. Thus, the astroglia are an essential part of an integrated concept of the BBB, not only in terms of their morphogenetic capabilities, but also as transformers of neuronal activation into BBB receptive signals.

Part III is devoted to hormonal and enzymatic control of the brain vessels. The initial article by *Bernhard Reuss* centers on the role of fibroblasts growth factors (FGFs) in regard to their influence on growth and differentiation of brain microvascular endothelium, and thus their involvement in establishing, maintaining and restoring the BBB. In particular, FGF-2 has come into focus as an important cofactor together with secretory products of astrocytes (TGF-β_1, GDNF) involved in the induction of certain specific barrier properties of brain microvessels. Also, the effect of FGF-2 to influence the synthesis and phosphorylation of the intermediate filament protein GFAP of astrocytes, which has been proven to be important for inducing the BBB phenotype, is covered here, to provide insight of the importance of the FGF family for the stabilization and preservation of BBB features.

This issue of factors that influence the functional properties of the BBB by humoral and enzymatic inputs is further considered by four articles. First, the contribution of *Weihing Pan, Shulin Xiang, Hong Tu* and *Abba Kastin* presents a general overview on the interaction of cytokines with the BBB. In particular, the mechanism of cytokine transport across the endothelial cells is described, as is their action on these cells, which results in altered endothelial function, cytotoxicity or cell proliferation. This article is followed by a description of insulin transport through the barrier and the effect of insulin on the BBB in concert with proinflammatory cytokines, authored by *William Banks* and *Wee Shion Lim*. It also addresses the important clinical issue of insulin resistance and its impact on the inflammatory susceptibility of the brain vasculature, an issue that is of profound socio-economic importance in view of the increasing number of obese people. The reviews on humoral effects on the BBB is completed by *Jean Bernard Dietrich*, who gives a detailed account of the interaction of glucocorticoid hormones and estrogens on the endothelium of the BBB. The focus of this contribution lies on the influence of glucocorticoids and estrogens with regard to their expression of endothelial adhesion molecules that are pivotal for the transendothelial migration of inflammatory cells through the BBB. In this context, their use as therapeutic tools for the treatment of autoimmune diseases is discussed. Part III on hormone and enzyme interaction ends with the chapter of *Dorothee Krause* and *Christina Lohmann* on metalloproteinases and the brain microvasculature. Besides the structural sealing of the BBB endothelium through tight junctions and its complement of specific transporters and carriers, the vascular wall of brain vessels is endowed with a battery of enzymes constituting a kind of "enzymatic barrier" to the passage of peptides across the BBB. These enzymes are proteinases and have been detected in variable amounts at the BBB. Among them are the metalloproteinases, which are involved in the cleavage of peripheral peptides as well as centrally released peptides, in remodelling the extracellular matrix during angiogenesis and in facilitating perivascular penetration of emigrating blood-borne cells, including tumor cells. The complexity of the metalloproteinases (MPs), and in particular the matrix bound subgroup (MMPs), for the function of the BBB are thoroughly reviewed in this paper.

Part IV centers on different approaches that have been taken to establish *in vitro* systems for culturing the BBB complex. It further includes a strategy that has recently been developed aimed to predict BBB properties by *in silico* approaches. *In vitro* models are of paramount importance to test the transport of drugs through the BBB. The use of alternate tissue culture models are helpful in some ways, but can only be regarded as approximative approaches. Thus, considerable effort has been channeled to develop reliable culture systems to mimic the BBB *in vitro*. The chapter by *Romeo Cecchelli, Caroline Coisne, Lucie Dehouck, Florence Miller, Marie-Pierre Dehouck, Valérie Buée-Scherrer* and *Bénédicte Dehouck* gives an overview on our present knowledge of *in vitro* BBB models that at least fulfil some of the essential criteria that are required for a well differentiated BBB endothelium. In extension of this article, *Alla Zozulya, Christian Weidenfeller* and *Hans-Joachim Galla* describe their specific approach to utilize a coculture system treated with hydrocortisone to achieve monolayers with high endothelial resistance, the key feature of a tight barrier. Both articles provide a comprehensive source of information for those who are interested in model systems of the BBB. The chapter by *Luca Cucullo, Emily Oby, Kerri Hallene, Barbara Aumayr, Ed Rapp* and *Damir Janigro* introduces advanced models that take into account the three-dimensional structure of the vascular tube including shear-factors applied on the endothelium by fluid flow. This dynamic *in vitro* model of BBB (DIV-BBB) and its newly designed model (NDIV-BBB) display the most advanced *ex vivo* approach to the *in vivo* BBB situation with the advantage of large upscaling, an obligatory requirement for industrial application. *Gerhard Eckert* and *Christian Noe* introduce the readers to the new world of *in silico* screening of drugs (pharmacoinformatics) with respect to their ability to penetrate the BBB. These *in silico* methods gain increasing interest in order to economize the process of standard high-throughput screening. The applicability of computational methods on the BBB is a novel promising strategy that may allow discovery of target families and presumably novel bioactive molecules. The fact that the permeation of the BBB is a multifactorial process necessitates advanced computational methods for modelling approaches and opens new perspectives in BBB drug research.

Part V completes consideration of pharmacological aspects of the BBB, and gives access to some fundamental issues of BBB pharmacology. *Sandra Turcotte, Michel Demeule, Anthony Régina, Chantal Fournier, Julie Jodoin, Albert Moghrabi* and *Richard Béliveau* introduce one of the most relevant systems that prevents significant accumulation of many hydrophobic molecules and drugs in the brain: the multidrug resistance transporter P-glycoprotein (P-gp). This efflux transporter is a member of the ATP-binding cassette group of transporters (ABC), which represent the largest family of transmembrane proteins. The existence of the P-gp at the BBB is one of the main causes of failure in chemotherapy, because of its ability to translocate xenobiotics against a concentration gradient across the plasma membrane. Thus, the P-gp plays an important role in brain protection at the BBB site, but by its nature provides a considerable hindrance for successful treatment of a variety of brain diseases. *Nicolas Bodor* and *Peter Buchwald* provide a survey on general aspects of targeting neuropharmacologicals by chemical delivery systems (CDSs). CDSs, as the authors state, represent a rational drug design approach that exploits sequential metabolism not only to deliver, but also to specifically target drugs to their site of action. The authors present a spectrum of approaches intended to deliver drugs, particularly bioactive peptides, through CDSs to the brain tissue. The chapter by *Pieter J. Gaillard, Corine C. Visser* and *Albertus G. de Boer* is centered on certain delivery systems, which include the transferrin system, the insulin receptor, the low-density lipoproteins I and II (LRP1, LRP2) and the diphtheria toxin receptor. The authors present the enormous opportunity that these systems offer for the successful delivery of drugs to the brain, but also enumerate the pitfalls that these systems face and their still limited applicability in clinical therapeutics.

Two major chapters herald part VI, both of which deal with vascular perfusion. *Albert Gjedde* introduces the field of blood-brain transfer and metabolism of oxygen, a field that sets the basics for functional imaging covered by the article of *Gerald A. Dienel*. The rationale to include these chapters within this handbook on the BBB is multi-faceted. As quoted by *A. Gjedde*, delivery of oxygen to brain tissue differs in major respects from the delivery of oxygen to other tissues. This is not a direct consequence of the properties of the BBB in its narrow sense, but reflects the specific demands that the vascular system of the brain must fulfil in order to satisfy the energetic requirements of the working brain. "The absent recruitment of capillaries in states of activation of neurons as well as the general principle of topographic arrangement of the vessels" account for the differences of brain with regard to other tissues. This statement by *A. Gjedde* pin-points the morphological and physiological specificity of the brain microvasculature, a key feature in the context of an integrated concept of the BBB. This is followed directly by the chapter of *Gerald Dienel*, which covers the entire spectrum of modern brain imaging and provides a thorough overview of the basic physiological features on which modern brain imaging is founded. The "take-home message" of this chapter is that blood flow, metabolism and cellular function are inseparable aspects of brain activities and the responsiveness of the BBB to neuronal activity represents a hallmark in coupling both sites of the active brain: blood flow and brain work.

The last and most extensive part VII includes a collection of articles on clinical afflictions of the BBB. It starts with a chapter on the impact of inflammation on the BBB by *Pedro Faustmann's* and *Claus Haase's* contribution on inflammatory responses of the blood-brain interface. The brain has long been considered to be an immunoprivileged part of the body. Under pathological conditions such as inflammation, trauma and neurodegeneration blood-borne cells immigrate into the CNS and changes of the permeability of the BBB occur. The mechanisms how these cells enter the brain and which proteins in the cerebrospinal fluid are disease-related are key issues, which represent central questions currently being addressed in neurology, neuropathology and neuroimmunology. The sections on inflammatory attacks on the BBB is followed by *Marilyn J. Cipolla's* chapter on stroke that summarizes the pathophysiological sequelae to the event of a stroke, i.e. development of brain edema, cellular regulation of cerebrovascular permeability, effects of stroke on the cytoskeleton of brain endothelium and reperfusion injury, just to mention a few highlights of this article.

A further important clinical issue is covered by the article of *Arshag D. Mooradian*, which is centered on diabetes and its consequences for the BBB. "The CNS complications of diabetes have not widely appreciated, because of the most overt complication, namely stroke", states *A. Mooradian*, pointing to the more subtle effects on the CNS that extends beyond clinically appreciated cerebrovascular accidents. These include: alterations in cerebral microvessels with poor autoregulation and blood distribution, altered BBB function, neurochemical changes, alterations in neurotransmitter receptor activity and contributing factors such as hypoglycemic reactions. All these changes in the CNS are discussed from the perspective of BBB function.

The chapter by a group of parasitologists contributing to the chapter by *Danny A. Milner* (*Mahalia S. Desruisseaux, Louis M. Weiss, Herbert B. Tanowitz* and *Adam Moss*) focuses on a group of infections with global impact: human parasitic diseases. The involvement of the BBB during parasitic infections has been elaborated in this conceptual context for the first time in this handbook. The chapter by *Milner et al.* offers a taxonomic guideline for those readers who are interested in parasitic infections which affect, disrupt and/or destroy the BBB. From a clinical point of view, these infections are of great relevance since they are more commonly fatal than parasitic diseases which do not destroy the BBB.

The final article by *Tailoi Chang-Ling* introduces the current understanding of the blood-retina interfaces. It describes both the features the retinal interface shares in common with the CNS blood vessels as well as its specific structural features. The unique morphology of the retina as an exposed part of the brain and the transparency of the ocular media has made it an exceptionally accessible target not only for *in vivo* detection of the vascular physiology but also for monitoring cardiovascular pathology, including changes to arterioles and venular walls due to arteriosclerosis and diabetic retinopathy. We feel that this is an appropriate last chapter, insofar as the article recapitulates many of the concepts regarding both healthy and diseased BBB that are encountered during the excursion through this handbook.

References

Davson, H. **1967**, *Physiology of the Cerebrospinal Fluid*, Churchill, London.
Dermietzel, R., Krause, D. **1991**, Molecular anatomy of the blood-brain barrier as defined by immunocytochemistry, *Int. Rev. Cytol.* 127, 57–103.
Ehrlich, P. **1885**, *Das Sauerstoffbedürfnis des Organismus. Eine farbanalytische Studie* Hirschwald, Berlin.
Goldmann, E. **1913**, Vitalfärbung am Zentralnervensystem. *Abh. K. Preuss. Akad. Wiss. Phys. Med.* 1, 1–60.
Goldstein, G. W., Betz, A. L. **1986**, The blood-brain barrier. *Sci. Am.* 255, 74–83.
Janzer, R. C., Raff, M. C. **1987**, Astrocytes induce blood-brain barrier properties in endothelial cells. Nature 325, 253–257.
Nag, S. **2003**, *The Blood-Brain Barrier*, Humana Press, Totowa.
Nimmerjahn, A., Kirchhoff, F., Helmchen, F. **2005**, Resting microglial cells are highly dynamic surveillants of brain parenchyma in vivo. *Science* 308, 1314–1318.
Pardrige, W. M. **1993**, *The Blood-Brain Barrier: Cellular and Molecular Biology*, Raven Press, New York.
Paulson, O. B., Moos Knudsen, G., Moos, T. **1999**, *Blood-Brain Barrier Systems* Munksgaard, Copenhagen.

Part I
Ontogeny of the Blood-Brain Barrier

1
Development of the Blood-Brain Interface

Britta Engelhardt

1.1
Introduction

The blood-brain barrier (BBB) is composed of a continuous layer of highly specialized vascular endothelial cells. The BBB maintains central nervous system (CNS) homeostasis by preventing the entry of substances that might disturb proper function of the neurons. Uncontrolled paracellular diffusion of hydrophilic molecules into the CNS is inhibited by a complex network of tight junctions between the endothelial cells. Transcellular passage of molecules into the CNS is prohibited by the lack of endothelial fenestrae and an extremely low pinocytotic activity of these endothelial cells. In order to meet the high metabolic needs of the neurons "behind" the barrier in spite of its restrictive capacity, specific transport systems are selectively expressed in the CNS microvascular endothelial cells, which mediate the directed transport of nutrients from the blood into the CNS or the removal of toxic metabolites out of the CNS. Recent years have dramatically advanced our knowledge about the growth factors and their receptors specifically acting on the developing vascular endothelium including the CNS vasculature. Despite this increased knowledge, all we know about the molecules involved in inducing BBB characteristics in endothelial cells during embryogenesis and maintaining them in the adult is that they are provided by the CNS microenvironment. In this chapter, the current knowledge of the cellular and molecular mechanisms involved in the induction, development and maintenance of the BBB is summarized.

1.2
Pioneering Research on the Blood-Brain Barrier

Discovery of the BBB is usually ascribed to the work of the German immunologist Paul Ehrlich. However, although he discovered in the 1880s that certain dyes, when injected into the vascular system of experimental animals, were rapidly taken up by all tissues with the exception of the brain and spinal cord [1], he did not asign this phenomenon to the existence of a barrier between the blood and the brain but rather interpreted his findings as a lack of affinity of the nervous system for these dyes. Only afterwards, when Edwin E. Goldman, an associate of Ehrlich, showed that the very same dyes, when injected into the cerebrospinal fluid, stained brain and spinal cord tissue but not any tissue outside of the CNS [2], was it concluded that these dyes were prevented from either getting access from the blood to the CNS or vice versa from the CNS to the blood circulation, suggesting the existence of a vascular barrier between the blood and the CNS [3]. The term "blood-brain barrier" was coined by Lewandowsky [4]. Based on the observation made by him and others that neurotoxic agents affected brain function solely when directly injected into the brain, but not when injected into the vascular system, he concluded the existence of a BBB [4, 5]. Until 1967, it was not clear whether the structural basis for the BBB was located at the level of endothelial or rather glial cells. Thomas Reese and Morris Karnowsky [6] were able to show by means of high resolution electron microscopy and the development of electron-dense tracers that intravenously injected horseradish peroxidase (HRP) was able to penetrate between endothelial cells in heart and muscle but not in the brain, where its diffusion was blocked at the level of tight junctions between the brain endothelial cells [6, 7]. Interestingly, exogenous peroxidase was also localized in some micropinocytotic vesicles within endothelial cells but none was found beyond the vascular endothelium, suggesting that vesicular HRP also did not reach the brain.

It should be noted that, with the exception of the elasmobranch fishes (which like many invertebrates [8] have a BBB at the level of glial cells [9]), in all vertebrates the BBB is localized at the level of the endothelial tight junctions.

Within the CNS there are some areas with neurohemal or neurosecretory functions (i.e. their neurons monitor hormonal stimuli and other substances within the blood or secrete neuroendocrines into the blood) which lack a vascular barrier [10]. As they are strategically localized in the midline of the ventricular system, they are collectively referred to as circumventricular organs (CVOs). The capillaries within the CVOs are fenestrated, allowing free diffusion of proteins and solutes between the blood and the CVOs. A barrier is, however, established by a complex network of tight junctions connecting specialized ependymal cells (tanycytes) at the border of the CVOs [10, 11]. Yet another structure, where endothelial cells do not form a barrier within the CNS, is the choroid plexus [12]. The choroid plexus is a villous structure consisting of an extensive capillary network enclosed by a single layer of cuboidal epithelium. It extends from the ventricular surface into the lumen of the ventricles. Its major known function is the secretion of cerebrospinal fluid. The choroid plexus capillaries are fenestrated like those in the CVOs, while tight junctions surround the apical regions of the choroid plexus epithelium, forming the blood-cerebrospinal fluid barrier.

Resembling the situation within the choroid plexus, barriers in the vertebrate retina, an external part of the brain, show a similar organization. Whereas the intraretinal blood vessels form a blood-retina barrier (BRB), the vessels of the choroidea are fenestrated and the barrier is formed by the retinal pigment epithelium.

1.3
The Mature Blood-Brain Interface

The fully differentiated BBB is composed of at least four different cell types, which all contribute to the regulation of the BBB. Embedded within the basal membrane of the highly specialized endothelial cells, a high number of pericytes can be found (about one pericyte per 2–3 endothelial cells). More than 99% of the abluminal surface of the CNS capillaries is invested by astrocytic endfeet [13] leaving only a small distance of 20 nm between the astrocytic foot process and the endothelial cell surfaces. It should be noted that nerve terminals derived from intracerebral and extracerebral neurons have also been reported in close proximity to brain endothelial cells [14, 15]. The space between the endothelial cells/pericytes and the astrocytes therefore forms the interface between the blood and the brain. Interestingly, in this strategic location, cells of the immune system can be found, namely perivascular macrophages or dendritic cells. Whether they have a direct influence in the proper function of the BBB remains to be investigated. Although the endothelial cells form the barrier proper, the interaction with the abovementioned cells adjacent to the endothelium is required for a proper barrier function.

The most prominent feature of the BBB is the presence of complex tight junctions between CNS endothelial cells, which establish a high electrical resistance across the endothelial barrier (about $2000\ \Omega\ cm^{-2}$) (see also Chapter 4) [16]. This seems to be due to the fact that, unlike cell junctions found between endothelial cells in peripheral blood vessels, the cell junctions in the brain are more extensive and seem to form an unbroken belt between the endothelial cells. Therefore, measurement of the transelectrical resistance (TER) across brain endothelium has become a well accepted criterion for BBB maturity or of barrier characteristics established by brain endothelial cells in vitro. BBB tight junctions can morphologically be distinguished from tight junctions between endothelial cells in the periphery, especially by freeze fracture electron microscopy. Whereas in BBB endothelial cells tight junction particles are preferentially associated with the protoplasmic face (the P-face), in nonbarrier endothelial cells tight junctional particles clearly predominate at the exocytoplasmic face (the E-face) [17, 18]. Interestingly during BBB development in the rat, it was observed that, besides the complexity of tight junctions, their P-face association increases [19]. As cultured BBB endothelial cells demonstrate a correlation of the P-face association of tight junction particles with the barrier function of the BBB endothelium [20], P-face association of tight junctions might be a measure for BBB maturation, at least in mammals. It should be noted that, in submammalian species, brain endothelial cells generally show tight junctions with high P-face association [21] and during BBB development in the chicken only tight junction/strand complexity is observed to be upregulated [19, 22, 23]. As elegant freeze-fracture immunogold staining studies have demonstrated, the localization of transmembrane tight junction proteins within the particle strands on P- and E-face freeze-fracture preparations [24–26], particle distribution in freeze-fracture analysis is most likely a function of the cytoplasmic anchoring of transmembrane tight junction proteins, which might be different in BBB and nonBBB endothelia [20]. P-face association of tight junction particles has become an accepted morphological criterion of endothelial barrier properties in mammals. It remains to be shown whether the above mentioned species differences in tight junction/morphology correlate with different barrier properties.

Occludin was the first integral membrane protein found to be exclusively localized within tight junctions, including the BBB (Fig. 1.1) [27]. However, mice carrying a null mutation in the occludin gene are viable and develop morphologically normal tight junctions in most tissues, including the brain [28]. Thus, although occludin localizes to tight junction particles in freeze-fracture preparations, it is not essential for proper tight junction formation. In contrast to occludin, the claudins, which exhibit no sequence homology to occludin, comprise an entire gene family of integral membrane tight junction proteins, with more than 20 members to date; and they have been shown to be sufficient for the formation of tight junction strands [25]. Claudins are not randomly distributed throughout the tissues. Besides the endothelial specific claudin-5, claudin-3 was shown to be localized in endothelial tight junction in the CNS of mice and man [29]. Additionally, claudin-12 has been described in CNS endothelium [30]. The presence of claudin-1 in brain endothelial cells is still controversial (see [29, 31, 32]). Part of the confusion has been caused by the use of an anti-claudin-1 antibody that was subsequently shown to cross-react with claudin-3. The present

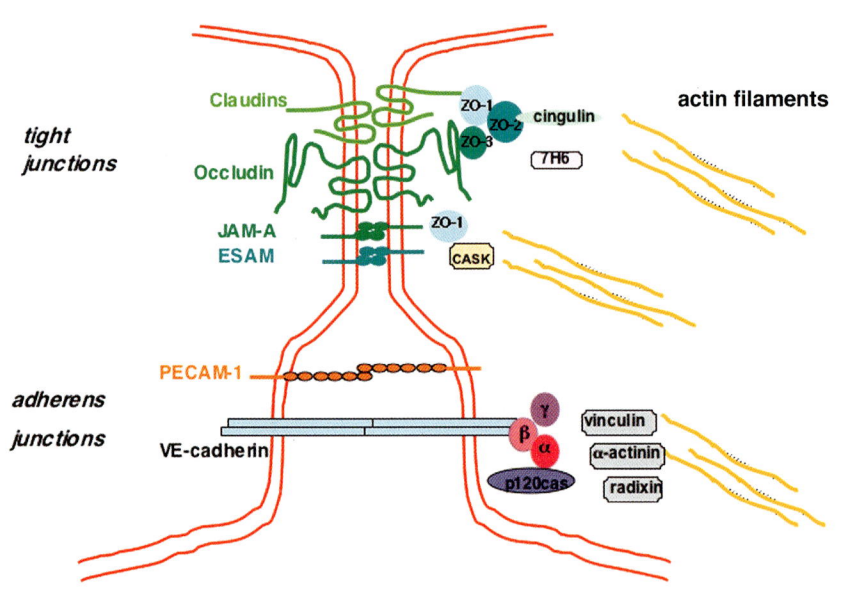

Fig. 1.1 BBB tight junctions. Schematic overview of the molecular composition of the cell-to-cell junctions between CNS microvascular endothelial cells and their putative linkage to the endothelial cytoskeleton.

availability of specific antibodies and molecular biology techniques should help to clarify the exact claudin makeup of BBB tight junctions in the near future.

With regard to BBB biology it is exciting that, in the absence of occludin, transfection of claudin-3 into fibroblasts (which lack tight junctions) induces P-face-associated (i.e. BBB-like) tight junctions [33]. In contrast, transfection of fibroblasts with the endothelial cell-specific claudin-5 induces E-face-associated tight junctions in the absence of occludin [24]. These observations suggest that different claudins induce structurally different tight junctions and that at the BBB claudin-3 and claudin-5 might be responsible for the presence of P-face- and E-face-associated tight junctions particles, respectively. That individual claudins within the BBB tight junctions fulfill specific functions regarding the regulation of the paracellular permeability has been shown by the establishment of mice deficient for claudin-5, which die as neonates about 10 h after birth due to a size-selective loosening of the BBB for molecules smaller than 800 Da [30]. Thus, each claudin seems to regulate the diffusion of a group of molecules of a certain size.

Additionally, members of the immunoglobulin supergene family have been described to be localized within tight junctions, including those of the BBB. Specifically, junctional adhesion molecule A (JAM-A [34]) and the endothelial cell-selective adhesion molecule (ESAM [35]) have been investigated in this context. JAM-A deficient mice have been established and described to be viable [36]. JAM-A deficient mice did not demonstrate significant alterations in organ development or morphology nor in vascular patterning or permeability, suggesting that JAM-A is not essential for proper BBB development. However, JAM-A has been demonstrated to regulate endothelial cell motility probably by providing an anchor to the cytoskeleton and might therefore fulfill a regulatory function at the BBB [37]. The functional involvement of ESAM in BBB tight junctions remains to be established.

The integral membrane proteins of the tight junctions, including those of the BBB [38], are linked to the cytoskeleton via cytoplasmic multidomain scaffolding proteins of the peripheral membrane associated guanylate kinase (MAGUK) family, such as ZO-1, ZO-2 and ZO-3 (Fig. 1.1; for reviews, see [39, 40]). Besides providing the cytoskeletal anchorage of transmembrane tight junction proteins, the MAGUKs seem to be important for the correct spacial distribution of the individual transmembrane tight junction proteins within the tight junction by binding their cytoplasmic domains via different domains [41, 42]. As ZOs have been reported to shuttle between tight junctions and the nucleus, they may also regulate gene expression at the BBB.

In cerebral endothelial cells, nonoccluding adherens junctions are found intermingled with tight junctions [43]. In adherens junctions, the endothelial specific integral membrane protein VE-cadherin [44] is linked to the cytoskeleton via catenins, which belong to the family of armadillo proteins (Fig. 1.1) [45]. In endothelial cells, expression and localization of β-catenin, γ-catenin and $p120^{cas}$ have been described to be crucial for the functional state of adherens junctions, including those in the brain [46, 47]. VE-cadherin has been shown to mediate contact inhibition of endothelial cell growth. The contribution of VE-cadherin and the catenins in maintaining the integrity of the BBB remains to be investigated, as expression of VE-cadherin and the catenins in the mature BBB was reported to be low [48, 49].

Finally, PECAM-1 has been demonstrated to localize to endothelial cell contacts outside of either tight junctions or adherens junctions, including the brain. In mice deficient for PECAM-1, however, no primary defect in BBB integrity has been reported [50]. However, in the case of chronic inflammation (such as experimental autoimmune encephalomyelitis (EAE), a mouse model for multiple sclerosis), mice deficient for PECAM-1 demonstrated earlier extravasation of mononuclear cells as compared to wild-type control animals [51]. Furthermore, PECAM-1 deficient mice showed a prolonged and exaggerated vascular permeability of CNS vessels during EAE, suggesting that PECAM-1 is a negative regulator of leukocyte migration across CNS microvessels and a positive regulator of BBB maintenance.

As indicated before, the junctional systems are connected to the endothelial cytoskeleton. The importance of the cytoskeleton in the establishment and maintenance of the BBB has become evident by mice lacking the actin-binding protein dystrophin. These so called mdx mice exhibit an increase in brain vascular permeability due to the disorganization of the a-actin cytoskeleton in endothelial cells and astrocytes, which leads to the altered subcellular localization of junctional proteins in the endothelium as well as the water channel aquaporin-4 (AQP4) in astrocytic endfeet [52].

The physical BBB established by specialized tight junctions and a low number of pinocytotic vesicles implies the necessity for specific transport systems which can ensure the transport of "nutrition" to the brain parenchyma. This biochemical BBB is established by transport systems of the BBB which can be grouped into three types: carrier-mediated transport, active efflux transport and receptor-mediated transport. Members of each family have been cloned at the BBB [53].

The glucose transporter Glut-1, belonging to the carrier-mediated transporters, is specifically expressed on brain endothelial cells with higher localization at the abluminal versus the luminal membrane [54]. The receptor-mediated transporters of the BBB that have been cloned to date include the leptin receptor (OBR) and the transferrin receptor. The transferrin receptor can be observed on endothelial cells forming a BBB but not on those within the choroid plexus or the CVOs [55]. From the group of active efflux transporters, expression of P-glycoprotein was shown to be required for the BBB and seems to ensure the rapid removal of ingested toxic, lipophilic metabolites from the neuroectoderm [56].

In addition to the physical and biochemical barrier, the specialized endothelial cells of the BBB provide a metabolic barrier by the expression of a number of enzymes which modify lipophilic endogenous and exogenous molecules, which otherwise could bypass the physical or biochemical barrier and negatively affect neuronal function (for detailed reviews, see [53, 57, 58]).

Although the above-described characteristics have originally been described solely for one particular segment of the vascular tree (namely the capillaries), more recently the term BBB has been used more widely, especially by immunologists to describe the unique characteristics of the CNS microvessels in the context of CNS inflammation and leukocyte extravasation across the BBB. The assumption that the unique characteristics of CNS capillary endothelial cells extend to the level of the endothelial cells lining the postcapillary venules, the vascular segment where leukocyte extravasation takes place, bears some danger as it is well known that there are functional and molecular differences between endothelial cells in different segments of the vascular tree. However, as both the unique molecular composition of tight junctions and the expression of several of the BBB specific transport systems have been described in microvascular endothelial cells of the CNS, the precise vascular localization of the functional term BBB might be extended beyond the capillary segments to the CNS microvessels [29, 59].

In any case, endothelial cells comprising the barrier characteristics in the CNS are unique and clearly distinguishable from any other endothelial cell in the body. The determination of how much variability there is in BBB characteristics along the vascular tree within the CNS awaits analysis. To understand how brain endothelial cells acquire the unique features of the BBB, it is therefore crucial to understand the development of the brain vasculature.

1.4
Development of the CNS Vasculature

During vertebrate embryogenesis, the development of the CNS vasculature begins when angioblasts, which originate from the lateral splanchnic mesoderm, enter the head region and form the perineural vascular plexus by de novo formation of blood vessels, a process called vasculogenesis. In a 2-day-old chick embryo and a 9-day-old rodent embryo, the perineural vascular plexus covers the entire surface of the neural tube [60–62]. None of the precursors that form the perineural vascular plexus nor the intraneural vessels are derived from the neuroectoderm [63, 64]. Thus, the vascular system within the CNS does not develop by vasculogenesis. Rather, vascular sprouts from the perineural plexus invade the proliferating neuroectoderm at day 4 of embryonic development in the chicken and at day 10 in rodents [60]. This mechanism whereby new vessels are formed from pre-existing vessels is called angiogenesis [61, 64, 65]. During brain angiogenesis, sprouting vessels grow radially into the neuroectodermal tissue, elongate, give rise to manifold branches and finally anastomose with adjacent sprouts, forming an undifferentiated network of capillaries near the ventricular zone of the developing brain [60, 66]. The onset of angiogenesis in the brain occurs at precisely reproducible stages of embryonic development (i.e. day 4 in chicken, day 10 in rodents) and follows a stereotypical temporal and spatial pattern, with a peak of angiogenic activity in early postnatal stages, establishing a reproducible pattern of cortical vascularization in mammals [60, 67].

The precise spatio-temporal orchestration of vascular development in the brain prompted Werner Risau in the 1980s to hypothesize that a paracrine mechanism operates by which angiogenic factors produced by the embryonic brain could activate specific receptors expressed on endothelial cells, thus leading to migration and proliferation of these cells. The first growth factors with angiogenic activities isolated from embryonic chicken brain were identified as acidic (aFGF) and basic fibroblast growth factors (bFGF) [68]. Although aFGF and bFGF (FGF-1, FGF-2) meet the requirements to be potent inducers of endothelial cell proliferation in vitro and of angiogenesis in vivo, developmental studies on their expression pattern in the brain and on their corresponding receptor on endothelial cells did not correlate well with the spatio-temporal pattern of brain angiogenesis [69–71]. In fact, expression of FGFs remains high, even in the adult brain when angiogenesis has ceased [72–74].

After vascular endothelial growth factor (VEGF) was identified, it became clear that this was the candidate paracrine factor stimulating specifically endothelial cell proliferation and sprouting via its high affinity receptors VEGF receptor 1 (VEGFR1, flt-1) and VEGF receptor 2 (VEGFR2, flk-1/KDR) [75]. The analysis of null or conditional alleles for VEGF and its endothelial tyrosine kinase receptors has since demonstrated that VEGF is required for the formation, remodeling and survival of embryonic and early postnatal blood vessels, including the brain, in a dose-dependent manner (for a summary, see [76]).

In the brain, VEGF is produced by cells in the ventricular layer forming a concentration gradient decreasing towards the neuroectodermal surface. The vessels originating from the perineural plexus therefore grow into the CNS along the concentration gradient towards these VEGF-producing cells (for a summary, see [77]). Besides the VEGF gradient, the ingrowth of blood vessels into the brain in a strictly radial manner might be supported by radial glial cells, which provide a preformed scaffold that the endothelial cells can follow [78]. A similar association of blood vessel growth guided by glial cells has been established for the vascularization of the retina [79] and could then also be traced to tracks formed by matrix-binding isoforms of VEGF [80, 81].

The question how a single growth factor could possibly accomplish functions like proliferation and migration of endothelial cells as well as vascular remodeling becomes more understandable when one considers that VEGF is expressed in several isoforms, which are produced by alternative splicing from a single gene. These isoforms differ by the absence or presence of domains that confer the ability to bind heparin and heparan sulfate proteoglycans in the extracellular matrix of cultured cells in vitro and therefore distribute differentially in the environment of VEGF-secreting cells in vivo [82]. Using mice either solely expressing a VEGF isoform lacking a heparin-binding domain or solely expressing a VEGF isoform binding to heparin, Ruhrberg and colleagues [83] elegantly demonstrated that differential diffusion rates and localization of different VEGF isoforms in the extracellular space in fact provide cues for regulating the vascular branch pattern during embryonic development, including brain angiogenesis.

Although VEGF also seems to provide the most important angiogenic stimulus in the brain, it has been recognized that the precise "wiring" of the vascular system, i.e. establishment of the correct vascular network and patterning, does not occur without an ordered series of guidance decisions which are brought about by a number of guidance molecules initially discovered for axons in the nervous system. In this context, the involvement of membrane-bound ligand receptor systems (such as the delta-notch, including the notch target genes Hey1 and Hey2 [84], the ephrinB-EphB families [85, 86], the robo-receptors and slit ligands [87], the netrins [88] and the semaphorins and their receptors of the plexin family [89]) has been demonstrated to play an important role in vascular development, including the brain.

Additional growth factor-receptor systems have been implicated in the stabilization and maturation of the laid down and developing vascular network, including the CNS. In particular, Angiopoietin-1 and -2 (Ang-1, Ang-2) and their common receptor Tie-2 have been shown to be involved in vascular sprouting and remodeling, particularly in the adhesion to the ECM and in the recruitment of perivascular cells [90]. A common feature of Tie-2 and Ang-1 deficient embryos is a defect in the association of endothelial cells with the underlying extracellular matrix and with perivascular cells. Platelet-derived growth factor BB (PDGF-BB) has been shown to be actively involved in the vascularization of the brain by its relevance for pericyte recruitment, both in general and also including the BBB. Interestingly, mice deficient for PDGF-BB show a dramatic decrease in pericyte investment of brain vessels, leading to the formation of lethal microaneurysm during late embryogenesis [91].

Additionally the TGFβ pathway, ubiquitously important during embryogenesis, has been shown to have specific effects in angiogenesis through the receptors Alk-1, Alk-5 and endoglin and their intracellular effectors Smad5 and Smad6 (for a review, see [92]).

Besides all these mechanisms, some evidence points to an involvement of the Wnt family of growth factors in vascular development. In canonical Wnt signaling, binding of the growth factor, which accumulates in the ECM [93, 94], to the frizzled receptor leads to the stabilization of cytoplasmic β-catenin, which involves proteins like disheveled and glycogene synthase kinase 3b (GSK-3b), leading to target gene transcription via a bipartite transcription factor formed by β-catenin and Lef/TCF [95–98].

In endothelial cells, little is known about the role of Wnt signaling during vasculogenesis or angiogenesis. However, the endothelial cell-specific deletion of the β-catenin gene was shown to lead to embryonic lethality around mid-gestation, due to vascular fragility, placenta and heart defects [99, 100], which at least for the heart defect could be linked to the transcriptional activity of β-catenin [99].

Regarding the development of the BBB, Wnt signaling might be of particular interest, as it has been shown that in particular brain endothelial cells are able to undergo canonical Wnt signaling. In vitro it could be demonstrated that primary mouse brain endothelial cells respond to Wnt-1 stimulation with activation of the canonical Wnt pathway involving β-catenin [101] and that endothelial cells growing into the neuroectoderm are positive for a Wnt signaling reporter, whereas other angiogenic vessels in the embryo are negative for Wnt signaling at all developmental stages investigated so far [102].

Support for a specific role of canonical Wnt signaling in brain vascularization comes also from some genetic diseases, like familial exudative vitroretinopathy (FEVR), in which the putative Wnt-receptor frizzled-4 (FZD4) is mutated, leading to a lack of vascularization in the peripheral retina [103]. The FZD4 deficient mouse demonstrates a regression of vessels in the cerebellum, suggesting a role of the Wnt pathway for vascular maintenance in the CNS [104]. Although the exact role of Wnt signaling during brain vascularization remains to be elucidated, the results so far have to be seen in the light of previous reports, which claim that β-catenin protein expression is high during brain vascularization and becomes downregulated after BBB establishment [47].

Even though many of the molecular components involved in brain angiogenesis have now been identified, their exact mechanisms of action are still not fully understood. One major drawback for understanding brain angiogenesis is that these mechanisms also apply outside the CNS. Therefore, it seems unlikely that they are specifically involved in BBB differentiation to create highly specialized endothelial cells. However, the biggest remaining obstacle to the understanding of the roles of these proteins for the development of the vascular bed in the CNS is the fact that mutations in their genes invariably lead to lethal phenotypes during early embryogenesis before BBB differentiation starts.

1.5
Differentiation of the Blood-Brain Barrier

1.5.1
Permeability

When vascular sprouts enter the neural tube early during brain angiogenesis, vessels show a simple sinusoidal morphology and are characterized by their large diameter, irregular shape and the presence of diaphragmatic fenestrae. These vessels are permeable to small hydrophilic substances but probably not to macromolecules (see below). Along with their permeable phenotype, endothelial cells exhibit rudimentary tight junctions, indicated by substantial junctional clefts [105]. As development proceeds, the vessels lose their fenestrae, become smaller and thinner-walled and more regular in shape [54, 106].

In early fetal brain capillaries, the tight junction strands of endothelial cells are short and have low complexity, a pattern which changes dramatically during development, as the strands become longer and interconnected (i.e. complex) and the outer leaflets of adjacent membranes within junctional contacts seem to be fused in so called "membrane kisses" [19, 107]. As initially described, the P-face association of particles in freeze fracture analysis is another parameter to evaluate the tight junction maturity in mammals. Indeed, it could be demonstrated that, in general, the density of particles within tight junction strands increases late during embryonic development (after day 18 in the rat) and after birth a significant increase in the P-face association could be observed, representing a transition to the adult conformation of tight junctions [19].

An important controversy exists with regard to the development of the mammalian BBB [61, 108, 109]. Studies performed in analogy to the early studies of Ehrlich, in which proteins or dyes were injected into the vascular system, showed that plasma proteins entered the fetal brain in significantly higher amounts than they entered adult brain [110–112] whereas others did not [113]. The former studies performed in the mouse and the chick even demonstrated that tightening of the barrier occurs as a gradual process, independent from vascular proliferation and starting late during embryogenesis when angiogenesis is not yet complete [114]. These studies were criticized based on the argument that, due to the small size of the fetus and its total blood volume, the amount of injected tracer as well as the injected volume were "enormous" and caused an artificial leakage of the developing BBB [108]. Based on immunohistochemical studies for native plasma proteins in fetal brains, which avoided the manipulation of small embryos and showed that there is very little plasma protein in fetal and newborn brain, it was argued that the barrier is present to plasma proteins as early as blood vessels form in the brain [115]. This controversy appears to be based on the lack of standardized quantitative BBB permeability measurements from the early fetal stage through later neonatal development to maturity, which might have simplified a comparison between the different species investigated. Comparison of BBB maturation between different species is extremely difficult due to the differences in their rates of brain development and especially since birth is not a reliable marker by which to stage BBB development [116]. Also, the independent maturation of the blood-cerebrospinal fluid barrier at the level of the choroid plexus epithelial cells is rarely considered in these measures [117]. In any case, endothelial barrier permeability to a small molecule, α-amino isobutyric acid (AIB), was shown to decrease in fetuses only late during gestation in sheep [116] and not until 17 days after birth in rabbits [118]. Additionally, permeability to ions as measured by electrical resistance of small pial blood vessels is several times higher in the fetus than in the adult and drops just before birth [119, 120]. Taken together, it seems that, depending on the species, small molecules access the fetal and newborn brain more readily than they access the adult.

It should be stressed at this point, though, that BBB tightness is not just "switched on" at a specific time-point during brain angiogenesis but rather occurs regionally when angiogenesis is still ongoing and increases gradually [121].

1.5.2
Transport Systems and Markers

During BBB development, capillary brain endothelial cells acquire a characteristic set of transport systems, considered to be related to the development of the physical barrier (for detailed reviews, see [53, 57, 61, 77, 122–124].

The glucose transporter Glut-1, establishing a carrier-mediated transport of glucose, is one of the earliest BBB markers as it is already expressed in brain endothelial cells of the first vascular sprouts entering the developing neural tube. At that time in embryogenesis, neuroepithelial cells also express Glut-1 which is, however, rapidly downregulated whereas expression in brain endothelial cells increases [125]. At first, Glut-1 is equally distributed between the luminal and the abluminal membranes of endothelial cells [54]. As development proceeds, expression of Glut-1 on brain endothelial cells increases in response to the increasing demands of the developing brain for glucose, with higher localization at the abluminal versus the luminal membrane.

Also, probably in response to the metabolic needs within the developing brain, the upregulation of the transferrin receptor, belonging to the receptor-mediated transporters, can be observed on endothelial cells forming a BBB but not on those within the choroid plexus or the CVOs [55].

Furthermore, the active efflux transporter P-glycoprotein becomes expressed in endothelial cells early during brain angiogenesis [126]. Expression of P-glycoprotein is required for the differentiation of the BBB [56] and seems to ensure the rapid removal of ingested toxic lipophilic metabolites from the neuroectoderm before the BBB has fully differentiated. Early expression of P-glycoportein at the BBB might be important to protect the developing brain from maternally ingested lipophilic molecules, since the placental barrier is ineffective against lipophilic molecules. It has been generally accepted that P-glycoprotein is localized to the luminal membrane of brain endothelial cells [127]. This viewpoint has, however, been challenged by the findings that, in human and primate brain, P-glycoprotein was immunolocalized to astrocytic endfeet rather than the luminal surface of cerebral endothelial cells [128]. It could well be that P-glycoprotein and other active efflux transporters are expressed in endothelial cells and pericytes and astrocytes working in concert to prevent entry of bloodborne lipophilic molecules into the brain parenchyme.

The upregulation of other markers of the BBB, such as the nonreceptor tyrosine kinase lyn [129] and the Ig-superfamily member HT7 can be phenomenologically correlated to the development of BBB vessels, but their function in CNS endothelial cells still needs to be elucidated [130–132].

In contrast to the genes which become upregulated, the panendothelial cell antigen MECA-32 becomes specifically downregulated in mouse brain endothelia during BBB maturation [133]. As a consequence, MECA-32 antigen is absent on the mature cerebral endothelium, whereas it remains present on vessels outside of the CNS and the capillaries within the CVOs. Based on sequence comparison using the BLAST algorithm, the mouse MECA-32 antigen is the mouse orthologue of the rat PV-1 gene product and the human plasmalemma vesicle-associated protein. PV-1 was shown to be specifically localized to the diaphragms of fenestrated endothelia [134, 135]. This might explain the specific downregulation of the MECA-32 antigen during the maturation of the BBB, as brain vessels lose their fenestrations. Therefore, it seems unlikely that MECA-32 antigen is involved in the paracellular barrier function of brain endothelial cell involving adherens and tight junctions, but rather in the formation of diaphragmed fenestrations directly responsible for the efflux of bloodborne molecules. This seems to be true for brain and nonbrain endothelia lacking these type of fenestrations, also explaining why the vasculature of cardiac and skeletal muscle largely lacks MECA-32 in resting, noninflamed conditions [136].

1.5.3
Extracellular Matrix

Agrin is a heparan sulfate proteoglycan that is required for the development of postsynaptic specializations at the neuromuscular junction. Absence of functional α-dystrophin in Duchenne muscular dystrophy (DMD) and in one of its animal models, the mdx mouse, leads to a reduction of α- and β-dystroglycan, members of a dystrophin-associated glycoprotein complex (DGC) in muscle, which bind to agrin but not to laminin [52]. As the mdx mouse displays a disturbed BBB due to a disorganized endothelial and astrocytic cytoskeleton, it is tempting to speculate that agrin is involved in the differentiation of BBB impermeability by establishing a symmetric polarization of endothelial cells and astrocytes. This is supported by the observation that, during chick and rat development, agrin accumulates in the brain microvascular basal lamina at the time when the BBB becomes less permeable to small tracers [137].

1.5.4
Putative Inductive Mechanisms

Although several aspects of BBB phenotypic development in brain capillary endothelial cells have been monitored, the crucial question concerning the induction of this differentiation process remains to be elucidated. It is obvious that, when brain angiogenesis starts, endothelial cells get in contact not only with various neuroectodermal cells like neuroblasts and with various glial cells and/or their precursors [78], but also with mesodermal cells such as pericytes or microglial cells. That endothelial cells do not show a predetermination to the BBB phenotype was elegantly shown by chick-quail xenografts in which vessels of the coelimic cavity of the embryonic chick acquired BBB characteristics when growing into a developing transplanted quail brain [63, 138]. These observations provided direct evidence that the development of BBB characteristics in endothelial cells is not predetermined but rather induced by the microenvironment of the nervous system during embryogenesis.

Due to the close apposition of astrocytic endfeet to the vessel wall, astrocytes and their precursors have immediately been implicated in BBB differentiation and/or maintenance (Fig. 1.2b) [139, 140]. It was demonstrated, in vivo, that astrocytes are capable of inducing some BBB characteristics in vessels of the anterior eye chamber [141]. However, the importance of astrocytes in that particular experimental system was questioned by observations that iris vessels were impermeable to dyes, even in the absence of astrocytes [142].

It is widely accepted instead that, in vitro, astrocytes or conditioned medium derived from astrocytes induce BBB-like characteristic in endothelial cells [20, 143, 144]. For many years, little has been known about the nature of the astrocyte-derived signals promoting brain endothelial maturation, although a recent publication by Lee et al. [32] shed some light on the astrocyte-endothelial interaction.

Fig. 1.2 a and b

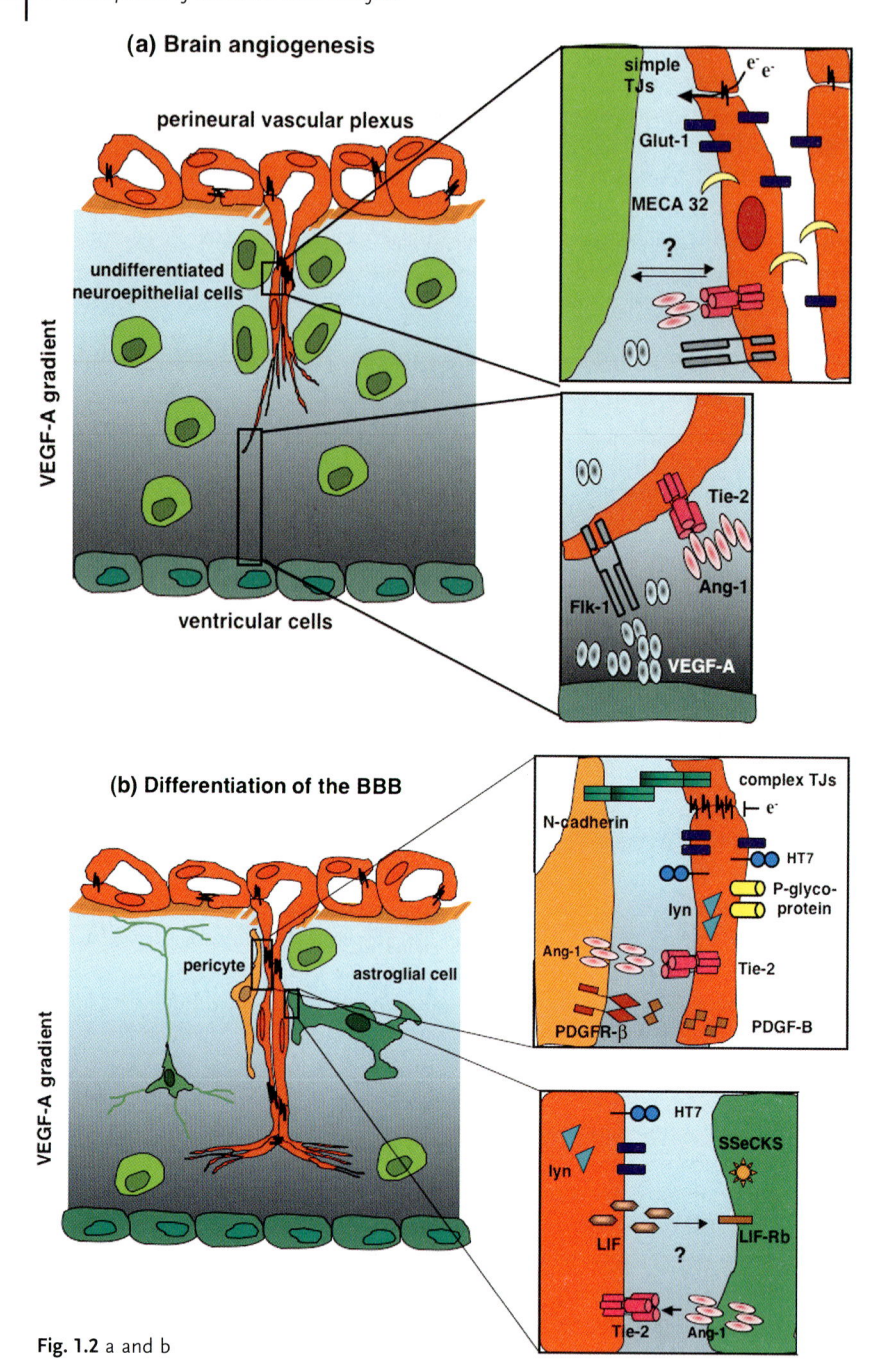

(c) The cellular elements of the mature BBB

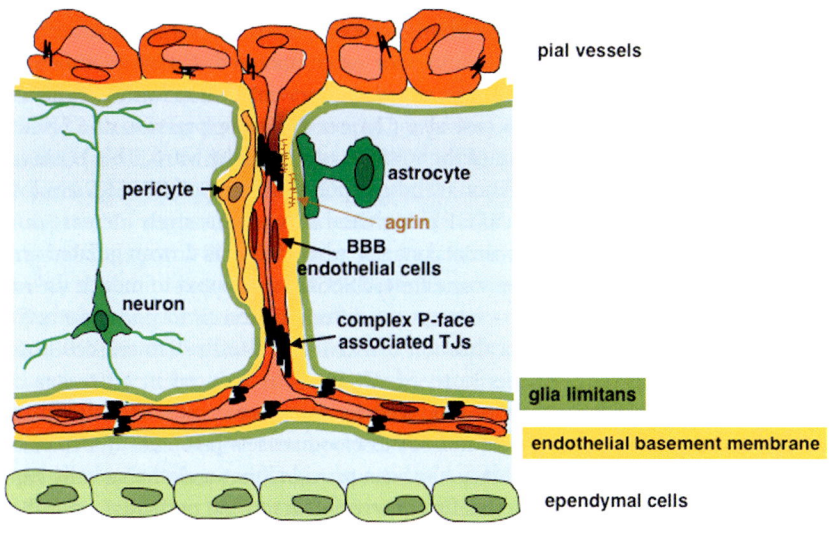

labels on figure: pial vessels; pericyte; astrocyte; agrin; BBB endothelial cells; neuron; complex P-face associated TJs; glia limitans; endothelial basement membrane; ependymal cells

Fig. 1.2 Three steps in BBB differentiation. (a) Brain angiogenesis: vascular sprouts radially invade the embryonic neuroectoderm towards a concentration gradient of VEGF-A, produced by neuroectodermal cells located in the ventricular layer. VEGF-A binds to its endothelial receptor, the receptor tyrosine kinase flk-1/KDR. The endothelial cell-specific receptor tyrosine kinase Tie-2 and its ligand Ang-1 are involved in angiogenic sprouting early during embryogenesis. The cerebral endothelial cells express Glut-1 and the MECA-32 antigen. The tight junctions (simple TJs) are permeable to small molecules.
(b) Differentiation of the BBB: the phenotype of cerebral endothelial cells changes such that they lose expression of the MECA-32 antigen and start to express the HT7 antigen. Glut-1 is now enriched on the abluminal surface of the endothelium. De novo expression of P-glycoprotein and the nonreceptor tyrosine kinase lyn can be observed. The tight junctions become complex TJs, P-face-associated and thus also tight for small polar molecules. Phenotypic changes of endothelial cells are accompanied by their close contact with pericytes and astroglial cells. The molecular mechanisms involved in the interaction between pericytes and endothelial cells are partially characterized and have been shown to be important for vessel maturation within the CNS. Recruitment of pericytes along the differentiating BBB vessels is ensured by several mechanisms:
(1) PDGF-B produced by endothelial cells binds to its receptor PDGFR-β on pericytes, (2) N-cadherin enriched on the respective membranes facing the neighboring cell type interact with each other and (3) Ang-1 expressed by pericytes binds to the endothelial receptor tyrosine kinase Tie-2. Only recently, some light has been shed on the molecular interactions between endothelial cells and astroglial cells in the developing CNS: (a) endothelial cells produce LIF, which induces the maturation of astrocytes via the LIF-Rb and (b) due to the presence of vessels, the oxygen level increases and endothelial cells produce PDGF-B, both leading to an upregulation of SSeCKS in astrocytes. In turn, SSeCKS upregulates Ang-1 expression in astrocytes, which acts as an endothelial differentiation marker and positively influences the membrane localization of junction protein as ZO-1 and claudin-1.
(c) The cellular elements of the mature BBB: despite the fact that the cerebral endothelial cells form the barrier proper, close contact with pericytes, astrocytes and maybe neuronal cells is required for the maintenance of the BBB. The presence of the heparan sulfate proteoglycan agrin in the endothelial basal membrane might be important for polarization of the cells involved in establishing the BBB. The precise molecular mechanisms involved in the supposedly paracrine crosstalk between the cellular elements required for BBB maintenance in the mature CNS remain unknown to date. Modified from an original figure in [170].

46 Dejana, E. **1996**, Endothelial adherens junctions: implications in the control of vascular permeability and angiogenesis, *J. Clin. Invest.* 98, 1949–1953.

47 Liebner, S., Gerhardt, H., Wolburg, H. **2000**, Differential expression of endothelial beta-catenin and plakoglobin during development and maturation of the blood-brain and blood-retina barrier in the chicken, *Dev. Dyn.* 271, 86–98.

48 Breier, G., et al. **1995**, Molecular cloning and expression of murine VE-cadherin in early developing cardiovascular system (in press).

49 Vorbrodt, A. W., Dobrogowska, D. H. **2004**, Molecular anatomy of interendothelial junctions in human blood-brain barrier microvessels, *Folia Histochem. Cytobiol.* 42, 67–75.

50 Duncan, G. S., et al. **1999**, Genetic evidence for functional redundancy of platelet/endothelial cell adhesion molecule-1 (PECAM-1): CD31-deficient mice reveal PECAM-1-dependent and PECAM-1-independent functions, *J. Immunol.* 162, 3022–3030.

51 Graesser, D., et al. **2002**, Altered vascular permeability and early onset of experimental autoimmune encephalomyelitis in PECAM-1-deficient mice, *J. Clin. Invest.* 109, 383–392.

52 Nico, B., et al. **2003**, Severe alterations of endothelial and glial cells in the blood-brain barrier of dystrophic mdx mice, *Glia* 42, 235–251.

53 Pardridge, W. M. **2005**, Molecular biology of the blood-brain barrier, *Mol. Biotechnol.* 30, 57–70.

54 Bolz, S., et al. **1996**, Subcellular distribution of glucose transporter (GLUT-1) during development of the blood-brain barrier in rats, *Cell Tissue Res.* 284, 355–365.

55 Kissel, K., et al. **1998**, Immunohistochemical localization of the murine transferrin receptor (TfR) on blood-tissue barriers using a novel anti-TfR monoclonal antibody, *Histochem. Cell Biol.* (in press).

56 Schinkel, A. H., et al. **1994**, Disruption of the mouse mdr1a P-glycoprotein gene leads to a deficiency in the blood-brain barrier and to increased sensitivity to drugs, *Cell* 77, 491–502.

57 Stewart, P. A. **2000**, Development of the blood-brain barrier, in *Morphogenesis of Endothelium*, eds. W. Risau, G. M. Rubanyi, Harwood, Amsterdam, pp 109–122.

58 el-Bacha, R. S., Minn, A. **1999**, Drug metabolizing enzymes in cerebrovascular endothelial cells afford a metabolic protection to the brain, *Cell Mol. Biol.* 45, 15–23.

59 Song, L., Pachter, J. S. **2003**, Culture of murine brain microvascular endothelial cells that maintain expression and cytoskeletal association of tight junction-associated proteins, *In Vitro Cell Dev. Biol. Anim.* 39, 313–320.

60 Bär, T. **1980**, The vascular system of the cerebral cortex, *Adv. Anat. Embryol. Cell Biol.* 59, 1–62.

61 Risau, W., Wolburg, H. **1990**, Development of the blood-brain barrier, *Trends Neurosci.* 13, 174–178.

62 Dermietzel, R., Krause, D. **1991**, Molecular anatomy of the blood-brain barrier as defined by immunocytochemistry, *Int. Rev. Cytol.* 127, 57–109.

References

63 Stewart, P. A., Wiley, M. J. **1981**, Developing nervous tissue induces formation of blood-brain barrier characteristics in invading endothelial cells: a study using quail-chick transplantation chimeras, *Dev. Biol.* 84, 183–192.

64 Noden, D. M. **1991**, Development of craniofacial blood vessels, in *Development of the Vascular System*, eds. R. N. Feinberg, G. K. Sherer, R. Auerbach, Karger, Basel, pp 1–24.

65 Risau, W. **1997**, Mechanisms of angiogenesis, *Nature* 386, 671–674.

66 Roncali, L., et al. **1986**, Microscopical and ultrastructural investigations on the development of the blood-brain barrier in the chick embryo optic tectum, *Acta Neuropathol.* 70, 193–201.

67 Kuban, K. C., Gilles, F. H. **1985**, Human telencephalic angiogenesis, *Ann. Neurol.* 17, 539–548.

68 Risau, W., Gautschi-Sova, P., Bohlen, P. **1988**, Endothelial cell growth factors in embryonic and adult chick brain are related to human acidic fibroblast growth factor, *EMBO J.* 7, 959–962.

69 Folkman, J., Klagsbrun, M. **1987**, Vascular physiology: a family of angiogenic peptides (news), *Nature* 329, 671–672.

70 Folkman, J., et al. **1988**, A heparin-binding angiogenic protein – basic fibroblast growth factor – is stored within basement membrane, *Am. J. Pathol.* 130, 393–400.

71 Risau, W., Gautschi-Sova, P., Böhlen, P. **1988**, Endothelial cell growth factors in embryonic and adult chick brain are related to human acidic fibroblast growth factors, *EMBO J.* 7, 959–962.

72 Schnurch, H., Risau, W. **1991**, Differentiating and mature neurons express the acidic fibroblast growth factor gene during chick neural development, *Development* 111, 1143–1154.

73 Emoto, N., et al. **1989**, Basic fibroblast growth factor (FGF) in the central nervous system: identification of specific loci of basic FGF expression in the rat brain, *Growth Factors* 2, 21–29.

74 Claus, P., Grothe, C. **2001**, Molecular cloning and developmental expression of rat fibroblast growth factor receptor 3, *Histochem. Cell Biol.* 115, 147–155.

75 Ferrara, N., Henzel, W. J. **1989**, Pituitary follicular cells secrete a novel heparin-binding growth factor specific for vascular endothelial cells, *Biochem. Biophys. Res. Commun.* 161, 851–858.

76 Gale, N. W., Yancopoulos, G. D. **1999**, Growth factors acting via endothelial cell-specific receptor tyrosine kinases: VEGFs, angiopoietins, and ephrins in vascular development, *Genes Dev.* 13, 1055–1066.

77 Engelhardt, B., Risau, W. **1995**, The development of the blood-brain barrier, in *New Concepts of a Blood-Brain Barrier*, eds. J. Greenwood, D. Begley, M. Segal, Plenum, London.

78 Bass, T., et al. **1992**, Radial glial interaction with cerebral germinal matrix capillaries in the fetal baboon, *Exp. Neurol.* 118, 126–132.

79 Fruttiger, M., et al. **1996**, PDGF mediates a neuron-astrocyte interaction in the developing retina, *Neuron* 17, 1117–1131.

80 Gerhardt, H., et al. **2003**, VEGF guides angiogenic sprouting utilizing endothelial tip cell filopodia, *J. Cell Biol.* 161, 1163–1177.

81 Stone, J., Dreher, Z. **1987**, Relationship between astrocytes, ganglion cells and vasculature of the retina, *J. Comp. Neurol.* 255, 35–49.

82 Houck, K. A., et al. **1992**, Dual regulation of vascular endothelial growth factor bioavailability by genetic and proteolytic mechanisms, *J. Biol. Chem.* 267, 26031–26037.

83 Ruhrberg, C., et al. **2002**, Spatially restricted patterning cues provided by heparin-binding VEGF-A control blood vessel branching morphogenesis, *Genes Dev.* 16, 2684–2698.

84 Fischer, A., et al. **2004**, The notch target genes Hey1 and Hey2 are required for embryonic vascular development, *Genes Dev.* 18, 901–911.

85 Adams, R. H. **2002**, Vascular patterning by Eph receptor tyrosine kinases and ephrins, *Semin. Cell Dev. Biol.* 13, 55–60.

86 Shawber, C. J., Kitajewski, J. **2004**, Notch function in the vasculature: insights from zebrafish, mouse and man, *Bioessays* 26, 225–234.

87 Park, K. W., et al. **2003**, Robo4 is a vascular-specific receptor that inhibits endothelial migration, *Dev. Biol.* 261, 251–267.

88 Eichmann, A., Makinen, T., Alitalo, K. **2005**, Neural guidance molecules regulate vascular remodeling and vessel navigation, *Genes Dev.* 19, 1013–1021.

89 Deutsch, U. **2004**, Semaphorins guide PerPlexeD endothelial cells, *Dev. Cell.* 7, 1–2.

90 Davis, S., Yancopoulos, G. D. **1999**, The angiopoietins: yin and yang in angiogenesis, *Curr. Top. Microbiol. Immunol.* 237, 173–185.

91 Lindahl, P., et al. **1997**, Pericyte loss and microaneurysm formation in PDGF-B-deficient mice, *Science* 277, 242–245.

92 Goumans, M. J., Lebrin, F., Valdimarsdottir, G. **2003**, Controlling the angiogenic switch: a balance between two distinct TGF-β receptor signaling pathways, *Trends Cardiovasc Med.* 13, 301–307.

93 Papkoff, J., Brown, A. M., Varmus, H. E. **1987**, The int-1 proto-oncogene products are glycoproteins that appear to enter the secretory pathway, *Mol. Cell Biol.* 7, 3978–3984.

94 Papkoff, J., Schryver, B. **1990**, Secreted int-1 protein is associated with the cell surface, *Mol. Cell Biol.* 10, 2723–2730.

95 Gumbiner, B. M. **1995**, Signal transduction by β-catenin, *Curr. Opin. Cell Biol.* 7, 634–640.

96 Behrens, J., et al. **1996**, Functional interaction of β-catenin with the transcription factor LEF-1, *Nature* 382, 638–642.

97 Zhurinsky, J., Shutman, M., Ben-Ze'ev, A. **2000**, Plakoglobin and beta-catenin: protein interactions, regulation and biological roles, *J. Cell Sci.* 113, 3127–3139.

98 Miller, J. R., Moon, R. T. **1996**, Signal transduction through β-catenin and specification of cell fate during embryogenesis, *Genes Dev.* 10, 2527–2539.

99 Liebner, S., et al. **2004**, Beta-catenin is required for TGF-beta-mediated endothelial to mesenchymal transformation during heart cushion development in the mouse, *J. Cell Biol.* (in press).

100 Cattelino, A., et al. **2003**, The conditional inactivation of the β-catenin gene in endothelial cells causes a defective vascular pattern and increased vascular fragility, *J. Cell Biol.* 162, 1111–1122.

101 Wright, M., et al. **1999**, Identification of a Wnt-responsive signal transduction pathway in primary endothelial cells, *Biochem. Biophys. Res. Commun.* 263, 384–388.

102 Maretto, S., et al. **2003**, Mapping Wnt/beta-catenin signaling during mouse development and in colorectal tumors, *Proc. Natl Acad. Sci. USA* 100, 3299–3304.

103 Robitaille, J., et al. **2002**, Mutant frizzled-4 disrupts retinal angiogenesis in familial exudative vitreoretinopathy, *Nat. Genet.* 32, 326–330.

104 Xu, Q., et al. **2004**, Vascular development in the retina and inner ear: control by norrin and frizzled-4, a high-affinity ligand-receptor pair, *Cell* 116, 883–895.

105 Cassella, J. P., Lawrenson, J. G., Firth, J. A. **1997**, Development of endothelial paracellular clefts and their tight junctions in the pial microvessels of the rat, *J. Neurocytol.* 26, 567–575.

106 Stewart, P. A., Hayakawa, E. M. **1994**, Early ultrastructural changes in blood-brain barrier vessels of the rat embryo, *Dev. Brain Res.* 78, 25–34.

107 Schulze, C., Firth, J. A. **1992**, Interendothelial junctions during blood-brain barrier development in the rat: morphological changes at the level of individual tight junctional contacts, *Dev. Brain Res.* 1–11.

108 Saunders, N. R., Dziegielewska, K. M., Mollgard, K. **1991**, Letter to the editor: the importance of the blood-brain barrier in fetuses and embryos, *Trends Neurosci.* 14, 14.

109 Risau, W., Wolburg, H. **1991**, The importance of the blood-brain barrier in fetuses and embryos – reply, *Trends Neurosci.* 14, 15.

110 Risau, W. **1991**, Induction of blood-brain barrier endothelial cell differentiation, *Ann. NY Acad. Sci.* 633, 405–419.

111 Wakai, S., Hirokawa, N. **1978**, Development of the blood-brain barrier to horseradish peroxidase in the chick embryo, *Cell Tissue Res.* 195, 195–203.

112 Stewart, P. A., Hayakawa, E. M. **1987**, Interendothelial junctional changes underlie the developmental 'tightening' of the blood-brain barrier, *Dev. Brain Res.* 32, 271–281.

113 Olsson, Y., et al. **1968**, Blood-brain barrier to albumin in embryonic new born and adult rats, *Acta Neuropathol.* 10, 117–122.

114 Risau, W., Hallmann, R., Albrecht, U. **1986**, Differentiation-dependent expression of protein in brain endothelium during development of the blood-brain barrier, *Dev. Biol.* 117, 537–545.

115 Mollgard, K., et al. **1988**, Synthesis and localization of plasma proteins in the developing human brain. Integrity of the fetal blood-brain barrier to endogenous proteins of hepatic origin, *Dev. Biol.* 128, 207–221.

116 Stonestreet, B.S., et al. **1996**, Ontogeny of blood-brain barrier function in ovine fetuses, lambs, and adults, *Am. J. Physiol.* 271, R1594–R1601.

117 Dziegielewska, K.M., et al. **2001**, Development of the choroid plexus, *Microsc. Res. Tech.* 52, 5–20.

118 Tuor, U.I., Simone, C., Bascaramurty, S. **1992**, Local blood-brain barrier in the newborn rabbit: postnatal changes in alpha-aminoisobutyric acid transfer within medulla, cortex, and selected brain areas, *J. Neurochem.* 59, 999–1007.

119 Butt, A.M., Jones, H.C., Abbott, N.J. **1990**, Electrical resistance across the blood-brain barrier in anaesthetized rats: a developmental study, *J. Physiol.* 429, 47–62.

120 Keep, R.F., et al. **1995**, Developmental changes in blood-brain barrier potassium permeability in the rat: relation to brain growth, *J. Physiol.* 488, 439–448.

121 Robertson, P.L., et al. **1985**, Angiogenesis in developing rat brain: an in vivo and in vitro study, *Dev. Brain Res.* 23, 219–223.

122 Betz, A.L., Goldstein, G.W. **1986**, Specialized properties and solute transport in brain capillaries, *Annu. Rev. Physiol.* 48, 241–250.

123 Broadwell, R.D. **1989**, Transcytosis of macromolecules through the blood-brain barrier: a cell biological perspective and critical appraisal, *Acta Neuropathol.* 79, 117–128.

124 Pardridge, W.M. **1988**, Recent advances in blood-brain barrier transport, *Annu. Rev. Pharmacol. Toxicol.* 28, 25–39.

125 Dermietzel, R., et al. **1992**, Pattern of glucose transporter (Glut 1) expression in embryonic brains is related to maturation of blood-brain barrier tightness, *Dev. Dyn.* 193, 152–163.

126 Qin, Y., Sato, T.N. **1995**, Mouse multidrug resistance 1a/3 gene is the earliest known endothelial cell differentiation marker during blood-brain barrier development, *Dev. Dyn.* 202, 172–180.

127 Stewart, P.A., Beliveau, R., Rogers, K.A. **1996**, Cellular localization of P-glycoprotein in brain versus gonadal capillaries, *J. Histochem. Cytochem.* 44, 679–685.

128 Golden, P.L., Pardridge, W.M. **2000**, Brain microvascular P-glycoprotein and a revised model of multidrug resistance in brain, *Cell Mol. Neurobiol.* 20, 165–181.

129 Achen, M.G., et al. **1995**, The non-receptor tyrosine kinase Lyn is localised in the developing murine blood-brain barrier, *Differentiation* 59, 15–24.

130 Albrecht, U., et al. **1990**, Correlation of blood-brain barrier function and HT7 protein distribution in chick brain circumventricular organs, *Brain Res.* 535, 49–61.

131 Gerhardt, H., Liebner, S., Wolburg, H. **1996**, The pecten oculi of the chicken as a new in vivo model of the blood-brain barrier, *Cell Tissue Res.* 285, 91–100.

132 Bertossi, M., et al. **2002**, Developmental changes of HT7 expression in the microvessels of the chick embryo brain, *Anat. Embryol.* 205, 229–233.

133 Hallmann, R., et al. **1995**, Novel mouse endothelial cell surface marker is suppressed during differentiation of the blood-brain barrier, *Dev. Dyn.* 202, 325–332.

134 Stan, R. V., et al. **1999**, Isolation, cloning, and localization of rat PV-1, a novel endothelial caveolar protein, *J. Cell Biol.* 145, 1189–1198.

135 Stan, R. V., Kubitza, M., Palade, G. E. **1999**, PV-1 is a component of the fenestral and stomatal diaphragms in fenestrated endothelia, *Proc. Natl Acad. Sci. USA* 96, 13203–13207.

136 Leppink, D. M., et al. **1989**, Inducible expression of an endothelial cell antigen on murine myocardial vasculature in association with interstitial cellular infiltration, *Transplantation* 48, 874–877.

137 Barber, A. J., Lieth, E. **1997**, Agrin accumulates in the brain microvascular basal lamina during development of the blood-brain barrier, *Dev. Dyn.* 208, 62–74.

138 Ikeda, E., Flamme, I., Risau, W. **1996**, Developing brain cells produce factors capable of inducing the HT7 antigen, a blood-brain barrier-specific molecule, in chick endothelial cells, *Neurosci. Lett.* 209, 149–152.

139 Phelps, C. H. **1972**, The development of glio-vascular relationships in the rat spinal cord, *Z. Zellforsch.* 128, 555–563.

140 Goldstein, G. W. **1988**, Endothelial cell-astrocyte interactions: a cellular model of the blood-brain barrier, *Ann. NY Acad. Sci.* 529, 31–39.

141 Janzer, R. C., Raff, M. C. **1987**, Astrocytes induce blood-brain barrier properties in endothelial cells, *Nature* 325, 253–257.

142 Small, R. K., et al. **1993**, Functional properties of retinal Muller cells following transplantation to the anterior eye chamber, *Glia* 7, 158–169.

143 Rubin, L. L., et al. **1991**, A cell culture model of the blood-brain barrier, *J. Cell Biol.* 115, 1725–1735.

144 Meresse, S., et al. **1989**, Bovine brain endothelial cells express tight junctions and monoamine oxidase activity in long-term culture, *J. Neurochem.* 53, 1363–1371.

145 Coats, S. R., et al. **2002**, Ligand-specific control of src-suppressed C kinase substrate gene expression, *Biochem. Biophys. Res. Commun.* 297, 1112–1120.

146 Bjarnegard, M., et al. **2004**, Endothelium-specific ablation of PDGFB leads to pericyte loss and glomerular, cardiac and placental abnormalities, *Development* 131, 1847–1857.

147 Mi, H., Haeberle, H., Barres, B. A. **2001**, Induction of astrocyte differentiation by endothelial cells, *J. Neurosci.* 21, 1538–1547.

148 Sims, D. E. **1986**, The pericyte – a review, *Tissue Cell* 18, 153–174.

149 Lindblom, P., et al. **2003**, Endothelial PDGF-B retention is required for proper investment of pericytes in the microvessel wall, *Genes Dev.* 17, 1835–1840.

150 Suri, C., et al. **1996**, Requisite role of angiopoietin-1, a ligand for the TIE2 receptor, during embryonic angiogenesis, *Cell* 87, 1171–1180.

151 Oh, S. P., et al. **2000**, Activin receptor-like kinase 1 modulates transforming growth factor-beta 1 signaling in the regulation of angiogenesis, *Proc. Natl Acad. Sci. USA* 97, 2626–2631.

152 Yang, X., et al. **1999**, Angiogenesis defects and mesenchymal apoptosis in mice lacking SMAD5, *Development* 126, 1571–1580.

153 Larsson, J., et al. **2001**, Abnormal angiogenesis but intact hematopoietic potential in TGF-beta type I receptor-deficient mice, *EMBO J.* 20, 1663–1673.

154 Oshima, M., Oshima, H., Taketo, M. M. **1996**, TGF-beta receptor type II deficiency results in defects of yolk sac hematopoiesis and vasculogenesis, *Dev. Biol.* 179, 297–302.

155 Li, D. Y., et al. **1999**, Defective angiogenesis in mice lacking endoglin, *Science* 284, 1534–1537.

156 Dickson, M. C., et al. **1995**, Defective haematopoiesis and vasculogenesis in transforming growth factor-beta 1 knock out mice, *Development* 121, 1845–1854.

157 Gerhardt, H., Wolburg, H., Redies, C. **2000**, N-cadherin mediates pericytic-endothelial interaction during brain angiogenesis in the chicken, *Dev. Dyn.* 218, 472–479.

158 Gerhardt, H., et al. **1999**, N-cadherin expression in endothelial cells during early angiogenesis in the eye and brain of the chicken: relation to blood-retina and blood-brain barrier development, *Eur. J. Neurosci.* 11, 1191–1201.

159 Miwa, N., et al. **2000**, Involvement of claudin-1 in the beta-catenin/Tcf signaling pathway and its frequent upregulation in human colorectal cancers, *Oncol. Res.* 12, 469–476.

160 Bauer, H. C., Bauer, H. **2000**, Neural induction of the blood-brain barrier: still an enigma, *Cell Mol. Neurobiol.* 20, 13–28.

161 McCarty, J. H., et al. **2002**, Defective associations between blood vessels and brain parenchyma lead to cerebral hemorrhage in mice lacking alpha integrins, *Mol. Cell Biol.* 22, 7667–7777.

162 Zhu, J., et al. **2002**, Abstract beta integrins are required for vascular morphogenesis in mouse embryos, *Development* 129, 2891–2903.

163 Engelhardt, B., Conley, F. K., Butcher, E. C. **1994**, Cell adhesion molecules on vessels during inflammation in the mouse central nervous system, *J. Neuroimmunol.* 51, 199–208.

164 Barber, A. J., Lieth, E. **1997**, Agrin accumulates in the brain microvascular basal lamina during development of the blood-brain barrier, *Dev. Dyn.* 208, 62–74.

165 Rascher, G., et al. **2002**, Extracellular matrix and the blood-brain barrier in glioblastoma multiforme: spatial segregation of tenascin and agrin, *Acta Neuropathol.* 104, 85–91.

166 Kniesel, U., Wolburg, H. **2000**, Tight junctions of the blood-brain barrier, *Cell Mol. Neurobiol.* 20, 57–76.

167 Coisne, C., et al. **2005**, Mouse syngeneic in vitro blood-brain barrier model: a new tool to examine inflammatory events in cerebral endothelium, *Lab. Invest.* 85, 734–746.

168 Farrell, C. L., Risau, W. **1994**, Normal and abnormal development of the blood-brain barrier, *Microsc. Res. Tech.* 27, 495–506.

169 Alt, C., et al. **2005**, Gene and protein expression profiling of the microvascular compartment in experimental autoimmune encephalomyelitis in C57Bl/6 and SJL mice, *Brain Pathol.* 15, 1–16.

170 Engelhardt, B. **2003**, Development of the blood-brain barrier, *Cell Tissue Res.* 314, 119–129.

2
Brain Angiogenesis and Barriergenesis

Jeong Ae Park, Yoon Kyung Choi, Sae-Won Lee, and Kyu-Won Kim

2.1
Introduction

Brain vasculature begins with an initial network of pia vessels, called as the primary perivascular plexus that is derived from the differentiation of angioblasts [1, 2]. These brain vessels eventually differentiate into the complex microvessel network, which requires at least two distinct processes: brain angiogenesis and barriergenesis [2, 3] (see Chapter 1). Brain angiogenesis involves endothelial proliferation and migration, endothelial assembling, vessel branching, and sprouting from the preexisting vessels, while barriergenesis involves remodeling and maturation into the blood-brain barrier (BBB) properties [1, 3–8]. Therefore, brain angiogenesis and barriergenesis occur throughout the embryonic and postnatal stages of the developing brain.

Intensive investigation into the development of brain vasculature has revealed the importance of signaling through the coordinated interaction with brain microenvironment. In the embryonic brain, active proliferation and migration of neuroglial progenitors from the subventricular zone (SVZ) to the cortex region result in the limiting of oxygen delivery by diffusion from the perivascular plexus. Subsequently, an oxygen gradient is generated and neuroglia-derived factors are released into the brain [8, 9]. The plexus may induce successive waves of angiogenic vessels toward the oxygen gradient and the factors. Thus, angiogenic vessels run between rows of migrating neuroglia, which can lead neuroglia to neuronal/astrocytic differentiation [10, 11]. Then, the new blood vessels will deliver oxygen and make contact with neuroglial progenitor cells. Accordingly, in the oxygenated brain, the endothelial cells (ECs) lining angiogenic vessels lie under the influence of the neural/glial environment, where they acquire a selective permeability barrier, the BBB [1, 3, 6, 12–15]. Therefore, elucidation of the cellular and molecular mechanisms of brain angiogenesis and barriergenesis in response to oxygen tension will provide an insight to delineate the developmental process of the BBB.

Blood-Brain Interfaces: From Ontogeny to Artificial Barriers.
Edited by R. Dermietzel, D.C. Spray, M. Nedergaard
Copyright © 2006 WILEY-VCH Verlag GmbH & Co. KGaA, Weinheim
ISBN: 3-527-31088-6

2.2
Brain Angiogenesis

Brain vessels are formed by angiogenic processes from primary perivascular plexus in the pia. In early embryonic brain, neuroglial cells from SVZ grow and migrates beyond the limit of oxygen diffusion from the perivascular plexus of the pia vessels [3]. This situation probably generates local hypoxia, which acts as a strong stimulus for the induction of angiogenesis. The generated hypoxia may trigger the growth of angiogenic vessels by hypoxia-inducible factor (HIF-1) [7, 16]. HIF-1 is a transcription factor that contributes to a variety of developmental and physiological events as well as several diseases [17, 18]. HIF-1 regulates a number of genes responsive to low cellular oxygen tension, especially angiogenesis. Examples include vascular endothelial growth factor (VEGF) and its receptors, platelet-derived growth factor-β (PDGF-β), basic fibroblast growth factor (bFGF), and erythropoietin (EPO) [19, 20].

2.2.1
Hypoxia-Regulated HIF-1 in the Development of the Brain

The existence of hypoxia in the developing brain has been found by using a hypoxia-specific marker, pimonidazole hydrochloride, which is converted by hypoxia-activated nitroreductase into a reactive intermediate to form adducts covalently with cellular component in hypoxic regions [21, 22]. In the early embryonic stage, hypoxia is detected in the mesenchymal region and gradually spreads into neural tubes. Moreover, the hypoxic region is well matched to that of HIF-1α immunoreactivity. In embryonic (E) days E13 and E18 in mice, hypoxic regions seem to move to the olfactory lobe, some connective tissues of craniofacial regions, marginal layer, and the ventricular neuroepithelia of the cortex, where neuroglial cells proliferate and generate locally avascular regions [21]. At postnatal (P) day P3, hypoxic regions are spread throughout almost the whole cerebral cortex (Fig. 2.1). As angiogenic vessels run into all brain regions, hypoxic regions are likely to disappear and then rarely be detected in the cerebral cortex of P21, when brain angiogenesis ceases. In addition, it has been reported that mouse brain exposed to lower oxygen concentrations also accumulates HIF-1α [23, 24].

Based on above observations, the appearance of hypoxic regions spatiotemporally occurring during brain development seems consistent with the idea that neuroglial cells proliferate and migrate away from the existing blood vessels, creating an oxygen gradient in the brain tissue [21, 24]. The hypoxic regions stimulate the expression of HIF-1α to increase the oxygen supply through new blood vessel formation [25, 26]. Therefore, hypoxia is implicated as a critical factor for vascularization in the developing brain through HIF-1α activation.

Further supporting the role of hypoxia in the developing brain, other lines of investigation have suggested that systemic inactivation of HIF-1α alters vascular as well as brain development. In HIF-1$\alpha^{-/-}$ mice, abnormal neural development

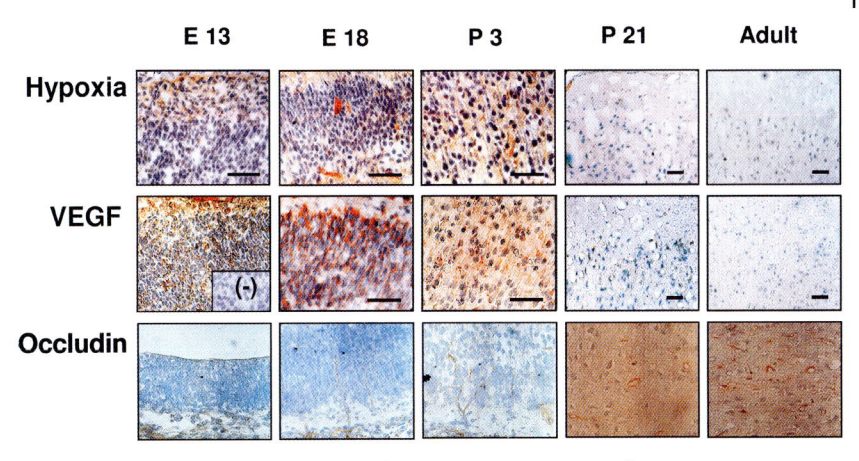

Fig. 2.1 Immunohistochemical staining of hypoxic regions, VEGF, and occludin in developing rat brain. Positive immunoreactivity is represented by the brown color of the DAB. Scale bars: 100 μm. (Adapted from [49], with permission).

and a defect in cephalic angiogenesis have been observed [22, 27]. More direct support for a hypoxic role has been investigated through the study of conditional knockout mice, with specifically targeted HIF-1α genes in the neural precursor cells of the developing brain [28]. Neural cell-specific ablations of HIF-1α, although ECs express HIF-1α, exhibit not only a reduction in neural cells but also a vascular regression. The defective vessels in the mutant are likely due to the disturbance of the hypoxic response from HIF-1α deficient neural cells. Therefore, hypoxia occurring in the development of the brain is likely a strong angiogenic signal to induce the formation of an embryonic vascular network [16, 17].

Given the widespread HIF-1 expression in the developing brain that experienced an oxygen gradient, the understanding of HIF signaling pathways can provide a clue for the investigation of hypoxia-induced angiogenesis associated with brain development.

2.2.2
Hypoxia-Inducible Factor

HIF-1 is composed of two subunits, an oxygen-sensitive HIF-1α subunit and an oxygen-insensitive HIF-1β subunit [29]. HIF-1α is stable under hypoxia, whereas it is rapidly degraded under normoxia. Each subunit has the basic helix-loop-helix (bHLH) and the PER-ARNT-SIM (PAS) domain. The N-terminal of HIF-1α contains bHLH and PAS that are required for dimerization and DNA binding. The C-terminal regions are required for degradation and transactivation, which include the oxygen-dependent degradation (ODD) domain, two transactivation domains (N-TAD, C-TAD), and an inhibitory domain (ID) [17].

HIF-1 activity is unlikely to be regulated in the transcriptional level, because of the rapid degradation of HIF-1α in oxygenated cells [17–19]. Rather, HIF-1α is regulated by processes involving stability and subcellular localization [30–32]. From the viewpoint of the availability of oxygen, the regulatory mechanisms of HIF-1 have been extensively studied [19]. Recently, insights into the oxygen-regulatory function of HIF-1α have come from studies on posttranslational modifications, such as ubiquitination, hydroxylation, acetylation, and sumolylation.

Under normoxic conditions, HIF-1α is subject to rapid degradation by the process of the pVHL-mediated ubiquitin protease pathway. The ODD domain within HIF-1α interacts with the pVHL that is part of an E3 ubiquitin ligase complex for HIF-1α polyubiquitination, letting HIF-1α be exposed to the pVHL-degradation pathway [31, 33, 34]. In detail, the association of HIF-1 with pVHL is triggered by HIF prolyl hydroxylase (PHD) that hydroxylates the two prolyl residues, Pro402 and Pro564, which are located within the ODD domain of HIF-1α [35–39]. In addition, factor inhibiting HIF-1 (FIH-1) exerts hydroxylation of an asparagine residue (Asn803) located in the C-terminal transactivation domain (C-TAD) [40–42]. Thereafter, FIH-1 downregulates transcriptional activation by apparently interfering with the recruitment of coactivator p300. These hydroxylases are dioxygenases requiring oxygen and 2-oxoglutarate, which transfer one oxygen atom to prolyl and asparagyl residues, creating hydroxylated amino acid, and the other oxygen atom reacts with 2-oxoglutarate, generating succinate and CO_2 [43]. The hydroxylase activity is likely to link oxygen availability. Hence, the posttranslational hydroxylation plays a key role in the direct regulation of HIF-1α stability.

While hydroxylation takes center stage in the regulation of HIF-1α stability, acetylation seems to be another important modification. Acetylation is mediated by ARD1, which has been identified as a protein interacting with the HIF-1α ODD domain [44]. Interestingly, ARD1 acetylates Lys532 residue within the ODD domain. The acetylation of Lys532 increases the proteosomal degradation of HIF-1α, whereas the mutation of Lys532 decreases the interaction with VHL and stabilizes HIF-1α. It is intriguing that mouse N-terminal acetyltransferase 1 (mNAT-1) is expressed in the developing brain, which may combine with ARD1 to form a functional acetyltransferase, suggesting an important role in dividing neuroglia [45]. In addition, recent analysis of the yeast two-hybrid assay reveals that Tid-1L [a mouse homologue of *Drosophila* tumor suppressor l(2)tid] interacting with pVHL accelerates the binding between HIF-1α and pVHL, which leads to reduced HIF-1α protein levels.

Under hypoxic condition, degradation of HIF-1α is prevented, allowing HIF-1α to accumulate and be stabilized within the nucleus. Then, HIF-1α dimerizes with HIF-1β to become an active transcription factor. HIF-1α and HIF-1β dimers recognize HIF-responsive elements (HREs) within the promoter of hypoxia-responsive target genes. Recent reports on the stability and transcriptional activity of HIF-1 come from the small ubiquitin-related modifier (SUMO) study. The modification by SUMO-1, sumoylation, is known to regulate transcriptional

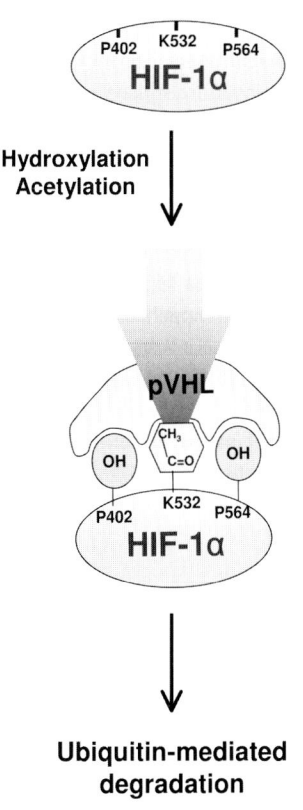

Fig. 2.2 Posttranslational modification of HIF-1α. Under normoxia, HIF-1α is subject to rapid degradation by the pVHL-mediated ubiquitin-proteasome pathway. Hydroxylation on Pro402 and Pro564 and acetylation on Lys532 within the ODD domain of HIF-1α potentiate the interaction with pVHL.

activities and the primary function of which prevents proteins from proteosomal degradation [46]. SUMO-1 and HIF-1α are concomitantly increased in hypoxic neurons and brain [47, 48]. Ectopic expression of SUMO-1 demonstrates that SUMO-1 increases HIF-1α stability; and coimmunoprecipitation and colocalization studies reveal the possibility of sumoylation regulating HIF-1α in the brain.

It is noteworthy that lysine residues in the ODD domain of HIF-1α are likely to serve the change of HIF-1α stability in a different oxygen tension through the process of ubiquitination, acetylation, and sumoylation. Until now, however, it is unclear which lysine residue of the ODD domain is in charge of HIF-1α ubiquitination. An overall schematic diagram of a molecular mechanism of HIF-1α regulation is represented in Fig. 2.2. To answer more detailed questions on how HIF-1α is involved in the BBB development, further studies on coordinated interactions between posttranslational modifications, including acetylation/deacetylation, hydroxylation, sumoylation, and ubiquitination, for HIF-1 activity seem warranted in the brain.

The Eph/ephrin family is the largest family of RTK characterized. The Eph/ephrin families are membrane-bound proteins that function as a receptor-ligand pair. Eph/ephrin has been known to regulate cell repulsion, adhesion, and migration, suggesting the regulatory function of axon guidance through chemorepulsion. In addition, vascular remodeling regulated by Eph/ephrin has been found in embryonic vasculature. Ephrin B2 is expressed in arterial ECs, whereas its receptor is reciprocally expressed in embryonic veins, providing the evidence for a distinction between arterial and venous endothelium. The expression of ephrin may be regulated by the Notch receptor, sonic hedgehog (Shh), and TGFβ, all of which are involved in angiogenesis, suggesting reciprocal interaction in angiogenic processes. In addition, the role of Eph/ephrin in angiogenic remodeling of primitive capillary network structure to more complex vascular network has been documented through targeted disruption of either ephrin B2 or EphB4. In addition to endothelial expression, ephrin B2 is expressed in smooth muscle cells and pericytes surrounding vessels as development proceeds [72, 73]. Eph/ephrin has also emerged as a critical regulator in postnatal vascular remodeling and angiogenesis in retinopathy prematurity.

Neuropilins (NRPs) are extracellular receptors for VEGF and semaphorin, and developmentally regulate both axon and vascular migration [70, 74]. NRP1 is identified as a VEGF receptor and simultaneously acts as a receptor for semaphorin families that is implicated in the neuronal guidance. Coexpressed with VEGFR2, NRP1 enhances the binding of VEGF to VEGFR2, by NRP1 forming a complex with VEGFR2. The role of NRP in the endothelial function has been found in studies of knockout mice that lacked the maturation and remodeling of the vascular network [75]. Likewise, emerging evidence shows that the slit/robo families, known as axonal guidance cues, are also involved in the blood vessel guidance [70, 76]. Netrin-1, a neuronal guidance factor, has been shown to act as an angiogenic factor [77].

Implication of Notch signaling in vascular development is suggested by genetic defects [78]. There are four Notch receptors and five ligands (Jagged-1, -2, Delta-1, -3, -4), which are membrane proteins. All the receptors and ligands have been expressed in the vascular system, including capillaries, vascular smooth muscles, and pericytes. Studies of endothelial function have shown that Notch signaling is negatively involved in angiogenesis by blocking β-integrated adhesion and sprouting. The β-integrin signaling is known to be necessary for VEGF-mediated endothelial sprouting. A further link has been found showing that VEGF induces the expression of Notch-1 and Delta-4.

Even if many factors involved in neuronal patterning are now being identified, their roles in vasculature are still unclear. Considering the strong parallels between these two systems, it is likely that the factors are regulated in a complex spatiotemporal interaction to shape the blood vessel network.

2.3
Oxygenation in the Brain: Brain Barriergenesis

Soon after ECs form primitive blood vessels by angiogenesis, commencement of blood flow is likely to provide sufficient oxygen tension and vascularize brain tissue. Once oxygenation in the brain occurs in a sufficient quantity, ECs lining the vessels seem to be able to develop the barrier phenotype with more complex tight junctions (TJs), according to the requirement of the brain microenvironment [3, 6, 49, 79–83]. Therefore, the remodeling process of the primitive blood vessels, barriergenesis, builds the functional blood vessels with the BBB.

TJs in the BBB are composed of transmembrane proteins, including occludin, claudin, junctional adhesion molecule, and VE-cadherin. In addition to transmembrane proteins, TJs also include several accessory proteins that are necessary for structural support: zonula occludens (ZO) proteins, 7H6, cingulin, and so forth. TJ assembly is seemingly regulated by interactions between multiple protein kinases, including PKC subtypes [84, 85]. In fact, TJ proteins are intimately associated with cytoskeletal architecture [86]. Besides complex TJs, BBB characteristics include several key features. The BBB structure expresses GLUT as a glucose transport, aquaporin as a water channel, and p-glycoprotein as an efflux transporter. All these proteins of the BBB regulate extremely low permeability, restriction of a free exchange of molecules by specific transport systems, regulation of homeostasis of the central nervous system (CNS), and prevention of the CNS from the extravascular environment [83]. More importantly, the BBB is enhanced by the presence of differentiated astrocytes and pericytes [6, 10, 12, 81, 82, 87]. These cells can reinforce barriergenesis through their direct interaction with ECs and/or secreted factors and finally lead to mature, functional blood vessels.

2.3.1
Cellular and Molecular Responses Following Brain Oxygenation

In earlier developmental stages, the primitive and discontinuous blood vessels invade the brain in response to a hypoxia- and neuroglia-induced factor gradient. Then, sufficient oxygen supplied by the invading vessels gives a great effect on the maturation of both neuroglia and ECs [5, 80, 88]. Several factors secreted from angiogenic ECs, including leukemia inhibitory factor-1 (LIF-1), have been shown to induce astrocyte differentiation [11].

In ECs, TJ proteins including ZO-1 and occludin increase in the brain tissue (Figs. 2.1 and 2.3). In addition, confocal microscopy studies in the adult and developing human brain demonstrate that occludin and claudin-5 are expressed and assembled at the junctional areas [89]. In contrast to occludin expression for TJ formation, hypoxic regions and VEGF expression are slightly detected in the embryonic stage, spread through the whole cerebral cortex at the postnatal stage, and then gradually disappear (Fig. 2.1). Besides, the significance of oxy-

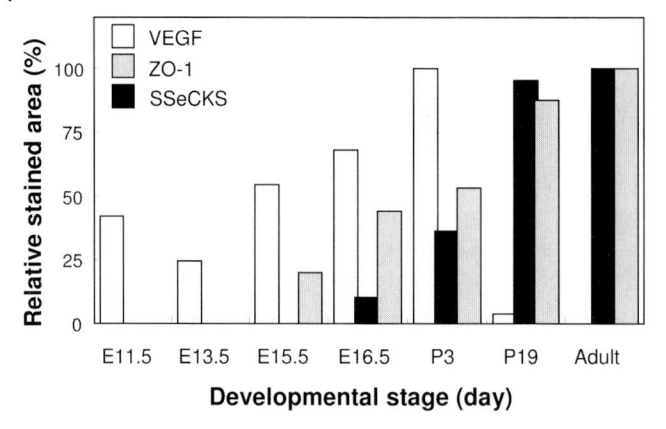

Fig. 2.3 Immunohistochemical staining of SSeCKS, VEGF, and ZO-1 in developing mouse brain. Immunoreactivities were quantified by analyzing each stained area relative to stained areas of adult brain (SSeCKS and ZO-1) or of P3 brain (VEGF). (Adapted from [79], with permission).

gen tension in several in vitro studies has suggested that posthypoxic reoxygenation in cerebral endothelial cells regulates their permeability and TJ proteins [90]. Thus, it seems that the oxygen gradient from hypoxia to oxygenation might regulate brain angiogenesis and BBB maturation during brain development.

2.3.2
Role of src-Suppressed C Kinase Substrate in the Induction of Barriergenesis

Given that ECs are surrounded by perivascular astrocytes, oxygen diffusion from blood vessels may induce oxygen-regulatory factors in the perivascular astrocytes. A very attractive approach to search for oxygen regulatory factors has been accomplished using a cDNA representative differential assay (RDA) in rat primary cultured astrocytes. As a result, the src-suppressed C kinase substrate (SSeCKS; rat homologue of human AKAP12) from astrocytes under posthypoxic reoxygenation has been identified as an oxygen regulatory gene [79].

SSeCKS is known to be a potential tumor suppressor and to function as a multivalent scaffolding protein for PKC, PKA, calmodulin, cyclins, and β-adrenergic receptors [91–93]. SSeCKS can control actin-based cytoskeletal architectures [94]. In fact, SSeCKS expression is relatively ubiquitous, with highest expression in the gonad, smooth and cardiac muscle, lung, brain, and heart [95]. In adult brain, the expression of SSeCKS has been detected in the cerebral and cerebellar white matter [96]. Interestingly, exposure of rat primary astrocytes to reoxygenation following hypoxia leads to increased SSeCKS expression, compared with hypoxic exposure, suggesting that SSeCKS is regulated by oxygen tension [79]. Oxygen-induced SSeCKS in perivascular astrocytes seems to influence the adjustment of ECs directly or indirectly. The indirect effect of SSeCKS on ECs shows that treatment with conditioned media from SSeCKS-overexpressing astrocytes reduces the endothelial capability for migration and tube formation, suggesting an antiangiogenic activity. Antiangiogenic effects of SSeCKS-mediated signals seem to be related to the report that SSeCKS is a potential inhibitor for metastatic tumors and controls cell cycle progression by regulating the expression and localization of cyclin D [93, 97].

In addition, SSeCKS-conditioned media from astrocytes not only increase the expression of TJ proteins, ZO-1, ZO-2, claudin-1, claudin-3, and occludin, but also decrease the permeability of human brain microvascular endothelial cells (HBMECs), suggesting that SSeCKS may play a key role in barriergenesis [79]. Confocal microscopic analysis demonstrates that SSeCKS-conditioned media from astrocytes enhance the linear distribution of ZO-1 at the HBMEC margin, where ECs contact. The linear distribution of ZO-1 is turned into a discontinuous and broken margin after downregulating SSeCKS expression. In developing brain tissue, SSeCKS filamentous staining in surrounding blood vessels stained with ZO-1 is taken to show the interaction between SSeCKS-expressing astrocytes and ZO-1-expressing ECs. This interaction is in line with the idea that reoxygenation during later stages of brain development increases SSeCKS in perivascular astrocytes, which in turn upregulates the expression of TJ proteins in neighboring ECs. Thus, SSeCKS is involved in stabilizing the BBB structure in the brain.

The cerebral cortical expression of SSeCKS and TJ proteins progressively increases in late embryogenesis. In adult mice, these proteins are abundantly and stably expressed. In contrast, the expression of VEGF is high from E11 to P3, then gradually decreases and is almost undetected in adult brain (Fig. 2.3). In addition, SSeCKS is co-localized with GFAP; and SSeCKS-expressing astrocytes closely interact with ECs lining blood vessels. An additional report is that platelet-dependent growth factor-BB (PDGF-BB) potently modulates SSeCKS expression in cultured smooth muscle cells [98]. The report implies a possibility that SSeCKS expression might be regulated by PDGF-BB and then affect BBB maturation. Even if a detailed understanding on the role of SSeCKS for BBB formation and integrity is still a prerequisite, it is known that SSeCKS-mediated signals in astrocytes induce the maturation and stabilization of permeable vessels by inhibiting brain angiogenesis and enhancing the expression of TJ proteins. In addition, the constitutive expression of SSeCKS in the adult brain may provide a stabilizing signal for BBB integrity under physiological conditions.

2.3.3
Barriergenic Factors in Perivascular Astrocytes and Pericytes Following Brain Oxygenation

Considering the importance of the brain endothelial barrier, it is likely that the list of factors involved in the signaling cascade might be large. There are numerous factors associated with maturation of vessels. Angiopoietin-1 (ang-1)/Tie2 seems partly to be responsible for the stabilization and maturation of vessels via interactions between the matrix, pericytes, and ECs [99, 100]. Ang-1/Tie2 is observed in both adult and embryonic stages and is upregulated during vascular formation [101, 102]. Mice lacking or overexpressing ang-1 have revealed that ang-1 is responsible for recruiting and sustaining periendothelial support cells and for contributing to the formation of leakage-resistant blood vessels [103, 104]. In addition, Tie2 deficiency normally produces the initial phases of angiogenesis and blood vessel formation, including sprouting [105]. However, blood vessels throughout the embryo are absent from the formation of normal hierarchical networks, suggesting a role of ang-1/Tie2 during the stages of EC maturation by cell-cell interaction.

In vitro studies show that exposure of cultured astrocytes to posthypoxic reoxygenation results in a higher expression and secretion of ang-1 and a lower expression of VEGF than does hypoxic exposure of astrocytes [49]. The secretion of ang-1 from perivascular astrocytes can bind its receptor, Tie-2, onto ECs. Direct exposure of cultured ECs to recombinant ang-1 protein not only markedly increases the expression of TJ proteins but also decreases its permeability (personal observations). Similarly, it has been documented that ang-1 opposes the permeability action of VEGF [106]. Moreover, ang-1 suppresses VEGF-induced vascular permeability in the skin and diabetic retinopathy [107]. Additionally, when BBB breakdown occurs, lesion and perilesional vessels decrease ang-1, supporting the leakage-resistant role of ang-1.

Other lines of investigation have demonstrated that brain pericytes are important for the control of EC growth and migration and the integrity of microvascular capillaries [108]. Consistent with the role of ang-1 for vascular remodeling processes, vascular pericytes secrete ang-1 in response to oxygenation. The pericyte-derived ang-1 induces occludin expression in brain capillary ECs through Tie2 activation *in vitro*, which results in barriergenesis. Hence, it may be possible that ang-1 secreted from astrocytes and pericytes stabilizes ECs through an increase in TJ proteins (see Chapter 5).

As a mechanism for SSeCKS-mediated barrier formation, it has been investigated whether ang-1 might be a barrier factor [79]. Overexpression of SSeCKS in rat primary astrocytes markedly reduces the expression of VEGF. In contrast, SSeCKS overexpression leads to an increased expression of ang-1, related to the maturation of preexisting vessels. Simultaneously, ang-1-neutralized, SSeCKS-conditioned media not only markedly reduce the expression of TJ proteins but also strongly increase vessel permeability. Therefore, SSeCKS is likely to enhance the barrier function of ECs through ang-1 signaling. These findings indicate that SSeCKS is likely to, at least in part, trigger vessel maturation and stabilization during BBB development by upregulation of TJ proteins mediated by ang-1.

Accumulated evidence shows that PDGF-β plays a critical role in the recruitment of pericytes to newly formed vessels [109]. It has been reported that PDGF-β is expressed by sprouting endothelium and pericyte/vSMC progenitors, suggesting a paracrine mode of interaction between the two cell types. Moreover, endothelium-specific PDGF-β knockout leads to vSMC/pericyte deficiency, but no obvious effect on the vasculature. From these results, the endothelial PDGF-β signal controls pericyte recruitment to angiogenic maturation of vessels.

Exposure of astrocytes to oxygenation following hypoxia leads thrombospondin-1 (TSP-1) to increase and sustain for a while [49]. Through this finding, it can be envisaged that the oxygenation signal in the brain microenvironment induces TSP-1. The increase in TSP-1, in concert with a decrease in VEGF, seems to promote a developmental switch from angiogenic to a differentiated, quiescent state of ECs. TSP-1 has been known to be a major activator of TGF-β1, which is a multifunctional cytokine expressed in ECs and mural cells and which promotes vessel maturation by stimulating extracellular matrix (ECM) production and by inducing the differentiation of mesenchymal cells to mural cells. Furthermore, it was recently reported that TGF-β1 upregulates the TJ and P-glycoprotein of brain microvascular ECs. Therefore, this suggests that TSP-1-dependent TGF-β1 in the brain may keep the BBB functioning. Besides, glia cell-derived neurotropic growth factor (GDNF), in the TGF-β family, is likely to be involved in BBB maturation [110].

Adrenomedullin (AM) has been identified as a vasodilator peptide and a new member of the calcitonin gene-related peptide (CGRP) family. AM is a peptide hormone with multifunctional biological properties, participating in the regulation of vascular tone, inflammation, and other physiological events of the vasculature. Northern blot and radioimmunoassay have revealed the AM distribution in the central nervous system, many neurons, ECs, and perivascular glia cells. The endothelial expression of AM plays a role in the induction and maintenance of BBB properties [111]. Moreover, interleukin-6 and bFGF are also involved in the regulation of the BBB [112].

2.4
Perspectives

Overall, many reports support the idea that oxygen tension and neuroglia-derived factors contribute to brain angiogenesis and barriergenesis. For the whole process, although an exception may exist, a change in oxygen tension is likely to cooperate with a variety of neuroglia-derived factors released in the microenvironment of the developing brain. These combined effects may contribute to the structural and functional integrity of the vascular system (Fig. 2.4). Endothelial, glial, pericyte, and neuronal cells at the level of individual cells are sure to respond to a variety of signals derived from oxygen tension. Therefore, further studies are required to lineate how complex signals mutually interact with and affect the proliferation and differentiation of the cells. These molecular and cell-based researches will be crucial to understand the development and maintenance of the BBB [71].

IGF1 and VEGF production in the hippocampus is unknown – in contrast to the songbird HVC, where gonadal steroids induce both IGF1 and VEGF (Jiang et al., 1998; Louissaint et al., 2002).

These observations indicate that, in the adult mammalian brain, as well as in its avian counterpart, neural progenitor mobilization may be dynamically regulated by the gliovascular unit. Yet besides these classic growth factor-signaled pathways linking angiogenesis and neurogenesis, several more primitive pathways appear to permit the vascular regulation of progenitor cell turnover and daughter cell fate. Among these, purinergic and nitric oxide signaling, two distinct but interacting pathways for rapid communication among contiguous cells, seem especially important in regulating the mobilization and fate of tissue progenitor cells. Both are intimately associated with and regulated by the local microvasculature, and as such can provide the means by which local microvascular cells can activate local progenitor cells.

3.6
Purinergic Signaling to Neural Progenitors Cells: the Gliovascular Unit as a Functional Entity

Purine nucleotides can act as extracellular transmitters, and have been implicated in interactions between neurons and glia, as well as between neural, blood, and inflammatory cells (Burnstock, 2002b). Purinergic signaling has been implicated in the proliferation of a variety of undifferentiated cell phenotypes (Burnstock, 2002a,b; Sanches et al., 2002; Wang et al., 1992). In the normal brain, ATP is released by astrocytes, for which it serves as an active signaling moiety and is responsible for the propagation of intercellular calcium waves among syncytially coupled astrocytes (Arcuino et al., 2002; Cotrina et al., 1998, 2000; Nedergaard et al., 2003). Acting through P2Y receptors, ATP can promote the proliferation of radial cells and astrocytes (Weissman et al., 2004), as well as brain microvascular cells (Burnstock 2002b; Wang et al., 1992). Moreover, ATP has been identified as a mitogen for v-myc immortalized neural stem cells (Ryu et al., 2003). Purines can signal through the Ras/Raf/MEK/MAPK pathway (Tu et al., 2000), indicating the potential for significant cross-talk with FGF, EGF/TGFα, PDGF, and erbB-driven neural mitogenic pathways, among others. These observations would suggest the possibility that astrocytic purines, and ATP in particular, might participate in the control of neurogenesis, suggesting a means by which glial activity might regulate neural progenitor cell mobilization. This might provide a basis for several studies that have reported both subventricular zone and hippocampal astrocytes as uniquely enabling neurogenesis by resident progenitor cells (Lim and Alvarez-Buylla, 1999; Song et al., 2002).

ATP is a short-acting molecule, both temporally and spatially, in that it is rapidly degraded by extracellular ectonucleotidases. Two ectonucleotidases, nucleotide triphosphate diphosphohydrolases 1 and 2 (NTDPases 1, 2), have been reported to modulate the local processing and clearance of ATP so as to modulate the local levels of ATP to ADP, the most potent ligands for P2Y receptor activation, and their metabolites, AMP and adenosine. NTDPase-1 catalyzes ATP hydrolysis to ADP and AMP, and ADP's to adenosine; it is selectively expressed in the brain by endothelial and smooth muscle cells (Braun et al., 2000), though ubiquitously so. In contrast, NTDPase 2 strongly localizes to areas of subventricular and hippocampal neurogenesis in the adult brain, within which it colocalizes with astrocytes (Braun et al., 2003). By this means, astrocytes expressing relatively high levels of ATP in regions of stem cell expansion may utilize NTDPase-2 to locally generate ADP, which may drive local neural stem cell expansion and neurogenesis. These cells may in effect partner with adjacent microvascular pericytes to then use NTPase-1 to rapidly clear the bioactive ATP and ADP, thereby delimiting these sites of stem cell expansion.

Although it is unclear whether purine receptor activation selectively favors the generation of any one phenotype, it is important to note the colocalization of these foci of purinergic signaling with areas of expression of noggin (Lim et al., 2000), a potent inhibitor of the pro-gliogenic bone morphogenetic proteins (Mabie et al., 1997). Persistent noggin expression in the subependyma (Lim et al., 2000) and hippocampus (Chmielnicki et al., 2004) appears to inhibit daughter cells from BMP-dependent gliogenesis, permitting their maintenance as undifferentiated progenitors, and hence accessibility to neurotrophic differentiation agents such as BDNF (Chmielnicki et al., 2004).

These observations suggest that multiple influences may converge in niches for cell genesis, to specifically permit neurogenesis. Angiogenesis, local restrictions on gliogenesis, and microvascular neurotrophic influences, such as BDNF and IGF1, might then be viewed as necessary concomitants for biasing new daughter cells to neurogenesis. Indeed, such permuted interactions may essentially define the locations of adult neurogenic niches, by specifying the distributions of ongoing progenitor cell mobilization and neurogenesis (Goldman, 2003).

3.7
Nitric Oxide is a Local Modulator of Progenitor Cell Mobilization

The net effects of peptidergic cytokines and purine signals, among others, on neural stem and progenitor cells depend upon the concurrent actions of other local cues. Of particular interest in this regard are recent observations that nitric oxide (NO), as released by both endothelial cells and neurons, is a powerful regulator of mitotic neurogenesis (Cheng et al., 2003; Packer et al., 2003). In normal brain, NO appears to be a negative regulator of neurogenesis, such that NO synthase inhibitors promote the rapid expansion of progenitor cells of both the olfactory subependyma and the hippocampal subgranular zone (Packer et al., 2003; Sun et al., 2005). Specifically, NOS inhibition may be sufficient to release NO's tonic suppression of progenitor cell turnover, thereby allowing the rapid reentry of these otherwise quiescent cells into the cell cycle. Conceivably, the resultant rapid mobilization of parenchymal progenitor cells might yield different daughter phenotypes in a context-dependent fashion, as newly generated daughters respond to regionally specified cues. By this logic, since BDNF is a powerful trigger for neuronal differentiation, those cells mobilized by NOS suppression, that then encounter BDNF might be expected to differentiate as neurons. This is precisely what is observed in the striatal and olfactory subependyma, as well as in the hippocampus, in which nNOS inhibition is associated with increased mitotic neurogenesis. In contrast, since BDNF-associated neuronal differentiation may be modulated by NO signaling (Chen et al., 2005; Cheng et al., 2003), NOS inhibition might also serve to impede BDNF's potentiation of terminal differentiation (Cheng et al., 2003). Nonetheless, mice deficient in neuronal NOS exhibited markedly elevated mitotic neurogenesis within both the olfactory subependyma and hippocampus, again pointing to the net suppressive actions of NO on progenitor cell mobilization in the adult brain (Packer et al., 2003).

Interestingly, a recent study has come to different conclusions with respect to the normal and pathological roles of eNOS signaling, in that mice deficient in eNOS were noted to exhibit diminished subependymal and hippocampal neurogenesis, and correspondingly deficient compensatory neurogenesis following ischemia, relative to wild-type mice (Chen et al., 2005). These observations suggest that eNOS and nNOS may have very different effects in the regulation of progenitor cell mobilization. Since they generate the same product, the differences noted between nNOS and eNOS inactivation may be due to the different paracrine pathways influenced by NO of endothelial and neuronal origin, and to the distinct spatial and temporal patterns of signal integration experienced by progenitor cells in eNOS- and nNOS-deficient environments.

3.8
Parenchymal Neural Progenitor Cells May Reside Among Microvascular Pericytes

Besides the neural stem and progenitor cells of the ventricular zone and subgranular layer, abundant glial progenitor cells pervade both the gray and white matter of the adult brain. Yet despite abundant studies of their distribution and lineage potential, the anatomic relationships of glial progenitor cells in the adult brain parenchyma remains unclear. Scolding et al. (1998, 1999) identified OPCs in the adult human white matter on the basis of both A2B5 and PDGFα receptor expression; and each has been quantified as representing roughly 3% of all white matter cells. Nunes et al. (2003) arrived at similar estimates, both in histological sections and upon A2B5 antigen-based FACS, as have others. Each described adult human OPCs as small and highly ramified, appearing similar to microglia, but none of these studies assessed the positions of these cells relative to the capillary microvasculature. Interestingly, these nominally glial progenitor cells of the adult brain express the NG2 chondroitin sulfate proteoglycan, by which they are operationally defined in rodents. Yet in humans, microglia (Pouly et al., 1999) and pericytes as well as OPCs can express NG2 (Ozerdam et al., 2001, 2002). Thus, given their antigenic and morphological similarities, it seems likely that some cells previously characterized as ramified microglia are instead parenchymal glial progenitors and vice versa, and similarly, that some putative adventitial pericytes might have been microglia or glial progenitors. Conceivably, the perivascular NG2$^+$ pool could then be composed of distinct populations of OPCs, microglia, and pericytes that adventitiously coexpress the same chondroitin sulfate proteoglycan. As such, some fraction of perivascular cells previously considered as smooth muscle pericytes might instead be resident progenitor cells; indeed, these morphologically and antigenically similar phenotypes might even be lineally related (Yamashima et al., 2004).

The identity of perivascular NG2$^+$ cells garnered particular interest following the observation that glial progenitor cells retain the possibility for neurogenesis, once removed from the tissue environment (Belachew et al., 2003; Kondo and Raff, 2000; Nunes et al., 2003). As such, the parenchymal glial progenitor cell pool may represent a reservoir of potential new neurons; and, to the extent that these cells might be perivascular, they may comprise an especially responsive pool of potentially neurogenic parenchymal progenitor cells. In this regard, Belachew et al. (2003) reported the generation of new neurons from resident NG2$^+$ progenitors of the hippocampus, specifically highlighting their generation from the tissue parenchyma. Okano and colleagues then reported the production of new granule neurons from the vascular adventitia of the adult monkey hippocampus, following transient ischemia (Yamashima et al., 2004). In the latter study, BrdU$^+$ cells were found in proliferative clusters within the vascular adventitia, within which a third of the cells expressed the neural progenitor marker musashi 1. Following transient ischemia, not only did the BrdU-labeling index rise, but so did the amount of BDNF expressed by SGZ pericytes, which appeared to potentiate neuronal differentiation by the newly generated daughter cells (Yamashima et al., 2004).

Together, these studies suggested that at least some fraction of parenchymal neural progenitor cells may be associated with the vascular walls, within which they may be identified or misidentified as NG2$^+$ perivascular cells, and that in some settings, these cells may be recruited not only for gliogenesis, but also for local neurogenesis. However, these studies have also highlighted the need for rigorous distinction between parenchymal glial progenitor cells, pericytes, and microglia.

3.9
The Role of the Vasculature in Post-Ischemic Mobilization of Progenitor Cells

Compensatory replacement of striatal neurons from resident progenitors was identified in experimental stroke models by several groups, who described neuronal recruitment into the neostriatum after focal ischemic injury (Arvidsson et al., 2002; Jin et al., 2003; Parent et al., 2002). Similarly, Nakafuku and coworkers described compensatory replacement of hippocampal pyramidal neurons – which, despite their apparent dissimilarity from striatal medium spiny cells, comprise another periventricular subcortical neuronal pool (Nakatomi et al., 2002). In each case, the trigger for the mobilization and striatal migration of resident progenitor cells appears to have been local hypoxic ischemia, possibly mediated by post-ischemic increases in VEGF, BDNF, and other neurotrophic and angiogenic cytokines (Sun et al., 2003), coupled with the parenchymal migration of the newly generated cells along the adventitial surfaces of local capillaries. Indeed, it seems likely that the response of the brain's persistent germinal layers, the subventricular and subgranular zones, to local ischemia may be to recapitulate and expand the local microvascular niches for cell genesis, differentiation, and migration. By this means, the microvascular contributions to progenitor cell maintenance and mobilization in the adult brain may be seen as important not only to the brain's normal homeostasis, but as critical components of its response to injury – microvascular influences might essentially establish the gain by which the extent and nature of compensatory neurogenesis and gliogenesis are regulated.

Acknowledgments

Work mentioned in the Goldman Laboratory is funded by NIH grants R01NS29813, R01NS33106, R01NS52534, and R01NS39559, and by the National Multiple Sclerosis Society.

References

Ahmed, S., Reynolds, B. A., Weiss, S. **1995**, BDNF enhances the differentiation but not the survival of CNS stem cell-derived neuronal precursors, *J. Neurosci.* 15, 5765–5778.

Arcuino, G., Lin, J., Takano, T., Liu, C., Jiang, L., Kang, J., Nedergaard, M. **2002**, Intercellular calcium signaling mediated by point-source bursts of ATP, *Proc. Natl Acad. Sci. USA* 99, 9840–9845.

Arvidsson, A., Collin, T., Kirik, D., Kokaia, Z., Lindvall, O. **2002**, Neuronal replacement from endogenous precursors in the adult brain after stroke, *Nat. Med.* 8, 963–970.

Belachew, S., Chittajallu, R., Aguirre, A. A., Yuan, X., Kirby, M., Anderson, S., Gallo, V. **2003**, Postnatal NG2 proteoglycan-expressing progenitor cells are intrinsically multipotent and generate functional neurons, *J. Cell Biol.* 161, 169–186.

Benraiss, A., Chmielnicki, E., Lerner, K., Roh, D., Goldman, S. A. **2001**, Adenoviral brain-derived neurotrophic factor induces both neostriatal and olfactory neuronal recruitment from endogenous progenitor cells in the adult forebrain, *J. Neurosci.* 21, 6718–6731.

Braun, N., Sevigny, J., Robson, S., Enjyoji, K., Guckelberger, O., Hammer, K., Di Virgilio, F., Zimmermann, H. **2000**, Assignment of ectonucleoside triphosphate diphosphohydrolase-1/CD39 expression to microglia and vasculature of the brain, *Eur. J. Neurosci.* 12, 4357–4366.

Braun, N., Sevigny, J., Mishra, S., Robson, S., Barth, S., Gerstberger, R., Hammer, K., Zimmermann, H. **2003**, Expression of the ecto-ATPase NTPDase2 in the germinal zones of the developing and adult rat brain, *Eur. J. Neurosci.* 17, 1355–1364.

Breier, G., Albrecht, U., Sterrer, S., Risau, W. **1992**, Expression of vascular endothelial growth factor during embryonic angiogenesis and endothelial cell differentiation, *Development* 114, 521–532.

Burnstock, G. **2002a**, Potential therapeutic targets in the rapidly expanding field of purinergic signalling, *Clin. Med.* 2, 45–53.

Burnstock, G. **2002b**, Purinergic signaling and vascular cell proliferation and death, *Arterioscler. Thromb. Vasc. Biol.* 22, 364–373.

Chen, J., Zacharek, A., Zhang, C., Jiang, H., Li, Y., Roberts, C., Lu, M., Kapke, A., Chopp, M. **2005**, Endothelial nitric oxide synthase regulates brain-derived neurotrophic factor expression and neurogenesis after stroke in mice, *J. Neurosci.* 25, 2366–2375.

Cheng, A., Wang, S., Cai, J., Rao, M., Mattson, M. **2003**, Nitric oxide acts in a positive feedback loop with BDNF to regulate neural progenitor cell proliferation and differentiation in the mammalian brain, *Dev. Biol.* 258, 319–333.

Chmielnicki, E., Benraiss, A., Economides, A. N., Goldman, S. A. **2004**, Adenovirally expressed noggin and brain-derived neurotrophic factor cooperate to induce new medium spiny neurons from resident progenitor cells in the adult striatal ventricular zone, *J. Neurosci.* 24, 2133–2142.

Cotrina, M. L., Lin, J. H., Nedergaard, M. **1998**, Cytoskeletal assembly and ATP release regulate astrocytic calcium signaling, *J. Neurosci.* 18, 8794–8804.

Cotrina, M. L., Lin, J. H., Lopez-Garcia, J., Naus, C., Nedergaard, M. **2000**, ATP-mediated calcium signaling, *J. Neurosci.* 20, 2835–2844.

Engerman, R. L., Pfaffenbach, D., Davis, M. D. **1967**, Cell turnover of capillaries. *Lab. Invest.* 17, 738–743.

Goldman, S. A. **2003**, Glia as neural progenitor cells, *Trends Neurosci.* 26, 590–596.

Goldman, S. A., Nottebohm, F. **1983**, Neuronal production, migration, and differentiation in a vocal control nucleus of the adult female canary brain, *Proc. Natl Acad. Sci. USA* 80, 2390–2394.

Greenberg, D., Jin, D. **2005**, From angiogenesis to neuropathology, *Nature* 438, 954–959.

Hidalgo, A., Barami, K., Iversen, K., Goldman, S. A. **1995**, Estrogens and non-estrogenic ovarian influences combine to promote the recruitment and decrease the turnover of new neurons in the adult female canary brain, *J. Neurobiol.* 27, 470–487.

Jiang, W., McMurtry, J., Niedzwiecki, D., Goldman, S. A. **1998**, IGFI is a radial cell-derived neurotrophin that promotes neuronal recruitment into the adult songbird brain, *J. Neurobiol.* 36, 1–15.

Sun, Y., Jin, K., Xie, L., Childs, J., Mao, X. O., Logvinova, A., Greenberg, D. A. **2003**, VEGF-induced neuroprotection, neurogenesis, and angiogenesis after focal cerebral ischemia, *J. Clin. Invest.* 111, 1843–1851.

Sun, Y., Jin, K., Childs, J. T., Xie, L., Mao, X. O., Greenberg, D. A. **2005**, Neuronal nitric oxide synthase and ischemia-induced neurogenesis, *J. Cereb. Blood Flow Metab.* 25, 485–492.

Tu, M. T., Luo, S. F., Wang, C. C., Chien, C. S., Chiu, C. T., Lin, C. C., Yang, C. M. **2000**, P2Y(2) receptor-mediated proliferation of C(6) glioma cells via activation of Ras/Raf/MEK/MAPK pathway, *Br. J. Pharmacol.* 129, 1481–1489.

Wang, D. J., Huang, N. N., Heppel, L. A. **1992**, Extracellular ATP and ADP stimulate proliferation of porcine aortic smooth muscle cells, *J. Cell Physiol.* 153, 221–233.

Weinstein, B. **2005**, Vessels and nerves: marching to the same tune. *Cell* 120, 299–302.

Weissman, T., Riquelme, P., Ivic, L., Flint, A., Kriegstein, A. **2004**, Calcium waves propagate through radial glial cells and modulate proliferation in the developing neocortex, *Neuron* 43, 647–661.

Yamashima, T., Tonchev, A. B., Vachkov, I. H., Popivanova, B. K., Seki, T., Sawamoto, K., Okano, H. **2004**, Vascular adventitia generates neuronal progenitors in the monkey hippocampus after ischemia, *Hippocampus* 14, 861.

Part II
The Cells of the Blood-Brain Interface

4
The Endothelial Frontier

Hartwig Wolburg

4.1
Introduction

The original finding of Paul Ehrlich [1] that an infused dye did not stain brain tissue, together with the complementary observation of his pupil Ernst Goldmann that the very same dye, if applied into the cerebrospinal fluid, did stain brain tissue, led to the concept of a biological barrier between blood and brain (Fig. 4.1). Due to the free access of dye from brain ventricle to brain tissue, it was concluded that there is no cerebrospinal fluid-brain barrier. However, the staining of circumventricular organs and the choroid plexus in Goldmann's experiment applying the dye into the general circulation (Goldmann experiment I) and the avoidance of staining of these organs in the experiment applying the dye into the cerebrospinal fluid (Goldmann experiment II) suggested the existence of a barrier between the cerebrospinal fluid and the blood (Fig. 4.1). The cellular basis of the barrier was unclear for decades. Today, we know that in most vertebrates the barrier is located within the endothelium (endothelial blood-brain barrier (BBB) in Fig. 4.1; only in elasmobranchs, the BBB is located in astrocytes) and in the epithelial choroid plexus cells and the tanycytes of the circumventricular organs (glial blood-cerebrospinal fluid barrier (BCSFB) in Fig. 4.1). The structures essentially responsible for the restriction and control of the paracellular flux between both epithelial and endothelial cells was identified as tight junctions. Originally, these intercellular contacts were studied exclusively by means of morphological methods. From the middle of the 1980s onward, the molecular organization of the tight junctions was unraveled step by step, but primarily in epithelial cells. Today, we overlook a multitude of molecules concerned with both the formation and the regulation of barrier properties, but we are far from understanding the molecular network which establishes the transcellular barrier. Endothelial cells are seemingly more complicated regarding barrier regulation than epithelial cells, because epithelial cells, but not endothelial cells, develop barrier properties in vitro (thus after isolation from their microenvironment) which are close at their physiological level in vivo. We have to take into account that the endothelial tight junc-

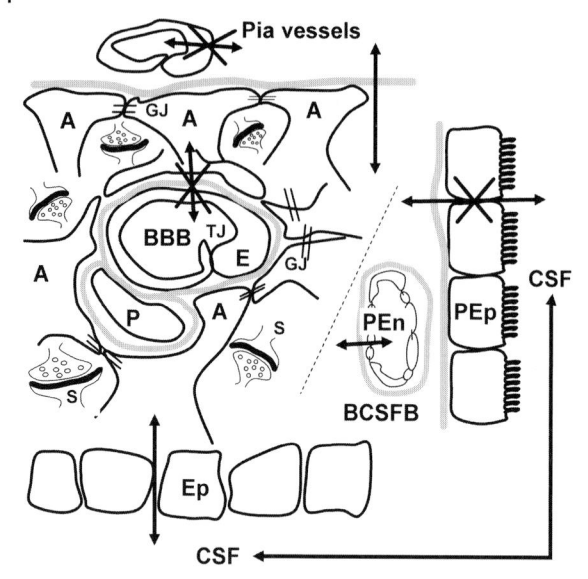

Fig. 4.1 Topological scheme of the BBB. Gray lines mark the basal laminae (*glia limitans superficialis et perivascularis*). A = Astrocytes, BBB = blood-brain barrier, BCSFB = blood-cerebrospinal fluid barrier, CSF = cerebrospinal fluid, E = endothelium, GJ = gap junction, P = pericyte, PEn = choroid plexus endothelial cell, PEp = choroids plexus epithelial cell, S = synapses, TJ = tight junctions. Crossed or uncrossed arrows mean impermeability or permeability across barriers, respectively.

tions are elements responsive to the brain microenvironment, including the surrounding basal lamina as well as the "second line of defense" which consists of pericytes, astrocytes, and microglia. In this chapter, we will focus on the structure and function of the brain microvascular endothelial cells and then, in the last paragraph, touch on the great importance of "peri-endothelial" factors for the establishment of the BBB.

4.2
The Brain Capillary Endothelial Cell

Mature BBB capillaries in the mammalian brain are mainly characterized by the small height of the endothelial cells (Figs. 4.2 and 4.3 A), the interendothelial tight junctions [2] (for recent reviews, see [3, 4]; Figs. 4.3 to 4.6), the small number of caveolae at the luminal surface of the cell [5], and the high number of endothelial mitochondria [6]. In addition, the subendothelial pericytes are completely surrounded by a basal lamina (Fig. 4.3 B), phagocytic perivascular cells, and astrocytic processes belonging to the set of elements directly adjacent to the cerebral vasculature [7, 8].

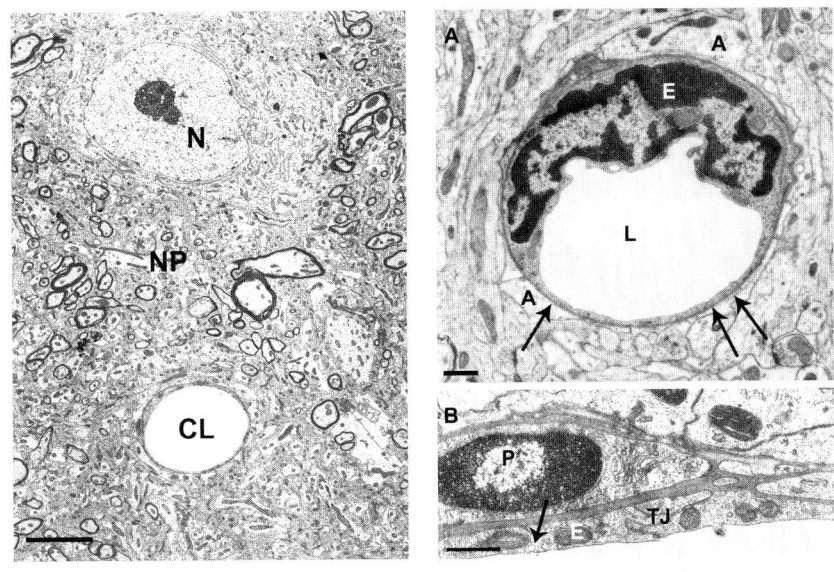

Fig. 4.2

Fig. 4.3

Fig. 4.2 Electron microscopical ultrathin section of the rat brain. CL=capillary lumen, lined by the endothelium, N=neuron, NP=neuropil. Bar: 5 µm.

Fig. 4.3 (A) Ultrathin section of a mouse brain capillary. The arrows point to the subendothelial basal lamina separating the pericyte from both the endothelial cell and the astrocyte. A=Astrocytic endfeet, E=endothelial cell, L=lumen, P=pericyte. Bar: 1 µm. (B) Ultrathin section of a rat brain capillary lined by endothelial cells (E) interconnected by tight junctions (TJ). The arrow points to a caveola at the luminal surface of the endothelial cell. P Pericyte completely surrounded by basal laminae. Bar: 0.5 µm.

The microvascular endothelial cells are doubtless most important in the restriction of BBB-related permeability. During brain angiogenesis and differentiation of the BBB, specific molecules have to be expressed in brain endothelial cells [9, 10]. For example, specific expression of the non-receptor Src family tyrosine kinase *lyn* has been demonstrated early during brain angiogenesis [11]. Recently, it could be shown that *lyn* is activated by hypotonic stress and phosphorylates a tyrosine of the transient receptor potential channel family member TRPV4 as a tonicity sensor [12]. Another important endothelial gene product encodes P-glycoprotein [13], which is required for the differentiation of the BBB [14] and seems to ensure the rapid removal of toxic metabolites from the neuroectoderm before the BBB has fully differentiated [15]. In the developing chicken CNS, it has been shown that angiogenic vessels invading the neuroectoderm express N-cadherin between endothelial cells and pericytes. With the onset of barrier differentiation, N-cadherin labeling decreased, suggesting that transient N-cadherin expression in endothelial and perivascular cells may represent an

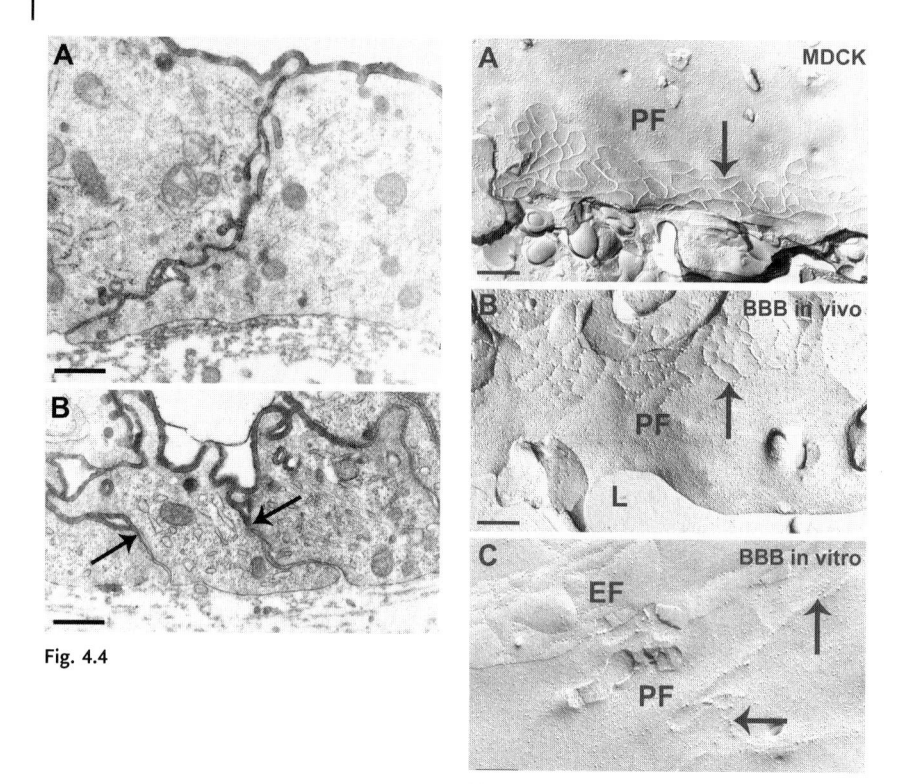

Fig. 4.4 Ultrastructural investigation of the passage of wheat germ agglutinin-horseradish peroxidase across bovine brain capillary endothelial cells, 12 days after coculture with rat astrocytes and two consecutive days after discontinued (A) or continued (B) coculture. In (A), the loss of astrocytic mediators is reflected by the free penetration of the tracer through the intercellular cleft into the subendothelial space. In (B), the intercellular cleft is occluded by tight junctions, which do not allow penetration of the tracer [49]. Bars: 1 µm.

Fig. 4.5 Freeze-fracture replicas of tight junctions. (A) MDCK II cells have tight junctions, nearly completely associated with the P-face (PF). The arrow points to tight junctional strands. Image produced in cooperation with Dr. G. Fricker (Heidelberg, Germany). (B) The endothelial cells of bovine brain capillary endothelial cells have a high ratio of PF-associated tight junctions, which is higher than that of all other endothelial cells outside the brain. The arrow points to tight junctional strands. L=Lumen.
(C) Bovine brain capillary endothelial cells in culture show a loss of PF-associated tight junctional particles. Most particles are associated with the E-face (EF). Arrows point to tight junction ridges which are only poorly occupied with particles. Bars: 0.2 µm.

initial signal which may be involved in the commitment of early blood vessels to express BBB properties [16]. The early adhesion between endothelial cells and pericytes might be the result of the release of chemotactic factors by endothelial cells to induce the migration of pericytes towards the endothelial cell wall and subsequent maturation of the vessels by an increased production of extracellular matrix components elicited by the action of activated TGF-β and other proteins [17]. Amongst these, platelet-derived growth factor (PDGF)-B, a high affinity ligand for the receptor tyrosine kinase PDGF-Rβ present on pericytes is produced by endothelial cells during development. PDGF-B has been shown to be involved in vascularization of the brain as disruption of the PDGF-B gene leads to pericyte loss, endothelial hyperplasia, and lethal microaneurysm formation during late embryogenesis [18]. The fine structural investigation of endothelial cells in these PDGF-B- and PDGF-Rβ-deficient mice showed a malformation of brain endothelial cells characterized by the folding of the luminal surface [19]. Interestingly, this increase in the luminal surface is also a typical feature of the blood vessels in the pecten oculus, which is a convolute of vessels within the vitreous body of the avian eye [20]. In these vessels, the pericytes die by apoptosis during development [21]. Thus, both the physiological loss of pericytes in the pecten oculi and the pathological loss of pericytes in the PDGF-Rβ-deficient mouse lead to a characteristic alteration of the shape of endothelial cells, suggesting a role for pericytes in the morphogenesis of microvessels.

Establishing the barrier is accompanied by further changes in the phenotype of the brain endothelial cells, such as upregulation of the HT7-antigen/basigin [22–24] or downregulation of the MECA-32 antigen [25]; and the expression of specific transporters and metabolic pathways can also be observed [26]. Concerning the development of the barrier function of brain capillaries [10, 27–32] it has become likely that BBB tightness is not just "switched on" at a specific time-point during brain angiogenesis, but rather the tightening of the barrier occurs as a gradual process which is independent from vascular proliferation and begins late during embryogenesis, when angiogenesis is not complete [33]. However, it should be stressed that an unequivocal correlation between tight junction structure and measured permeability has not been shown so far [34, 35]. The molecular mechanisms involved in barrier maturation are poorly understood yet. From transplantation studies showing that vessels derived from the coelomic cavity gain BBB characteristics when growing into an ectopic brain transplant [36], it is known that the development of BBB characteristics in endothelial cells is not pre-determined but rather is induced by the neuroectodermal microenvironment.

4.3
Endothelial Structures Regulating Transendothelial Permeability

4.3.1
Tight Junctions

4.3.1.1 Morphology of Tight Junctions

In endothelial cells, tight junctions as specialized contact zones were already known from ultrathin sections (Figs. 4.3 B and 4.4) [2, 4, 5] and, in epithelial cells, their morphology was described in detail by Farquhar and Palade [37]. Around the same time when the endothelial nature of the BBB was detected by Reese and Karnovsky [38] and classically described in the comprehensive study of Brightman and Reese [2], a novel technique was developed which was suitable to visualize the morphology of cytoplasmic membranes and intercellular contacts and to corroborate the then novel fluid mosaic model of membranes [39]. This technique was the freeze-fracture technique, and soon after its inauguration, tight junctions were the issue of many-fold freeze-fracture descriptions (as summarized by [40] and, more recently, by [41]), including endothelial cell tight junctions of the BBB (Fig. 4.5) [42–48].

If sectioned transversally, the tight junction appears as a system of fusion ("kissing") points, each of which represents a sectioned strand. These strands form the real obstacle proper within the intercellular pathway, which are under close control of the brain microenvironment, in particular the astrocytes (see Sect. 4.4 of this chapter and, e.g. [49]; Fig. 4.4). Two parameters can be visualized by freeze-fracture electron microscopy: the complexity of strands and the association of particles with the inner (P-face) or outer (E-face) lipidic leaflet of the membrane. The complexity of the tight junction network was recognized to be related to the transepithelial electrical resistance [50]. Epithelial tight junctions are mostly associated with the P-face, forming a network of strands and leaving grooves at the E-face which are occupied by only a few particles (Fig. 4.5 A) [51, 52]. After ATP depletion, MDCK cells suffer from deterioration of the paracellular barrier ("gate") function, which is accompanied by a reorganization of the actin cytoskeleton [53, 54] and a decreased P-face association of the tight junctions. Thus, the degree of particle association to the P-face seems directly to correlate with the observed transepithelial resistance.

The tight junction structure of the brain capillary endothelial cells was investigated by Nagy et al. [47], again using the freeze-fracture method. They found that the brain endothelial tight junctions were the most complex in the whole vasculature of the body. This was in nice correlation to the hypothesis of Claude [50], according to which there is a logarithmic relationship between the number of tight junction strands and the transcellular electrical resistance. However, as pointed out already for epithelial tight junctions, the association of the tight junction particles with the P-face or E-face of the membrane has been described to be a further parameter of the quality of the endothelial permeability barrier in the brain [48]. The BBB tight junctions are unique among all endothelial tight junctions in that their P-face association is as high as, or even slightly higher than their E-face association (Fig. 4.5 B). Interestingly, the P-face/E-face ratio of BBB tight junctions continuously increases during development [29]. In cell culture, the freeze-fracture morphology of BBB endothelial cells is similar to non-BBB endothelial cells (Fig. 4.5 C), indicating that the association of the strand particles with the membrane leaflets reflects the quality of the barrier and is under the control of the brain microenvironment.

4.3.1.2 Molecular Biology of Tight Junctions

The molecular biology of tight junctions has become extremely complex in recent years. As claimed by us elsewhere [3, 55, 56], most data have been found for epithelial cells, probably because the regulation of BBB endothelial cell tight junctions is considerably more complex than that in epithelial cells. Generally, the molecular components identified at tight junctions can be separated into different classes based on their structures and functions. First, there are the integral membrane proteins occludin and the members of the claudin family, and, as detected more recently, Ig-superfamily members such as the junction-adhesion molecules of the JAM group and the endothelial cell-selective adhesion molecule ESAM (see below). Second, there are adaptor proteins which are distinguished according to their function to be first- or second-order adaptors. First-order adaptors are based on their direct association with the integral tight junction proteins via PDZ domains and include, for example, ZO-1, ZO-2, and ZO-3 (see below). Second-order adaptors are based on their indirect association with the integral tight junction proteins and include, for example, cingulin or the newly described cingulin-related JACOP (see below). Furthermore, there are other proteins involved in signaling cascades, such as G-proteins, regulators of G-protein signaling, and small GTPases which have been reviewed elsewhere [56, 67, 68].

In the past ten years, the knowledge of the molecular composition and regulation of the tight junctions has rapidly extended [3, 4, 57–69]. Occludin and the claudin family are the most important membranous components, both of which are proteins with four transmembrane domains and two extracellular loops (Fig. 4.6).

Occludin Occludin was the first tight junctional transmembrane molecule discovered [57]. It was initially isolated from junction-enriched membrane fractions of chicken liver as a transmembranous tight junction protein of approx. 65 kDa, which exists in several isoforms. Occludin shows high interspecies variability between chicken and mammals [58], sharing less than 50% identity in amino acid sequence. In contrast, human, mouse and dog occludins are more closely related, showing approximately 90% identity. Beside the high content of tyrosine and glycine in the first extracellular loop (approx. 60%), the most conserved region of occludin comprises the carboxy terminal ZO-1 binding domain, an a-helical coiled-coil structure, putatively linking occludin to the cytoskeleton.

claudin expression in the tight junctions. In non-BBB endothelial cells, tight junctions are almost completely associated with the E-face and claudin-3 is rarely or not expressed. BBB endothelial cells cultured in vitro develop tight junctions which are associated with the E-face [48] and express less claudin-1 [90]. However, an antibody was used which is now known to recognize claudin-3 as well. Under pathological conditions such as malignant glioma or experimental allergic encephalomyelitis, claudin-1/-3 was found to be lost and/or the tight junctions were E-face-associated [91].

Immunoglobulin-Like Proteins at Tight Junctions Almost simultaneously with the identification of claudins, a junctional adhesion molecule (JAM) has been reported as the first member of the immunogloblin (Ig) superfamily to be present at tight junctions (Fig. 4.6) [95]. JAM, which is now called JAM-A, localizes at homotypic cell-cell contacts of endothelial and epithelial cells and is highly enriched at tight junctions [95, 96]. Two Ig-like proteins closely related to JAM-A, JAM-B, and JAM-C have been identified recently [97, 98]. JAM-B and JAM-C are restricted to endothelial cells and are largely absent from epithelial cells. Although ultrastructural analyses for JAM-B and JAM-C in endothelial cells are still missing, the co-localization of JAM-C with occludin and the zonula occludens protein-1 (ZO-1; see below) upon ectopic expression in MDCK epithelial cells suggests its localization at tight junctions [98]. Additional evidence for a role of JAMs in the formation of tight junctions is based on the observation that anti-JAM-A antibodies as well as soluble JAM-A negatively affect the formation of functional tight junctions after Ca^{2+} switch-induced cell-cell contact formation [96, 99] and on the identification of cytosolic proteins which associate with JAMs and which are implicated in the formation/function of tight junctions (see below). In this respect, it is interesting to note that JAM-A and JAM-B are expressed by Sertoli cells in the testis where they could be involved in the formation and/or maintenance of the blood-testis barrier [100].

Another role of JAMs in endothelial cells might be related to their predicted function in regulating leukocyte-endothelial cell interaction during inflammation, through homophilic and heterophilic interactions (for recent reviews, see [101–103]). Originally, blocking JAM-A was found to inhibit leukocyte diapedesis in vitro and during inflammation in vivo [104]. However, there is a considerable body of evidence that leukocyte diapedesis does not follow the paracellular route, but the transcellular route via a mechanism called emperipolesis (for overviews of both the classic transmigration studies and the recent literature, see [105, 106] and also Chapter 24). The transcellular mechanism as observed in experimental allergic encephalomyelitis (EAE) leaves tight junctions intact and implies a complex rearrangement of the luminal and abluminal membranes [107]. This does not exclude the possibility that, despite leaving tight junctions morphologically intact during transcellular transmigration, these junctions can molecularly be changed. In EAE, we demonstrated a selective loss of anti-claudin-3 immunoreactivity [91]. This molecular alteration is associated with an increase in vascular permeability, but not an opening of tight junctions for paracellular leukocyte diapedesis. If we therefore assume that transcellular migration of leukocytes delivers a signal to the endothelial junction – probably via the actin cytoskeleton – it is tempting to speculate that functional antibodies that "tickle" junctional molecules such as VE-cadherin, PECAM-1, the JAM family or CD99 might trigger these intracellular signaling cascades such that they increase or decrease the endothelial mechanisms required for either transcellular or paracellular migration.

More recently, four additional Ig superfamily members have been identified at tight junctions. These include the coxsackie and adenovirus receptor (CAR) [108], endothelial cell-selective adhesion molecule (ESAM) [109], junctional adhesion molecule (JAM) 4 [110] and the coxsackie and adenovirus receptor-like membrane protein (CLMP) [111]. They share with JAM-A, JAM-B, and JAM-C a similar organization, with two Ig-like domains. However, they are more closely related to each other than to the three JAMs and thus form a subfamily within the tight junction-associated Ig superfamily members [103]. Interestingly, CAR, ESAM, and JAM4 end in a type I PDZ domain-binding motif, whereas JAM-A, JAM-B, and JAM-C end in a type II motif, which suggests functional differences between the two subfamilies. The function of these four Ig superfamily members at tight junctions is not clear. CAR, JAM-4, and CLMP are predominantly expressed by epithelial cells, whereas ESAM is expressed exclusively in endothelial cells, including those in brain capillaries [109, 112]. Endothelial cells derived from ESAM-deficient mice display defects in endothelial tube formation, suggesting a role for ESAM in endothelial cell contact formation [113]. How this function relates to its specific localization at tight junctions is not yet clear.

Peripheral Membrane Components at Tight Junctions The transmembrane proteins associate in the cytoplasm with peripheral membrane components which form large protein complexes, the cytoplasmic "plaque". ZO-1, a 220-kDa phosphoprotein, was the first peripheral membrane component identified and characterized at tight junctions [114]. In cellular systems with less elaborate or no tight junctions at all, ZO-1 is found enriched in regions of the adherens junctions [115], where it may interact with components of the cadherin-catenin system [116, 117]. Since the discovery of ZO-1, many further components of peripheral tight junction proteins have been described (Fig. 4.6). One type of plaque protein consists of adaptors, proteins with multiple protein-protein interaction domains such as SH-3 domains, guanylate kinase (GUK) domains, and PDZ domains [118, 119]. The adaptor proteins include members of the membrane-associated guanylate kinase (MAGUK) [120] and membrane-associated guanylate kinase with an inverted orientation of protein-protein interaction domains (MAGI) families [121, 122]), such as ZO-1, -2, -3, MAGI-1, -2, -3, as well as proteins with one or several PDZ domains, such as PAR-3, PAR-6, and MUPP1 [66, 103, 122]. The adaptor proteins serve as scaffolds to organize the close proximity of the second type of plaque proteins, the regulatory and signaling proteins. These include small GTPases, their regulators, and the transcriptional regulator ZO-1-associated nucleic acid binding protein ZONAB (Fig. 4.6;

for further literature, see [56]). Furthermore, a new protein called JACOP (junction-associated coiled-coil protein) has been discovered. This protein is found in the tight junction complex of epithelial but also endothelial cells and is suggested to anchor especially the junctional complex to the actin-based cytoskeleton [123, 124]. JACOP has considerable sequence similarity to cingulin, another previously detected peripheral protein at tight junctions [125]. In many cases, the role of the regulatory and signaling proteins in tight junction biology is still poorly understood; and it is to be expected that they are involved in completely different aspects of tight junction biology. The proteins of the PAR-3/aPKC/PAR-6 complex are most likely involved in the regulation of tight junction formation and establishment of cell polarity, since overexpression of dominant-negative mutants of these proteins leads to delayed tight junction formation [126–128]. Both JAM-A and PAR-3 localize to cell-cell contacts of endothelial cells [129] and, therefore, it is conceivable that the formation of tight junctions is similarly regulated by JAM-A and the PAR-3/aPKC/PAR-6 complex. Information on the maturation state of cell-cell contacts is required for many cellular events which are regulated by cell density (e.g. proliferation) and transcription factors associated with the cytoplasmic plaque at tight junction provide a direct link between tight junctions and the nucleus [67]. Generally, many of these molecules have not been described explicitly in the BBB endothelial cells so far. Therefore, some of them are not included in Fig. 4.6. However, one can not exclude that in the future they have to be incorporated into the scenario of BBB regulation and maintenance.

Indeed, the vast majority of experiments addressing the role of tight junction-associated proteins for tight-junction biology were performed with epithelial cells. The predominant concern of tight junctions is related to the generation of cellular polarity. Regarding polarity, we have also to consider molecules which were originally detected in invertebrate systems but are now known in vertebrates as well and which will have a yet unknown impact in understanding junctions [56, 124, 130–132]. Our knowledge about the role of these tight junction-associated proteins in tight junction formation in endothelial cells, and in particular those in the BBB, is still limited. However, it is to be expected that the principal mechanisms underlying tight junction formation operate in both cellular systems in a similar way.

4.3.2
Caveolae

Barrier permeability is determined by both the tight junction-controlled paracellular and the caveolae-mediated transcellular permeability. In addition, there is receptor-mediated endocytosis, the morphological basis of which is provided by the clathrin-coated pits and vesicles. The literature on these vesicular structures is huge (see, for example, reviews [133–141]). Caveolae have been suggested as sites of endothelial transcytosis, endocytosis, signal transduction, and as docking

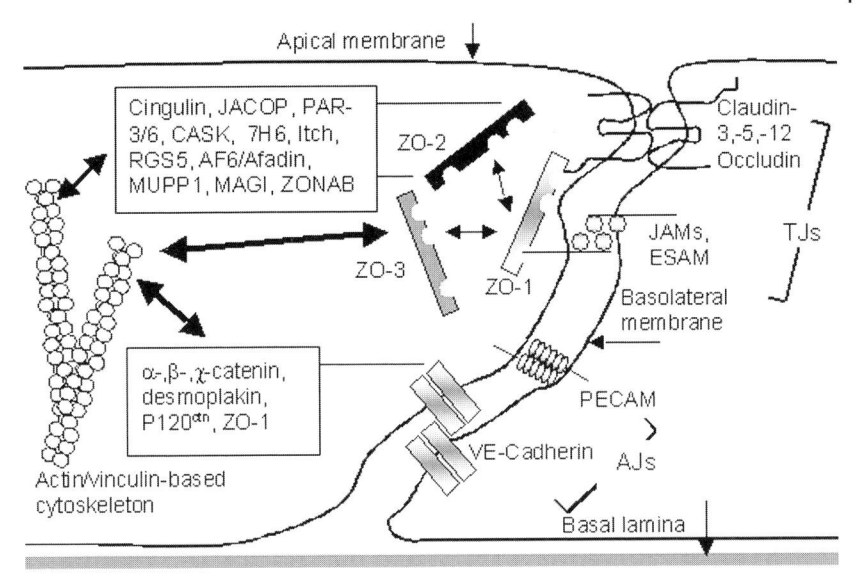

Fig. 4.6 Simplified scheme of the molecular composition of endothelial tight junctions. Occludin and the claudins are the most important membranous components, which are proteins with four transmembrane domains and two extracellular loops. The junctional adhesion molecules (JAMs) and the endothelial selective adhesion molecule (ESAM) are members of the immunoglobulin superfamily. Inside the cytoplasm, there are many associated proteins which partly contain PDZ domains binding the C-terminus of the intramembrane proteins (first-order adaptor proteins). Among those are the *zonula occludens* proteins 1–3 (ZO-1, ZO-2, ZO-3) and the calcium-dependent serin protein kinase (CASK). The three PDZ domains of these proteins are indicated by the pits in the molecular symbols. Among the second-order adaptor molecules, cingulin was described to be expressed in BBB endothelial cells and the junction-associated coiled-coil protein (JACOP) in other endothelial and epithelial cells. Signaling and regulatory proteins described in endothelial cells, but not in any case explicitly in BBB endothelial cells, are the multi-PDZ-protein 1 (MUPP1), the partitioning defective proteins-3 and -6 (PAR-3/6), the membrane-associated guanylate kinase with an inverted orientation of protein-protein interaction domains (MAGI), the ZO-1-associated nucleic acid binding protein (ZONAB), afadin/AF6, and the regulator of G-protein signaling, RGS5. All these first- and second-order adaptors and regulatory/signaling proteins control the interaction of the membranous components with the actin/vinculin-based cytoskeleton. Whereas tight and adhesion junctions in epithelial cells are strictly separated from each other, these junctions are intermingled in endothelial cells. The most important molecule of endothelial adhesion junctions is the vascular endothelial cadherin (VE-cadherin). Also, the platelet-endothelial cell adhesion molecule (PECAM) mediates homophilic adhesion. Linker molecules between adhesion junctions and the cytoskeleton are primarily the catenins (α-, β-, γ-catenin), but also desmoplakin and p120^{ctn}. For further information and for literature, see text.

sites for glycolipids and glycosylphosphatidylinositol-linked proteins (for literature, see [136, 142]). Among the components of the caveolar membranes are the receptors for LDL (not only present in coated pits), HDL, transferrin, insulin, albumin, ceruloplasmin and advanced glycation end-products (AGE), interleukin-1, vesicle-associated membrane protein-2 (VAMP-2), and caveolin-1/-2. Regarding the blood-brain barrier, caveolin-1 was not only detected in endothelial cells, but also in astrocytes and pericytes [143]. Caveolin-1 binds to cholesterol and fatty acids and forms high molecular weight oligomers. Among the molecules forming signaling complexes at caveolin-1 as a multivalent docking site are heterotrimeric G-proteins, members of the MAP kinase pathway, *src* tyrosine kinase, protein kinase C, and the endothelial NO synthase. All these molecular complexes are organized in lipid-based microdomains or rafts (as summarized in [140]). The involvement of caveolin-1 at least in NO and calcium signaling processes was impressively documented in the caveolin-1-deficient mouse [144]. The function of caveolae as organelles of endocytosis still remains controversial. Internalization of caveolae has been demonstrated to be regulated by phosphorylation [142]. However, caveolin-1-deficient mice seem to have an unaffected transendothelial transport [144]. Caveolin-1 was proposed not to be required for endocytosis but would stabilize endocytic raft domains decreasing endocytosis [145].

In the endothelium, the relationship of paracellular and transcellular permeability is of crucial importance for the regulation of overall transendothelial permeability (Fig. 4.7). It is well known that the vascular endothelial growth factor (VEGF) is identical to the permeability growth factor (VPF) [146]. VEGF plays a central role in triggering angiogenesis and vascular permeability. It was shown previously that endothelial cells under the influence of VEGF form fenestrations, caveolae, and caveolin-1- and VAMP-positive vesiculo-vacuolar organelles (VVOs) [133, 147, 148]. More recently, the *src*-suppressed C-kinase substrate (SSeCKS) in astrocytes has been reported to be responsible for the decreased expression of VEGF and increased release of the anti-permeability factor angiopoietin-1 (Ang-1) [149] (see also Chapter 2). In parallel, the authors demonstrated that SSeCKS overexpression increased the expression of tight junction molecules and decreased paracellular permeability in endothelial cells. Indeed, VEGF has been shown to induce the phosphorylation of occludin and ZO-1 which could result in both dissociation of caveolin from the junction [78] and targeting to the luminal membrane. The VEGF receptor 2, also known as Flk-1, is closely associated with caveolin-1, the main molecule of caveolae [150], which has also been shown to co-precipitate with occludin (Fig. 4.7) [151]. In addition, the monocyte chemoattractant protein 1 (MCP-1) was reported to alter the expression of both tight junction-related proteins and caveolin-1 [152]. Thus, tight junctions could play a role as a "sink" for caveolin-1, or, put the other way, occludin-bound caveolin-1 may be a stabilizator of tight junctions. Once dissociated from occludin, caveolin-1 could increasingly bind to the dystrophin-dystroglycan complex, in particular to NO synthase [153], which, at least in muscle cells, is associated with syntrophin [154]. However, any information about a connection between NOS and endothelial dystroglycan is lacking so far. In contrast,

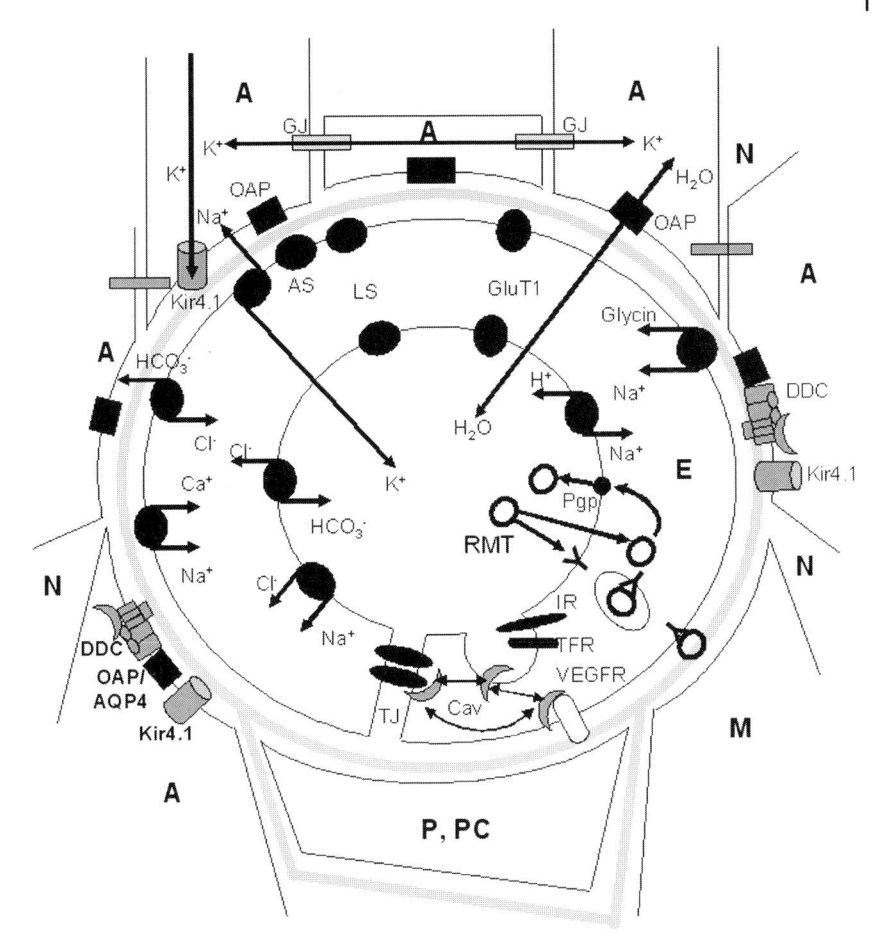

Fig. 4.7 Schematic view of the distribution of some transporters, receptors, and other molecules at the BBB. A=Astrocyte, E=endothelial cell, N=neuron, M=microglial cell, P=pericyte, PC=perivascular cell, AS=A-system of amino acid transport, as present in the abluminal membrane, LS=L-system of the amino acid transport, as present in both the luminal and abluminal endothelial membrane, RMT=receptor-mediated transport, TFR=transferrin receptor, IR=insulin receptor, Pgp=P-glycoprotein, DDC=dystrophin- dystroglycan complex, Kir4.1=inward rectifying potassium channel 4.1, VEGFR=vascular endothelial growth factor receptor, Cav=caveolin, OAP=orthogonal arrays of particles in the glial endfoot membrane, representing the site of the water channel protein aquaporin-4 (AQP4), GJ=gap junctions between astroglial cells, TJ=tight junctions between endothelial cells or cell processes. The gray line around endothelial and perivascular cells represents the basal laminae.

endothelial NO synthase is present in BBB endothelial cells and its activity is increased in permeable blood vessels [155, 156]. Also, NO donors were shown to disrupt the BBB [157]. Accordingly, the well known cytokine erythropoietin seems to protect the BBB against VEGF-induced permeability by reducing the level of NO synthase and by stabilizing tight junctions [158]. This relationship could suggest that tight junctions and vesicular transcytosis are interconnected inversely. However, as has been demonstrated by means of direct electron microscopical observations, the cleft index, as a morphological indicator of the tight junctional integrity, in capillaries of C6 or RG2 rat glioma was higher than in normal brain tissue. In contrast, the vesicular density was found not to be increased in the tumor. Interestingly, administration of bradykinin or leukotriene C4 was not able to alter the cleft index in both normal brain and brain tumor, but the vesicular density was increased in the tumor only [159], suggesting that vesicular transcytosis and tight junctions can be regulated separately.

4.3.3
Transporters in the Blood-Brain Barrier Endothelium

The tight junction-based paracellular impermeability of the brain capillary endothelial cells implies that the hydrophilic substances essentially needed for the metabolism of the brain require transport through the endothelial wall. Several independent carrier systems for the transport of hexoses (glucose, galactose), neutral, basic and acidic amino acids, monocarboxylic acids (lactate, pyruvate, ketone bodies), purines (adenine, guanine), nucleosides (adenosine, guanosine, uridine), amines (choline), and ions have been described (for recent overviews, see [160, 161]). Among these transporters, the glucose transporter is of special importance due to the fact that glucose is the main energy source of the brain (Fig. 4.7). The 55-kDa form of the glucose transporter isoform Glut1 as one of five members of a supergene family of sodium-independent glucose transporters is highly restricted to the capillary endothelial cells in the brain [160, 162, 163]). This restriction to endothelial cells and the asymmetrical distribution in the endothelial cell membranes (the density of transporter molecules is three to four times higher in the abluminal than in the luminal membrane [164–167]) are believed to reflect the onset of BBB function during development. In functional terms, the lower concentration of transporter molecules in the luminal membrane limits the intensity of glucose flux from the blood to endothelium; and the higher concentration in the abluminal membrane may reduce the endothelial glucose concentration in comparison to that in blood and secure a better efficiency of transport from endothelium to brain parenchyma [166].

In addition to carrier-mediated transport from blood to brain, receptor-mediated transport has also been investigated intensively. In BBB endothelial cells, there are receptors for the endocytosis of transferrin [168–170], LDL [171], immunoglobulin G [172], insulin, and insulin-like growth factor (Fig. 4.7) [173]. In the case of the insulin receptor and the insulin-like growth factor receptor, it should be stressed that the BBB of mice (the endothelial cells of which were specifically deficient for both receptors) was not compromised, indicating that neither insulin nor insulin-like growth factor are required for the development and integrity of the BBB [174].

The transfer of dopamine between brain and blood is restricted by the enzymatic conversion of L-DOPA to dopamine by the DOPA-decarboxylase within the endothelial cell ("metabolic BBB") [175]. Other enzymes such as γ-glutamyl-transpeptidase and antioxidant enzymes (glutathione peroxidase, glutathione reductase, catalase, superoxide dismutase) are involved in protecting the brain against oxidative stress. Thereby, the reduced glutathione form GSH reacts nonenzymatically, or catalyzed by glutathione peroxidase, with reactive oxygen species and forms the oxidized form glutathione disulfide (GSSG). By the glutathione reductase reaction, GSH can be regenerated from GSSG [176]. It has been shown that oxidative stress induced by cerebral ischemia disrupts the BBB and that antioxidants such as superoxide dismutase and catalase attenuate the BBB disruption [177, 178]. Presumably, the antioxidative capacity of the brain may be promoted by the homeostasis of glutathione metabolism in the brain and between brain and blood. In capillary endothelial cells and in astrocytes, a sodium-dependent GSH transporter has been described which may contribute to the delivery of GSH from blood to brain [179, 180].

There are still other antigens associated with the BBB. For example, in order to restrict free diffusion of lipophilic compounds from blood to brain, a control mechanism is required to export them effectively out of the brain endothelium. In this context, the P-glycoprotein which is a transmembrane protein conferring multidrug resistance is of special importance (see Chapter 19). P-glycoprotein and the multidrug resistance-associated protein Mrp1 [13] have been identified in brain endothelial cell membranes (Fig. 4.7). These proteins importantly are responsible for the active extrusion of nonpolar molecules out of endothelial cells and are therefore in the focus of research on drug delivery in the brain. Interestingly, P-glycoprotein was described to be co-expressed together with caveolin in brain microcapillary endothelial cells [181, 182]. However, only a fraction of P-glycoprotein may be associated with caveolin-1, suggesting that the expression of both proteins may be under the control of defined metabolic requirements [183]. Moreover, another important class of surface molecules mediating the interaction between blood cells and endothelium represents the adhesion molecules. The expression and upregulation of the intercellular and vascular cell adhesion molecules (ICAM-1 and VCAM-1, respectively) have been identified as essential steps for lymphocyte recruitment during inflammation (for a review, see [106]). The transendothelial migration of lymphocytes was shown to be dependent on Rho signaling: when activation of Rho as a consequence of leukocyte binding to the adhesion receptor ICAM-1 was blocked by C3-transferase, leukocyte transmigration was compromised [184]. Whereas the extracellular domain of endothelial ICAM-1 suffices to mediate T cell adhesion, the cytoplasmic domain is required to mediate transmigration of T cells, probably by inducing Rho-signaling within the endothelial cells [185]. The involvement of tight junctions during inflammation is a topic which arouses contradictory discussion in the field. It seems widely accepted that inflammatory cells invade the CNS by using the paracellular pathway [102]. However, there is a multitude of morphological evidence that suggests consideration of the transcellular transmigration as well (for reviews, see [56, 105, 106]). This does not rule out that the molecular composition of tight junctions can be altered during inflammation [186, 187].

4.4
Brief Consideration of the Neuroglio-Vascular Complex

The neuroglio-vascular complex is involved in the regulation of blood flow and nutrient supply within the CNS (Fig. 4.7). This regulation includes: (a) control of perfusion parameters differentially realized in specific brain regions according to local requirements, (b) maintenance of energy supply from blood to neuronal and synaptic metabolism via glial cells, and (3) the protection of the nervous parenchyma from alterations in blood composition, particularly from neurotoxic compounds including reactive oxygen species. The network of neuroglio-vascular interactions (the "neurovascular unit") is involved in manifold metabolic dependencies between neurons and glial cells on the one hand and between glial cells and vascular cells on the other hand (for recent comprehensive overviews on the neurovascular unit, see [188, 189]). The direct interface between the neuroglial compartment and the vascular compartment is established by the perivascular glial endfeet, forming the glial limiting border. It is generally accepted now that astrocytes play a decisive role in the maintenance of the barrier properties of the brain microcapillary endothelial cells. An interesting correlation exists between astroglial differentiation and BBB maturation. The astroglial differentiation can be morphologically recognized as the polarization of astrocytes, which arises concomitantly with the maturation of the BBB [190, 191] and is not maintained by reactive [192] or cultured astrocytes [193, 194].

Polarization of astrocytes means getting a molecular and structural heterogeneity of specific membrane domains of the astroglial surface. Where the processes of glial cells (so-called endfeet) contact the superficial or perivascular basal lamina (*glia limitans superficialis et perivascularis*), the glial membrane is studded with numerous square arrays or orthogonal arrays of intramembranous particles (OAPs) which can only be visualized by means of the freeze-fracture technique (for an overview of the literature, see [193]; Figs. 4.7 and 4.8). Where the contact of the glial cell membrane with the basal lamina is lost by bending away into deeper parenchymal regions of the neuropil, the density of OAPs is dramatically reduced. In highly polarized epithelial cells, the apical and basolateral membrane domains are separated by tight junctions; but this is not the case in highly polarized astrocytes. The establishment and maintenance of the glial polarity is suggested to be evolved by the extracellular matrix of the basal lamina (Fig. 4.8).

Regarding the OAPs, it is well known now that they contain at least the water channel protein aquaporin-4 (AQP4) (Fig. 4.7). Aquaporins mediate water move-

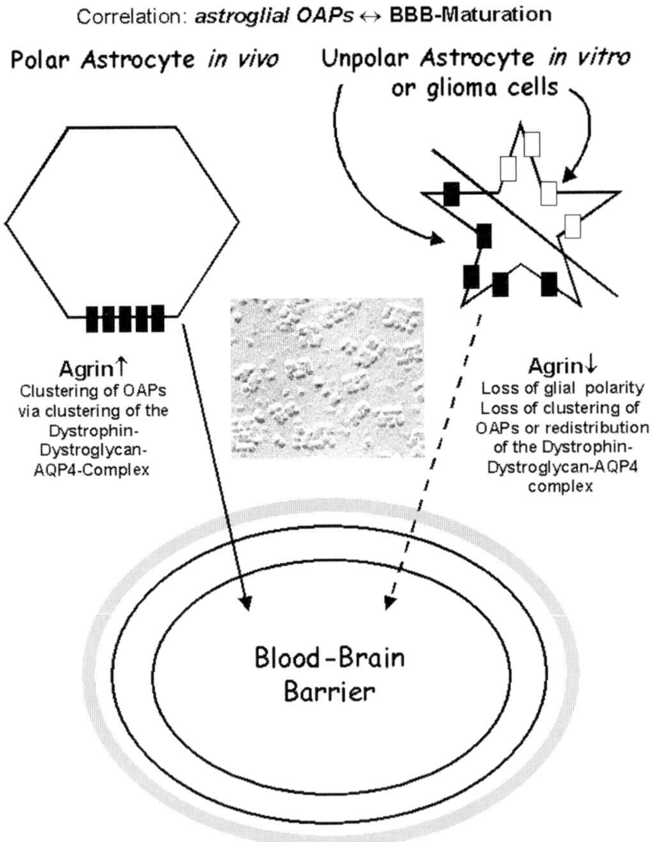

Correlation: *astroglial OAPs* ↔ BBB-Maturation

Polar Astrocyte *in vivo*

Unpolar Astrocyte *in vitro* or glioma cells

Agrin↑
Clustering of OAPs
via clustering of the
Dystrophin-
Dystroglycan-
AQP4-Complex

Agrin↓
Loss of glial polarity
Loss of clustering of
OAPs or redistribution
of the Dystrophin-
Dystroglycan-AQP4
complex

Blood-Brain Barrier

Fig. 4.8 Working hypothesis on glial involvement in the formation and maintenance of the BBB. When the BBB is fully established, the associated astroglial cells are highly polarized in terms of orthogonal arrays of particles (OAPs, symbolized as black rectangles; picture in the middle of the scheme is a freeze-fracture replica) and the water channel protein aquaporin-4 (AQP4). The development of astroglial polarity goes along with an increasing expression rate of the heparansulfate proteoglycan agrin within the perivascular basal lamina. In culture and under pathologic conditions such as brain tumor, agrin is down-regulated and the OAPs are redistributed (black rectangles), or lost, or AQP4 is no more recognizable in freeze-fracture replicas (white rectangles). This down-regulation may be causative for the redistribution of AQP4 across the whole surface of the glial cell, which can be described as a loss of glial polarity. This loss can constitutively be observed when the BBB is disturbed. Therefore, agrin as a molecule significantly involved in the clustering of the dystrophin/dystroglycan/AQP4 complex seems to play a pivotal role in the management of the endothelial BBB (see text).

ments between the intracellular, interstitial, vascular, and ventricular compartments which are under the strict control of osmotic and hydrostatic pressure gradients (see Chapter 9) [195–198]. The involvement of AQP4 in OAP formation was demonstrated by the absence of OAPs in the astrocytes of an AQP4-deficient mouse [199], by the formation of OAPs in chinese hamster ovary cells stably transfected with AQP4 cDNA [200], and by the immunogold fracture-labeling technique showing that AQP4 is a component of the arrays [201, 202]. Moreover, Nielsen et al. [203] were able to demonstrate by immunogold immunocytochemistry that the distribution of the AQP4-related immunoreactivity was identical to that of the OAPs.

Interestingly, the OAP-related polarity of astrocytes seems to correlate with the quality of the BBB (Fig. 4.8) [190, 193]. An indirect proof of this relationship is the observation that, under brain tumor conditions when the BBB is known to be leaky, the OAP-related polarity of glial cells was decreased (Fig. 4.8) [204]. In addition, the AQP4 content as detected by immunocytochemistry was increased [205, 206]. The apparent contradiction of upregulation of AQP4 and downregulation of AQP4-positive OAPs in glioma cells can be resolved by the assumption that, under glioma conditions, AQP4 exists separately from OAPs in the glial membrane and is no longer restricted to the glial endfeet membranes. Indeed, AQP4 exists as two isoforms (M1, with 323 amino acids, and M23, with 301 amino acids [207]). Only isoform M23 seems to assemble in OAPs, whereas isoform M1 does not [208]. This could mean that, under pathological conditions, the ratio between the M1 and M23 isoforms of AQP4 changes heavily in favor of M1. However, this seems not to be the case, again supporting the view that M23 would become unrecognizable in freeze-fracture replicas by dissociating from AQP4-based arrays (Warth and Wolburg, unpublished observation). What this would mean in functional terms is unknown so far. In any case, the restriction of AQP4 immunoreactivity at the endfoot membrane was maintained only where agrin was present in the perivascular basal lamina [206]. Agrin is an extracellular heparan sulfate proteoglycan which was originally characterized as the essential molecule for clustering acetylcholine receptors at the motor endplate [209, 210] but has also been described as being important within the CNS, particularly for the integrity of the BBB [211–213]. The agrin splicing variant Y0Z0 was reported to be specifically present in the endothelial cell basal lamina of CNS capillaries [214]. If agrin was absent from the basal lamina, AQP4 immunoreactivity was randomly found across the whole surface of the cell [215]. This would suggest that agrin is responsible for the restriction of AQP4 molecules at the glial endfoot membrane (Fig. 4.8).

However, agrin has no binding site to AQP4, but to α-dystroglycan [216]. α-Dystroglycan is a member of the dystrophin-dystroglycan complex (DDC) which is best investigated in the muscular system, but is also present in the CNS (Fig. 4.7) [154, 217]. Thus, α-dystroglycan has to be connected to AQP4. Indeed, they are found to be coexpressed in glial cells. Another member of the DDC is α1-syntrophin, which has a binding site to dystrophin and a PDZ domain binding to AQP4 [218]. In glioblastoma, the binding strength between α1-syntrophin and AQP4 seems to be strong enough to bind AQP4 and α1-syntrophin together during their redistribution across the surface of the glioma cell [206]. In contrast, dystrophin remains restricted at the endfoot membrane suggesting a cleavage of the dystrophin/α1-syntrophin/AQP4 complex. Concomitant with these observations, we found a loss of α-dystroglycan in the perivascular domain of glioma cells [206]. Thus, loss of agrin went along with loss of α-dystroglycan, redistribution of AQP4/α-syntrophin and loss of OAPs, including a severe reduction in the OAP-related polarity of glioma cells. All these observations together speak in favor of the suggestion that the OAPs consist of more than AQP4, but contain components of the DDC as well.

It has been shown that the truncated dystrophin isoform Dp71 is essential for the clustered localization of the weakly rectifying potassium channel Kir4.1 in retinal Müller (glial) cells [219]. In addition, the PDZ domain of α-syntrophin can also bind to Kir4.1 [220], which is normally restricted to the endfoot membrane in astrocytes and retinal Müller glial cells (Fig. 4.7) [221–223]. On the basis of colocalization of AQP4 and Kir4.1 in retinal Müller cells and due to the well known fact that water fluxes are driven by ion fluxes, it was hypothesized that K^+ clearance is coupled to water flux [189, 203, 224–227]. Accordingly in the α-syntrophin-deficient mouse, in which AQP4 is dislocalized across the glial surface, the K^+ clearance was delayed [228]. The authors argue that K^+ uptake would be facilitated if accompanied by water flux. In the hypoxic retina, Kir4.1 is downregulated in retinal Müller cells compromising the spatial buffering capacity; and as a consequence, intracellular K^+ concentration increases and consecutive water flux is causative for the observed cell swelling [226, 227]. Alternatively, the cell swelling can also be due to a failure to release water at the endfoot membrane, because the molecular complex consisting of Kir4.1 and AQP4 may dissociate. In human glioblastoma tissue, we were able to observe a redistribution of both anti-AQP4- and anti-Kir4.1-immunoreactivities across the surface of the glioma cell [229]. The observation that Kir4.1, but not AQP4, was already redistributed in low-grade astrocytomas suggested different targeting or cluster mechanisms for Kir4.1 and AQP4. An uncoupling of water transport from K^+ siphoning under different pathological conditions, such as brain contusion, bacterial meningitis, and brain tumors, has recently been postulated [230] and may be one reason for cell swelling. Interestingly, before the detection of aquaporins, OAPs were frequently discussed as morphologically correlating to potassium channels [193, 231–233]. However, we must not forget that this kind of molecular aggregation of different channels in one morphological structure is also established at the subpial endfoot membrane and not exclusively at the perivascular endfoot membrane, suggesting that the OAP-related molecular arrays have no exclusive significance for the BBB.

In the α-syntrophin-deficient mouse, Amiry-Moghaddam et al. [234] described an elongation of the survival time of animals under hyponatremic conditions in comparison to wild-type mice. The reason for the longer survival time might be the reduced formation of brain edema, which was interpreted as a consequence of the redistribution of AQP4 after its cleavage from α-syntrophin. However, as we were able to show, the cleavage of α-syntrophin from AQP4 seems not to be a prerequisite for the redistribution of AQP4 [206]. Rather, we believe that loss of agrin reduces the OAP/AQP4-related polarity of astrocytes. The loss of agrin might be caused by an activation of metalloproteinase-3, which has recently been reported to occur under conditions of cerebral ischemia [235]. Furthermore, this could lead to a redistribution of "free" AQP4 molecules outside the OAP structure, or an upregulation of the M1 isoform of AQP4 (see above), a downregulation of Kir4.1, and a consecutive failure of spatial buffering of K^+, followed by osmotic water influx and swelling. All these processes together might represent key events for both the development of an edema and the loss of the glial cell ability to maintain BBB properties within the endothelial cell. An unpolarized astrocyte has already been suggested to be unable to induce or maintain the complete set of BBB properties in endothelial cells (Fig. 4.8) [236].

4.5
Conclusions

It has been 120 years since the discovery of the BBB by Paul Ehrlich. In the following period, BBB research has focused on the morphological description of the barrier, using mainly conventional histological and electron microscopical methods as well as methods to demonstrate the tightness of the barrier against a variety of low and high molecular weight substances (for a review, see [237]). Tight junctions have been described as a network of protein particles, using freeze-fracture electron microscopy. But, it required almost a further 30 years and a great methodological advance in molecular biology to discover the protein components of the tight junctional complex and to identify the protein particles within the freeze-fracture replicas. Then, it took further time to recognize these tight junctional complexes as very dynamic structures. Now, it might be possible to go into more detail, to characterize the interplay of these molecules and their regulation by and association with cellular signaling cascades and cytoskeletal components. It has been shown that the endothelial barrier in the brain differs from epithelial barriers. The simple observation that epithelial, but not endothelial cells are able to form a high-resistance and low-permeability barrier in vitro sheds light on the importance of the brain microenvironment in the formation and maintenance of the barrier in vivo. This microenvironment consists of endothelial cells, pericytes, microglial cells, astrocytes, neurons, and the extracellular matrix in between, which itself forms a microcosmos of its own. All components may or may not operate together or at certain periods during development or during pathological derangements and in different combinations which are not recognizable so far. The greatest future challenge will be characterizing the mechanisms involved in BBB differentiation and pathology in order to understand the key principles of barrier formation in the CNS.

Acknowledgments

Personal work cited in this chapter was supported by grants from the Fortüne program of the University of Tübingen, the IZKF Tübingen, the Deutsche Krebshilfe, and the Deutsche Forschungsgemeinschaft. Drs. A. Lippoldt and K. Ebnet contributed to this work.

References

1 P. Ehrlich **1885**, *Das Sauerstoff-Bedürfnis des Organismus. Eine farbenanalytische Studie*, PhD thesis, Herschwald, Berlin, pp 69–72.
2 M. W. Brightman, T. S. Reese **1969**, *J. Cell Biol.* 40, 648–677.
3 H. Wolburg, A. Lippoldt **2002**, *Vasc. Pharmacol.* 38, 323–337.
4 A. W. Vorbrodt, D. H. Dobrogowska **2003**, *Brain Res. Rev.* 42, 221–242.
5 A. Peters, S. L. Palay, H. de F. Webster **1991**, *The Fine Structure of The Nervous System: Neurons and Their Supporting Cells*, Oxford University, Oxford.
6 B. L. Coomber, P. A. Stewart **1985**, *Microvasc. Res.* 30, 99–115.
7 D. N. Angelov, M. Walther, M. Streppel, O. Guntinas-Lichius, W. F. Neiss **1998**, *Adv. Anat. Embryol. Cell Biol.* 147, 1–87.
8 H. Gerhardt, C. Betsholtz **2003**, *Cell Tissue Res.* 314, 15–23.
9 B. Engelhardt **2003**, *Cell Tissue Res.* 314, 119–129.
10 D. Virgintino, M. Errede, D. Robertson, C. Capobianco, F. Girolamo, A. Vimercati, M. Bertossi, L. Roncali **2004**, *Histochem. Cell Biol.* 122, 51–59.
11 M. G. Achen, M. Clauss, H. Schnürch, W. Risau **1995**, *Differentiation* 59, 15–24.
12 H. Xu, H. Zhao, W. Tian, K. Yoshida, J. B. Roullet, D. M. Cohen **2003**, *J. Biol. Chem.* 278, 11520–11527.
13 Y. Qin, T. N. Sato **1995**, *Dev. Dyn.* 202, 172–180.
14 A. H. Schinkel, J. J. M. Smit, O. van Tellingen, J. H. Beijnen, E. Wagenaar, L. van Deemter, C. A. A. M. Mol, M. A. van der Walk, E. C. Robanus-Maandag, H. P. J. te Riele, A. J. M. Berns, P. Borst **1994**, *Cell* 77, 491–502.
15 D. J. Begley **2004**, *Curr. Pharmaceut. Design* 10, 1295–1312.
16 H. Gerhardt, S. Liebner, C. Redies, H. Wolburg **1999**, *Eur. J. Neurosci.* 11, 1191–1201.
17 J. Folkman, P. A. D'Amore **1996**, *Cell* 87, 1153–1155.
18 P. Lindahl, B. R. Johansson, P. Leveen, C. Betsholtz **1997**, *Science* 277, 242–245.
19 M. Hellström, H. Gerhardt, M. Kalén, X. Li, U. Eriksson, H. Wolburg, C. Betsholtz **2001**, *J. Cell Biol.* 153, 543–553.
20 H. Wolburg, S. Liebner, A. Reichenbach, H. Gerhardt **1999**, *Int. Rev. Cytol.* 187, 111–159.
21 H. Gerhardt, G. Rascher, J. Schuck, U. Weigold, C. Redies, H. Wolburg **2000**, *Glia* 31, 131–143.
22 B. Schlosshauer, K.-H. Herzog **1990**, *J. Cell Biol.* 110, 1261–1274.

23 H. Seulberger, F. Lottspeich, W. Risau **1990**, *EMBO J.* 9, 2151–2158.
24 H. Seulberger, C. M. Unger, W. Risau **1992**, *Neurosci. Lett.* 140, 93–97.
25 R. Hallmann, D. N. Mayer, E. L. Berg, R. Broermann, E. C. Butcher **1995**, *Dev. Dyn.* 202, 325–332.
26 W. M. Pardridge **1991**, *Semin. Cell Biol.* 2, 419–426.
27 S. Wakai, N. Hirokawa **1978**, *Cell. Tissue Res.* 195, 195–203.
28 W. Risau **1991**, *Ann. N.Y. Acad. Sci.* 633, 405–419.
29 U. Kniesel, W. Risau, H. Wolburg **1996**, *Dev. Brain Res.* 96, 229–240.
30 P. A. Stewart **2000**, in *Morphogenesis of Endothelium*, eds. W. Risau, G. M. Rubanyi, Harwood, Amsterdam, pp 109–122.
31 K. M. Dziegielewska, J. Ek, M. D. Habgood, N. R. Saunders **2001**, *Microsc. Res. Tech.* 52, 5–20.
32 B. Engelhardt **2003**, *Cell Tissue Res.* 314, 119–129.
33 W. Risau, R. Hallmann, U. Albrecht **1986**, *Dev. Biol.* 117, 537–545.
34 K. Mollgard, D. H. Malinowska, N. R. Saunders **1980**, *Nature* 264, 293–294.
35 N. R. Saunders, G. W. Knott, K. M. Dziegielewska **2000**, *Cell. Mol. Neurobiol.* 20, 29–40.
36 P. A. Stewart, M. J. Wiley **1981**, *Dev. Biol.* 84, 183–192.
37 M. G. Farquhar, G. E. Palade **1963**, *J. Cell Biol.* 17, 375–412.
38 T. S. Reese, M. J. Karnovsky **1967**, *J. Cell Biol.* 34, 207–217.
39 S. J. Singer, G. L. Nicolson **1972**, *Science* 175, 720–731.
40 L. A. Staehelin **1974**, *Int. Rev. Cytol.* 39, 191–283.
41 H. Wolburg, S. Liebner, A. Lippoldt **2003**, *Methods Mol. Med.* 89, 51–66.
42 R. Dermietzel **1975a**, *Cell Tissue Res.* 164, 45–62.
43 R. Dermietzel **1975b**, *Cell Tissue Res.* 164, 309–329.
44 K. Mollgard, N. R. Saunders **1975**, *J. Neurocytol.* 4, 453–468.
45 E. Tani, S. Yamagata, Y. Ito **1977**, *Cell Tissue Res.* 176, 157–165.
46 R. R. Shivers **1979**, *Brain Res.* 169, 221–230.
47 Z. Nagy, H. Peters, I. Hüttner **1984**, *Lab. Invest.* 50, 313–322.
48 H. Wolburg, J. Neuhaus, U. Kniesel, B. Krauss, E.-M. Schmid, M. Öcalan, C. Farrell, W. Risau **1994**, *J. Cell Sci.* 107, 1347–1357.
49 S. Hamm, B. Dehouck, J. Kraus, K. Wolburg-Buchholz, H. Wolburg, W. Risau, R. Cecchelli, B. Engelhardt, M.-P. Dehouck **2004**, *Cell Tissue Res.* 315, 157–166.
50 P. Claude **1978**, *J. Membrane Biol.* 39, 219–232.
51 A. Martinez-Palomo, I. Meza, G. Beaty, M. Cereijido **1980**, *J. Cell Biol.* 87, 736–745.
52 J. L. Madara, K. Dharmsathaphorn **1985**, *J. Cell Biol.* 101, 2124–2133.
53 L. J. Mandel, R. Bacallao, G. Zampighi **1993**, *Nature* 361, 552–555.
54 R. Bacallao, A. Garfinkel, S. Monke, G. Zampighi, L. J. Mandel **1994**, *J. Cell Sci.* 107, 3301–3313.
55 U. Kniesel, H. Wolburg **2000**, *Cell. Mol. Neurobiol.* 20, 57–76.
56 H. Wolburg, A. Lippoldt, K. Ebnet **2005**, Tight junctions and the blood-brain barrier, in *Tight Junctions*, ed. L. Gonzales-Mariscal, Landes Bioscience, Georgetown.

57 M. Furuse, T. Hirase, M. Itoh, A. Nagafuchi, S. Yonemura, S. Tsukita **1993**, *J. Cell Biol.* 123, 1777–1788.
58 Y. Ando-Akatsuka, M. Saitou, T. Hirase, M. Kishi, A. Sakakibara, M. Itoh, S. Yonemura, M. Furuse, M. Tsukita **1996**, *J. Cell Biol.* 133, 43–48.
59 M. Furuse, K. Fujita, T. Hiiragi, K. Fujimoto, S. Tsukita **1998**, *J. Cell Biol.* 141, 1539–1550.
60 K. Morita, M. Furuse, K. Fujimoto, S. Tsukita **1999**, *Proc. Natl. Acad. Sci. USA* 96, 511–516.
61 S. Tsukita, M. Furuse, M. Itoh **1999**, *Curr. Opin. Cell Biol.* 11, 628–633.
62 S. Tsukita, M. Furuse, M. Itoh **2001**, *Nat. Rev. Mol. Cell Biol.* 2, 285–293.
63 M. Heiskala, P. A. Peterson, Y. Yang **2001**, *Traffic* 2, 92–98.
64 J. D. Huber, R. D. Egleton, T. P. Davis **2001**, *Trends Neurosci.* 24, 719–725.
65 F. D'Atri, S. Citi **2002**, *Mol. Membr. Biol.* 19, 103–112.
66 L. Gonzalez-Mariscal, A. Betanzos, P. Nava, B. E. Jaramillo **2003**, *Progr. Biophys. Mol. Biol.* 81, 1–44.
67 K. Matter, M. S. Balda **2003**, *Nat. Rev. Mol. Biol.* 4, 225–236.
68 E. Dejana **2004**, *Nat. Rev. Mol. Cell Biol.* 5, 261–270.
69 K. Turksen, T.-C. Troy **2004**, *J. Cell Sci.* 117, 2435–2447.
70 M. Saitou, M. Furuse, H. Sasaki, J.-D. Schulzke, M. Fromm, H. Takano, T. Noda, S. Tsukita **2000**, *Mol. Biol. Cell* 11, 4131–4142.
71 A. Traweger, R. Fuchs, I. A. Krizbai, T. M. Weiger, H. C. Bauer, H. Bauer **2000**, *J. Biol. Chem.* 277, 10201–10208.
72 A. Sakakibara, M. Furuse, M. Saitou, Y. Ando-Akatsuka, S. Tsukita **1997**, *J. Cell Biol.* 137, 1393–1401.
73 Y. Chen, C. Merzdorf, D. L. Paul, D. A. Goodenough **1997**, *J. Cell Biol.* 138, 891–899.
74 M. S. Balda, C. Flores-Maldonado, M. Cereijido, K. Matter **2000**, *J. Cell Biochem.* 78, 85–96.
75 D. Huber, M. S. Balda, K. Matter **2000**, *J. Biol. Chem.* 275, 5773–5778.
76 T. Hirase, S. Kawashima, E. Y. Wong, T. Ueyama, Y. Rikitake, S. Tsukita, M. Yokoyama, J. M. Staddon **2001**, *J. Biol. Chem.* 276, 10423–10431.
77 L. DeMaio, Y. S. Chang, T. W. Gardner, J. M. Tarbell, D. A. Antonetti **2001**, *Am. J. Physiol.* 281, H105–H113.
78 D. A. Antonetti, A. J. Barber, L. A. Hollinger, E. B. Wolpert, T. W. Gardner **1999**, *J. Biol. Chem.* 174, 23463–23467.
79 M. Furuse, H. Sasaki, S. Tsukita **1999**, *J. Cell Biol.* 147, 891–903.
80 M. Furuse, K. Furuse, H. Sasaki, S. Tsukita **2001**, *J. Cell Biol.* 153, 263–272.
81 L. C. Mitic, C. M. Van Itallie, J. M. Anderson **2000**, *Am. J. Physiol.* 279, G250–G254.
82 S. Tsukita, M. Furuse **2000**, *J. Cell Biol.* 149, 13–16.
83 K. Morita, H. Sasaki, M. Furuse, S. Tsukita **1999**, *J. Cell Biol.* 147, 185–194.
84 C. Rahner, L. L. Mitic, J. M. Anderson **2001**, *Gastroenterology* 120, 411–422.
85 D. F. B. Simon, Y. Lu, K. A. Choate, H. Velazquez, E. Al-Sabban, M. Praga, G. Casari, A. Bettinelli, G. Colussi, J. Rodriguez-Soriano, D. McCredie, D. Milford, S. Sanjad, R. P. Lifton **1999**, *Science* 285, 103–106.

86 T. Hirase, J. M. Staddon, M. Saitou, Y. Ando-Akatsuka, M. Itoh, M. Furuse, K. Fujimoto, S. Tsukita, L. L. Rubin **1997**, *J. Cell Sci.* 110, 1603–1613.
87 S. Tsukita, M. Furuse **1999**, *Trends Cell Biol.* 9, 268–273.
88 T. Inai, J. Kobayashi, Y. Shibata **1999**, *Eur. J. Cell Biol.* 78, 849–855.
89 N. Sonoda, M. Furuse, H. Sasaki, S. Yonemura, J. Katahira, Y. Horiguchi, S. Tsukita **1999**, *J. Cell Biol.* 147, 195–204.
90 S. Liebner, A. Fischmann, G. Rascher, F. Duffner, E.-H. Grote, H. Wolburg **2000**, *Acta Neuropathol.* 100, 323–331.
91 H. Wolburg, K. Wolburg-Buchholz, J. Kraus, G. Rascher-Eggstein, S. Liebner, S. Hamm, F. Duffner, E.-H. Grote, W. Risau, B. Engelhardt **2003**, *Acta Neuropathol.* 105, 586–592.
92 T. Nitta, M. Hata, S. Gotoh, Y. Seo, H. Sasaki, N. Hashimoto, M. Furuse, S. Tsukita **2003**, *J. Cell Biol.* 161, 653–660.
93 T. Soma, H. Chiba, Y. Kato-Mori, T. Wada, T. Yamashita, T. Kojima, N. Sawada **2004**, *Exp. Cell Res.* 300, 202–212.
94 H. Wen, D. D. Warty, M. C. Marcondes, H. S. Fox **2004**, *Mol. Cell Biol.* 24, 8408–8417.
95 I. Martin-Padura, S. Lostaglio, M. Schneemann, L. Williams, M. Romano, P. Fruscella, C. Panzeri, A. Stoppacciaro, L. Ruco, A. Villa, D. Simmons, E. Dejana **1998**, *J. Cell Biol.* 142, 117–127.
96 Y. Liu, A. Nusrat, F. J. Schnell, T. A. Reaves, S. Walsh, M. Pochet, C. A. Parkos **2000**, *J. Cell Sci.* 113, 2363–2374.
97 D. Palmeri, A. van Zante, C. C. Huang, S. Hemmerich, S. D. Rosen **2000**, *J. Biol. Chem.* 275, 19139–19145.
98 M. A. Aurrand-Lions, L. Duncan, L. Du Pasquier, B. A. Imhof **2000**, *Curr. Top. Microbiol. Immunol.* 251, 91–98.
99 T. W. Liang, R. A. DeMarco, R. J. Mrsny, A. Gurney, A. Gray, J. Hooley, H. L. Aaron, A. Huang, T. Klassen, D. B. Tumas, S. Fong **2000**, *Am. J. Physiol. Cell Physiol.* 279, C1733–C1743.
100 G. Gliki, K. Ebnet, M. Aurrand-Lions, B. A. Imhof, R. H. Adams **2004**, *Nature* 431, 320–324.
101 C. Johnson-Léger, B. A. Imhof **2003**, *Cell Tissue Res.* 314, 93–105.
102 W. A. Muller **2003**, *Trends Immunol.* 24, 327–334.
103 K. Ebnet, A. Suzuki, D. Ohno, D. Vestweber **2004**, *J. Cell Sci.* 117, 19–29.
104 A. DelMaschio, A. D. Luigi, I. Martin-Padura, M. Brockhaus, T. Bartfai, P. Fruscella, L. Adorini, G. V. Martino, R. Furlan, M. G. De Simoni, E. Dejana **1999**, *J. Exp. Med.* 190, 1351–1356.
105 C. V. Carman, T. A. Springer **2004**, *J. Cell Biol.* 167, 377–388.
106 B. Engelhardt, H. Wolburg **2004**, *Eur. J. Immunol.* 34, 2955–2963.
107 H. Wolburg, K. Wolburg-Buchholz, B. Engelhardt **2005**, *Acta Neuropathol.* 109, 181–190.
108 C. J. Cohen, J. T. Shieh, R. J. Pickles, T. Okegawa, J. T. Hsieh, J. M. Bergelson **2001**, *Proc. Natl. Acad. Sci. USA* 98, 15191–15196.

109 I. Nasdala, K. Wolburg-Buchholz, H. Wolburg, A. Kuhn, K. Ebnet, T. Brachtendorf, U. Samulowitz, B. Kuster, B. Engelhardt, D. Vestweber, S. Butz **2002**, *J. Biol. Chem.* 277, 16294–16303.
110 S. Hirabayashi, M. Tajima, I. Yao, W. Nishimura, H. Mori, Y. Hata **2003**, *Mol. Cell Biol.* 23, 4267–4282.
111 E. Raschperger, U. Engstrom, R. F. Pettersson, J. Fuxe **2004**, *J. Biol. Chem.* 279, 796–804.
112 K. Hirata, T. Ishida, K. Penta, M. Rezaee, E. Yang, J. Wohlgemuth, T. Quertermous **2001**, *J. Biol. Chem.* 276, 16223–16231.
113 T. Ishida, R. K. Kundu, E. Yang, K. Hirata, Y. D. Ho, T. Quertermous **2003**, *J. Biol. Chem.* 278, 34598–34604.
114 B. R. Stevenson, J. D. Siliciano, M. S. Mooseker, D. A. Goodenough **1986**, *J. Cell Biol.* 103, 755–766
115 M. Itoh, A. Nagafuchi, S. Yonemura, T. Kitaniyasuda, S. Tsukita **1993**, *J. Cell Biol.* 121, 491–502.
116 M. Itoh, A. Nagafuchi, S. Moroi, S. Tsukita **1997**, *J. Cell Biol.* 138, 181–192.
117 A. K. Rajasekaran, M. Hojo, T. Huima, E. Rodriguez-Boulan **1996**, *J. Cell Biol.* 132, 451–464.
118 B. Z. Harris, W. A. Lim **2001**, *J. Cell Sci.* 114, 3219–3231.
119 T. Pawson, P. Nash **2003**, *Science* 300, 445–452.
120 J. M. Anderson **1996**, *Curr. Biol.* 6, 382–384.
121 I. Dobrosotskaya, R. K. Guy, G. L. James **1997**, *J. Biol. Chem.* 272, 31589–31597.
122 Y. Hamazaki, M. Itoh, H. Sasaki, M. Furuse, S. Tsukita **2002**, *J. Biol. Chem.* 277, 455–461.
123 M. S. Balda, M. D. Garrett, K. Matter **2003**, *J. Cell Biol.* 160, 423–432.
124 H. Ohnishi, T. Nakahara, K. Furuse, H. Sasaki, S. Tsukita, M. Furuse **2004**, *J. Biol. Chem.* 279, 46014–46022.
125 S. Citi, H. Sabanay, J. Kendrick-Jones, B. Geiger **1989**, *J. Cell Sci.* 93, 107–122.
126 Y. Nagai-Tamai, K. Mizuno, T. Hirose, A. Suzuki, S. Ohno **2002**, *Genes Cells* 7, 1161–1171.
127 A. Suzuki, T. Yamanaka, T. Hirose, N. Manabe, K. Mizuno, M. Shimizu, K. Akimoto, T. Izumi, T. Ohnishi, S. Ohno **2001**, *J. Cell Biol.* 152, 1183–1196.
128 L. Gao, G. Joberty, I. Macara **2002**, *Curr. Biol.* 12, 221–225.
129 K. Ebnet, M. Aurrand-Lions, A. Kuhn, F. Kiefer, S. Butz, K. Zander, M. K. Meyer zu Brickwedde, A. Suzuki, B. A. Imhof, D. Vestweber **2003**, *J. Cell Sci.* 116, 3879–3891.
130 E. Knust **2002**, *Mol. Membrane Biol.* 19, 113–120.
131 E. Knust, O. Bossinger **2002**, *Science* 298, 1955–1959.
132 S. A. Van de Pavert, A. Kantardzhieva, A. Malysheva, J. Meuleman, I. Versteeg, C. Levelt, J. Klooster, S. Geiger, M. W. Seeliger, P. Rashbass, A. Le Bivic, J. Wijnholds **2004**, *J. Cell Sci.* 2004, 4169–4177.
133 A. M. Dvorak, D. Feng **2001**, *J. Histochem. Cytochem.* 49, 419–431.

134 J. Couet, M.M. Belanger, E. Roussel, M.-C. Drolet **2001**, *Adv. Drug Delivery Rev.* 49, 223–235.
135 R.-V. Stan **2002**, *Microsc. Res. Tech.* 57, 350–364.
136 M. Simionescu, A. Gafencu, F. Antohe **2002**, *Microsc. Res. Tech.* 57, 269–288.
137 M. Bendayan **2002**, *Microsc. Res. Tech.* 57, 327–349.
138 B. Rippe, B.-I. Rosengren, O. Carlsson, D. Venturoli **2002**, *J. Vasc. Res.* 39, 375–390.
139 L. Pelkmans, A. Helenius **2002**, *Traffic* 3, 311–320.
140 R.G. Parton, A.A. Richards **2003**, *Traffic* 4, 724–738.
141 S. Nag **2003**, in *The Blood-Brain Barrier, Biology and Research Protocols*, ed. S. Nag, Humana Press, Totowa, NJ, pp. 3–36.
142 R.G. Parton, B. Joggerst, K. Simons **1994**, *J. Cell Biol.* 127, 1199–1215.
143 D. Virgintino, D. Robertson, M. Errede, V. Benagiano, U. Tauer, L. Roncali, M. Bertossi **2002**, *Neuroscience* 115, 145–152.
144 M. Drab, P. Verkade, M. Elger, M. Kasper, M. Lohn, B. Lauterbach, J. Menne, C. Lindschau, F. Mende, F.C. Luft, A. Schedl, H. Haller, T.V. Kurzchalia **2001**, *Science* 293, 2449–2452.
145 I.R. Nabi, P.U. Le **2003**, *J. Cell Biol.* 161, 673–677.
146 D.R. Senger, S.J. Galli, A.M. Dvora, C.A. Perruzzi, V.S. Harvey, H.F. Dvorak **1983**, *Science* 219, 983–985.
147 W.G. Roberts, G.E. Palade **1995**, *Cancer Res.* 57, 765–772.
148 S. Esser, K. Wolburg, H. Wolburg, G. Breier, T. Kurzchalia **1998**, *J. Cell Biol.* 140, 947–959.
149 S.-W. Lee, W.-J. Kim, Y.K. Choi, H.S. Song, M.J. Son, I.H. Gelman, Y.-J. Kim, K.-W. Kim **2003**, *Nat. Med.* 9, 900–906.
150 L. Labreque, I. Royal, D.S. Surprenant, C. Patterson, D. Gingras, R. Bélivuae **2003**, *Mol. Biol. Cell* 14, 334–347.
151 A. Nusrat, C.A. Parkos, P. Verkade, C.S. Foley, T.W. Liang, W. Innis-Whitehouse, K.K. Eastburn, J.L. Madara **2000**, *J. Cell Sci.* 113, 1771–1781.
152 L. Song, J.S. Pachter **2004**, *Microvasc. Res.* 67, 78–89.
153 T. Michel **1999**, *Braz. J. Med. Biol. Res.* 32, 1361–1366.
154 D.J. Blake, S. Kröger **2000**, *Trends Neurosci.* 23, 92–99.
155 J.C. De la Torre, G.B. Stefano **2000**, *Brain Res. Rev.* 34, 119–136.
156 S. Nag, P. Picard, D.J. Stewart **2001**, *Lab. Invest.* 81, 41–49.
157 W.G. Mayhan **2000**, *Brain Res.* 866, 101–108.
158 O.M. Martinez-Estrada, E. Rodriguez-Millán, E. González-de Vicente, M. Reina, S. Vilaró, M. Fabre **2003**, *Eur. J. Neurosci.* 18, 2538–2544.
159 K. Hashizume, K.L. Black **2002**, *J. Neuropathol. Exp. Neurol.* 61, 725–735.
160 G.E. Mann, D.L. Yudilevich, L. Sobrevia **2003**, *Physiol. Rev.* 83, 183–252.
161 A.G. De Boer, I.C. Van der Sandt, P.J. Gaillard **2003**, *Annu. Rev. Pharmacol. Toxicol.* 2003, 629–656.
162 H. Bauer, U. Sonnleitner, A. Lametschwandtner, M. Steiner, H. Adam, H.C. Bauer **1995**, *Dev. Brain Res.* 86, 317–325.
163 F. Maher, S.J. Vannucci, J.A. Simpson **1994**, *FASEB J.* 8, 207–212.

164 C. L. Farrell, W. M. Pardridge **1991**, *Proc. Natl Acad. Sci. USA* 88, 5779–5783.
165 S. Bolz, C. L. Farrell, K. Dietz, H. Wolburg **1996**, *Cell Tissue Res.* 284, 355–365.
166 D. H. Dobrogowska, A. W. Vorbrodt **1999**, *J. Histochem. Cytochem.* 47, 1021–1029.
167 I. A. Simpson, S. J. Vannucci, M. R. DeJoseph, R. A. Hawkins **2001**, *J. Biol. Chem.* 276, 12725–12729.
168 W. A. Jefferies, M. R. Brandon, S. V. Hunt, A. F. Williams, K. C. Gatter, D. Y. Mason **1984**, *Nature* 312, 162–163.
169 M. W. B. Bradbury **1997**, *J. Neurochem.* 69, 443–454.
170 J. R. Burdo, D. A. Antonetti, E. B. Wolpert, J. R. Connor **2003**, *Neuroscience* 121, 172, 883–890.
171 S. Méresse, J. C. Delbart, J. C. Fruchart, R. Cechelli **1989**, *J. Neurochem.* 53, 340–345.
172 B. V. Zlokovic, D. S. Shundric, M. B. Segal, M. V. Lipovac, J. B. Mackic, H. Davson **1990**, *Exp. Neurol.* 107, 263–290.
173 K. R. Duffy, W. M. Pardridge **1987**, *Brain Res.* 420, 32–38.
174 T. Kondo, A. Hafezi-Moghadam, K. Thomas, D. D. Wagner, C. R. Kahn **2004**, *Biochem. Biophys. Res. Comm.* 317, 315–320.
175 G. W. Goldstein, A. L. Betz **1986**, *Sci. Am.* 255, 70–79.
176 R. Dringen **2000**, *Progr. Neurobiol.* 62, 649–671.
177 P. Lagrange, I. A. Romero, A. Minn, P. A. Revest **1999**, *Free Radic. Biol. Med.* 27, 667–672.
178 G. W. Kim, A. Lewén, J.-C. Copin, B. D. Watson, P. H. Chan **2001**, *Neuroscience* 105, 1007–1018.
179 R. Kannan, R. Mittur, Y. Bao, T. Tsuruo, N. Kaplowitz **1999**, *J. Neurochem.* 73, 390–399.
180 M. R. Kannan, R. Chakrabarti, D. Tang, K. J. Kim, N. Kaplowitz **2000**, *Brain Res.* 852, 374–382.
181 D. Virgintino, D. Robertson, M. Errede, V. Benagiano, F. Girolamo, E. Maiorano, L. Roncali, M. Bertoss **2002**, *J. Histochem. Cytochem.* 50, 1671–1676.
182 J. Jodoin, M. Demeule, L. Fenart, R. Cecchelli, S. Farmer, K. J. Linton, C. F. Higgins, R. Béliveau **2003**, *J. Neurochem.* 87, 1010–1023.
183 M. Demeule, J. Jodoin, D. Gingras, R. Beliveau **2000**, *FEBS Lett.* 466, 219–224.
184 P. Adamson, S. Etienne, P. O. Couraud, V. Calder, J. Greenwood **1999**, *J. Immunol.* 162, 2964–2973.
185 S. Etienne, P. Adamson, J. Greenwood, A. D. Strosberg, S. Cazaubon, P. O. Couraud **1998**, *J. Immunol.* 161, 5755–5761.
186 D. Huber, M. S. Balda, K. Matter **2000**, *J. Biol. Chem.* 275, 5773–5778.
187 H. Wolburg, K. Wolburg-Buchholz, J. Kraus, G. Rascher-Eggstein, S. Liebner, S. Hamm, F. Duffner, E.-H. Grote, W. Risau, B. Engelhardt **2003**, *Acta Neuropathol.* 105, 586–592.

188 C. Iadecola **2004**, *Nat. Rev. Neurosci.* 5, 347–360.
189 M. Simard, M. Nedergaard **2004**, *Neuroscience* 129, 877–896.
190 B. Nico, A. Frigeri, G. P. Nicchia, F. Quondamatteo, R. Herken, M. Errede, D. Ribatti, M. Svelto, L. Roncali **2001**, *J. Cell Sci.* 114, 1297–1307.
191 J. Brillault, V. Berezowski, R. Cecchelli, M. P. Dehouck **2002**, *J. Neurochem.* 83, 807–817.
192 S. Saadoun, M. C. Papadopoulos, D. C. Davies, S. Krishna, B. A. Bell **2002**, *J. Neurol. Neurosurg. Psychiatry* **2002**, 72, 262–265.
193 H. Wolburg **1995**, *J. Brain Res.* 36, 239–258.
194 G. P. Nicchia, A. Frigeri, G. M. Liuzzi, M. P. Santacroce, B. Nico, G. Procino, F. Quondamatteo, R. Herken, L. Roncali, M. Svelto **2000**, *Glia* 31, 29–38.
195 G. P. Nicchia, B. Nico, L. M. A. Camassa, M. G. Mola, N. Loh, R. Dermietzel, D. C. Spray, M. Svelto, A. Friegeri **2004**, *Neuroscience* 129, 935–945.
196 M. C. Papadopoulos, S. Krishna, A. S. Verkman **2002**, *Mount Sinai J. Med.* 69, 242–248.
197 J. Badaut, F. Lasbennes, P. J. Magistretti, L. Regli **2002**, *J. Cer. Blood Flow Metab.* 22, 367–378.
198 M. Amiry-Moghaddam, O. P. Ottersen **2003**, *Nat. Rev. Neurosci.* 4, 991–1001.
199 J.-M. Verbavatz, T. Ma, R. Gobin, A. S. Verkman **1997**, *J. Cell Sci.* 110, 2855–2860.
200 B. Yang, D. Brown, A. S. Verkman **1996**, *J. Biol. Chem.* 271, 4577–45806.
201 J. E. Rash, T. Yasumura, C. S. Hudson, P. Agre, S. Nielsen **1998**, *Proc. Natl Acad. Sci. USA* 95, 11981–11986.
202 J. E. Rash, K. G. V. Davidson, T. Yasumura, C. S. Furman **2004**, *Neuroscience* 129, 915–934.
203 S. Nielsen, E. A. Nagelhus, M. Amiry-Moghaddam, C. Bourque, P. Agre, O. P. Ottersen **1997**, *J. Neurosci.* 17, 171–180.
204 J. Neuhaus **1990**, *Glia* 3, 241–251.
205 S. Saadoun, M. C. Papadopoulos, D. C. Davies, S. Krishna, B.A. Bell **2002**, *J. Neurol. Neurosurg. Psychiatry* 72, 262–265.
206 A. Warth, S. Kröger, H. Wolburg **2004**, *Acta Neuropathol.* 107, 311–318.
207 J. S. Jung, R. V. Bhat, G. M. Preston, W. B. Guggino, J. M. Baraban, P. Agre **1994**, *Proc. Natl Acad. Sci. USA* 91, 13052–13056.
208 C. S. Furman, D. A. Gorelick-Feldman, K. G. V. Davidson, T. Yasumura, J. D. Neely, P. Agre, J. E. Rash **2003**, *Proc. Natl Acad. Sci. USA* 2003, 13609–13614.
209 U. J. McMahan **1990**, *Cold Spring Harb. Symp. Quant. Biol.* 55, 407–418.
210 G. Bezakova, M. A. Ruegg **2003**, *Nat. Rev. Mol. Cell Biol.* 4, 295–308.
211 A. J. Barber, E. Lieth **1997**, *Dev. Dyn.* 208, 62–74.
212 T. M. Berzin, B. D. Zipser, M. S. Rafii, V. Kuo-Leblanc, G. D. Yancopoulos, D. J. Glass, J. R. Fallon, E. G. Stopa **2000**, *Neurobiol. Aging* 21, 349–355.
213 M. A. Smith, L. G. W. Hilgenberg **2002**, *NeuroReport* 13, 1485–1495.
214 D. M. Stone, K. Nikolics **1995**, *J. Neurosci.* 15, 6767–6778.
215 G. Rascher, A. Fischmann, S. Kröger, F. Duffner, E.-H. Grote, H. Wolburg **2002**, *Acta Neuropathol.* 104, 85–91.

216 S. H. Gee, F. Montanaro, M. H. Lindenbaum, S. Carbonetto **1994**, *Cell* 77, 675–686.
217 T. Haenggi, A. Soontornmalai, M. C. Schaub, J.-M. Fritschy **2004**, *Neuroscience* 129, 403–413.
218 J. D. Neely, M. Amiry-Moghaddam, O. P. Ottersen, S. C. Froehner, P. Agre, M. E. Adams **2001**, *Proc. Natl Acad. Sci. USA* 98, 14108–14113.
219 N. C. Connors, P. Kofuji **2002**, *J. Neurosci.* 22, 4321–4327.
220 N. C. Connors, M. E. Adams, S. C. Froehner, P. Kofuji **2004**, *J. Biol. Chem.* 279, 28387–28392.
221 P. Kofuji, P. Ceelen, K. R. Zahs, L. W. Surbeck, H. A. Lester, E. A. Newman **2000**, *J. Neurosci.* 20, 5733–5740.
222 K. Higashi, A. Fujita, A. Inanobe, M. Tanemoto, K. Doi, T. Kubo, Y. Kurachi **2001**, *Am. J. Physiol.* 281, C922–C931.
223 P. Kofuji, E. A. Newman **2004**, *Neuroscience* 129, 1045–1056.
224 T. Pannicke, I. Iandiev, O. Uckermann, B. Biedermann, F. Kutzera, P. Wiedemann, H. Wolburg, A. Reichenbach, A. Bringmann **2004**, *Mol. Cell. Neurosci.* 26, 493–502.
225 A. Bringmann, A. Reichenbach, P. Wiedemann **2005**, *Ophthal. Res.*, in press.
226 E. A. Nagelhus, Y. Horio, A. Inanobe, A. Fujity, F.-M. Haug, S. Nielsen, Y. Kurachi, O. P. Ottersen **1999**, *Glia* 26, 47–54.
227 E. A. Nagelhus, T. M. Mathiisen, O. P. Ottersen **2004**, *Neuroscience* 129, 905–913.
228 M. Amiry-Moghaddam, A. Williamson, M. Palomba, T. Eid, N. C. De Lanerolle, E. A. Nagelhus, M. E. Adams, S. C. Froehner, P. Agre, O. P. Ottersen **2003**, *Proc. Natl Acad. Sci. USA* 100, 13615–13620.
229 A. Warth, M. Mittelbronn, H. Wolburg **2005**, *Acta Neuropathol.* 109, 418–426.
230 S. Saadoun, M. C. Papadopoulos, S. Krishna **2003**, *J. Clin. Pathol.* 56, 972–975.
231 W. Risau, H. Wolburg **1990**, *Trends Neurosci.* 13, 174–178.
232 H. Wolburg, W. Risau **1995**, in *Neuroglia*, eds. H. Kettenmann, B. R. Ransom, Oxford University Press, Oxford, pp. 763–776.
233 E. A. Newman **1987**, *J. Neurosci.* 7, 2423–2432.
234 M. Amiry-Moghaddam, R. Xue, F.-M. Haug, J. D. Neely, A. Bhardwaj, P. Agre, M. E. Adams, S. C. Froehner, S. Mori, O. P. Ottersen **2004**, *FASEB J.* 10, 1096/fj.03-0869 fje.
235 S. Solé, V. Petegnief, R. Gorina, Á. Chamorra, A. M. Planas **2004**, *J. Neuropathol. Exp. Neurol.* 63, 338–349.
236 H. Wolburg **1995**, in *Neuron-Glia Interrelations During Phylogeny. II. Plasticity and Regeneration,* eds. A. Vernadakis, B. I. Roots, Humana Press, Totowa, NJ pp. 479–510.
237 S. Nag (ed.) **2003**, *The Blood-Brain Barrier. Biology and Research Protocols* Humana Press, Totowa, NJ

5
Pericytes and Their Contribution to the Blood-Brain Barrier

Markus Ramsauer

5.1
Introduction

The simplistic view of the blood-brain barrier (BBB), which defines this barrier as an exclusively endothelial entity, must be reconsidered. In particular, cerebral pericytes constitute a neglected component of the metabolic BBB complex. These cells are ideally positioned to create an interface between the circulatory system and the brain parenchyma.

Pericytes, originally described in 1870 by Rouget, are perivascular cells that are found adjacent to capillaries, share a common basement membrane with the cerebral endothelium, and are distinctively shaped with many cytoplasmic processes that encircle capillaries. These cells are derived developmentally from the mesoderm and are indicated to possess morphological and functional differences within the same tissue or even within a single capillary bed. Pericytes are difficult to define, plastic, and have the capacity to differentiate into other mesenchymal cell types. Thus, these cells in the brain may serve multiple functional roles and among these are macrophage activity, modulation of blood flow and vascular permeability, and regulation of vessel growth and stability.

Contraction (and reciprocal relaxation) appears to be the way that pericytes influence microvascular blood flow. Furthermore, pericytes in the central nervous system (CNS) express macrophage functions and are actively involved in immune responses operating as a "second line of defense" at the BBB. And finally, they play a regulatory role in BBB differentiation, endothelial cell tight junction formation, and brain angiogenesis. The regulation of endothelial cells has been suggested to control vessel growth and contribute to vascular stability. Pericytes have been reported to provide: (a) BBB-specific enzymes, (b) potential modulators of endothelial permeability, (c) stabilizing effects on microvessel walls, and (d) a promoting activity on angiogenic processes and capillary sprouting.

A host of different cell factors and signaling agents appear to be involved with these cellular functions, some effecting the pericyte and others produced by this cell. What is emerging is a realization of the reciprocal dependence of the peri-

Blood-Brain Interfaces: From Ontogeny to Artificial Barriers.
Edited by R. Dermietzel, D.C. Spray, M. Nedergaard
Copyright © 2006 WILEY-VCH Verlag GmbH & Co. KGaA, Weinheim
ISBN: 3-527-31088-6

cyte and its nextdoor neighbor, the endothelial cell, and the resulting complex biological feedback mechanisms.

5.2
Pericyte Structure and Positioning

The BBB regulates the exchange of blood solutes with the interstitial cerebral fluid and consists of a physical and metabolic barrier. The BBB is composed of the microvascular endothelium lining blood vessels surrounded by pericytes and astrocytes, which tightly clasp cerebral microvessels by their endfeet (Fig. 5.1a). Due to their strategic position, pericytes create an interface between the circulatory system and the surrounding tissue (CNS; [1]). To date, pericytes have been defined primarily by their periendothelial cell location in the microvasculature. A variety of terms are used as synonyms for pericytes: microvascular smooth muscle cells, perivascular macrophages, (intra)mural cells, myofibroblasts, and perivascular cells. However the term pericytes best describes their anatomic relevance (*peri*: around, *cyto*: cell).

Pericytes are intimately associated with endothelial cells. An intercellular distance of less than 20 nm is often noted between them [2]. Pericytes are enveloped in a basement membrane, which is continuous with the basal lamina of the capillary tubes. The thin basement membrane provides mechanical support and a barrier function. There is in vitro evidence that both pericytes and endothelial cells contribute to the production of this basement membrane [3, 4]. Being contained within the basement membrane distinguishes them from smooth muscle cells, which are separated by the basement membrane from endothelial cells and do not establish direct contact. In mature vessels, the most reliable identification is still electron microscopy, revealing the complete surrounding of the pericytes by the endothelial basement membrane. However, in embryogenesis and during angiogenic sprouting, this method of identification cannot be applied, since the basement membrane is not fully developed [5].

Each pericyte possesses a large cell body with a prominent nucleus and a small amount of surrounding cytoplasm [2]. In the CNS, pericytes have an oval to elongated cell body arranged parallel to the vessel long axis. The shape and contour of the cell body is determined by the vessel outline and the overlying basement membrane. Protruding from the soma are several long primary processes, which also run parallel to the long axis of the blood vessel; and orthogonally oriented smaller secondary processes radiate along the length of the primary process and encircle the capillary wall. This provides the appearance that pericytes cradle or girdle the blood vessel [6]. Early studies by Zimmermann [7] revealed the presence of up to 90 processes with a width of 0.3–0.8 µm per 100 µm of capillary length. Interestingly, the distance between cell bodies is up to 50 µm in the brain, suggesting an interdigitation of processes among neighboring pericytes. Processes from multiple pericytes can cover a single endothelial cell and a single pericyte can extend processes to more than one capillary

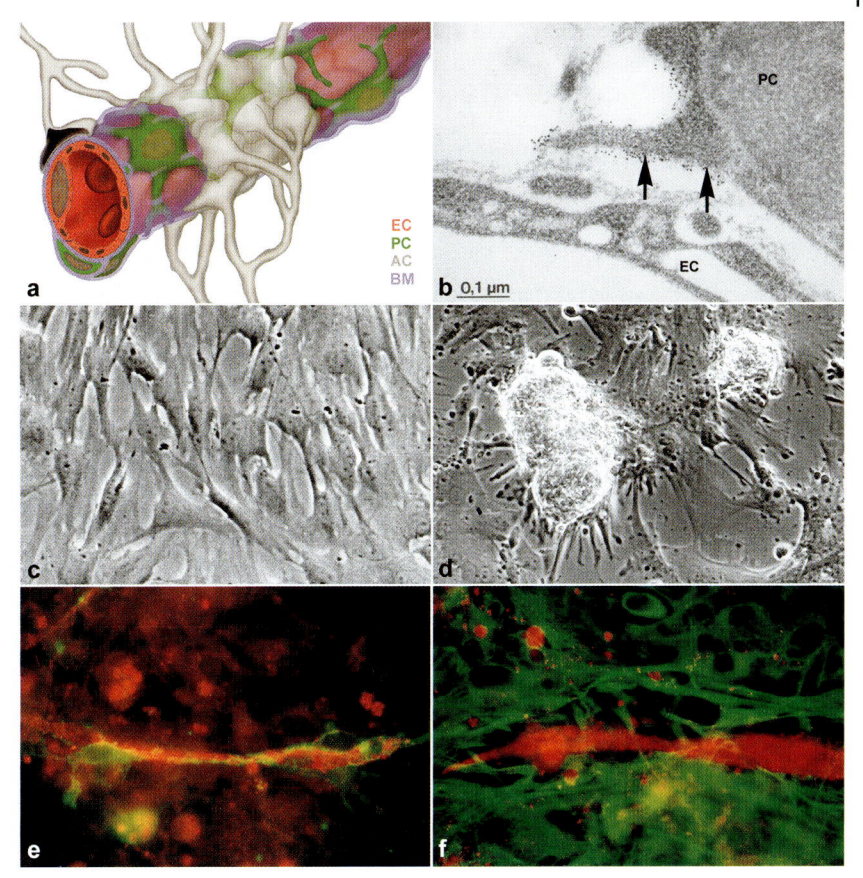

Fig. 5.1 (a) This schematic view recapitulates the in vivo situation of the BBB (EC=endothelial cells, PC=pericytes, AC=astrocytes, BM=basement membrane). (b) Immunogold transmission electron micrograph from a cross-section of an isolated cerebral microvessel, utilizing pericytic aminopeptidase N. The pericyte (PC), identified by the ensheathing basement membrane, shows immunogold beads on its abluminal plasma membrane, whereas the endothelial cell (EC) is devoid of immunoreactivity. (c, d) Pericytes in vitro grow not contact-inhibited with overlapping processes (c) and exhibit noticeable retraction with formation of numerous multicellular nodules (d). (e, f) Double-immunofluorescence of mixed endothelial-pericyte cultures in triculture with primary astrocytes. Endothelial cells are visualized by immunostaining for vW F VIII-ra (red). Capillary-like structures closely associated with pAPN-positive pericytes (green, e) are surrounded by GFAP-expressing astrocytes (green, f).

within the microcirculation [1]. Their processes penetrate the basement membrane to directly contact the underlying endothelium and, in a reciprocal manner, endothelial processes penetrate into the pericytes [8]. Also, reciprocal invaginations may occur in endothelial cells or pericytes, filled by processes of the other cell. These have been described as heterologous "peg and socket" contacts [9]. Next to an array of adhesion plaques that exhibit membrane specializations between pericytes and endothelial cells [2], the two also form gap junctions [10, 11]. Gap junctions provide direct connections between the cytoplasm of the two cells and allow the passage of ionic currents and small molecules [12]. These contact sites are thought to provide anchorage and to support transmission of a mechanical contractile force from the pericytes to the endothelium [2, 13].

The pericyte coverage of the abluminal endothelial surface is only partial, varies extensively between the capillary beds of different tissues, and may exhibit vessel variation [14]. The highest pericyte density has been described for neural tissues and is stated to be almost in parity in brain and retina. Pericyte to endothelial ratios are 1:3 in the brain and only outnumbered in the retina (1:5). For comparison, pericyte to endothelial ratio is stated to be 1:100 in striated muscle [15]. A comparison of pericyte coverage (pericyte plasma membrane length in contact with the vascular circumference, versus the outer circumference of the endothelial cell tube) in the rat's retina and brain revealed a ratio of 0.41 for the retinal microvessels and ranged from 0.22 to 0.30 for five regions of the cerebral cortex [16]. The high number of pericytes may reflect their contribution to the blood-retina and blood-brain barrier. The number of pericytes, however, varies significantly among different-sized vessels, being found on arterioles, venules, and particularly capillaries [1]. On either side of a capillary bed, there is a continuum of cells, from "true" pericytes surrounding capillaries, via "intermediate" cells resembling both smooth muscle cells and pericytes at the interface between arteriolar or venular capillaries and arterioles or venules, to "true" smooth muscle cells surrounding terminal arterioles, venules, and larger vessels. The smooth muscle cells provide structural support to the large vessels and are, especially in the arterioles, important regulators of blood flow due to their contractile behavior. Pericytes are more abundant on post-capillary venules and relatively sparse on capillaries. The distribution of pericytes could reflect postarteriolar hydrostatic pressures [17]. In addition to this selective positioning in the vessel wall, there are structural differences among pericytes on the arterial and venous sides of capillary beds. Pericytes in "true" capillaries are often highly elongated with extensive slender processes that wrap many times around the endothelial tube, whereas in the post-capillary venules the pericytes are stubby with thick and short radial processes [15].

5.3
Pericyte Markers

So far no pericyte-specific marker has been identified, such as the von Willebrand factor VIII-related antigen (vWFVIII-ra) for endothelial cells, which makes pericytes positive identification often difficult. However, a number of pericyte markers have been described which can be used to differentiate them from other cell types.

Cerebral pericytes express surface markers: high molecular weight melanoma-associated antigen (HMW-MAA; called nerve/glial antigen 2, NG2, in the mouse; [18, 19]) and differentiation antigen Thy 1.1, found on thymus-derived lymphocytes, is also expressed by brain pericytes and astrocytes [20]. A monoclonal antibody (3G5) directed against a surface glycolipid has been shown to be a useful marker for pericytes [21]. Rat pericytes also express the integrin-subunit CD11b (aM) and low levels of the marker ED2 [22, 23]. Although they share markers with cells of macrophage lineage, they do not express CD45 and isolectin IB4 (GSA; [23]). Krause et al. [24] reported the characterization of a monoclonal antibody that recognizes a 140-kDa protein specific for pericytes of the rat. Later, the protein was identified as a specific isoform of aminopeptidase N (CD 13) in cerebral pericytes (pAPN; [25]). The lack of pAPN expression in brain regions devoid of a tight endothelium, like the circumventricular organs, suggested that this ectopeptidase is a member of the metabolic complement of the BBB which is involved in neuropeptide degradation [26]. pAPN was recently revealed to be a late BBB marker in rodent brain development, which occurs around day E18 of brain angiogenesis [27] and is expressed at the extracytoplasmic side of the pericyte plasma membrane (Fig. 5.1b). Cerebral pericytes express vimentin as an intermediate filament but not desmin, which is found in peripheral pericytes and smooth muscle cells [28]. Pericytes contain both smooth muscle and nonsmooth muscle isoforms of actin and myosin [29]. In situ, pericytes from "true" capillaries (mid-capillaries) are reported to be virtually devoid of smooth muscle isoforms, whereas pericytes from pre-capillary arterioles and post-capillary venules contain mainly smooth muscle isoforms [30]. The expression of these markers varies, depending on the species-, tissue- or development-related contexts. For example, in chicken embryos, pericytes surrounding the angiogenic vasculature of the brain can easily be visualized by their expression of smooth muscle actin (SMA) [31]; but this is not the case in mice or rats [32].

5.4
Pericytes in Culture

Most of the data about pericytes is generated from in vitro experiments. Pericyte isolation and culture techniques are usually modifications of a protocol worked out by Gitlin and D'Amore [33]. This involves mechanical disruption of tissue, enzymatic digestion, and collection of capillary fragments that contain both endothelial cells and pericytes. The vascular cells can be differentiated by their preferential growth on different culture substrata. Endothelial cells have been noted to prefer coatings with some matrix component [34]. When grown under supplemented culture conditions (10% fetal bovine serum) on uncoated plastic, the proliferation of pericytes is favored over the other cell species. Pericytes, unlike endothelial cells, generally grow at similar rates regardless of the substrates [15]. Fetal bovine serum contains sufficient growth factors, such as platelet-derived growth factor (PDGF), to stimulate pericyte proliferation and to yield almost pure cultures of pericytes. Since endothelial cells are inhibited in growth by proliferating pericytes [35, 36], the growth of endothelial cells is rare and transient in primary cultures. Proliferation of pericytes is favored and they eventually become the dominant cell type.

Cultured pericytes appear as large, spreading, stellate cells with highly irregular edges (Fig. 5.1c). Pericytes in culture display numerous long filamentous processes and are characterized by prominent intracellular fibers of actin (stress fibers). They grow in small islets, proliferate slowly with a prolonged lag phase, and do not establish true monolayers. Once in the log phase of growth, the doubling time is in the order of 3–4 days. Their morphology differentiates them from endothelial cells ("cobblestone-like" morphology), fibroblasts (long, spindle-shaped, with extended filopodia), and smooth muscle cells ("hill and valley" growth). Their growth is not contact-inhibited and, after reaching confluence, pericytes form multilayers and then nodules by a process involving the sudden retraction of a multilayer area into a small nodule or cell aggregate, leaving empty patches in the substratum, re-attachment of the nodule on the substratum, followed by cell migration out of the nodule and cell proliferation (Fig. 5.1d) [37]. Nodule formation is associated with changes in the synthesis and distribution of matrix molecules and extracellular calcification occurs in the nodules [38]. Thus, pericytes in vitro can differentiate along the osteogenic pathway.

Culture conditions may influence pericyte phenotype and marker expression. Alpha-smooth muscle actin (a-SMA) protein is expressed in pericytes in vitro but not in brain microvascular pericytes in vivo [39]. The expression and organization of a-SMA can be modulated by the presence of transforming growth factor-beta 1 (TGF-β1) [40], endothelin-1 (ET1) [41], and extracellular matrix molecules [42]. Such factor-dependent phenotypic changes may be the basis for pericyte functional heterogeneity [28].

In order to gain more information on the inductive mechanisms involved in the expression of pAPN, a prominent pericyte-specific BBB enzyme, and to obtain purified cultures of cerebral pericytes, we designed a solid-phase isolation technique, which makes use of the monoclonal antibody directed against the pAPN [43]. pAPN has been detected to constitute a major component of the pericytic protease complement and can be regarded as a specific marker for the pericytic phenotype in brain. The isolation protocol consisted of an immunoabsorption of antibody-labeled pericytes to immunomagnetic beads. Expression of pAPN was high in acute isolated pericytes, as revealed by immunocytochemistry. In this study, we showed that the pAPN associated with the BBB is down-regulated in pericytic cultures. This observation is in accordance with previous data describing comparable in vitro effects for BBB-specific enzymes of endothelial or pericytic origin, such as γ-glutamyl transpeptidase (GGT) [44]. Although purified pericytes deprived of endothelial cells did not reveal a re-expression effect, pericytes that were kept in contact with endothelial cells were able to acquire a pAPN-positive phenotype, indicating that endothelial cells constitute an essential requirement for the in vitro re-expression of pAPN. The in vitro history of the pericytes is essential for the re-inducibility of pAPN. Thus, our experimental design provides stringent in vitro evidence that BBB differentiation requires the cooperation of different vascular cellular components and that the achievement of functionally differentiated cerebral pericytes is closely related to endothelial cells. Because of their close association with endothelial cells in vivo, the physiological functions of pericytes are best studied in the context of their interactions with endothelial cells.

5.5
Contractility and Regulation of Blood Flow

It is usually stated that pericytes have a contractile function, although this has been extensively scrutinized and debated for many years. In fact, originally, Rouget [45] regarded pericytes as a variety of smooth muscle cells and this view was adapted later by Zimmermann [7]. Many researchers today presume that the pericytes are contractile elements related to smooth muscle cells and involved in the regulation of blood flow through the microvasculature, although definitive evidence of their contractility remains elusive.

The first line of such evidence came from the indication of contractile proteins in the pericytic cells. Pericytes exhibit a cytoskeleton similar to vascular smooth muscle cells. Early ultrastructural studies reported the presence of actin- and myosin-like filaments in rat brain pericytes [46]. More recent publications describe the similarities between pericyte and smooth muscle cell actins [29] and (tropo-)myosins [47, 48]. In the pericytes of small capillaries, nonmuscle isomyosin is the predominant form, whereas the smooth muscle isomyosin is present in a very low concentration. The reverse relationship is found in pericytes associated with larger capillaries and post-capillary venules. At the subcellular level, actin is localized in stress fibers and microfilament bundles [49, 50]. Nehls and colleagues [28] reported that muscle-specific actin appeared to be absent or scarce, even in the pericytes of pre- and post-capillary segments. A systematic study has shown that the alpha-actin mRNA is expressed in brain pericytes in tissue culture, but the immunoreactive alpha-actin protein is not expressed in brain microvascular pericytes in vivo [39]. The nonphysiological conditions of tissue culture may cause dedifferentiation of pericytes towards a primitive, smooth muscle-like phenotype. These authors hypothesize that only the SMA-expressing capillaries participate in blood flow control and the midcapillaries serve a different function.

Functional evidence for the contractile ability of the pericytes comes from studies utilizing in vitro techniques, where their contraction of collagen or silicone substrata has been directly observed [51–53]. However, contraction of collagen and silicone rubber in vitro does not necessarily equate to contraction in vivo and the latter is much more difficult to prove morphometrically. But especially the assessment of cell contraction in vitro allowed the investigation of vasoactive compounds which alter or regulate pericyte contractile tone. Pericytes have been indicated to possess both cholinergic and adrenergic receptors and their response leads to relaxation or produces contraction [6]. It has been demonstrated that the contractile response of pericytes could be altered by certain vasoactive agents; for example histamine and serotonin contracted the pericytes [54]. The most attention received concerns the endothelium-derived vasorelaxants nitric oxide and prostacyclin, the vasoconstrictors angiotensin II and endothelin-1, and their corresponding receptors expressed on pericytes [41, 55, 56]. Through a paracrine mechanism, endothelial cells could control contraction in pericytes, which suggests the involvement of cerebrovascular auto-regulation.

Given their intimate association with vessels, their shape, distribution, and morphology, it is compelling to seriously consider possible functions that would involve contractility analogous to the smooth muscle cells of larger vessels. At the level of the capillary, extremely subtle changes in the internal diameter could be functionally very effective at altering capillary blood flow.

In the mature vascular system, the endothelium is supported by mural cells, which express characteristics specific to their localization, for example pericytes of the pre- and post-capillary segments display gradual transitions to smooth muscle. Therefore, at least two pericyte populations appear to exist: smooth muscle-like transitional pericytes and nonmuscle-like mid-capillary pericytes [30, 39, 40]. The marked similarity to vascular smooth muscle cells has led to the theory that pericytes and vascular smooth muscle cells represent phenotypic variants of a continuous population of mural cells. Depending on external stimuli, pericytes have been suggested to give rise to smooth muscle cells and vice versa [28]. Thus, pericytes may function as the progenitors for smooth muscle cells in cases of vessel enlargement or remodeling.

5.6
Macrophage Function

Pericytes in different tissues have been suggested to serve different functional roles; and diversity or heterogeneity within the pericytes of an individual tissue has often been noted. The potential phagocytotic role of pericytes has only been well investigated in the CNS, leading to the proposal that pericytes are to be considered part of the BBB [57]. A structural feature often used in the assessment of macrophages is the presence of lysosomes. Pericytes have been shown to contain notable lysosomes and inclusion bodies, which are strongly reactive for acid phosphatase [58]. Lysosomal inclusions are common in pericytes of the CNS, but are only occasionally observed in other tissues [2].

In further support of the theory that pericytes act as phagocytic cells, the cells ingest substances after systemic injection from the circulation [59], after intraventricular administration [60], and after direct introduction into the extracellular fluid [61]. Based on this activity, it has been suggested that pericytes serve to clean the extracellular fluid. The phagocytic capacity of CNS pericytes has been well documented in numerous conditions of trauma [62–65]. In addition, in vitro ingestion of various materials by the cells, including polystyrene beads, can be directly observed [23]. It appears, however, that pericytes may be limited to the pinocytosis of proteins such as horseradish peroxidase or trypan blue, as no phagocytosis of carbon particles by pericytes (but perivascular cells not completely surrounded by basement membrane) was seen in a recent critical study [66].

With respect to macrophage function, pericytes have been reported to possess numerous marker components of macrophages under some conditions: these include CR3 complement receptor, leukocyte-common antigen CD4, Fc receptor, and classes I and II major histocompatibility complex molecules (MHC class I and II) [67]. Therefore, CNS pericytes may be actively involved in the regulation of leukocyte transmigration, antigen presentation, and T-cell activation, critical aspects of macrophage function [5]. They constitutively express low levels of vascular cell adhesion molecule-1 (VCAM-1) and intercellular adhesion molecule-1 (ICAM-1) [68], which have potential costimulatory activity in MHC class II-dependent antigen presentation. Like the adhesion molecules, T-cell adherence is increased by TNF-α treatment. Another component which has been demonstrated to up-regulate pericytes is interferon-γ [69]. While the pericytes express some macrophage properties, the cells are also capable of up-regulating this activity, like other tissue macrophages. Pericytes do not display full activity in the normal CNS tissue and have been viewed as a "second line of defense" at the boundary between blood and brain [26].

Maxwell and Kruger [62] reported that, following irradiation of the brain, pericytes undergo mitosis and penetrate the brain parenchyma. Similar anatomical investigations led some authors to speculate that pericytes transform into microglial cells [64, 70]. The origin of microglia from elements detached from the walls of the vessels has been sustained throughout the decades, but it has been denied as well. They are distinguished from pericytes by morphology, location, and functional properties. However, while clearly being different cell types, microglia and pericytes do appear to share the same origin and overall functional role. A recent study has shown a strong correlation between the numbers of microglia and pericytes in different tissue regions, suggesting a transformational relationship and, on the basis of electron microscopic images, proposed that pericytes break out of the basal lamina, which involves astrocytes, and become perivascular microglia [71]. At this point, the exact relationship between pericytes and microglial cells still remains incompletely resolved. Furthermore, a subgroup of pericyte-like cells, based upon their not being completely enclosed within the vascular basal lamina, has been distinguished from pericytes. These cells reside on the surface of the basal lamina, between it and the glia limitans [66]. Also, the relationship between pericytes and the pericyte-like perivascular cells is unknown. This makes investigation of a release or liberation of pericytes from the basement membrane even harder, as the classic definition of a pericyte as a cell embedded in the capillary basement membrane cannot be applied in these situations.

Overall, in vivo and in vitro, many structural and functional properties of macrophages are exhibited by pericytes and they appear to be intermediate in terms of their macrophage potential.

5.7
Regulation of Homeostasis and Integrity

Important aspects of vascular homeostasis are related to the control of vascular tone and perfusion by way of endothelial cell-secreted contracting or relaxing factors and control of the leukocyte traffic from the blood compartment to the interstitium [15]. In addition, an active role of pericytes in the regulation of the BBB was highlighted by their expression of degrading enzymes such as butyrylcholinesterase (BChE) [72], γ-glutamyltranspeptidase (GGT) [44, 73], glutamyl aminopeptidase (EAP) [56], and aminopeptidase N [25]. Pericytes thus have been regarded as a "second line of defense", which helps to maintain homeostasis of the CNS, particularly during breakdown of the endothelial BBB.

With respect to homeostasis, to maintain a selective permeability barrier for plasma constituents and to regulate their passage from the intravascular compartment to the interstitium is very important. In this context, the role of pericytes in BBB establishment should be reconsidered. Recent in vitro studies have shown that pericytes and endothelial cells of CNS origin have mutual influence on each other's biosynthetic and replicative behavior. Pericytes also influence the BBB at the level of the endothelial tight junction and transendothelial vesiculation.

BBB differentiation was originally thought to be a process induced and directed by astrocytes [74, 75]. However, astrocytes may not be the only cells to contribute regulating factors, since the development of endothelial cell tight junctions is not altered in conditions associated with astrocyte-endothelial cell decoupling [76] or in areas lacking astrocytes [77]. The group of Dore-Duffy found increased pericyte coverage in glial fibrillary acidic protein (GFAP)-deficient mice characterized by reduced cytoarchitectural integrity of the astroglial layer surrounding the microvasculature [5].

Early in the embryonic CNS development of the mouse, invasion of the intraneural section by the perineural mesenchymal endothelial cells is accompanied by pericytes [78]. The subsequent appearance of tight junctions and the decrease in pinocytotic vesicles are also associated with pericyte coverage. Others have shown that growth and maturation of the microvasculature is dependent on the establishment of endothelial cell-pericyte gap junctions [11]. Furthermore, there is also recent evidence indicating that pericytes are able to mimic the astrocytic inductive influence. The absence of pericytes in platelet-derived growth factor-B (PDGF-B)- and PDGF receptor-beta (PDGFR-β)-deficient mouse embryos leads to defects in endothelial junction formation and to the formation of numerous cytoplasmic folds at the luminal surface of the endothelium [79]. This suggests that pericytes control endothelial differentiation in vivo. Most in vitro studies of capillary permeability have focused on endothelial cell monolayers and ignored the second cell that forms the capillary wall: the microvascular pericyte. Later, a model to study the permeability characteristics of endothelial cells, pericytes, and co-cultures of both cell types using semipermeable culture inserts was described [80]. Indeed, the addition of pericytes to endothelial monolayers increased their barrier effect to macromolecules and electrical resistance. Other transfilter co-culture experiments using conditionally immortalized adult rat brain pericytes revealed that the pericyte-derived angiopoietin-1 (Ang-1) induces occludin expression, an integral tight junction protein [81, 82]. The results of Sundberg et al. [83] provide strong evidence that Ang-1 is expressed by pericytes in vitro and in vivo and that the role proposed for Ang-1 in vessel maturation in development can be extended to vessel maturation after angiogenesis in adult tissues. Only recently it was shown in another co-culture model that brain pericytes induce and enhance the barrier function and P-glycoprotein (P-gp) functional activity of brain endothelial cells [84]. This pericyte-induced up-regulation of BBB properties was mediated, at least in part, through continuous production of transforming growth factor-beta 1 (TGF-β1).

Another, and more speculative, possible function of pericytes is that they may be modulators of leakage from microvessels [2]. Pericytes are most numerous on venules, the most leaky microvessels. Pericytes and their basal lamina appear to block the extravasation of macromolecules and blood-borne elements and provide mechanical support. Pericytes synthesize and deposit various extracellular matrix proteins, such as laminin, fibronectin, collagen (types I, III, IV, X), tenascin, and thrombospondin [37]. Behaving as a mechanical part of the vessel wall was one of the originally stated functions of pericytes. Maintenance and structural rigidity of the microvessel wall could also be inferred by the presence of stress fibers within pericytes, the enveloping design of pericytes, and pericytic contacts with endothelial cells that provide anchorage and are also thought to support transmission of the mechanical contractile force from the pericytes to the endothelium. Additionally, pericyte loss observed in PDGF-deficient mice results in compromised vascular stability; and these mice also develop capillary microaneurysms [85].

All in all, the pericyte is in a prime location to be involved with changes in microvascular permeability and to contribute to stable and mature capillaries.

5.8
Angiogenesis and Stability

Development of the vascular system is a complex process, governed by vascular endothelial growth factor (VEGF), that begins with the clustering of endothelial precursors, also known as angioblasts, into tube-like endothelial structures (vasculogenesis) [86]. Vascularization of neural tissues, like the brain, occurs by angiogenesis [74]. The preexistent vessel tubes hereby remodel and form the more complex architecture of the vascular system. Vessel assembly, patterning, and maintenance are complex and highly regulated processes. Developmental studies suggest that the endothelial cells in the nascent tubes may govern the acquisition of additional vessel layers [87]. As the next few paragraphs explain, there is much more to the positioning and supportive roles of pericytes than passive maintenance of the vessel wall.

5.8.1
PDGF-B and Pericyte Recruitment

Mural cells become associated with the forming vessels at later stages of development, which has led to the suggestion that the endothelial cells may govern vessel layer acquisition [88]. Accumulated evidence shows that PDGF-B plays a critical role in the recruitment of pericytes to newly formed vessels. In the developing mouse embryo, PDGF-B is produced by the sprouting endothelium, whereas PDGFR-β is present on developing pericytes [32, 85, 89], suggesting a paracrine mode of interaction between the two cell types. Targeted disruption of PDGF-B or PDGFR-β genes both lead to virtually identical phenotypes: perinatal lethality due to widespread microvascular leakage and hemorrhage [90, 91] caused by a severe deficit in pericytes [85]. Moreover, endothelium-specific PDGF-B knockout leads to pericyte deficiency [92]. In support of this theory, it has been demonstrated in an in vitro co-culture system that proliferating endothelial cells secrete PDGF-B, which acts as a chemoattractant and mitogen for undifferentiated mural cell precursors and induces a shape change from polygonal to spindle-shaped, reminiscent of smooth muscle cells in culture [93, 94].

It is well documented that pericytes become associated with endothelial cells while the vasculature is forming [78]. In fact, pericytes in culture have a select affinity for the endothelia [95]. The migration of pericytes to capillary-like structures was observed to be faster than that of astrocytes and pericytes covered these structures more extensively than astrocytes. We completed this in vitro BBB system by adding astrocytes to these mixed cultures of endothelial cells and pericytes [96]. Under these tri-culture conditions, endothelial cells reorganized into capillary-like structures with closely associated pericytes (Fig. 5.1 e, f), which allowed a definition of the heterocellular interactions involved in BBB development. The assembly of heterocellular complexes seems to be an essential step in the process of vessel maturation, because functional differentiation of cerebral pericytes (pAPN-expression) was first acquired when association with endothelial cells occurred [43] and only the presence of pericytes guaranteed stability of the astrocyte-induced capillary-like structures and prevented regression. Taken together, tri-culturing of endothelial cells, pericytes, and astrocytes provides a reliable in vitro model by which cerebral angiogenesis and the underlying well balanced mechanistic effects can be mimicked. This view is fostered mainly by data from retinal angiogenesis, suggesting that a newly formed vascular network remains immature and "plastic" until pericytes are recruited [97].

The presence of pericytes in angiogenic sprouts suggests that they may be functional in the angiogenic process [98]. While pericytes appear to be normally induced in the absence of PDGF-B or PDGFR-β, the subsequent co-migration and proliferation of the pericyte population is impaired [32]. Detailed analysis of the microvasculature in PDGF-B and PDGFR-β mutants shows that angiogenic sprouting, at least in certain locations such as in the embryonic brain, proceeds relatively normally in the absence of pericytes and leads to a normal number of branch points and a normal microvessel density in this tissue [79]. However, blood vessels devoid of pericytes exhibit several other defects, such as irregular diameter, increased permeability, and rupturing microaneurysms at late gestation. The temporal correlation between failure of pericyte recruitment and endothelial hyperplasia suggests that the lack of pericytes directly regulates endothelial cell number [79].

Taken together, the available evidence therefore strongly implies that the endothelial cell-derived PDGF-B signal controls pericyte recruitment to angiogenic vessels and that pericyte growth during vessel formation may be positively regulated by PDGF-B.

5.8.2
TGF-β1 and Differentiation

The above data suggest that the nature of the microvessel may be altered as pericytes attach to maturing capillaries. Pericytes are suspected to be responsible for the arrest of vessel formation and the induction of vessel maturation.

Some groups were able to demonstrate that pericytes are capable of rendering endothelial cells quiescent [36, 99]. This pericyte-induced endothelial cell growth arrest and inhibited migration are mediated via contact and activation of TGF-β1 [100]. Both endothelial cells and pericytes appear to synthesize a latent or inactive form of TGF-β1 when grown separately, but the activation of the protein may require the presence of both cells and physical contact, as implied from in vitro studies [99, 101, 102]. It is the activated TGF-β1 that then inhibits endothelial cell proliferation and migration. Thus, TGF-β1 in part constitutes a pericyte-derived signal for the endothelial cells. This factor also inhibits the proliferation of pericytes in culture [103], like it does endothelia, and induces at least in part the expression of contractile properties in these cells [40, 94].

These findings are supported by observations of mice null for TGF-β1 [104] or its receptor TGF-β receptor type II (TGF-βR2) [105] and co-receptor endoglin

[106] that display defective endothelial-mesenchymal interactions and show defects in vascular wall structures.

Thus, pericytes do provide a negative regulation of endothelial growth through TGF-β1, which is actually activated through participation of the endothelial cell itself.

5.8.3
Ang-1 and Maturation

Somewhere in the remodeling process lies the contribution of the angiopoietins. The angiopoietins Ang-1 and Ang-2 have been identified as ligands of the endothelial receptor tyrosine kinase Tie-2 [107–109], which controls vascular remodeling and maturation more than induction of angiogenesis. The largely complementary phenotypes of Ang-1- or Tie-2-deficient mice [108, 110] and Ang-2-overexpressing mice [107] have led to an antagonistic model in which Ang-1 acts as a Tie-2-activating agonist and Ang-2 acts as a Tie-2-inhibiting antagonist. These mice exhibit aberrant vascular development, characterized by abnormal interactions between endothelial cells and their supporting cells.

A clue to the role of the angiopoietins is provided by examination of the expression patterns of the angiopoietins and their receptor. All endothelial cells express Tie-2 during development and in adult tissues, while perivascular cells are the primary source of Ang-1 [108, 109]. That pericytes contribute to the production of Ang-1 has also been shown in vitro [81–83]. Recently, it was found that Ang-1, acting via the Tie-2 receptor on endothelial cells, is essential for the remodeling and stabilization of the embryonic vasculature [111]. The authors demonstrated that, whereas blocking the function of the PDGFR-β in the developing retinal vasculature led to mural cell-deficient vessels that were poorly remodeled and leaky, administration of recombinant modified Ang-1 restored the vascular structure of the larger vessels in the absence of the mural cells. Ang-1 is also known as an anti-permeability factor and prevents vascular leakage [112]. It has been reported that administration of Ang-1 reduces BBB leakage in the ischemic brain [113].

Ang-2 has been identified as a functional antagonist of Ang-1 that binds to Tie-2 without inducing signal transduction [107]. Ang-2 expression is restricted to endothelial cells at sites of vascular remodeling. This pattern of expression led to a hypothesis that Ang-1 is involved in vessel stabilization whereas Ang-2 is a destabilizing factor. The expression level of Ang-2 is induced under hypoxic conditions in brain, whereas Ang-1 expression is unchanged [114]. Ang-2 function is considered to be contextually determined by the presence of other cytokines, for example, Ang-2 has a pro-angiogenic effect in the presence of angiogenic factors such as VEGF and induces vessel regression in the absence of angiogenic cytokines [115].

All these observations clearly demonstrate a strong interdependence between the endothelial cell and its intimate partner (the pericyte). The tightly controlled reciprocal signaling cues will determine the cells' abundance, positions, phenotype and functions. These create a balance leading to a mature, quiescent and stable state blood vessel.

5.9
Conclusion

In essence, cerebral pericytes are emerging as an important member of the cellular complex that constitutes the BBB. The pericytes are complex cells, with metabolic, signaling and mechanical roles to support the endothelial cells. Endothelial cell-pericyte relationships are better understood and appear to involve complex biological feedback mechanisms between endothelial cells and pericytes. Interactions between these two cell types are important for the maturation, remodeling, and maintenance of the vascular system via the secretion of growth factors or modulation of the extracellular matrix. An important role for pericytes in pathology, and neuropathology in particular, has been indicated in hypertension, diabetic microangiopathy, Alzheimer's disease, multiple sclerosis, and CNS tumor formation [116–121]. Putting the pericyte functions in context with the requirements of the environment may be the key to understanding pericyte nature.

References

1 K. K. Hirschi, P. A. D'Amore **1996**, *Cardiovasc. Res.* 32, 687.
2 D. E. Sims **1986**, *Tissue Cell* 18, 153.
3 L. J. Mandarino, N. Sundarraj, J. Finlayson, H. R. Hassell **1993**, *Exp. Eye Res.* 57, 609.
4 M. P. Cohen, R. N. Frank, A. A. Khalifa **1980**, *Invest. Ophthalmol. Vis. Sci.* 19, 90.
5 R. Balabanov, P. Dore-Duffy **1998**, *J. Neurosci. Res.* 53, 637.
6 H. K. Rucker, H. J. Wynder, W. E. Thomas **2000**, *Brain Res. Bull.* 51, 363.
7 K. W. Zimmermann **1923**, *Z. Anat. Entwickl. Ges.* 68, 29.
8 R. G. Tilton, E. J. Miller, C. Kilo, J. R. Williamson **1985**, *Invest. Ophthalmol. Vis. Sci.* 26, 68.
9 T. S. Leeson **1979**, *Can. J. Ophthalmol.* 14, 21.
10 P. Cuevas, J. A. Gutierrez-Diaz, D. Reimers, M. Dujovny, F. G. Diaz, J. I. Ausman **1984**, *Anat. Embryol. (Berl.)* 170, 155.
11 K. Fujimoto **1995**, *Anat. Rec.* 242, 562.
12 D. M. Larson, M. P. Carson, C. C. Haudenschild **1987**, *Microvasc. Res.* 34, 184.
13 L. Diaz-Flores, R. Gutierrez, H. Varela, N. Rancel, F. Valladares **1991**, *Histol. Histopathol.* 6, 269.
14 D. E. Sims **1991**, *Can. J. Cardiol.* 7, 431.
15 D. Shepro, N. M. Morel **1993**, *FASEB J.* 7, 1031.

16 R. N. Frank, S. Dutta, M. A. Mancini **1987**, *Invest. Ophthalmol. Vis. Sci.* 28, 1086.
17 D. E. Sims **2000**, *Clin. Exp. Pharmacol. Physiol.* 27, 842.
18 U. Ozerdem, W. B. Stallcup **2003**, *Angiogenesis* 6, 241.
19 D. J. Ruiter, R. O. Schlingemann, J. R. Westphal, M. Denijn, F. J. Rietveld, R. M. De Waal **1993**, *Behring Inst. Mitt.* 1993, 258.
20 W. Risau, B. Engelhardt, H. Wekerle **1990**, *J. Cell Biol.* 110, 1757.
21 R. C. Nayak, A. B. Berman, K. L. George, G. S. Eisenbarth, G. L. King **1988**, *J. Exp. Med.* 167, 1003.
22 M. B. Graeber, W. J. Streit, G. W. Kreutzberg **1989**, *J. Neurosci. Res.* 22, 103.
23 R. Balabanov, R. Washington, J. Wagnerova, P. Dore-Duffy **1996**, *Microvasc. Res.* 52, 127.
24 D. Krause, B. Vatter, R. Dermietzel **1988**, *Cell Tissue Res.* 252, 543.
25 J. Kunz, D. Krause, M. Kremer, R. Dermietzel **1994**, *J. Neurochem.* 62, 2375.
26 D. Krause, J. Kunz, R. Dermietzel **1993**, *Adv. Exp. Med. Biol.* 331, 149.
27 R. Dermietzel, D. Krause **1991**, *Int. Rev. Cytol.* 127, 57.
28 V. Nehls, D. Drenckhahn **1993**, *Histochemistry* 99, 1.
29 I. M. Herman, P. A. D'Amore **1985**, *J. Cell Biol.* 101, 43.
30 V. Nehls, D. Drenckhahn **1991**, *J. Cell Biol.* 113, 147.
31 H. Gerhardt, H. Wolburg, C. Redies **2000**, *Dev. Dyn.* 218, 472.
32 M. Hellstrom, M. Kalen, P. Lindahl, A. Abramsson, C. Betsholtz **1999**, *Development* 126, 3047.
33 J. D. Gitlin, P. A. D'Amore **1983**, *Microvasc. Res.* 26, 74.
34 J. Folkman, C. C. Haudenschild, B. R. Zetter **1979**, *Proc. Natl Acad. Sci. USA* 76, 5217.
35 J. C. Swinscoe, E. C. Carlson **1992**, *J. Cell Sci.* 103, 453.
36 A. Orlidge, P. A. D'Amore **1987**, *J. Cell Biol.* 105, 1455.
37 A. M. Schor, A. E. Canfield, A. B. Sutton, T. D. Allen, P. Sloan, S. L. Schor **1992**, *EXS* 61, 167.
38 A. M. Schor, A. E. Canfield, P. Sloan, S. L. Schor **1991**, *In Vitro Cell Dev. Biol.* 27A, 651.
39 R. J. Boado, W. M. Pardridge **1994**, *J. Neurosci. Res.* 39, 430.
40 M. M. Verbeek, I. Otte-Holler, P. Wesseling, D. J. Ruiter, R. M. de Waal **1994**, *Am. J. Pathol.* 144, 372.
41 M. P. Dehouck, P. Vigne, G. Torpier, J. P. Breittmayer, R. Cecchelli, C. Frelin **1997**, *J. Cereb. Blood Flow Metab.* 17, 464.
42 P. M. Newcomb, I. M. Herman **1993**, *J. Cell Physiol.* 155, 385.
43 M. Ramsauer, J. Kunz, D. Krause, R. Dermietzel **1998**, *J. Cereb. Blood Flow Metab.* 18, 1270.
44 W. Risau, A. Dingler, U. Albrecht, M. P. Dehouck, R. Cecchelli **1992**, *J. Neurochem.* 58, 667.
45 C. Rouget **1873**, *Arch. Physiol. Norm. Pathol.* 5, 603.
46 Y. J. Le Beux, J. Willemot **1978**, *Anat. Rec.* 190, 811.
47 N. C. Joyce, M. F. Haire, G. E. Palade **1985**, *J. Cell Biol.* 100, 1387.
48 N. C. Joyce, M. F. Haire, G. E. Palade **1985**, *J. Cell Biol.* 100, 1379.

49 D. DeNofrio, T.C. Hoock, I.M. Herman **1989**, *J. Cell Biol.* 109, 191.
50 O. Skalli, M.F. Pelte, M.C. Peclet, G. Gabbiani, P. Gugliotta, G. Bussolati, M. Ravazzola, L. Orci **1989**, *J. Histochem. Cytochem.* 37, 315.
51 A.B. Dodge, H.B. Hechtman, D. Shepro **1991**, *Cell Motil. Cytoskeleton* 18, 180.
52 C. Kelley, P. D'Amore, H.B. Hechtman, D. Shepro **1987**, *J. Cell Biol.* 104, 483.
53 A.M. Schor, S.L. Schor **1986**, *Microvasc. Res.* 32, 21.
54 C. Kelley, P. D'Amore, H.B. Hechtman, D. Shepro **1988**, *J. Muscle Res. Cell Motil.* 9, 184.
55 U. Chakravarthy, T.A. Gardiner **1999**, *Prog. Retina Eye Res.* 18, 511.
56 D.P. Healy, S. Wilk **1993**, *Brain Res.* 606, 295.
57 R.D. Broadwell, M. Salcman **1981**, *Proc. Natl Acad. Sci. USA* 78, 7820.
58 C.R. Farrell, P.A. Stewart, C.L. Farrell, R.F. Del Maestro **1987**, *Anat. Rec.* 218, 466.
59 K. Kristensson, Y. Olsson **1973**, *Acta Neurol. Scand.* 49, 189.
60 M. Mato, S. Ookawara, K. Kurihara **1980**, *Am. J. Anat.* 157, 329.
61 P.A. Cancilla, R.N. Baker, P.S. Pollock, S.P. Frommes **1972**, *Lab. Invest.* 26, 376.
62 D.S. Maxwell, L. Kruger **1965**, *Exp. Neurol.* 12, 33.
63 B. Jeynes **1985**, *Stroke* 16, 121.
64 J. Boya **1976**, *Acta Anat. (Basel)* 95, 598.
65 B. van Deurs, **1976**, *J. Ultrastruct. Res.* 56, 65.
66 S. Kida, P.V. Steart, E.T. Zhang, R.O. Weller **1993**, *Acta Neuropathol. (Berl.)* 85, 646.
67 W.E. Thomas **1999**, *Brain Res. Brain Res. Rev.* 31, 42.
68 M.M. Verbeek, J.R. Westphal, D.J. Ruiter, R.M. de Waal **1995**, *J. Immunol.* 154, 5876.
69 K. Vass, H. Lassmann **1990**, *Am. J. Pathol.* 137, 789.
70 S. Mori, C.P. Leblond **1969**, *J. Comp. Neurol.* 135, 57.
71 R.A. Monteiro, E. Rocha, M.M. Marini-Abreu **1996**, *J. Submicrosc. Cytol. Pathol.* 28, 457.
72 D.Z. Gerhart, L.R. Drewes **1987**, *Cell Tissue Res.* 247, 533.
73 A. Frey, B. Meckelein, H. Weiler-Guttler, B. Mockel, R. Flach, H.G. Gassen **1991**, *Eur. J. Biochem.* 202, 421.
74 P.A. Stewart, M.J. Wiley **1981**, *Dev. Biol.* 84, 183.
75 F.E. Arthur, R.R. Shivers, P.D. Bowman **1987**, *Brain Res.* 433, 155.
76 C.B. Jaeger, A.R. Blight **1997**, *Exp. Neurol.* 144, 381.
77 P.A. Felts, K.J. Smith **1996**, *Neuroscience* 75, 643.
78 H.C. Bauer, H. Bauer, A. Lametschwandtner, A. Amberger, P. Ruiz, M. Steiner **1993**, *Brain Res. Dev. Brain Res.* 75, 269.
79 M. Hellstrom, H. Gerhardt, M. Kalen, X. Li, U. Eriksson, H. Wolburg, C. Betsholtz **2001**, *J. Cell Biol.* 153, 543.
80 C.J. Dente, C.P. Steffes, C. Speyer, J.G. Tyburski **2001**, *J. Surg. Res.* 97, 85.

81 T. Asashima, H. Iizasa, T. Terasaki, K. Hosoya, K. Tetsuka, M. Ueda, M. Obinata, E. Nakashima **2002**, *Eur. J. Cell Biol.* 81, 145.
82 S. Hori, S. Ohtsuki, K. Hosoya, E. Nakashima, T. Terasaki **2004**, *J. Neurochem.* 89, 503.
83 C. Sundberg, M. Kowanetz, L. F. Brown, M. Detmar, H. F. Dvorak **2002**, *Lab. Invest.* 82, 387.
84 S. Dohgu, F. Takata, A. Yamauchi, S. Nakagawa, T. Egawa, M. Naito, T. Tsuruo, Y. Sawada, M. Niwa, Y. Kataoka **2005**, *Brain Res.* 1038, 208.
85 P. Lindahl, B. R. Johansson, P. Leveen, C. Betsholtz **1997**, *Science* 277, 242.
86 J. Folkman, M. Klagsbrun **1987**, *Science* 235, 442.
87 K. K. Hirschi, P. A. D'Amore **1997**, *EXS* 79, 419.
88 J. E. Hungerford, C. D. Little **1999**, *J. Vasc. Res.* 36, 2.
89 L. Holmgren, A. Glaser, S. Pfeifer-Ohlsson, R. Ohlsson **1991**, *Development* 113, 749.
90 P. Soriano **1994**, *Genes Dev.* 8, 1888.
91 P. Leveen, M. Pekny, S. Gebre-Medhin, B. Swolin, E. Larsson, C. Betsholtz **1994**, *Genes Dev.* 8, 1875.
92 M. Enge, M. Bjarnegard, H. Gerhardt, E. Gustafsson, M. Kalen, N. Asker, H. P. Hammes, M. Shani, R. Fassler, C. Betsholtz **2002**, *EMBO J.* 21, 4307.
93 K. K. Hirschi, S. A. Rohovsky, L. H. Beck, S. R. Smith, P. A. D'Amore **1999**, *Circ. Res.* 84, 298.
94 K. K. Hirschi, S. A. Rohovsky, P. A. D'Amore **1998**, *J. Cell Biol.* 141, 805.
95 T. Minakawa, J. Bready, J. Berliner, M. Fisher, P. A. Cancilla **1991**, *Lab. Invest.* 65, 32.
96 M. Ramsauer, D. Krause, R. Dermietzel **2002**, *FASEB J.* 16, 1274.
97 L. E. Benjamin, I. Hemo, E. Keshet **1998**, *Development* 125, 1591.
98 V. Nehls, K. Denzer, D. Drenckhahn **1992**, *Cell Tissue Res.* 270, 469.
99 Y. Sato, D. B. Rifkin **1989**, *J. Cell Biol.* 109, 309.
100 M. S. Pepper **1997**, *Cytokine Growth Factor Rev.* 8, 21.
101 Y. Sato, R. Tsuboi, R. Lyons, H. Moses, D. B. Rifkin **1990**, *J. Cell Biol.* 111, 757.
102 A. Antonelli-Orlidge, K. B. Saunders, S. R. Smith, P. A. D'Amore **1989**, *Proc. Natl Acad. Sci. USA* 86, 4544.
103 Q. Yan, E. H. Sage **1998**, *J. Cell Biochem.* 70, 70.
104 M. C. Dickson, J. S. Martin, F. M. Cousins, A. B. Kulkarni, S. Karlsson, R. J. Akhurst **1995**, *Development* 121, 1845.
105 M. Oshima, H. Oshima, M. M. Taketo **1996**, *Dev. Biol.* 179, 297.
106 D. Y. Li, L. K. Sorensen, B. S. Brooke, L. D. Urness, E. C. Davis, D. G. Taylor, B. B. Boak, D. P. Wendel **1999**, *Science* 284, 1534.
107 P. C. Maisonpierre, C. Suri, P. F. Jones, S. Bartunkova, S. J. Wiegand, C. Radziejewski, D. Compton, J. McClain, T. H. Aldrich, N. Papadopoulos, T. J. Daly, S. Davis, T. N. Sato, G. D. Yancopoulos **1997**, *Science* 277, 55.
108 C. Suri, P. F. Jones, S. Patan, S. Bartunkova, P. C. Maisonpierre, S. Davis, T. N. Sato, G. D. Yancopoulos **1996**, *Cell* 87, 1171.

109 S. Davis, T. H. Aldrich, P. F. Jones, A. Acheson, D. L. Compton, V. Jain, T. E. Ryan, J. Bruno, C. Radziejewski, P. C. Maisonpierre, G. D. Yancopoulos **1996**, *Cell* 87, 1161.
110 T. N. Sato, Y. Tozawa, U. Deutsch, K. Wolburg-Buchholz, Y. Fujiwara, M. Gendron-Maguire, T. Gridley, H. Wolburg, W. Risau, Y. Qin **1995**, *Nature* 376, 70.
111 A. Uemura, M. Ogawa, M. Hirashima, T. Fujiwara, S. Koyama, H. Takagi, Y. Honda, S. J. Wiegand, G. D. Yancopoulos, S. I. Nishikawa **2002**, *J. Clin. Invest.* 110, 1619.
112 G. Thurston **2002**, *J. Anat.* 200, 575.
113 Z. G. Zhang, L. Zhang, S. D. Croll, M. Chopp **2002**, *Neuroscience* 113, 683.
114 S. J. Mandriota, M. S. Pepper **1998**, *Circ. Res.* 83, 852.
115 M. Ramsauer, P. A. D'Amore **2002**, *J. Clin. Invest.* 110, 1615.
116 L. Claudio, C. S. Raine, C. F. Brosnan **1995**, *Acta Neuropathol. (Berl.)* 90, 228.
117 G. Allt, J. G. Lawrenson **2001**, *Cells Tissues Organs* 169, 1.
118 I. M. Herman, S. Jacobson **1988**, *Tissue Cell* 20, 1.
119 P. Ballabh, A. Braun, M. Nedergaard **2004**, *Neurobiol. Dis.* 16, 1.
120 M. M. Verbeek, I. Otte-Holler, D. J. Ruiter, R. M. de Waal **1999**, *Cell Mol. Biol.* 45, 37.
121 M. M. Verbeek, R. M. de Waal, J. J. Schipper, W. E. Van Nostrand **1997**, *J. Neurochem.* 68, 1135.

6
Brain Macrophages: Enigmas and Conundrums

Frederic Mercier, Sebastien Mambie, and Glenn I. Hatton

6.1
Introduction

Macrophages are cells from hematopoietic origin that migrate and reside in all organs and tissues throughout adulthood. The central nervous system (CNS) is not an exception and contains numerous macrophages in the meninges, choroid plexus, circumventricular organs, and in the whole CNS vasculature. Brain macrophages are described as sentinels responsible for the brain immune defense and for clearing cell and extracellular matrix debris inherent in the maintenance of neural tissue. A fascinating emerging possibility is that macrophages, which are efficient producers of extracellular matrix molecules and growth factors/cytokines, have an influence on, and perhaps regulate, neuronal and glial cell functions as well as the proliferation and differentiation of neural precursor cells and neural stem cells (NSCs) throughout adulthood. In addition, because they are highly mobile cells, macrophages can migrate to sites of physiological challenge or injury in order to more efficiently deliver signaling molecules and ultimately influence the morphological and functional plasticity of neural tissue.

First, we describe the various types of macrophages present in the adult brain, commenting on their distribution and ultrastructure. Macrophages have been, and often still are, confused with other resident cell types: pericytes, dendritic cells, and microglial cells. We show here that macrophages of the brain ventricle walls ultrastructurally resemble adult neuroblasts. Distinguishing these cells from one another is not easy, based upon only ultrastructural and location criteria. This is made even more difficult by cell "specific" markers such as CD cell surface antigens that can be up- and down-regulated with maturation, differentiation, and eventually trans-differentiation (differentiation from a given cell type to another cell type).

We review macrophage functions starting with the conventional roles attributed to these cells, such as macrophagic activity and immune defense. We focus on the striking diversity of active molecules produced by brain macrophages, including extracellular matrix molecules and growth factors, and then analyze how the molecules may interact in the basal lamina at the astrocytic/macro-

Blood-Brain Interfaces: From Ontogeny to Artificial Barriers.
Edited by R. Dermietzel, D.C. Spray, M. Nedergaard
Copyright © 2006 WILEY-VCH Verlag GmbH & Co. KGaA, Weinheim
ISBN: 3-527-31088-6

phage interface to ultimately regulate neural functions. Finally, we advance our view of the emerging possibilities that are suggested by recent literature, including the trans-differentiation of macrophages into neural cells.

6.2
Different Types and Locations of Brain Macrophages

Macrophages of the central nervous system (CNS) have various morphologies in the meninges, perivasculature, ventricle walls, ventricle lumen, choroid plexus, and circumventricular organs. Due to their high degree of morphological plasticity and because they migrate throughout and between these different locations, macrophages display different morphologies, even within compartments. Macrophages capable of phagocytic activity are primarily located in the CNS extraparenchyma (meninges, brain perivasculature, stroma of the choroid plexus and circumventricular organs), but microglial cells also function as phagocytes and are at least partly replaced by populations of macrophages that enter the neural tissue proper (parenchyma) after crossing the basal lamina located at the parenchymal/extraparenchymal interface (Hickey and Kimura 1988; Ling 1979; Ling and Wong 1993; Vallieres and Sawchenko 2003).

The CNS cell types that can be considered to be macrophagic include microglia, pericytes, perivascular macrophages (PVM, also termed Mato cells or fluorescent granular perithelial cells; Mato et al. 1980), dendritic cells, Kolmer cells, epiplexus and supraependymal cells (Ling 1979; Ling et al. 1998; Lu et al. 1993). All these cell types vary in morphology within a given tissue compartment, migrate from one compartment to another, do not express strict specific markers that may distinguish the cells from each other in a given CNS compartment, and convert from one type to another in the different compartments upon the influence of cytokines and growth factors. It is, thus, currently impossible to state whether the CNS macrophage-type cells constitute one cellular entity with various morphologies, diversity of marker-expression, and different functions, or are morphologically and functionally distinct cell types.

Even though, in lineage, they are most likely related to macrophages and may act phagocytically, we will not consider microglia and pericytes to be macrophages. The frequent confusion of pericytes with PVMs can be resolved ultrastructurally on the basis of their respective locations in the blood vessel wall, i.e. in tunica media for pericytes and in tunica adventitia for PVM (Bechmann et al. 2001a; Graeber et al. 1989). Because the media and adventitia of arteries and arterioles form sleeves that completely cover the endothelium, PVMs and pericytes/smooth muscle cells can be distinguished from each other in their respective layers (Fig. 6.1). To distinguish the two cell types along capillaries where the media and adventitia form incomplete layers is more difficult (Fig. 6.2A, B). This is why the existence of a capillary adventitial layer that consists of PVM has not been recognized in the past. Figure 6.3C shows a capillary with easily identifiable and ultrastructurally distinct pericyte and PVM. Smooth

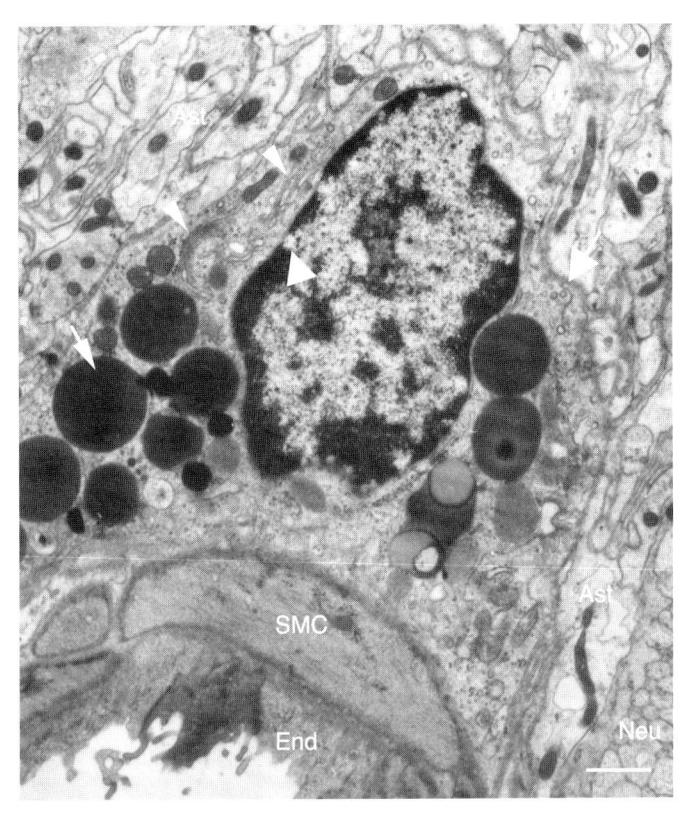

Fig. 6.1 Ultrastructure of arteriolar perivascular macrophages (PVMs). PVMs are located in the outer layer of the vasculature (adventitia), between astrocytes (Ast) and the smooth muscle cells (SMC) forming the media. Endothelial cells (End) form the intima. Basal laminae (large arrow) separate PVMs from Ast and SMC. PVMs possess large lysosomes (arrow), a developed Golgi apparatus (arrowheads), and clumped electron-dense heterochromatin (large arrowhead). The electron density of the PVM cell body is intermediate between electron-lucent astrocytes (ASM) and electron-dense smooth muscle cells (SMC) and endothelium (End). Neu: neuronal processes. Scale bar: 500 nm.

muscle cells and pericytes are always electron-dense (in the media: Figs. 6.1, 6.3C and 6.5B). Most PVMs show an electron-density that is intermediate between pericyte/smooth muscle cells and astrocytes (Fig. 6.1). Pial macrophages, however, which can be likened to PVMs, are often electron-dense (Fig. 6.3A). The adventitia represents an extension of the pia at the blood vessel surface. It has been established that macrophages migrate from the pia to the adventitial layer during inflammation (McKeever and Ballantine 1978). To our knowledge, there are no data suggesting that pial macrophages change electron-density when reaching the adventitia. Thus, the electron-lucent and electron-dense macrophages may either represent two different cell types or one single cell type that changes its ultrastructural characteristics.

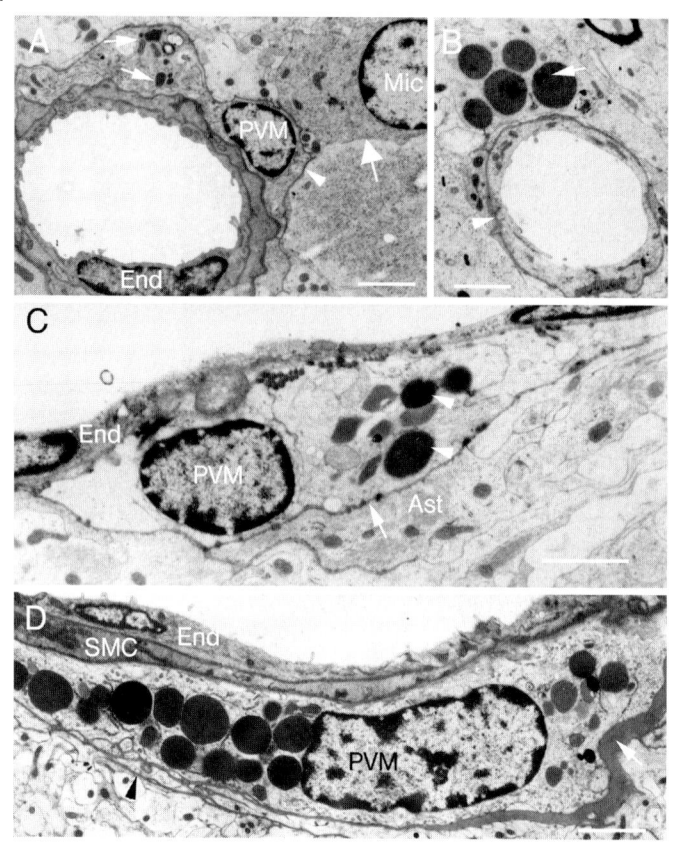

Fig. 6.2 Perivascular macrophages (PVMs) in the brain vasculature.
(A) Microglial cell (Mic) abutting a PVM that displays lysosomes (arrows). Unlike Mics, PVMs are on the vessel side of the basal laminae (arrowhead). The large arrow indicates the Mic interface outside of basal lamina. End: endothelial cell.
(B) Mic, identified by its large lysosomes (arrow), abutting a capillary. Arrowhead indicates subendothelial basal lamina.
(C) Venous PVMs are directly apposed to the surface of endothelial cells (End). A basal lamina (arrow) separates the PVM from surrounding astrocytes (Ast). Arrowheads indicate lysosomes.
(D) Arteriolar PVMs contain smooth muscle cells (SMC) separating endothelium from PVMs. Note the presence of numerous large lysosomes in the PVM. The basal lamina overlying this PVM is thick (white arrow) or elongates between Ast (black arrowhead). All scale bars: 2 μm.

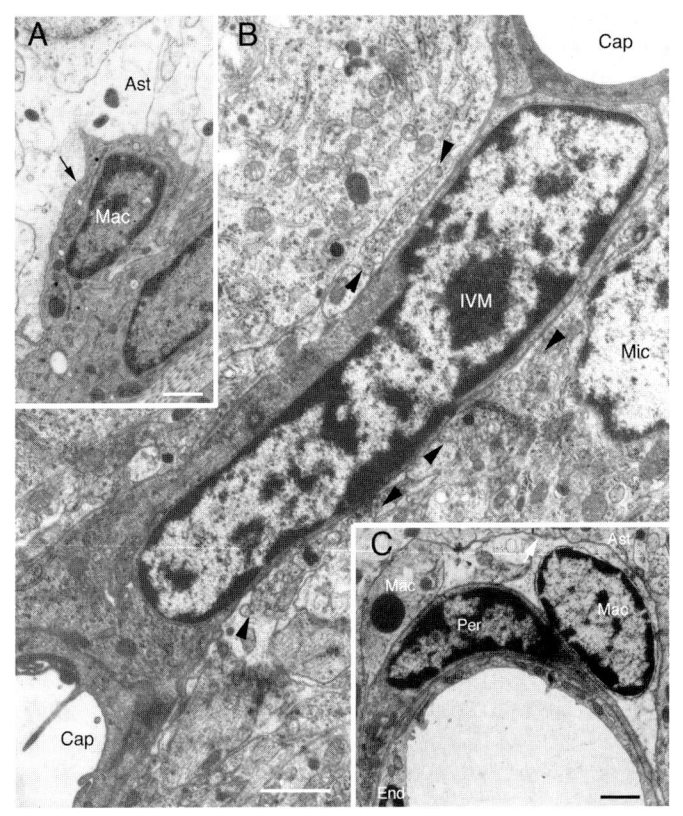

Fig. 6.3 Meningeal and intervascular macrophages.
(A) Pial macrophage (Mac) located just beneath the basal lamina (arrow) underlying the astrocyte (Ast) endfeet of the glia limitans. Scale bar: 2 µm.
(B) Intervascular macrophage (IVM) between two capillaries (Cap) of the hypothalamus. Arrays of axons (between arrowheads), which are separated from the IVM by thin astrocytic endfeet, are found along the IVM. A microglial cell (Mic) is closely associated with the IVM. Scale bar: 2 µm.
(C) Cytoarchitectonics of a capillary, showing the intima (End: endothelial cells), the media (Per = pericyte) and the adventitia (Mac = macrophage). Scale bar: 1 µm.

6.2.1
Macrophage Structure and Ultrastructure

6.2.1.1 Perivascular Macrophages

The tunica adventitia of brain blood vessels consists of pia-arachnoid tissue. Macrophages and fibroblasts are the primary cells encountered in this tissue, although the cell ratio varies from a high fibroblast density in the meninges to a high macrophage density along the brain vasculature. The arterial adventitia resembles the meninges proper, and possesses a Virchow-Robin space that is continuous with the sub-arachnoid space (Mercier, 2004; Mercier and Hatton,

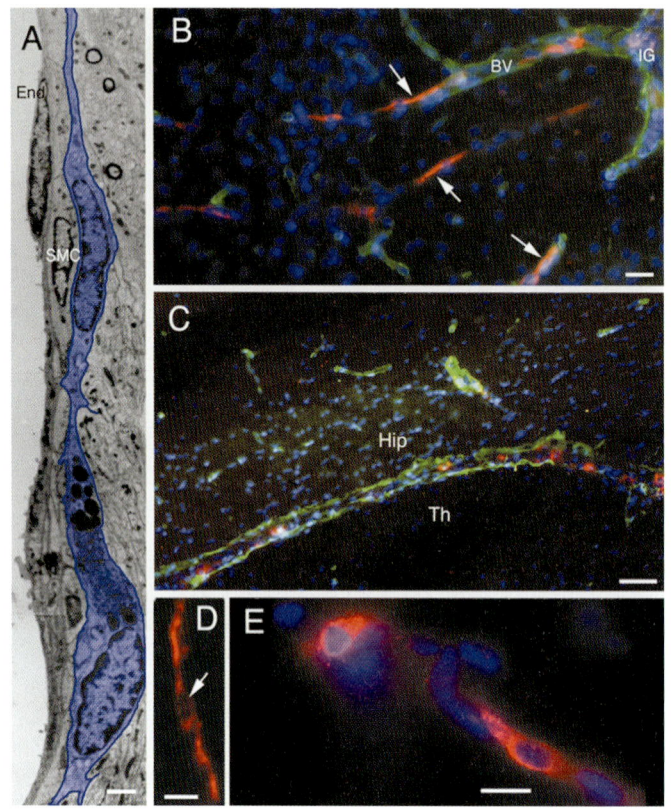

Fig. 6.4 Meningeal and PVM network.
(A) Electromicrograph showing the PVM network (highlighted in light purple) along an arteriole (with a lumen diameter of 13 μm) in the caudate putamen/corpus callosum interface. The basal laminae separating PVMs from smooth muscle cells (SMC) and astrocytes (Ast) is highlighted in dark blue. End: endothelium. Scale bar: 2 μm.
(B) Immunofluorescence micrograph showing a triple label for CD163 (visualizing PVMs, red), laminin (visualizing blood vessel basal laminae, green) and bisbenzimide (visualizing cell nuclei, blue).
PVMs (arrows) are seen on most blood vessels (BV). IG: indusium griseum. Scale bar: 20 μm.
(C) The *lamina affixa*, a meningeal tissue located between the thalamus (Th) and hippocampus (Hip), contain numerous PVMs (red). The lamina affixa/neural tissue interface is delineated by basal laminae (green). Laminin: green; bisbenzimide: blue. Scale bar: 50 μm.
(D) Typical bipolar PVM labeled with CD163. An arrow indicates the cell body. Scale bar: 10 μm. (E) PVMs labeled with CD163 (red). Bisbenzimide: blue. Scale bar: 10 μm.

2004). The arterial Virchow-Robin space disappears as arteriole lumens approach 15–20 μm in diameter. Most arterioles with a lumen diameter less than 15 μm have PVMs tightly apposed to both the basal lamina overlying the media and the sub-astrocytic basal lamina (Figs. 6.1, 6.2 C, D and 6.4 A). Contrary to common belief, brain capillaries, even those having a lumen diameter of 6–8 μm, possess an adventitia filled with PVMs (Figs. 6.2 A and 6.3 C).

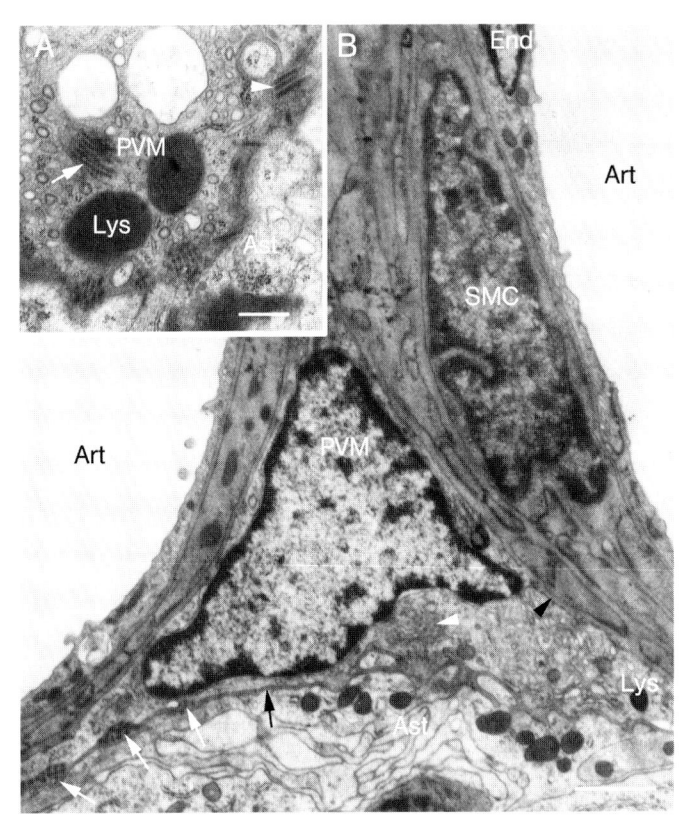

Fig. 6.5 Perivascular macrophages (PVMs) express collagen fibers.
(A) Aligned fibers of collagen with a diameter of 35–50 nm in the PVM cytoplasm (arrow) and the thick basal lamina overlying the PVM (arrowhead). The PVM contains characteristic large lysosomes (Lys).
Ast: astrocyte. Scale bar: 500 nm.

(B) PVM located at the interface of two merging arterioles (Art). 35–50 nm collagen fibers are visible in both the PVM cytoplasm (arrowhead) and the superficial basal lamina (arrows).
Ast = astrocyte; End = endothelial cell; Lys = lysosome; SMC = smooth muscle cell.
Scale bar: 1 μm.

Figure 6.1 shows the ultrastructure of a typical macrophage in the tunica adventitia of a brain blood vessel. PVMs are located between astrocytes and smooth muscle cells in large blood vessels, or astrocytes and pericytes in capillaries. Because pericytes only partly cover the capillary endothelium surface (approximately 30%), capillary PVMs can also be found directly inserted between astrocytes and endothelial cells. There are no fundamental ultrastructural differences between venous (Fig. 6.2C), arteriolar (Figs. 6.1 and 6.2D) and capillary (Fig. 6.2A) PVMs. The PVMs display large lysosomes, some of which have a diameter of more than 1 μm (Fig. 6.1, arrow). The cytoplasm of PVMs is often filled with densely packed organelles, including Golgi apparatus, mitochondria, rough reticulum endoplasmic, and characteristic clusters of free ribosomes

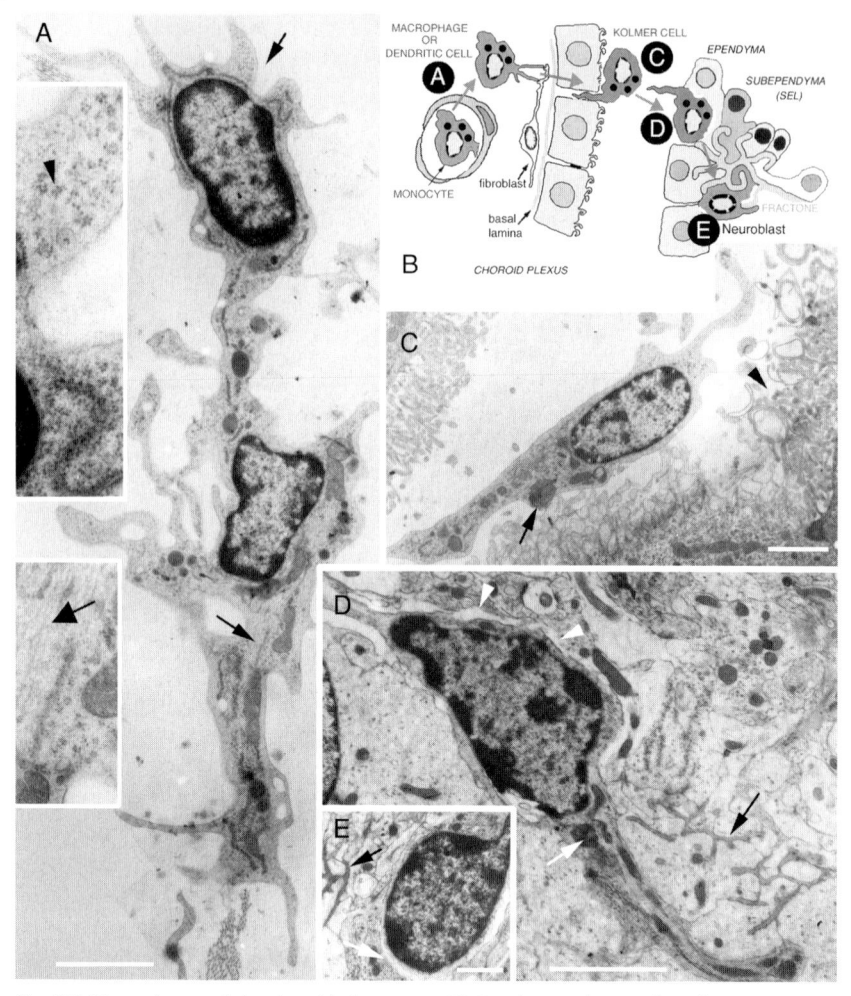

Fig. 6.6 Macrophages of the choroid plexus (CP) and ventricle walls.
(A) Bi-nucleated dendritic cell in the CP stroma showing numerous tortuous cell processes. The insets (4.0×, top; 2.5×, bottom) show areas indicated by arrows in A, showing that dendritic cells possess clusters of free ribosomes (arrowhead) and intermediate filaments (large arrow).
(B) Schema depicting the migratory pathway of monocyte-derived macrophages from the CP to the ventricle wall. The locations of the ultramicrographs shown in the figure are reported in the schema as their respective letters.

(C) Bipolar epiplexus cell (Kolmer macrophage) in the lateral ventricle. Arrow: lysosome. Arrowhead: CP epithelium.
(D) Bipolar subependymal macrophage identified by a lysosome (arrow) and surrounding electron-lucent bands (arrowheads). Black arrow: subependymal basal lamina (fractone).
(E) Neuroblasts of the lateral ventricle wall display characteristics of migrating macrophages: clumps of electron-dense heterochromatin, narrow cytoplasm, and surrounding electron-lucent bands. Black arrow: fractone. All scale bars: 2 µm.

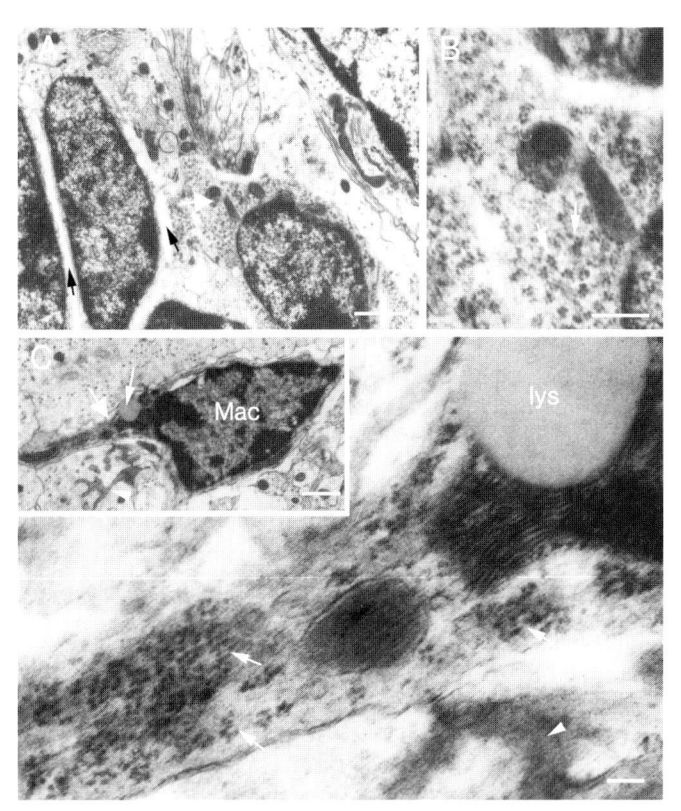

Fig. 6.7 Ultrastructural similarities between macrophages and pre-migrating neuroblasts. (A) Pre-migrating neuroblasts in the subependymal layer of the lateral ventricle are surrounded by electron-lucent lateral bands (black arrows). Scale bar: 2 μm.
(B) 4× magnified field of the neuroblast cytoplasm indicated by a white arrow in A, showing clusters of free ribosomes (arrows). Scale bar: 200 nm.

(C) Subependymal macrophage (Mac) identified by the presence of a large lysosome (small arrow). Arrowhead: fractone. Scale bar: 1 μm.
(D) 10× magnified field of the Mac process indicated by a large arrow in C, showing clusters of free ribosomes (arrows). Lys=lysosome. Arrowhead: fractone. Scale bar: 100 nm.

(Fig. 6.7C). PVM nuclei are ovoid and display a characteristic electron-dense heterochromatin with indentations (Fig. 6.1, large arrowhead; Fig. 6.2). Like pericytes, PVMs are encircled by a continuous basal lamina (Fig. 6.1, large arrow; Fig. 6.2A, arrowhead; Fig. 6.3C, arrow), except where PVMs contact one another to form a network (Fig. 6.4A; Mercier et al. 2002). PVMs are often coated with a very thick (Fig. 6.2D, arrow) or complex-shaped (Fig. 6.2D, arrowhead) basal lamina. PVMs also connect adjacent blood vessels as intervascular macrophages (Fig. 6.3B; see also Mercier et al. 2003).

6.2.1.2 Meningeal Macrophages

Macrophages are abundant throughout the meninges, at the surface of the brain, and in the projections located between major brain structures, such as the lamina affixa between hippocampus and thalamus. Pial macrophages, which contact the basal lamina underlying the glial limitans, are electron-dense and possess few large lysosomes (Fig. 6.3 A). Arachnoid macrophages contact fibroblast processes to form a network (Mercier et al. 2002), but whether the arachnoid macrophages form a network on their own is difficult to assess by transmission electron microscopy without serial sectioning and three-dimensional reconstruction over a large tissue volume. Along large blood vessels of the brain, PVMs can be considered as meningeal macrophages, bathing with fibroblasts (Mercier, 2004) in the Virchow-Robin/subarachnoid space.

6.2.1.3 Dendritic Cells

According to recent studies, rat meninges and choroid plexus stroma contain numerous dendritic cells that are distinguished from the ED2 (also termed CD163) macrophages by their expression of the cell surface antigens OX6 and OX62, markers of major histocompatibility complex (MHC) class II (McMenamin, 1999; McMenamin et al. 2003). Dendritic cells were previously characterized in the CNS but were reported as rare in normal rodents, only being present in substantial numbers after acute injury or experimental autoimmune encephalomyelitis (EAE; Matyszak and Perry 1996; Matyszak et al. 1997). Dendritic cells, which originate from bone marrow after maturation in lymphoid organs, are considered as "sentinels" of the immune system. Dendritic cells are antigen-presenting cells (APC), express MHC class II antigens, and initiate T-cell responses. Macrophages do not express MHC class II in the normal brain, but only in activated macrophage subpopulations (McMaster and Williams 1979). Dendritic cells form networks in the choroid plexus and meninges (McMenamin 1999) as they do in peripheral organs and tissues (Steinman 1991).

Morphologically, dendritic cells are pleiomorphic, i.e. dendriform, displaying numerous processes. One dendritic cell of the choroid plexus stroma is shown in Fig. 6.6 A. The cell has an electron-dense heterochromatin that is typical of the macrophage family, with indentations. The numerous processes of variable diameter make this cell distinguishable from a regular macrophage. Although it is believed that dendritic cells and macrophages may originate from different circulating monocyte precursors, the cells can convert to one another (Ardavin 2003; Dauer et al. 2003). Dendritic cells intervene in inflammation (Fischer and Reichmann 2001; Matyszak and Perry 1996). However, because these cells have only recently been characterized in brain, their specific function in brain is unknown. Interestingly, dendritic cells exist in the hypophysis (Sato and Inoue 2000), where they intervene in endocrine regulation (Hoek et al. 1997).

6.2.1.4 Ventricular Macrophages

Different cell populations reside or transit in the brain ventricles. Ventricular cells have been identified by their location at the surface of the choroid plexus (epiplexus cells, also termed Kolmer cells), or ependyma (supraependymal cells; Scott 1999). Intraventricular, free motile cells are also found along the entire ventricular neuroaxis (Ling 1976). Most of these cells are macrophages (Bleier et al. 1975, 1982; Ling 1979; Ling and Wong 1993, 1982; Lu et al. 1993), although macrophages often cluster at the ependymal surface with T-lymphocytes and immature neurons expressing gamma aminobutyric acid (GABA) and serotonin (Del Brio et al. 1992; Harandi et al. 1986; Hirunagi et al. 1989). It has been demonstrated that epiplexus macrophages, free motile intraventricular macrophages, and supraependymal macrophages are the same cells at different locations along a migratory pathway extending from the choroid plexus stroma to the ventricle wall (Ling 1979; Ling and Wong 1993). Using carbon (Indian ink) and rhodamine labeling techniques, the authors have shown that circulating monocytes first enter the choroid plexus stroma, where they mature into macrophages. Then, the macrophages cross the choroid epithelium, transit in the lumen of the lateral ventricle as Kolmer cells, attach to the ependymal surface as supraependymal cells, and infiltrate the ventricle wall. In neonate rodents at least, some of the infiltrated cells mature into microglia (Ling 1979; Ling and Wong 1993). Brain infiltration of monocytes and their maturation into macrophages have been recognized in adult animals (for reviews, see Hickey 2001; Ransohoff et al. 2003), using models of bone marrow radiation chimeras (Lassmann et al. 1993) and bone marrow grafts from enhanced green fluorescent protein (EGFP) mice (Vallieres and Sawchenko 2003). All data support the view that brain macrophages are constantly replaced by blood-derived cells throughout life. Most macrophages stay within the meninges, perivascular layer, and choroid plexus. However, in agreement with the theory of Rio-Hortega (1932), a fraction of these macrophages enter the neural compartment to replenish microglia. It is however not clear whether all microglia derive from macrophages or whether all microglia and macrophages derive from separate precursors (Walker 1999). It is also discussed, using ultrastructural similarity, that infiltrated ventricular macrophages may represent a source of neural stem cells or neural progenitor cells (further explained in a later section).

6.2.2
Immunotyping by Cell Surface Antigens

There is a considerable heterogeneity in the immunophenotype of brain macrophages, dendritic cells and microglia (Graeber et al. 1989). These cells express several cell-surface antigens termed CD, ED, or OX, which greatly vary among species, organs, location in the tissue, and the degree of physiological or pathological activation. The common confusion of macrophages, pericytes, and microglia (Guillemin and Brew 2004; Kida et al. 1993; Thomas 1999) has led to several mistakes. In addition, some markers thought to be specific for other cells have been later found in macrophages. For example, the lymphocyte markers CD4 and CD8 are expressed by healthy and diseased brain macrophages (Perry and Gordon 1987; Perry et al. 1987; Popovich et al. 2003). CD14 is present in both macrophages and microglia (Beschomer et al. 2002); and numerous markers have been characterized in diseased brain (AIDS, auto-immune encephalomyelitis, Alzheimer's disease, Parkinson's disease). Also, it is important to remember that macrophages are scavenger cells that phagocytose necrotic or apoptotic cells. By phagocytosis or pinocytosis, macrophages can absorb cell markers of foreign origin. It has been shown, for example, that macrophages can absorb myeloperoxidase activity from ingested neutrophils (Graeber et al. 1990; Wendling et al. 1991). However, in the rat, CD163 is a reliable marker of PVMs (Fig. 6.4 B, D, E) and meningeal macrophages (Fig. 6.4 C; Angelov et al. 1996).

6.2.3
Macrophages Contact Basal Laminae

All brain macrophages, except those located in the ventricle lumen, contact a basal lamina by their cell body or processes. PVMs are totally enclosed by basal laminae (Fig. 6.1). Macrophages of the choroid plexus stroma contact the basal lamina underlying the epithelial cells; and those of the ventricle walls contact the fractones, the specialized subependymal basal laminae (Mercier et al. 2003). Pial macrophages directly contact the basal lamina interfacing the pia and glia limitans (Fig. 6.3A). Macrophages of the arachnoid cell barrier layer (at the interface between dura and arachnoid) also contact a basal lamina (Vandenabeele et al. 1996). In addition, the punctate basal laminae that exist throughout the arachnoid (Mercier 2004; Mercier and Hatton 2001) often contact macrophages or fibroblasts.

It is likely that macrophages participate in the synthesis and deposition of extracellular matrix (ECM) molecules composing the basal lamina. Collagens IV, XV, and XVIII and laminins β-1 and γ-1, which are the principal basal lamina components, are produced by fibroblasts, but the absence of fibroblasts along capillaries and in the subependymal layer (SEL) of the entire ventricular system suggests that basal laminae are there produced by other cells. The characterization of collagen fibrils within and at the surface of PVMs strongly suggests that PVMs produce collagens (Fig. 6.5). It is known that macrophages are the primary producers of the metalloproteinases MMP-2 and MMP-9, which are the principal enzymes involved in the degradation of basal laminae. Macrophages also produce heparan sulfate proteoglycans (HSPG) present in basal laminae, such as perlecan, a crucial binder and activator of heparin-binding growth factors. The heparin-binding growth factors are among the most powerful signaling molecules, intervening in cell proliferation, differentiation, and migration. But these factors are usually inactive or weakly active when free in the extracellular space and need to be activated and presented by the HSPG of the basal laminae to reach their full potential. Thus, the production of HSPG by macrophages and the interactions of these HSPG with growth factors in front of macrophages within basal laminae are likely very important for the control of cell renewal, cell differentiation or maturation, and any sort of morphological plasticity.

6.2.4
Network of Macrophages Through the Brain

PVMs contact each other via their elongated processes to form a network all along the brain vasculature. The PVM network has been previously characterized both at the light microscopic (Cuff et al. 1996) and ultrastructural levels (Mercier et al. 2002). Although the PVM network has been primarily described in arterioles, both venous (Fig. 6.2 C) and capillary (Fig. 6.3 C), PVMs belong to the macrophage perivascular network. The ultramicrograph in Fig. 6.4 A shows the PVM network in an arteriole of the lateral ventricle wall. The PVM network can also be visualized by labeling PVMs with the specific cell surface antigen CD163 (Fig. 6.4 B–E), although it is not established that CD163 stains all PVMs. Immunolabeling for CD163 visualizes both the cell body and process surfaces of PVMs (Fig. 6.4 D, E), but the process terminals either do not express CD163 or are not detected with the sensitivity of the antibody in our immunochemistry protocol. Thus, the PVM visualized by CD163 immunolabeling appears as dashed lines along the vasculature (Fig. 6.4 B). Macrophages present in meningeal infoldings with the brain, like those of the lamina affixa, are also labeled with CD163 (Fig. 6.4 C). The PVMs of the superficial meninges also form a network that can be visualized by transmission electron microscopy and immunolabeled with CD163 (not shown).

PVMs and meningeal macrophages are constantly repopulated from bone marrow (Hickey et al. 1992) with a turnover rate of 3–4 weeks (Bechmann et al. 2001 a). These cells are known to be efficient APCs, even in comparison with microglial cells (Fabriek 2005 a; Hickey and Kimura 1988), and to participate in the brain immune defense and inflammation (Schiltz and Sawchenko 2003; Williams et al. 2001). Although PVM and meningeal macrophages express numerous cytokines, growth factors, and extracellular matrix molecules, their specific role(s) remain to be determined.

The PVM network (Cuff et al. 1996) runs along the adventitial layer of the vasculature into the CSF compartment (Virchow-Robin space) that is continuous with the subarachnoid space of the meninges. In the meninges, macrophages also form a network that is interwined with a fibroblast network connected by functional gap junctions (Mercier and Hatton 2001; Spray et al. 1991).

The gap junction protein Connexin (Cx)43 appears to be a key player in the development of progenitor cells and is expressed by macrophages, neutrophils and mast cells (Oviedo-Orta and Evans, 2004). Moreover, gap junctional communication has been found between particular macrophages (foam cells of atherosclerotic lesions; Polacek et al. 1993). Thus, it is also possible that brain macrophages communicate to each other via gap junctions throughout the CNS. Interestingly, macrophages inhibit gap junctional communication and downregulate Cx43 expression in cultured astrocytes (Rouach et al. 2002).

6.3
Migration of Brain Macrophages

Macrophages are highly mobile cells that can move throughout the meninges, perivascular layer (Virchow-Robin space), and ventricles. They are particularly mobile after brain injury or bacterial infection, converging at the injury site, where they secrete pro-inflammatory cytokines and ultimately induce an inflammation. Brain macrophages are further assisted in this task by monocytes and lymphocytes infiltrating from the circulation. Inflammation leads to a serious disturbance of the cytokine and growth factor microenvironment (see Chapter 11), which then may lead to repair or partial repair, but often induces astrogliosis and neuronal death. Meningeal macrophages can also cross the glia limitans, as shown in the suproptic nucleus after dehydration (Mercier 2004). The purpose of this migration into the brain parenchyma is unexplained but may be related to morphological changes associated with neuroendocrine regulation (Beagley and Hatton 1992; Hatton 1997).

6.4
Fast Renewal of Brain Macrophages

Macrophages have a short life span, being replaced every 3–5 weeks. It has been shown that macrophages are replenished from circulating monocytes, themselves of bone marrow origin (Hess et al. 2004; Hickey and Kimura 1988; Hickey et al. 1992). Monocytes are attracted by chemotaxy to cross the vessel wall. Macrophage chemo-attracting protein-1 (MCP-1) is one of the principal chemokines implicated in this process, termed extravasation. MCP-1 also induces migration of neural stem cells (Widera et al. 2004). Once past the endothelium and eventual smooth muscle cell layer, the monocytes mature into PVMs or dendritic cells. Circulating dendritic precursors also exist; and the conversion of PVMs into dendritic cells also occurs in the perivascular layer. Some PVMs and choroid plexus macrophages also continue their voyage to reach the brain parenchyma, where they mature into microglia. However, both macrophages and microglial cells are also capable of self-renewal in their respective compartments and thus they are replenished only partly by infiltration of new monocytes from the circulation. However, microglia replace themselves at a much slower rate than macrophages. The entry of monocytes in the adult brain occurs mainly in the choroid plexus and arterioles in both brain and meninges (Ransohoff et al. 2003). Because the macrophages have a short life, one may wonder whether the molecules and debris absorbed by phagocytosis are released in the neural microenvironment. Experiments based on the injection of carbon particles (Indian Ink) into brain have demonstrated that carbon particles are kept inside brain macrophages for at least 2 years, a result that is not consistent with the short life of macrophages (Hickey 2001). A possible explanation is that carbon particles can be successively re-absorbed by new macrophages after the death of the old ones.

6.5 Functions

6.5.1 Known Functions of Brain Macrophages

PVMs and other macrophages are primary soldiers in the brain's immune defenses. These cells use conventional weapons such as phagocytic activity, which consists in the absorption and digestion of pathogens and cellular debris. Macrophages also act as APCs, cooperating with T lymphocytes to direct the specific immune response. Macrophages produce less conventional weapons that directly benefit neural cell functions, secreting signaling molecules such as cytokines, growth factors, and ECMs, the last of which influence growth factor activity in the neural environment. The resident CNS macrophages are not the only soldiers available to defend the brain after a challenge. As alluded to earlier, by a mechanism termed extravasation, which involves chemokines produced by vascular cells and resident PVMs, circulating monocytes enter the brain and accumulate at the sites of injury or physiological challenge. In the infiltrating pathways (vasculature, meninges, choroid plexus), the monocytes mature into macrophages, expand their lysosomal arsenal, and eventually become dendritic cells that produce MHC class II molecules, participating in the immune defense (Ransohoff et al. 2003). Ironically, infiltrating circulating monocytes, if infected by viruses, also serve as carriers for pathogens, as "trojan horses" that are welcomed by the brain, but they quickly betray their host after crossing the blood-brain barrier (BBB). Upon crossing the BBB, the HIV-infected macrophages insinuate through the neural tissue, a disastrous situation that leads to the destruction of neural cells and can ultimately trigger dementia (Williams et al. 2001). The infiltrating macrophages also participate in the progression of most neurodegenerative processes, accentuating the inappropriate composition of the altered neural environment. However, under physiological conditions, the infiltrating macrophages are beneficial to the CNS.

6.5.1.1 Phagocytosis

PVMs, other brain macrophages, and (to a lesser extent) microglia, via their numerous and sizeable lysosomes, have a large capacity for engulfing and digesting cellular, matrix, and pathogen debris. Macrophages phagocyte neural cells, vascular cells, and inflammatory cells that comprise macrophages themselves (Chan et al. 2003). To perform their phagocytic activity against pathogens, macrophages express receptors directed against bacterial and fungal components, such as peptidoglycan, teichoic acid, lipopolysaccharide (LPS), mannose (Galea et al. 2005), flagelin, and glucans, or viral components such as double-stranded RNA. Low pH, hydrolases, and peroxidases, and a lack of appropriate nutrients prevent the ingested parasitic organisms from replicating within the macrophage lysosomes. The phagocytic activity and release of hydrolases is linked with the immune function of macrophages (Cardella et al. 1974).

6.5.1.2 Immune Function

In all organs and tissues, including the brain, macrophages and dendritic cells act as APCs. Antigens that have been previously absorbed and digested are presented to the "attention" of the T lymphocytes, which then initiate the immune response. Macrophages produce growth factors and cytokines that create or enhance immune responses in collaboration with other immune cells. Among these growth factors, macrophage colony stimulating factor (M-CSF) and granulocyte macrophage (GM)-CSF induce chemotaxis, phagocytosis, proliferation, differentiation, and maturation of hematopoietic stem cells and leucocyte progenitors (Cairo et al. 1990). Interestingly, GM-CSF also participates in the differentiation of neural progenitor cells (Kim et al. 2004) and in brain inflammation (Franzen et al. 2004). Macrophages also produce vascular endothelial growth factor (VEGF), which increases vascular permeability and contributes to the afflux of lymphocytes and monocytes (Lee et al. 2002).

The presence of PVMs along the vascular adventitia of most blood vessels (Mercier et al. 2002) and their mobility (Cuff et al. 1996) make these cells particularly suitable for intervening rapidly at every level of the brain and initiating an appropriate immune response (Bechmann et al. 2001b; Mato et al. 1996). Migrating intraventricular macrophages also participate in the immune and inflammatory response of the neural tissue (Maxwell and McGadey 1988). Evidence that the brain microenvironment and macrophages/microglia intervene in the brain immune surveillance and defense has been intensively reviewed. The reader is invited to read the excellent reviews of Becher et al. (2000), Fabriek et al. (2005b), and Thomas (1999).

6.5.1.3 Production of Growth Factors, Cytokines, and Chemokines

Although not the only cells, macrophages produce several growth factors, including basic fibroblast growth factor (FGF-2), transforming growth factor-beta1 (TGF-β1), and tumor necrosis factor α (TNFα; Mato et al. 1998). In brain, FGF-2 is primarily produced by endothelial cells, ependymal cells, macrophages (Frautschy et al. 1991; Grotte et al. 2001), and choroid plexus cells (Johanson et al. 2001). Monocytes and macrophages also produce hormones that are identical to neuronally produced hormones, for example corticotropin releasing hormone (CRH), which locally activates the inflammatory response (De Souza, 1995). Macrophages produce inflammatory and anti-inflammatory cytokines, such as interleukin-1β (Angelov et al. 1998a,b; Bauer et al. 1993). It was previously thought that circulating cytokines have their effects on the brain only via circumventricular organs, which have a leaky, if any, BBB. It has been demonstrated that circulating cytokines induce the production of the same cytokines by PVMs, pia-arachnoid macrophages, and microglia. Because all these cells are inside the BBB, the cytokines released are directly accessible to glia and neurons. Cytokines have strong effects on brain function, generating fever, behavior changes, stress responses, and multiple effects due to their influence on the neuroendocrine system (for a review, see Mercier and Hatton 2004). Cytokines released by macrophages are also capable of inducing neuronal death by necrosis or apoptosis, astrocytosis, and the proliferation and differentiation of neural progenitor cells and NSCs. Brain macrophages are themselves activated by other immune cells, such as lymphocytes, which constantly co-enter the brain with monocytes during physiological (discussed in a later section) and pathological conditions. For example, lymphocytes produce growth hormone, which activates macrophages. In addition, brain macrophages attract circulating monocytes by producing the chemokine monocyte chemotactic protein-1 (MCP-1; Calvo et al. 1998). This is used by macrophages that have already been recruited into an inflammation site to further recruit other macrophages. In the neurohypophysis, macrophages also produce inducible nitric oxide synthase (iNOS; Gajkowska et al. 1999). Nitric oxide is known to serve as a neuromediator in brain.

6.5.1.4 Production and Degradation of the Extracellular Matrix

Macrophages are an important source of ECM molecules. Macrophages produce matrix metalloproteinase (MMP)-2 and MMP-9, two collagenases that are involved in basal lamina and ECM degradation. Macrophages also produce urokinase-type plasminogen activator (uPA), an enzyme used to clear apoptotic and necrotic cells, and cyclo-oxygenase-2 (COX-2; Elmquist et al. 1997). Macrophages also express iNOS (Mato et al. 1998). Nitric oxide, which is also supplied by perivascular nerve fibers, is released in the perivasculature, particularly abundantly in the arteries of the Circle of Willis, where it may intervene as a co-neurotransmitter (Edvinsson et al. 2001).

6.5.1.5 Repair After Injury

Until quite recently, it was thought that the CNS lacked any capability of self-repair after injury. Recent evidence, involving cell therapy with ensheathing glia, demonstrates the opposite: self-repair of the transected spinal cord in the adult rat (Ramon-Cueto 1998, 2000). In this model, anatomical recovery (axonal reconnection through 4 mm of missing tissue) preceded by several weeks the recovery of sensorimotor functions. Although the mechanisms of repair induced by injected ensheathing glia are not understood, it is likely that these cells release molecules that counterbalance an inhibitory spinal cord microenvironment resulting from the activity of local growth factors, cytokines, and ECM molecules (Becker and Becker 2002; Coumans et al. 2001; Krekoski et al. 2001). Jones et al. (2003) have shown that basal lamina molecules inhibit axonal growth. It is thought that the inflammation produced by fibroblasts and macrophages attracted by chemotaxis in the wounded area, and the associated glial scar, inhibit neural tissue repair. However, basal laminae also promote cytogenesis and repair throughout all organs and tissues in adulthood (Bunge 1987; Fujimoto et al. 1997; Gospodarowicz et al. 1981). It was recently demonstrated that macrophages produce trophic factors with both positive and negative effects on optic nerve regeneration (Yin et al. 2003). Macrophages also produce and release growth factors and ECM molecules promoting cytogenesis and cell differentiation (discussed in a later section). It is likely that connective tissue fibroblasts and macrophages and their adjacent basal laminae govern tissue plasticity everywhere in the body and thus both promote and inhibit cell proliferation, differentiation, and migration to an extent that is appropriate for the function of the tissue. It was recently shown that dendritic cells promote the repair of injured spinal cord (Mikami et al. 2004). The authors demonstrated that dendritic cells strongly induce the proliferation and survival of neural progenitor cell and neural stem cell activity.

6.5.2
Potential Functions of Brain Macrophages

6.5.2.1 Interactions with Meningeal/Vascular Cells, Neurons, and Astrocytes
Macrophages are strategically located at the interface between astrocytes and the connective tissue/CSF compartment. The CSF potentially carries information from all over the brain, and from peripheral tissues if the fluid that contacts macrophages is outside of the BBB (for example in the stroma of the choroid plexus). All macrophages have access to CSF, carriers of signaling molecules, and the basal lamina permits exchange of information with the other cell types. SEL macrophages/microglia also have access to CSF information by interstitial clefts that connect the ventricles to the extracellular space bathing the SEL cells (Brightman 1965, 2002). The astrocyte/macrophage contact occurs directly in the SEL or indirectly via the basal lamina located between the two cell types, at the pia/glia limitans interface or at the blood vessel surface.

Thus, macrophages and astrocytes can exchange information via signaling molecules that may be activated in the basal laminae. Eventually, macrophages can enter the neural tissue to further influence astrocytes and neurons. Migration of meningeal macrophages was observed in the supraoptic nucleus after dehydration (Mercier 2004). In the supraoptic nucleus, both macrophages located in pia and the infiltrated macrophages may produce growth factors and cytokines to signal astrocytes (and neurons?) of a homeostatic change (osmolarity change) and initiate morphological and functional changes in the neural tissue that will lead to osmotic regulation. It has been demonstrated that following dehydration, astrocytes retract processes from between adjacent neurosecretory neurons, which in turn have more opportunity to contact each other and coordinate the release of the neurohormones vasopressin and oxytocin (Hatton 2004). Another possible example of the influence of macrophages on neural tissue is given in the next section.

6.5.2.2 Do Macrophages Govern the Neural Stem Cell Niche in Adulthood?

Adult Neurogenesis Neurons and glia (both neural cells) are continuously generated in the brain of adult vertebrates (Altman 1962a,b; Ekstrom et al. 2001; Goldman and Nottebohm 1983; Lois and Alvarez-Buylla 1993; Luskin 1993). In mammals, the vast majority of neurogenesis occurs in the SEL of the lateral ventricles (Doetsch et al. 1997; Eriksson et al. 1998; Lois and Alvarez-Buylla, 1993) and dentate gyrus (Altman and Das 1965; Cameron et al. 1993; Seri et al. 2004). In the SEL of the lateral ventricle, the new neural cells are produced from self-renewing and pluripotent (with potential to produce neurons, astrocytes, oligodendrocytes) NSCs. These can be grown in culture as free-floating spherical aggregates, termed neurospheres, which contain new NSCs, neurons, astrocytes, and oligodendrocytes (Reynolds and Weiss 1992; Weiss et al. 1996). NSCs have been found throughout the entire ventricular system in adult mammals, but for unknown reasons, the SEL of the lateral ventricles is more mitotically active and supports the production of both neurons and glia, while the SEL of the other ventricles and spinal cord shows poor mitotic activity and favors gliogenesis.

NSCs are not clearly identified in vivo. Controversial data suggest that NSCs may be ependymocytes (Johansson et al. 1999), astrocytes (Doetsch et al. 1999), or radial glial cells originating from embryonic neuroepithelium (Mercle et al. 2004). Moreover, sufficient evidence has not been provided that neuroblasts, which are ultrastructurally identifiable unipotent neural precursor cells, derive from NSCs, although the prevalent theory suggests such a lineage (Alvarez-Buylla et al. 2001; Doetsch et al. 1999).

The Neurogenic Niche The concept of "microenvironment" or "niche" regulating NSCs proliferation and differentiation in the SEL was recently proposed (see Chapter 3). In support of this concept, it was shown that the fate of a precursor or stem cell depends on the niche in which it is housed. The induction mechanisms from the niche may even overrule the genetic cell commitment. For example, hippocampus-derived neuroblasts transplanted into the olfactory bulb differentiate into olfactory neurons and not hippocampus-type neurons (Suhonen et al. 1996). NSCs can even trans-differentiate into non-neural cells, for example into hematopoietic cells, after transplantation in bone marrow face antigen CD44 (Jones et al. 2000), a HSPG that binds the powerful neurogenic factors FGF-2 and HB-EGF (Bennett et al. 1995). Jones et al. (2000) further demonstrated that CD44 are intermediate affinity receptors for FGF-2 and HB-EGF, present growth factors to their specific receptors on adjacent cells, and upregulate the activity of these growth factors. These authors suggested that macrophages regulate the bioavailability of heparin-binding growth factors to trigger cell proliferation and differentiation at the appropriate time and location. Interestingly, Jones et al. (2000) also demonstrated in vitro that CD44 HSPG expression increases more than five-fold when differentiation from monocytes to macrophages is induced. If monocyte-derived macrophages of the CNS behave similarly, these cells may express HSPG after extravasation, once in contact with subpial, perivascular basal laminae, or fractones in the case of choroid plexus-derived supraependymal cells. Interestingly, Jonakait et al. (2000) demonstrated that macrophages express molecules that synergize with nerve growth factor and influence differentiation of progenitor cells. Based on our findings that: (1) SEL macrophages are systematically associated with fractones (Mercier et al. 2002, 2003), (2) perlecan is specifically found in fractones, and (3) heparin-binding neurogenic factors, including FGF-2, are specifically sequestered into fractones after intracerebrovascular injection (Mercier et al. unpublished observations), we propose that the mechanisms of HSPG/growth factor interactions are the means by which macrophages operate in neurogenesis during adulthood. As crucial sites of ECM/growth factor interactions, fractones may activate the neurogenic growth factors and initiate their presentation to the receptors of abutting NSCs and neural progenitor cells, or participate in the mechanisms of internalization of the growth factors into the cells for activation of mitosis and differentiation. Similar mechanisms have been previously described for the mitotic action of FGF-2, although not in the context of neural stem cell biology in adulthood (Allen and Rapraeger 2003; Reiland and Rapraeger, 1993).

Can Macrophages Trans-Differentiate into Neural Cells or Become Neural Stem Cells? It has been clearly demonstrated with both carbon and rhodamine labeling techniques that circulating monocytes enter the choroid plexus stroma, where they become macrophages. As macrophages, the cells cross the choroid epithelium, transit in the lumen of the lateral ventricle as Kolmer cells, attach to the ependymal surface as supraependymal cells, and infiltrate the lateral ventricle wall (Fig. 6.6 B). Some of these infiltrating cells mature into microglia in neonatal rodents (Ling 1979; Ling and Wong 1993). The choroid plexus-ventricle wall migration pathway has also been recognized in adult animals (for reviews, see Hickey 2001; Ransohoff et al. 2003), using models of bone marrow radiation chimeras (Lassmann et al. 1993) and bone marrow grafts from EGFP mice (Vallieres and Sawchenko 2003). All available data support the view that cell populations within the meninges, choroid plexus, as well as perivascular macrophages and microglia, are replaced by blood-derived cells throughout life. None of these studies support the view that bone marrow-derived cells trans-differentiate into neural cells after infiltration in the SEL, but the studies of Cogle et al. (2004)

and Mezey et al. (2000) appear to suggest just this. Indeed, several studies demonstrate that trans-differentiation from a cell type to another cell type is possible (Bjornson et al. 1999; Kondo et al. 2000). Our hypothesis is that supraependymal macrophages originating from circulating monocytes and transiting via the choroid plexus as Kolmer cells naturally infiltrate the lateral ventricle – the neurogenic niche – and have the ability to trans-differentiate into neural cells, or to acquire multipotency, i.e. to become NSCs. Supporting the possibility that monocytes potentially trans-differentiate into other cells after extravasation, it has been demonstrated in vitro that monocytes can transform into endothelial-like cells in the presence of angiogenic factors (Fernandez-Pujol et al. 2000; Haveman et al. 2003). Second, the neuroblasts present in the adult neurogenic niche, termed "A" cells in the nomenclature adopted by Doetsch et al. (1997), possess numerous ultrastructural features in common with macrophages. Figure 6.7A shows typical clusters of A cells present in the SEL at the caudate putamen/corpus callosum interface. The A cells display electron-lucent bands at their periphery, have a dense heterochromatin with some clustering, narrow perinuclear cytoplasm, and one or two large and short processes filled with numerous organelles, including a high density of clustered free ribosomes (Fig. 6.7B). Macrophages present in the SEL of the ventricular system possess similar ultrastructural characteristics, displaying one or two large processes (Figs. 6.6C and 6.7D), a dark heterochromatin, and the characteristic electron-lucent lateral bands (Figs. 6.6D and 6.7C) and dense clusters of free ribosomes encountered in no other cell type in brain, besides neuroblasts and cells of the macrophage family. Interestingly, dendritic cells in the choroid plexus stroma, as well as epiplexus cells also display these numerous clusters of free ribosomes in their processes (Fig. 6.6A, insets). The only noticeable difference between cells of the macrophage family and neuroblasts is the presence of few large lysosomes in macrophages (Figs. 6.6C, D and 6.7C, D).

The intriguing perspective is that macrophages may enter the neurogenic niche via the lateral ventricle lumen and trans-differentiate into neuroblasts, or even acquire neural stem cell properties upon influence of the neurogenic niche microenvironment. It is possible that macrophages lose their lysosomes upon trans-differentiation. In response to injury, microglia can express the hematopoietic stem cell marker CD34 (Labedy et al. 2005). That the proteoglycan NG2, a marker of oligodendrocyte precursor cells, is expressed by macrophages (Bu et al. 2001) is also going in the same direction.

6.5.2.3 Role of Macrophages in CNS Angiogenesis

Blood vessels constantly reconstitute in the adult CNS. New endothelial cells, smooth muscle cells, pericytes, and PVMs are generated to replace dead cells, which do not have a long lifespan (all these cells have to be replaced within weeks). New blood vessels are also generated during adulthood, in both brain and meninges. All components required for angiogenesis are therefore present in the CNS. PVMs appear ideally located to synthesize and release ECM molecules necessary for angiogenesis. Similar mechanisms implicating macrophages appear to initiate and regulate adult angiogenesis (Anghelina et al. 2004; Moldovan 2002 a, b). These authors suggest that macrophages are "architects of angiogenesis", scaffolding and preparing the orderly assemblage of newly generated endothelial cells. Other authors suggest that endothelial cells replenish from circulating bone marrow-derived monocytes and dendritic cells (Fernandez et al. 2000; Havemann et al. 2003). These authors have demonstrated in vitro that both monocytes and dendritic cells can transform into endothelial cells in the presence of angiogenic growth factors. The phenotypic overlap between monocytes and endothelial cells further supports the possibility of lineage relationships between the two cell types (Schmeisser et al. 2003). Again, the HSPG perlecan may be involved, activating the growth factors that lead to angiogenesis (Jiang et al. 2004). Whatever the case is, the paradigm that postnatal angiogenesis is exclusively caused by outgrowth of endothelial cells from preformed vessels has to be modified (Havemann et al. 2003).

6.5.2.4 Role of Macrophages in CNS Plasticity

We suggest that PVMs and pial macrophages behave similarly to SEL macrophages and govern CNS plasticity, i.e. orchestrate morphological and functional changes that permit astrocytes and neurons to adapt to physiological challenges, respond to any biological requirement for a new neural architecture, and respond to injury by attempting repair. PVMs produce growth factors and ECM molecules that may participate in the assembly and dynamics of basal laminae, which serve as substrate for binding and activation of growth factors and cytokines involved in the morphological modification required for proper CNS functioning. In every part of the CNS the cells facing the basal laminae and subadjacent macrophages are astrocytes (astrocytes of the glia limitans at the brain surface and at the border of every major brain structure, perivascular astrocytes, and SEL astrocytes). The distribution of basal laminae and their potential function in brain is described by Mercier and Hatton (2004). How astrocytes are implicated in CNS plasticity is reviewed by Hatton (2004) and Mercier (2004). Astrocytes are ideally located to serve as an obligatory information relay between meningeal cells (macrophages and fibroblasts) and neurons. This is facilitated by the neural architecture, neurons never directly contacting basal laminae, except in circumventricular organs where the total retraction of astrocyte endfeet facilitates the passage of neurohormones or neuromediators.

It is important to consider the possibility that fibroblasts may assist macrophages in the CNS extraparenchymal compartment in both providing information captured in the brain fluids and producing ECM molecules and growth factors appropriate to the initiation of plasticity. It has been shown that meningeal cells, both macrophages and fibroblasts, produce ECM molecules for the formation of the basal lamina (Sievers et al. 1993). Fibroblasts produce numerous collagens, growth factors (for a review, see Mercier and Hatton 2004) and morphogens such as semaphorins. Meningeal cells are involved in nerve regeneration (Shearer and Fawcett 2001) and in the formation and dynamics of the glia limitans (Struckhoff 1995). Fibroblasts are connected to each other by their processes (Mercier and Hatton 2001) via functional gap junctions (Spray et al. 1991) to form a network throughout the CNS extraparenchyma (meninges, meningeal separations between the brain major structures, perivasculature). Gap junctional inter-fibroblast communication likely facilitates the propagation of information available from the CSF and blood to inform the neighboring macrophages. Interestingly, it has been shown that fibroblast-produced collagen-1 fibrils appear to participate in the mediation of growth factor by HSPG. Perlecan promotes FGF-2 delivery in collagen-1 fibrils (Yang et al. 2005).

6.6 Conclusion: Macrophages as Architects of the CNS Throughout Adulthood

The association of CNS macrophages with basal laminae throughout the CNS, together with the production by these cells of ECM molecules that interact with proliferating and differentiating growth factors and cytokines suggest that macrophages are involved in tissue maintenance, cell replacement, and morphological/functional plasticity in adulthood. It is tempting to conclude that macrophages achieve their architectural functions by producing and secreting appropriate signaling molecules, attracting signaling molecules secreted by other cells, and supervising the ECM/growth factors interactions that take place in the adjacent basal laminae. Regulation of the signaling molecules' activity, presentation to neuronal and astrocytic receptors, and eventual intracellular internalization of the HSPG/growth factors complexes constitute one of the basic mechanisms by which the macrophages could influence CNS plasticity. We further suggest that PVMs and meningeal macrophages form a highly organized network throughout the CNS. Gap junction intercellular communication would allow the macrophage network to sense blood, CSF, and neurone-borne signals, analyze eventual biological deficit, tissue injury, or entry of pathogens into the CNS, produce signaling molecules, and ultimately elaborate the appropriate response to alleviate the challenge. The same organization and mechanisms may constantly serve neural plasticity, orchestrating the production, differentiation, migration, and orderly integration of new vascular and neural cells throughout the CNS, as well as controlling the growth and destruction of cell processes and synapses necessary for the dynamic activity of the CNS.

References

Allen BL, Rapraeger AC **2003**, Spatial and temporal expression of heparan sulfate in mouse development regulates FGF and FGF receptor assembly, *J. Cell Biol.* 163, 637–648.

Altman J **1962a**, Autoradiographic and histological studies of postnatal neurogenesis. IV. Cell proliferation and migration in the anterior forebrain, with special reference to persisting neurogenesis in the olfactory bulb, *J. Comp. Neurol.* 137, 433–458.

Altman J **1962b**, Are neurons formed in the brain of adult mammals? *Science* 135, 1127–1128.

Altman J, Das GD **1965**, Autoradiographic and histological evidence of postnatal hippocampal neurogenesis in rats, *J. Comp. Neurol.* 124, 319–335.

Alvarez-Buylla A, Garcia-Verdugo JM, Tramontin AD **2001**, A unified hypothesis on the lineage of neural stem cells, *Nat. Rev. Neurosci.* 2, 287–293.

Amoureux MC, Cunningham BA, Edelman GM, Crossin KL **2000**, N-CAM binding inhibits the proliferation of hippocampal progenitor cells and promotes their differentiation to a neural phenotype, *J. Neurosci.* 20, 3631–3640.

Angelov DN, Neiss WF, Streppel M, Walther M, Guntinas-Lichius O, Stennert E **1996**, ED2-positive perivascular cells act as neuronophages during delayed neuronal loss in the facial nucleus of the rat, *Glia* 16, 129–139.

Angelov DN, Walther M, Streppel M **1998a**, The cerebral perivascular cells, *Adv. Anat. Embryol. Cell Biol.* 147, 1–87.

Angelov DN, Walther M, Streppel M, Guntinas L, Van D, Stennert E, Neiss WF **1998b**, ED-2 positive perivascular phagocytes produce interleukin-1 beta during delayed neuronal loss in the facial nucleus of the rat, *J. Neurosci. Res.* 54, 820–827.

Anghelina M, Krishnan P, Moldovan L, Moldovan NI **2004**, Monocytes and macrophages from branched cell columns in Matrigel: implications for a role in neovascularization, *Stem Cells Dev.* 13, 665–673.

Ardavin C **2003**, Origin, precursors and differentiation of mouse dendritic cells, *Nat. Rev. Immunol.* 3, 582–591.

Bauer J, Berkenbosch F, Van D, Dijkstra CD **1993**, Demonstration of interleukin-1 beta in Lewis rat brain during experimental allergic encephalomyelitis by immunocytochemistry at the light and ultrastructural level, *J. Neuroimmunol.* 48, 13–21.

Beagley GH, Hatton GI **1992**, Rapid morphological changes in supraoptic nucleus and posterior pituitary induced by a single hypertonic saline injection, *Brain Res. Bull.* 28, 613–618.

Becher B, Prat A, Antel JP **2000**, Brain-immune connection: immuno-regulatory properties of CNS-resident cells, *Glia* 29, 293–304.

Bechmann I, Kwidinski E, Kovac AD **2001a**, Turnover of rat brain perivascular cells, *Exp. Neurol.* 168, 242–249.

Bechmann I, Priller J, Kovac A **2001b**, Immune surveillance of mouse brain perivascular spaces by blood-borne macrophages, *Eur. J. Neurosci.* 14, 1651–1658.

Becker GC, Becker T **2002**, Repellent guidance of regenerating optic axons by chondroitin sulfate glycoaminoglycans in zebrafish, *J. Neurosci.* 22, 842–853.

Bennett K, Jackson DG, Simon JC, Tankzos E, Peach R, Modrell B, Stamenkovic I, Plowman G, Aruffo A **1995**, *J. Cell Biol.* 107, 743–750.

Beschomer R, Nguyen TD, Gozalan F, Pedal I, Mattem R, Schluesener HJ, Meyerman R, Schwab JM **2002**, CD14 expression by activated parenchymal microglia/macrophages and infiltrating monocytes following human traumatic brain injury, *Acta Neuropathol.* 103, 541–549.

Bjornson CR, Rietze RL, Reynolds BA, Vescovi AL **1999**, Turning brain into blood: an hematopoetic fate adopted by adult neural stem cells in vivo, *Science* 283, 534–537.

Bleier R, Albrecht R, Cruce JA **1975**, Supraependymal cells of hypothalamic third ventricle: identification as resident macrophages of the brain, *Science* 189, 299–301.

Bleier R, Siggelkow I, Albrecht R **1982**, Macrophages of hypothalamic third ventricle. I. Functional characterization of supraependymal cells in situ, *J. Neuropathol. Exp. Neurol.* 41, 315–329.

Bragg DC, Hudson LC, Liang YH, Tompkins LB, Fernandes A, Meeker RB **2002**, Choroid plexus macrophages proliferate and release toxic factors in response to feline immunodeficiency virus, *J. Neurovirol.* 8, 225–239.

Brickman YG, Ford MD, Small DH, Barlett PF, Nurcombe V **1995**, Heparan sulfate mediate the binding of basic fibroblast growth factor to a specific receptor on neural precursor cells, *J. Biol. Chem.* 42, 24941–24948.

Brightman MW **1965**, The distribution within the brain of ferritin injected into cerebrospinal fluid compartments: II parenchymal distribution, *Am. J. Anat.* 117, 193–220.

Brightman MW **2002**, The brain interstitial clefts and their glial walls, *J. Neurocytol.* 31, 595–603.

Bu J, Akhtar N, Nishiyama A **2001**, Transient expression of the NG2 proteoglycan by a subpopulation of activated macrophages in a excitotoxic hippocampal lesion, *Glia* 34, 296–310.

Bunge RP **1987**, Tissue culture observations relevant to the study of axon-Schwann cell interactions during peripheral nerve development and repair, *J. Exp. Biol.* 132, 21–34.

Cairo MS, Vande Ven C, Toy C, Mauss D, Sheikh K, Kommareddy S, Modanlou H **1990**, Lymphokines: enhancement by granulocyte-macrophage and granulocyte colony-stimulating factors of neonatal myeloid kinetics and functional activation of polymorphonuclear leukocytes, *Rev. Infect. Dis.* 12, S492–S497.

Calvo CF, Yoshimura T, Gelman M, Mallat M **1998**, Production of monocyte chemotactic protein-1 by rat brain macrophages, *J. Exp. Med.* 188, 1359–1368.

Getchell TV, Shah DS, Parting JV, Subhedar NK, Getchell ML **2002**, Leukemia inhibiting factor mRNA expression is upregulated in macrophages and olfactory receptor neurons after target ablation, *J. Neurosci. Res.* 67, 246–254.

Gilbert SF **1994**, The cellular basis of morphogenesis, in Developmental Biology, ed. Gilbert SF, Sinauer Associates, Sunderland, Mass., pp. 77–112.

Gilbert SF **2001**, Continuity and change: paradigm shifts in neural induction, *Int. J. Dev. Biol.* 45, 155–164.

Goldman SA, Nottebohm F **1983**, Neuronal production, migration, and differentiation in a vocal control nucleus of the adult female canary brain, *Proc. Natl Acad. Sci. USA* 80, 2390–2394.

Gordon MY, Ryley GP, Watt SM, Greaves MS **1987**, Compartmentalization of a haematopoietic growth factor (GM-CSF) by glycosaminoglycans in the bone marrow microenvironment, *Nature* 326, 403–405.

Gospodarowicz D, Fujii DK, Giguere L, Savion N, Tauber JP, Vlodavsky I **1981**, The role of the basal lamina in cell attachment, proliferation and differentiation. Tumor cells vs normal cells, *Prog. Clin. Biol. Res.* 75A, 95–132.

Graeber MB, Streit WJ, Kreutzberg GW **1989**, Identity of ED2-positive perivascular cells in rat brain, *J. Neurosci. Res.* 22, 103–106.

Graeber MB, Streit WJ, Kiefer R, Schoen SW, Kreutzberg GW **1990**, New expression of myelomonocytic antigens by microglia and perivascular cells following lethal motor neuron injury, *J. Neuroimmunol.* 27, 121–132.

Gritti A, Parati EA, Cova L, Frolichsthal P, Galli R, Wanke E, Favarelli L, Morasutti DJ, Roisen F, Nickel DD, Vescovi AL **1996**, Multipotential stem cells from the adult mouse brain proliferated and self-renew in response to basic fibroblast growth factor, *J. Neurosci.* 16, 1091–1100.

Grotte C, Meisinger C, Claus P **2001**, In vivo expression and localization of the fibroblast growth factor system in the intact and lesioned rat peripheral nerve and spinal ganglia, *J. Comp. Neurol.* 434, 342–357.

Guillemin GJ, Brew BJ **2004**, Microglia, macrophages, and pericytes: a review of function and identification, *J. Leukoc. Biol.* 75, 388–397.

Harandi M, Didier M, Aguera M, Calas A, Belin MF **1986**, GABA and serotonin (5-HT) pattern in the supraependymal fibers of the rat epithalamus: combined radioautographic and immunocytochemical studies. Effects of 5-HT content on [^{3}H]GABA accumulation, *Brain Res.* 370, 241–249.

Hatton GI **1997**, Function-related plasticity in hypothalamus, *Annu. Rev. Neurosci.* 20, 375–397.

Hatton GI **2004**, Morphological plasticity of astroglial/neuronal interactions: functional implications, in *Glial-Neuronal Signaling*, eds. GI Hatton, V Parpura, Kluwer, Amsterdam, pp. 99–124.

Havemann K, Pujol BF, Adamkiewicz J **2003**, In vitro transformation of monocytes and dendritic cells into endothelial-like cells, in *Novel Angiogenic Mechanisms: Role of Circulating Progenitor Endothelial Cells*, ed. NI Moldovan, Kluwer Academic/Plenum, New York.

Hess DC, Abe T, Hill WD, Martin Studdard A, Carothers J, Masuya M, Fleming PA, Drake CJ, Ogawa M **2004**, Hematopoietic origin of microglial and perivascular cells in brain, *Exp. Neurol.* 186, 134–144.

Hickey WF **2001**, Basic principles of immunological surveillance of the normal central nervous system, *Glia* 36, 118–124.

Hickey WF, Kimura H **1988**, The perivascular microglial cells of the CNS are bone marrow derived and present antigens in vivo, *Science* 239, 290–292.

Hickey WF, Vass K, Lassmann H **1992**, Bone marrow derived elements in the central nervous system: an immunohistochemical and ultrastructural survey of rat chimeras, *J. Neuropathol. Exp. Neurol.* 51, 246–256.

Hienola A, Pekkhanen M, Raulo E, Vantolla P, Rauvala H **2004**, HB-GAM inhibits proliferation and enhances differentiation of neural stem cells, *Mol. Cell Neurosci.* 26, 75–88.

Hirunagi K, Uryu K, Fujioka T **1989**, Supraependymal cells and fibers in the third ventricle of the domestic chicken. A scanning electron microscopy study, *Z. Mikrosk. Anat. Forsch.* 103, 529–539.

Hoek A, Allaerts W, Leenen PJM, Schoemaker J, Drexhage HA **1997**, Dendritic cells and macrophages in the pituitary and the gonads. Evidence for their role in the fine regulation of the reproductive endocrine response, *Eur. J. Endocrinol.* 136, 8–24.

Jiang X, Multhaupt H, Chan E, Schaefer L, Schaefer RM, Couchman JR **2004**, Essential contribution of tumor-derived perlecan to epidermal tumor growth and angiogenesis, *Histochem. Cytochem.* 52, 1575–1590.

Jin K, Mao XO, Sun Y, Xie L, Jin L, Nishi E, Klagsbrun M, Greenberg DA **2002**, Heparin-binding epidermal growth factor like growth factor: hypoxia inducible expression in vitro and stimulation of neurogenesis in vitro and in vivo, *J. Neurosci.* 22, 5365–5373.

Johanson CE, Gonzalez AM, Stopa EG **2001** Water-imbalance-induced expression of FGF-2 in fluid-regulatory centers; choroid plexus and neurohypophysis, *Eur. J. Pediatr. Surg. Suppl.* 1, S37–S38.

Johansson CB, Momma S, Clarke DL, Risling M, Lendahl U, Frisen J **1999**, Identification of a neural stem cell in the adult mammalian central nervous system, *Cell* 96, 25–34.

Jonakait GM, Wen Y, Wan Y, Ni L **2000**, Macrophage cell-conditioned medium promotes cholinergic differentiation of undifferentiated progenitors and synergizes with nerve growth factor action in the developing basal forebrain, *Exp. Neurol.* 161, 285–296.

Jones LL, Sajed D, Tuszynski MH **2003**, Axonal regeneration through regions of chondroitin sulfate proteoglycan deposition after spinal cord injury: a balance of permissiveness and inhibition, *J. Neurosci.* 23, 9276–9288.

Jones M, Tussey L, Athanasou N, Jackson DG **2000**, Heparan sulfate proteoglycan isoforms of the CD44 hyaluronan receptor induced in human inflammatory macrophages can function as paracrine regulators of fibroblast growth factor action, *J. Biol. Chem.* 275, 7964–7974.

Kearns SM, Laywell ED, Kukekov VK, Steindler DA **2003**, Extracellular effects on neurosphere cell motility, *Exp. Neurol.* 182, 240–244.

Kida S, Steart PV, Zhang ET **1993**, Perivascular cells act as scavengers in the cerebral perivascular spaces and remain distinct from pericytes, microglia and macrophages, *Acta Neuropathol.* 85, 646–652.

Kiefer R, Schweitzer T, Jung S, Toyka KV, Harting HP **1998**, Sequential expression of transforming growth factor-beta1 by T-cells, macrophages, and microglia in rat spinal cord during autoimmune inflammation, *J. Neuropathol. Exp. Neurol.* 57, 385–395.

Kim JK, Choi BH, Park HC, Park SR, Kim YS, Yoon SH, Park HS, Kim EY, Ha Y **2004**, Effects of GM-CSF on the neural progenitor cells, *Neuroreport* 15, 2161–2165.

Kondo M, Sherer DC, Miyamoto T, King AG, Akashi K, Sugamura K, Weissman IL **2000**, Cell fate conversion of lymphoid-committed progenitors by instructive action of cytokines, *Nature* 407, 383–386.

Krekoski CA, Neubauer D, Zuo J, Muir D **2001**, Axonal regeneration into acellular nerve grafts is enhanced by degradation of chondroitin sulfate proteoglycans, *J. Neurosci.* 21, 6206–6213.

Kuhn HG, Winkler J, Kempermann G, Tha l LJ, Gage FH **1997**, Epidermal growth factor and fibroblast growth factor-2 have different effects on neural progenitors in the adult rat brain, *J. Neurosci.* 17, 5820–5829.

Labedy R, Wirenfeldt M, Dalmau I, Gregersen R, Garcia-Ovejero D, Babcock A, Owens T, Finsen B **2005**, Proliferating resident microglia express the stem cell antigen CD34 in response to acute neural injury, *Glia* 50, 121–131.

Lassmann H, Schmied M, Vass K, Hickey WF **1993**, Bone marrow derived elements and resident microglia in brain inflammation, *Glia* 7, 19–24.

Lee TH, Avraham H, Lee SH, Avraham S **2002**, Vascular endothelial growth factor modulates neutrophil transendothelial migration via up-regulation of interleukin-8 in human brain microvascular endothelial cells, *J. Biol. Chem.* 22, 10445–10451.

Lim DA, Tramotin AD, Trevejo JM, Herrera DG, Garcia-Verdugo JM, Alvarez-Buylla A **2000**, Noggin antagonizes BMP signaling to create a niche for adult neurogenesis, *Neuron* 28, 713–726.

Ling EA **1976**, Some aspects of ameboid microglia in the corpus callosum and neighbouring regions of neonatal rats, *J. Anat.* 121, 29–45.

Ling EA **1979**, Ultrastructure and origin of epiplexus cells in the telencephalic choroid plexus of postnatal rat studies by intravenous injection of carbon particles, *J. Anat.* 129, 479–492.

Ling EA, Wong WC **1993**, The origin and nature of ramified and amoeboid microglia: a historical review and current concepts, *Glia* 7, 9–18.

Ling EA, Kaur C, Lu J **1998**, Origin, nature, and some functional considerations of intraventricular macrophages with special reference to the epiplexus cells, *Microsc. Res. Tech.* 41, 43–56.

Lois C, Alvarez-Buylla A **1993**, Proliferating subventricular zone cells in the adult mammalian forebrain can differentiate into neurons and glia, *Proc. Natl Acad. Sci. USA* 90, 2074–2077.

Lu J, Kaur C, Ling EA **1993**, Intraventricular macrophages in the lateral ventricles with special reference to epiplexus cells: a quantitative analysis and their uptake of fluorescent tracer injected intraperitoneally in rays of different ages, *J. Anat.* 183, 405–414.

Luskin MB **1993**, Restricted proliferationand migration of postnatally generated neurons derived from the forebrain subventricular zone, *Neuron* 11, 173–189.

Makatsori E, Lamari FN, Theocharis AD **2003**, Large matrix proteoglycans, versican and perlecan, are expressed and secreted by human leukemic monocytes, *Anticancer Res.* 23, 3303–3309.

Mato M, Ookawara S, Kurihara K **1980** Uptake of exogenous substances and marked infoldings of the fluorescent granular pericyte in cerebral fine vessels, *Am. J. Anat.* 157, 329–332.

Mato M, Ookawara S, Sakamoto A, Aikawa E, Ogawa T, Mitsuhashi U, Masuzawa T, Suzuki H, Honda M, Yazaki Y, Watanabe E, Luoma J, Yla-Herttuali S, Fraser I, Gordon S, Kodama T **1996**, Involvement of specific macrophage-lineage cells surrounding arterioles in barrier and scavenger function in brain cortex, *Proc. Natl Acad. Sci. USA* 93, 3269–3274.

Mato M, Sakamoto A, Ookawara S, Takeuchi K, Suzuki K **1998**, Ultrastructural and immunohistochemical changes of fluorescent granular perithelial cells and the interaction of FGP cells to microglia after lipopolysaccharide administration, *Anat. Rec.* 251, 330–338.

Matyszak MK, Perry VH **1996**, The potential role of dendritic cells in immune-mediated inflammatory diseases in the central nervous system, *Neuroscience* 74, 599–608.

Matyszak MK, Townsend MJ, Perry VH **1997**, Ultrastructural studies of an immune-mediated inflammatory response in the CNS parenchyma directed against a non-CNS antigen, *Neuroscience* 78, 549–560.

Maxwell WL, McGadey J **1988**, Response of intraventricular macrophages after a penetrant cerebral lesion, *J. Anat.* 160, 145–155.

McKeever PE, Ballantine JD **1978**, Macrophages migration through the brain parenchyma to the perivascular space following particle ingestion, *Am. J. Pathol.* 93, 153–164.

McMaster WR, Williams AM **1979**, Identificaion of Ia glycoproteins in the rat thymus and purification of rat spleen, *Eur. J. Immunol.* 9, 426–433.

McMenamin PG **1999**, Distribution and phenotype of dendritic cells and resident tissue macrophages in the dura mater, leptomeninges, and choroid plexus of the rat brain as demonstrated in whole mount preparations, *J. Comp. Neurol.* 405, 553–562.

McMenamin PG, Wealthall RJ, Deverall M, Cooper SJ, Griffin B **2003**, Macrophages and dendritic cells in the rat meninges and choroid plexus: three dimensional localization by environmental scanning electron microscopy and confocal microscopy, *Cell Tissue Res.* 313, 259–269.

Mercier F **2004**, Astroglia as a modulation interface between meninges and neurons, in *Glial-Neuronal Signaling*, eds. GI Hatton, V Parpura, Kluwer, Amsterdam, pp. 125–162.

Mercier F, Hatton GI **2001**, Connexin 26 and bFGF are primarily expressed in subpial and subependymal layers in adult brain parenchyma: roles in stem cell proliferation and morphological plasticity? *J. Comp. Neurol.* 431, 88–104.

Mercier F, Hatton GI **2004**, Meninges and perivasculature as mediators of CNS plasticity, in Hertz L. (ed.) *Non-Neuronal Cells in the Nervous System: Function and Dysfunction*, Elsevier, Amsterdam, pp. 215–253.

Mercier F, Kitasako JT, Hatton GI **2002**, Anatomy of the brain neurogenic zones revisited: fractones and the fibroblast/macrophage network, *J. Comp. Neurol.* 451, 170–188.

Mercier F, Kitasako JT, Hatton GI **2003**, Fractones and other basal laminae in the hypothalamus, *J. Comp. Neurol.* 455, 324–340.

Mercle FT, Tramontin AD, Garcia-Verdugo JM, Alvarez-Buylla A **2004**, Radial glia give rise to adult neural stem cells in the subventricular zone, *Proc. Natl Acad. Sci. USA* 101, 17528–17532.

Mezey VA, Chandross KJ, Harta GN, Maki RA, McKercher SR **2000**, Turning blood into brain: cells bearing neuronal antigens generated in vivo from bone marrow, *Science* 290, 1779–1782.

Mikami Y, Okano H, Sakaguchi M, Nakamura M, Shimazaki T, Okano HJ, Kawakami Y, Toyama Y, Toda M **2004**, Implantation of dendritic cells in injured adult spinal cords results in activation of endogenous neural stem/progenitor cells leading to de novo neurogenesis and functional recovery, *J. Neurosci. Res.* 15, 453–465.

Miller JD, Cummings J, Maresh GA, Walker DG, Castillo GM, Ngo C, Kimata K, Kinsella MG, White TM, Snow AD **1997**, Localization of perlecan (or a perlecan-related macromolecule) to isolated microglia in vitro and to microglia/macrophages following infusion of beta-amyloid protein into rodent hippocampus, *Glia* 21, 228–243.

Moldovan NI **2002a**, Current priorities in the research of circulating pre-endothelial cells, *Adv. Exp. Med. Biol.* 522, 1–8.

Moldovan NI **2002b**, State of the art review on vascular stem cells and angiogenesis. Role of monocytes and macrophages in adult angiogenesis: a light at the tunnel's end, *J. Hematother. Stem Cell Res.* 11, 179–194.

Nordeen EJ, Nordeen KW **1989**, Estrogen stimulates the incorporation of new neurons into avian song nuclei during adolescence, *Dev. Brain Res.* 49, 27–32.

Oviedo-Orta E, Evans WH **2004**, Gap junctions and connexin-mediated communication in the immune system, *Biochem. Biophys. Acta* 1662, 102–112.

Perry VH, Gordon S **1987**, Modulation of CD4 antigen on macrophages and microglia in rat brain, *J. Exp. Med.* 166, 1138–1143.

Perry VH, Brown MC, Gordon S **1987**, The macrophage response to central and peripheral nerve injury. A possible role for macrophages in regeneration, *J. Exp. Med.* 165, 1218–1223.

Polacek D, Lal R, Volin MV, Davies PF **1993**, Gap junctional communication between vascular cells. Induction of connexin43 messenger RNA in macrophage foam cells of atherosclerotic lesions, *Am. J. Pathol.* 142, 593–606.

Popovich PG, van Rooijen N, Hickey WF, Preidis G, McGaughy V **2003**, Hematogenous macrophages express CD8 and distribute to regions of lesion cavitation after spinal cord injury, *Exp. Neurol.* 182, 275–287.

Ramon-Cueto A, Plant GW, Avila J, Bunge M **1998**, Long distance axonal regeneration in the transected adult rat spinal cord is promoted by olfactory ensheathing glia transplants, *J. Neurosci.* 18, 3803–3815.

Ramon-Cueto A, Cordeo MI, Santos-Benito FF, Avila J **2000**, Functional recovery of paraplegic rats and motor axon regeneration in their spinal cord by olfactory ensheathing glia, *Neuron* 25, 425–435.

Ransohoff RM, Kivisakk P, Kidd G **2003**, Three or more routes for leukocyte migration into the central nervous system, *Nat. Rev. Immunol.* 3, 569–581.

Rasika S, Nottebohm F, Alvarez-Buylla A **1994**, Testosterone increases the recruitment and/or survival of new high vocal center neurons in adult female canaries, *Proc. Natl Acad. Sci.* 91, 7854–7858.

Reape TJ, Wilson VJ, Kanczler JM, Ward JP, Burnand KG, Thomas CR **1997**, Heparin-binding epidermal growth factor-like growth factor mRNA and protein in human atherosclerotic tissue, *J. Mol. Cell Cardiol.* 29, 1639–1648.

Reiland J, Rapraeger AC **1993**, Heparan sulfate proteoglycan and FGF receptor target basic FGF to different intracellular destinations, *J. Cell Sci.* 105, 1085–1093.

Reynolds BA, Weiss S **1992**, Generation of neurons and astrocytes from isolated cells of the mammalian nervous system, *Science* 255, 1707–1710.

Rio-Hortega P del **1932**, Microglia, in *Cytology and Cellular Pathology of the Nervous System*, ed. W Penfield, Hoeber, New York, pp. 481–534.

Roberts R, Gallagher J, Spooncer E, Allen TD, Bloomfield F, Dexter TM **1988**, Heparan sulfate bound growth factors: a mechanism for stromal cells mediated haemopoiesis, *Nature* 322, 376–378.

Rouach N, Calvo CF, Glowinski J, Giaume C **2002**, Brain macrophages inhibit gap junctional communication and downregulate connexin 43 expression in cultured astrocytes, *Eur. J. Neurosci.* 15, 403–407.

Sato T, Inoue K **2000**, Dendritic cells in the rat pituitary gland evaluated by the use of monoclonal antibodies and electron microscopy, *Arch. Histol. Cytol.* 63, 291–303.

Schilz JC, Sawchenko PE **2003**, Signaling the brain in systemic inflammation: the role of perivascular cells, *Front. Biosci.* 8, 1321–1329.

Schingo T, Cregg C, Enwere E, Fujikawa H, Hassam R, Geary C, Cross JC, Weiss S **2003**, Pregnancy-stimulated neurogenesis in the adult female forebrain mediated by prolactin, *Science* 299, 117–120.

Schmeisser A, Graffy C, Daniel WG, Strasser RH **2003**, Phenotypic overlap between monocytes and vascular endothelial cells, in *Novel Angiogenic Mechanisms: Role of Circulating Progenitor Endothelial Cells*, ed. NI Moldovan, Kluwer Academic/Plenum, New York, pp. 59–74.

Scott DE **1999**, Post-traumatic migration and emergence of a novel cell line upon the ependymal surface of the third cerebral ventricle in the adult mammalian brain, *Anat. Rec.* 256, 233–241.

Seri B, Garcia-Verdugo JM, Collado-Morente L, McEwen BS, Alvarez-Buylla A **2004**, Cell types, lineage, and architecture of the germinal zone in the adult dentate gyrus, *J. Comp. Neurol.* 478, 359–378.

Shearer MC, Fawcett JW **2001**, The astrocyte/meningeal interface – a barrier to successful nerve regeneration, *Cell Tissue Res.* 305, 267–273.

Sievers J, Pehlemann FW, Gude S, Berry M **1993**, Meningeal cells organize the superficial glia limitans of the cerebrum and produce components of both the interstitial matrix and the basement membrane, *J. Neurocytol.* 23, 135–149.

Spray DC, Moreno AP, Kessler JA, Dermietzel R **1991**, Characterization of gap junctions between cultured leptomeningeal cells, *Brain Res.* 568, 1–14.

Steinman RM **1991**, The dendritic cell system and its role in immunogenicity, *Annu. Rev. Immunol.* 9, 271–296.

Struckhoff G **1995**, Cocultures of meningeal and astrocytic cells – a model for the formation of the glial-limiting membrane, *Int. J. Dev. Neurosci.* 13, 595–606.

Suhonen JO, Peterson DA, Ray J, Gage FH **1996**, Differentiation of adult hippocampus-derived progenitors into olfactory neurons in vivo, *Proc. Natl Acad. Sci. USA* 85, 141–145.

Thomas WE **1999**, Brain macrophages: on the role of the pericytes and perivascular cells, *Brain Res. Rev.* 31, 42–57.

Vallieres L, Sawchenko PE **2003**, Bone marrow-derived cells that populate the adult mouse brain preserve their hematopoietic identity, *J. Neurosci.* 23, 5197–5207.

Vallieres L, Campbell IL, Gage FH, Sawchenko PE **2002**, Reduced hippocampal neurogenesis in adult transgenic mice with chronic astrocytic production of interleukin-6, *J. Neurosci.* 22, 486–492.

Vandenabeele F, Creemers J, Lambrichts I **1996**, Ultrastructure of the human spinal arachnoid mater and dura mater, *J. Anat.* 189, 417–430.

Walker WS **1999**, Separate precursor cells for macrophages and microglia in mouse brain immunphenotypic and immunoregulatory properties of the progeny, *J. Neuroimmunol.* 94, 127–133.

Weiss S, Dunne C, Hewson J, Wohl C, Wheatley M, Peterson AC, Reynolds BA **1996**, Multipotent CNS stem cells are present in the adult mammalian spinal cord and ventricular neuroaxis, *J. Neurosci.* 16, 7599–7609.

Wendling D, Didier JM, Vuitton DA **1991**, The phagocyte oxidative metabolism function in ankylosing spondylitis, *Rheumatology* 11, 187–189.

Widera D, Holtkamp W, Entschladen F, Niggemann B, Zanker K, Kaltschmidt B, Kaltschmidt C **2004**, MCP-1 induces migration of adult neural stem cells, *Eur. J. Cell Biol.* 83, 381–387.

Williams K, Alvarez X, Lackner AA **2001**, Central nervous system perivascular cells are immunoregulatory cells that connect the CNS with the peripheral immune system, *Glia* 36, 156–164.

Wong G, Goldsmith Y, Turnley AM **2004**, Interferon-gamma but not TNF alpha promotes neuronal differentiation and neurite outgrowth of murine adult neural stem cells, *Exp. Neurol.* 187, 171–177.

Yang WD, Gomes RR, Alicknavitch M, Farach-Carson MC, Carson DD **2005**, Perlecan domain 1 promotes fibroblast growth factor-2 delivery in collagen-1 fibril scaffolds, *Tissue Eng.* 11, 76–99.

Yin YQ, Cui Q, Li YM, Irwin M, Fisher D, Harwey AR, Benowitz LI **2003**, Macrophage-derived factors stimulate optic nerve regeneration, *J. Neurosci.* 23, 2284–2293.

Zhang JM, Hoffmann R, Sieber-Blum M **1997**, Mitogenic and anti-proliferative signals for neural crest cells and the neurogenic action of TGF-β1, *Dev. Dyn.* 208, 375–386.

7
The Microglial Component

Ingo Bechmann, Angelika Rappert, Josef Priller, and Robert Nitsch

7.1
Microglia: Intrinsic Immune Sensor Cells of the CNS

7.1.1
Development

Microglial cells were first described by the Spanish neuroscientist del Rio-Hortega (1932). On the basis of the first selective stain for microglial cells, a weak silver carbonate method, del Rio-Hortega (1932) introduced microglial cells as a distinct entity belonging to the class of glial cells. Despite the longstanding debate in past decades over the origin of microglial cells, del Rio-Hortega's original observation of "fountains of microglia" in the tela choroidea and the pia appears to hold true, in that leptomeningeal mesenchymal cells are indeed found to enter the neuropil, where they transform into microglia. In addition, it is now clear that monocytes provide a second source of microglia, while convincing evidence for a neuroectodermal origin is lacking.

Microglial precursors populate the CNS parenchyma during early embryonic development. At that stage, they are referred to as fetal macrophages (Takahashi et al. 1989; Alliot et al. 1999a), which, like their counterparts in the blood, show a rounded morphology. Later, they develop short processes, which subsequently mature into a fully ramified shape characteristic of adult microglia. Their presence prior to vascularization of the neuroectoderm (for a review, see Kurz et al. 2004) indicates that they are not blood-borne, but rather derive from leptomeningeal cells which, in turn, derive from the yolk sac. At later stages, monocytes from the blood are also recruited and transform into microglia. Perinatally, microglial cells aggregate as clusters of so-called ameboid cells at specific locations in the brain, particularly in the corpus callosum (for a review, see Ling and Wong 1993). These microglial progenitor cells undergo substantial proliferation, migrate into the overlying cerebral cortex along fiber tracts, and differentiate into fully ramified microglia. In the adult, early studies using an injection of colloidal carbon to prelabel circulating monocytes demonstrated their recruitment and subsequent transformation into microglia following brain lesion (for a review, see Kaur et al. 2001). On the basis of such findings, Kaur et al. (2001) stated that: "circulating monocytes are the major source of brain macrophages in traumatic brain lesion. With the completion of their roles as scavanger cells at the site of injury, they will become microglia in the healing process. The whole process, therefore, recapitulates the microglial ontogeny in early development." In fact, this concept has been confirmed in modern models of bone marrow transplantation (Priller et al. 2001; Bechmann et al. 2005; see Chapter 6). In the adult brain, microglia make up about 5–20% of the entire central nervous system glial cell population; and one can thus roughly calculate that there are about as many microglial cells as neurons in the adult brain.

7.1.2
Microglial Activation

In the absence of pathology, microglial cells exhibit a ramified morphology with a small cell body and long, thin processes which define their territory (Fig. 7.1). In this state, they are known as "resting" microglia. Virtually any kind of brain pathology is accompanied by a morphologic transformation of these ramified cells into what is referred to as their "activated" form. Such activated cells show stubby processes, losing the small extensions typical of the "resting" form. Once activated, they are not easy to distinguish from infiltrating mononuclear cells/ macrophages, which, in turn, develop microglia-like ramifications once inside the brain (Priller et al. 2001; Bechmann et al. 2005) under the influence of astrocytic signals (Sievers et al. 1994; Hailer et al. 1998; for a review, see Bechmann and Nitsch 2004). The induction of microglial activation as indicated by their morphologic transformation has led to the helpful concept of microglia as "sensors" of pathologic events within the CNS (Kreutzberg 1996). While there are certainly uniform phenotypic changes during microglial activation from ramified, to ameboid, and finally phagocytic, a closer look reveals remarkable changes in surface antigen expression and cytokine release in different pathologies, which certainly reflect the different activating signals. For instance, the MHC-II complex is not induced in the course of retrograde (Liu et al. 2005) axonal degeneration, but is enhanced for up to several years after anterograde (Wallerian) axonal degeneration (Kosel et al. 1997). Interestingly, such MHC-II$^+$ microglia in zones of degeneration are highly ramified, challenging the concept that ramification always reflects a "resting" state (Bechmann et al. 2001c). In fact, in vitro, the cytokine GM-CSF drives ramification and differentiation of a microglial subpopulation towards a dendritic cell-like state, while other cells of the same culture do not differentiate in that direction (Fischer and Reichmann 2001; Santambrogio et al. 2001). One can thus assume that the functional plasticity of microglia is as heterogeneous as the catalogue of brain disorders, and that site-specific subpopulations may exist with different capacities to differentiate in a macrophage-like or dendritic cell-like direction.

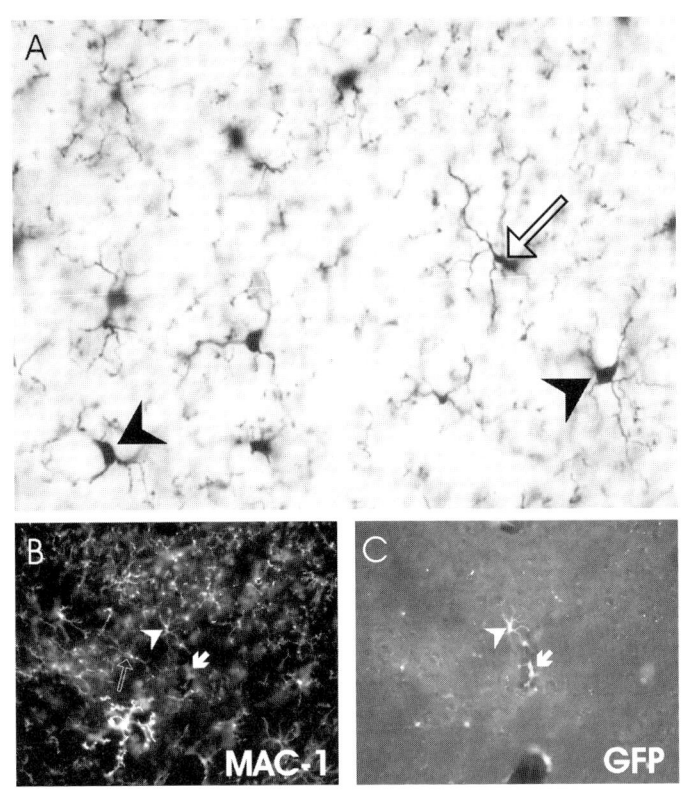

Fig. 7.1 (A) Microglial cells in the normal brain. Note the typical ramified morphology of what are regarded as "resting" microglia (open arrow). Juxtavascular microglia are found to engulf small vessels (arrowheads). (B, C) (double fluorescence) Supplementation of perivascular macrophages and microglia by bone marrow cells. The marrow was transduced to express green fluorescent protein (GFP), allowing clear-cut detection of blood-derived cells within the brain (Priller et al. 2001; C). These cells express the microglia/macrophage marker MAC-1 (B). A perivascular cell (white arrow) and a juxtavascular microglial cell (white arrowhead) are GFP-positive, while none of the highly ramified intraparenchymal microglia (black arrow, with white outline) appear to derive from blood.

7.1.3
Antigen Presentation/Cytotoxicity

In most circumstances, microglia are immature antigen-presenting cells (APCs; Carson et al. 1998). Only after a multi-step activation process involving the CD40-CD40 ligand, CD80-CD28 binding, as well as cytokine stimulation with GM-CSF and IFN-gamma (Matyszak et al. 1999), are they rendered competent APCs. Unless such strong activation signals are present, antigen presentation by microglia provides insufficient co-stimulation, thus inducing T cell anergy, an important mechanism of immune tolerance (for a review, see Kamradt and Mitchison 2001). The relative resistance of microglia to transformation into activated APCs and the resulting tolerogenic antigen presentation reflect the need to be "more tolerant" within the brain, compared with less vulnerable organs, in order to minimize secondary damage, an unavoidable side-effect of any immune response (for a review, see Kwidzinski et al. 2003). In fact, besides providing help by phagocytosing degenerated material (such as myelin with its growth-inhibiting epitopes) and secreting neuroprotective molecules, microglia can also become cytotoxic, e.g. through the release of NO, IL1-beta, and TNF-alpha (Stoll et al. 2002). This is particularly prominent upon binding of LPS to Toll-like receptor 4, leading to microglia-mediated death of neurons and oligodendrocytes (Lehnardt et al. 2002, 2003). This and similar activating pathways of innate immunity may in part underlie the devastating effects of bacterial meningitis, rendering not only the infectious agent, but also the host's microglial response against it, potential targets of therapeutic intervention.

7.2
Terminology: Subtypes and Their Location in Regard to Brain Vessels

7.2.1
Perivascular Macrophages

Since microglia share most markers with cells of the monocyte/macrophage lineage, immunocytochemical detection in situ identifies them together with extraparenchymal macrophages of the leptomeninges and the perivascular (Virchow-Robin) spaces. This latter population is sometimes referred to as perivascular microglia, although they are located outside the neuropil proper. As shown in Fig. 7.2, perivascular macrophages are also distinct from pericytes, which regularly engulf capillary endothelial cells (see Chapter 6). Perivascular macrophages accumulate degenerated myelin (Kosel et al. 1997) and seem to be particularly important for antigen presentation. In fact, during brain inflammation, T cells first reside within the perivascular spaces, where they are believed to be re-stimulated by these macrophages (Platten and Steinman 2005; see Fig. 7.3).

7.2.2
Juxtavascular and Other Microglia

The first layer of the neuropil proper is formed mainly by astrocytic endfeet building the *glia limitans perivascularis*, which is separated from the outer vessel wall or, where existent, the Virchow-Robin spaces by a basement membrane (for details of the perivascular space, see Fig. 7.2). Some of the microglial cells participate in forming the glia limitans (Lassmann et al. 1991) or engulf the smaller vessels of the brain (Fig. 7.1). These cells are known as juxtavascular microglia. Other microglial cells seem to have no direct contact with the vasculature and are simply referred to as microglia. In routine light microscopic sections used for neuropathol-

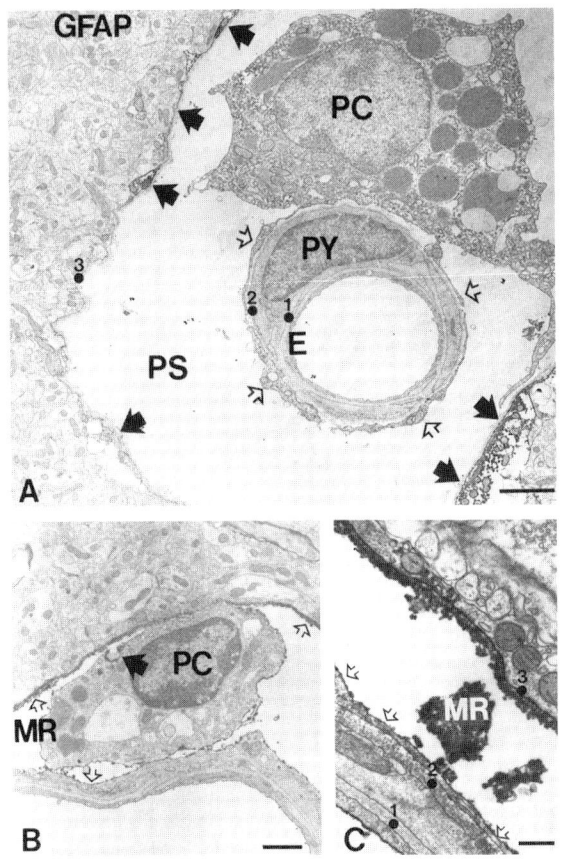

Fig. 7.2 (A) Location of perivascular cells (PC) in the perivascular (Virchow-Robin) spaces (PS). The dots mark the basement membranes. A first basement membrane (1) surrounds the endothelial cell (E), which in turn is surrounded by a pericyte (PY) and the second basement membrane (2). The third basement membrane (3) is located on the top of the glia limitans, which is visualized by GFAP-immunocytochemistry (arrows). The perivascular cells are located between the second and the third basement membrane and are often found to wrap small processes (open arrows) around brain vessels. Scale bar: 2 µm.
(B) Perivascular space after intraventricular injection of Mini Ruby (MR). Within minutes upon injection, the tracer can be found in the perivascular spaces (open arrows), demonstrating their connection to the subarachnoid space, where the CSF drains from the ventricles. Phagocytosis of the tracer by the perivascular macrophages (PC) is evident. Scale bar: 1 µm.
(C) Mini Ruby (MR) clusters confined to the perivascular space. Following injection of Mini Ruby, the tracer remains confined to the space between the second (2) and the third basement membrane (3). Mini Ruby clusters are attached to the processes of perivascular cells (open arrows), but cannot be found within pericytes or astrocytic processes of the glia limitans. Scale bar: 0.5 µm. (Reprinted from Bechmann et al. 2001b, copyright, with permission from Elsevier.)

7.4.2
Chemokines – an Overview

Chemokines are a diverse family (43 human chemokines are known today) of chemotaxis-inducing cytokines. They are small peptides (8–10 kDa) and have been found to be involved in the pathogenesis of many neuroinflammatory diseases, ranging from multiple sclerosis (MS) and stroke to HIV encephalopathy. The chemokine family is currently divided into four groups, depending on the number and spacing of the conserved cysteines in the protein sequence (Baggiolini et al. 1997; Luster 1998). The C-X-C family (also called alpha) has a single amino acid between the initial two cysteines. If the first two cysteines are adjacent to each other, the peptides are classified in the C-C family (also called beta), while the unique gamma chemokine or C chemokine (lymphotactin) has lost two of the four conserved cysteines (Cys-2, Cys-4). Finally, in the CX_3C family, the chemokines have three amino acid residues separating the first two cysteines in the sequence. The CX_3C chemokines are the only membrane-bound chemokines. Chemokines mediate their biological activities through 19 different G protein-coupled cell surface receptors. The chemokine receptors are named according to their chemokine subfamily classification. Chemokine receptors are promiscuous in that they can bind to more than one chemokine, and more than one chemokine can often bind to the same receptor.

7.4.3
GAG/Duffy

The presence of chemokine-binding sites on human brain microvessels (Andjelkovic et al. 1999) suggests that chemokines produced locally by microglia, astrocytes, and perivascular cells will either diffuse or be transported to the endothelial cell surface, where they will be immobilized for presentation to leukocytes. However, in vivo chemokines interact with sulfate sugars of the glycosaminoglycan (GAG) family, such as heparan sulfate and chondroitin sulfate, which can limit their dissemination. GAGs decorate proteins in the extracellular matrix and on the cell surface and not only interfere with chemokine diffusion, but also provide highly specific substrates for the presentation of chemokines to other cells, e.g. on the luminal surface of endothelial cells. Hence, they are appropriately positioned to activate leukocytes in close proximity (Middleton et al. 2002; Rot and von Andrian 2004). Tissue-derived chemokines can cross the endothelial cell barrier passively through intercellular junctions (Song and Pachter 2004) or can be transported (Middelton et al. 1997; Rot and von Andrian 2004). One molecule is expressed by endothelial cells and may contribute to chemokine transport: the Duffy antigen (also known as the Duffy antigen receptor for chemokines, or DARC), which binds numerous chemokines (Nibbs et al. 2003).

7.4.4
Chemokine Expression in the CNS

Various chemokine receptors of the different chemokine families are expressed by different types of brain cells, neurons, astrocytes, and microglia. However, with the exception of CX3CL1 (fractalkine), all chemokines found in brain tissue are expressed by glial cells and infiltrating leukocytes. Both astrocytes and juxtavascular microglia are in physical proximity to the endothelium. Pericytes and perivascular macrophages are widely distributed at the level of the BBB. Thus, these cells are in a prime position to influence the transmigration of leukocytes. In addition, they are established sources of soluble molecules that can enhance vascular permeability. Most information concerning the migration of lymphocytes into the neuropil has emerged from studies using experimental allergic encephalomyelitis (EAE), the animal model of multiple sclerosis (MS), and HIV infection. In a state of health, very few leukocytes infiltrate the brain (Ludowyk et al. 1992); but in disorders such as MS the "barrier" becomes permissive, resulting in intense infiltration of the CNS by T lymphocytes, whose subsequent activity appears to underlie the onset and progression of disease (Brown 2001). In patients with HIV encephalitides and MS or in EAE tissue of animals, several chemokines, such as CCL2 (monocyte chemoattractant protein, MCP-1), CCL3 (macrophage inflammatory protein 1a, MIP-1a), CCL5 (RANTES), and CXCL10 (interferon-inducible protein 10), can be detected in the CSF (Letendre et al. 1999; Sorensen et al. 1999). The corresponding receptors CCR5 and CXCR3 are expressed by activated T cells, which are increased in the peripheral blood, CSF, and lesions of MS patients (Misu et al. 2001). These findings lead to the idea that these receptors and their ligands are important for lymphocyte trafficking into the CNS during disease. In fact, all these chemokines are expressed by microglia (Kremlev et al. 2004; Rock et al. 2004).

7.4.5
CCL2 and CCR2

CCL2 and its receptor CCR2 have been implicated as key mediators of leukocyte entry into the CNS, since CCL2 expression has been found in virtually all forms of CNS insult, including acquired immunodeficiency syndrome (AIDS) with dementia (Cinque et al. 1998; Mengozzi et al. 1999; Weiss et al. 1999) and EAE (Ransohoff et al. 1993; Mahad and Ransohoff 2003). CCL2 is produced by macrophages, microglia, activated astrocytes, perivascular macrophages, and endothelial cells (Sozzani et al. 1995; Rollins 1996; Simpson et al. 1998; Hofmann et al. 2002; Babcock et al. 2003). It potently recruits monocytes and T cells into the brain (Sozzani et al. 1995; Rollins 1996). By using a coculture model of the BBB, Weiss and co-workers demonstrated that 90% monocytes and 10% lymphocytes transmigrate in response to CCL2, while the input population of mononuclear cells consisted of 90% lymphocytes and 10% monocytes (Weiss et al. 1998). Neutralizing antibodies specific for CCL2 ameliorated the progression of EAE disease (Karpus et al. 1997). Mice lacking either the corresponding receptor CCR2 or one of its ligands were relatively resistant to the development of EAE, despite having T cell responses equivalent to those observed in wild-type mice (Fife et al. 2000; Izikson et al. 2000; Huang et al. 2001). CNS tissues of CCL2-deficient mice were virtually devoid of monocytes at time-points equivalent to those at which wild-type mice developed neurological impairment (Huang et al. 2001). CCR2 was also essential for the accumulation of monocytes at the epicenter of a spinal cord contusion; and their absence in CCR2-deficient mice was associated with impaired clearance of tissue debris (Ma et al. 2002). In zones of axonal degeneration, CCR2 deficiency also completely blocked the invasion of leukocytes (Babcock et al. 2003), which normally takes place within the first days after injury (Bechmann et al. 2001c, 2005). Nevertheless, controversial data are found in humans. The CCL2 concentration in the CSF of patients with early active MS or during relapses is reduced (Sorensen et al. 2004). However, these data did not exclude a role of CCL2 in other phases of MS. Thus, further studies directly in MS brain tissue could clarify the role of CCL2 in MS.

7.4.6
CCL3 and CCL5

Of the CSF chemokines, only CCL3 and CCL5 are associated with enhanced migration of T lymphocytes from MS patients, an effect that is ascribed to the over-expression of CCR5 on these cells (Zang et al. 2000). CCL3 is expressed by perivascular macrophages (Hofmann et al. 2002). Neutralizing antibodies and targeted gene deletion in animal models of CNS inflammation have yielded conflicting results about the roles of these chemokines. Inactivation of CCR5 in acute viral encephalitis was effective in decreasing the numbers of infiltrating leukocytes (Glass et al. 2001). In contrast, loss of CCR5 activity in mouse models of MS was without effect; and in MS it may slow disease progression (Tran et al. 2000; Kantor et al. 2003). While neutralizing antibodies specific for CCL3 ameliorated the progression of EAE (Karpus et al. 1995), CCL3 knockout mice were entirely susceptible to EAE (Tran et al. 2000). Thus, these findings leave the specific functions of these chemokines in these disease models unresolved.

7.4.7
CXCR3 and CXCL10

Numerous studies utilizing animal models and transgenic approaches have supported a role for CXCL10 and its receptor CXCR3 in the trafficking of lymphocytes during acute CNS inflammation. Under inflammatory conditions, both CXCL10- and CXCR3-expressing T cells could be found in the CSF (Sorensen et al. 1999). CXCR3 expression is not specific to T cells in the brain, but can also be found on other populations outside the neuropil. However, if the ligand CXCL10 appears in the CSF, CXCR3-expressing T cells stop recirculating in the periphery (Trebst and Ransohoff 2001). In addition, CXCL10-CXCR3 is probably also involved in the transmigration into the CNS parenchyma: Sorensen could demonstrate a direct correlation between CXCL10 expression and the number of CXCR3-expressing cells (Sorensen et al. 2002). CXCL10 is produced by astrocytes, perivascular macrophages (Karpus and Ransohoff 1998), and microglia (Lokensgard et al. 2001), and may establish the gradient necessary for the transmigration of T cells into the CNS parenchyma. Neutralizing antibodies specific for CXCL10 may ameliorate the progression of disease by inhibiting mononuclear infiltration (Fife et al. 2001). In contrast, Narumi et al. (2002) reported an exacerbation of EAE progression by applying similarly utilized CXCL10-neutralizing antibodies. Their finding is supported by studies using CXCL10 knockout mice, which were entirely susceptible to EAE (Klein et al. 2004). Interestingly, CXCL10 expression in the CNS leads to an increase in leukocyte infiltration; however, these infiltrates are restricted to the meninges and perivascular compartments (Boztug et al. 2002; Trifilo and Lane 2003). This indicates that CXCL10 could be a signal which directs lymphocytes across the endothelial layer, but not beyond, i.e. across the glia limitans (see Fig. 7.2).

7.4.8
Microglia-Endothelial Cell Dialogue

All these results support the view that chemokines are crucial in lymphocyte trafficking during established CNS autoimmunity. Their individual role in attracting particular subsets of leukocytes such as T and B lymphocytes, monocytes, or NK cells, as well as the signals initiating chemokine expression during pathology have yet to be identified. Some controversy in the results probably reflects that chemokines and their receptors are promiscuous, given that by blocking one chemokine, others compensate for the effect (Klein et al. 2004). However, by using conditional and cell type-specific knockout strategies, a picture may eventually evolve of how microglia and perivascular macrophages attract particular subtypes of leukocytes into particular regions of the CNS (Rock et al. 2004). It now seems clear that the dialogue between microglia and brain endothelial cells plays a crucial role in the process.

7.4.9
Microglial Effects on Tight Junctions

Inflammation not only involves the recruitment of leukocytes, but also the opening of endothelial barriers in order to allow serum molecules such as antibodies and complement factors to gain access to areas of degeneration. Activated microglia produce cytokines and other toxic factors promoting neurodegeneration and activation of other glial cells, including astrocytes. It is likely that microglia-mediated astrocytic activation also affects BBB integrity through crosstalk with astrocytes and perivascular cells. In post mortem studies of HIV-1-associated dementia, the presence of damaged tight junctions, indicated by fragmentation or absence of immunoreactivity for occludin and ZO-1, correlated with levels of monocyte infiltration (Dallasta et al. 1999; Boven et al. 2000). The coincidence of the greatest number of abnormal junctions with the highest levels of microglial activation in a study of active MS suggests that some of the tight junction abnormalities result from the pathophysiological action of cytokines, matrix metalloproteinases, and other immune effectors present in the disordered milieu of the MS lesion (Kieseier et al. 1999; Plumb et al. 2002). This apparently also applies to perivascular macrophages: in experimental simian immunodeficiency virus encephalitis, the accumulation of these cells correlates with the fragmentation and decreased immunoreactivity of ZO-1 and occludin (Luabeya et al. 2000). Several lines of evidence indicate that locally produced microglia-derived cytokines, in particular TNF-α, are the most likely candidates for mediating tight junction disruption (Gloor et al. 2001; Poritz et al. 2004; Wang et al. 2005). TNF-α and interferon-γ induce a striking fragmentation of ZO-1 via F-actin rearrangement in cultures of microvascular endothelial cells (Blum et al. 1997). Indeed, high levels of these cytokines are present in the CNS of patients with HIV-1-associated dementia complex (Griffin 1997), and microglia are a potent source of TNF-α (Benveniste 1997; Persidsky et al. 2000). These findings could explain how microglial cells control the opening and closing of tight junctions, thus fostering permeability for molecules from the blood and possibly the infiltration of leukocytes.

7.5
Concluding Remarks

Sound data indicate that microglia, once activated by pathologic changes in their surrounding milieu, secrete chemokines and cytokines which impact endothelial cells at two levels: the expression of adhesion/chemoattractant molecules and the maintenance of tight junctions. While the former clearly is related to cell type-specific leukocyte recruitment, it is unclear as to how far tight junction changes are involved in initiating and facilitating the infiltration of leukocytes. In vitro data are difficult to interpret, since endothelial cells derive from various sites of the vascular arbor and there is a lack of sound studies specifically addressing changes in tight junctions at the level of postcapillary venules, where diapedesis takes place. Recently, trans-rather than paracellular diapedesis has been shown, using serial electron microscopic sections from venules of EAE brains (Wolburg et al. 2004), suggesting that diapedesis is independent of the maintenance of tight junctions. It is, however, clear that the permeability of the BBB is enhanced during neuroinflammation; and this may well allow serum molecules involved in the inflammatory cascade to gain access.

Microglia as early *sensors of pathological events* within the neuropil (Kreutzberg 1996) are likely to initiate innate and adaptive immune responses by cytokine and chemokine signaling. Be it directly or via astrocytes forming the glia limitans, this will induce endothelial expression of adhesion molecules, allowing leukocyte recruitment. There are also indications that microglia institute tight junctional changes, thereby fostering at least BBB permeability. Beyond this, liberation of cells from their tight junctions may provide a more general mechanism of inflammatory cascades by supporting migration into sites of injury. The current challenge is to identify the crucial signaling molecules from microglia to endothelial cells, which may provide promising targets for interfering with undesired (autoimmune) neuroinflammation.

References

Alliot F, Godin I, Pessac B **1999a**, Microglia derive from progenitors, originating from the yolk sac, and which proliferate in the brain, *Brain Res. Dev. Brain Res*. 117, 145–152.

Alliot F, Rutin J, Leenen PJ, Pessac B **1999b**, Pericytes and periendothelial cells of brain parenchyma vessels co-express aminopeptidase N, aminopeptidase A, and nestin, *J. Neurosci. Res*. 58, 367–378.

Andjelkovic AV, Spencer DD, Pachter JS **1999**, Visualization of chemokine binding sites on human brain microvessels, *J. Cell Biol*. 145, 403–412.

Babcock AA, Kuziel WA, Rivest S, Owens T **2003**, Chemokine expression by glial cells directs leukocytes to sites of axonal injury in the CNS, *J. Neurosci*. 23, 7922–7930.

Baggiolini M, Dewald B, Moser B **1997**, Human chemokines: an update, *Annu. Rev. Immunol*. 15, 675–705.

Bechmann I **2005**, Failed central nervous system regeneration: a downside of immune privilege? *Neuromolecular Med*. 7, 217–228.

Bechmann I, Nitsch R **2001**, Plasticity following lesion: help and harm from the immune system, *Restor. Neurol. Neurosci*. 19, 189–198.

Bechmann I, Nitsch R **2004**, Interaction of glial cells with monocytes, in *Neuroglia*, eds. H Kettenman, BR Ramson, Oxford University Press, Oxford.

Bechmann I, Priller J, Kovac A, Bontert M, Wehner T, Klett FF, Bohsung J, Stuschke M, Dirnagl U, Nitsch R **2001a**, Immune surveillance of mouse brain perivascular spaces by blood-borne macrophages, *Eur. J. Neurosci*. 14, 1651–1658.

Bechmann I, Kwidzinski E, Kovac AD, Simburger E, Horvath T, Gimsa U, Dirnagl U, Priller J, Nitsch R **2001b**, Turnover of rat brain perivascular cells, *Exp. Neurol*. 168, 242–249.

Bechmann I, Peter S, Beyer M, Gimsa U, Nitsch R **2001c**, Presence of B7-2 (CD86) and lack of B7-1 (CD80) on myelin phagocytosing MHC-II-positive rat microglia is associated with nondestructive immunity in vivo, *FASEB J*. 15, 1086–1088.

Bechmann I, Goldmann J, Kovac AD, Kwidzinski E, Simburger E, Naftolin F, Dirnagl U, Nitsch R, Priller J **2005**, Circulating monocytic cells infiltrate layers of anterograde axonal degeneration where they transform into microglia, *FASEB J.* (online 25 Jan 2005).

Benveniste EN **1997**, Role of macrophages/microglia in multiple sclerosis and experimental allergic encephalomyelitis, *J. Mol. Med.* 75, 165–173.

Blum MS, Toninelli E, Anderson JM, Balda MS, Zhou J, O'Donnell L, Pardi R, Bender JR **1997**, Cytoskeletal rearrangement mediates human microvascular endothelial tight junction modulation by cytokines, *Am. J. Physiol.* 273, H286–H294.

Boven LA, Middel J, Verhoef J, De Groot CJ, Nottet HS **2000**, Monocyte infiltration is highly associated with loss of the tight junction protein zonula occludens in HIV-1-associated dementia, *Neuropathol. Appl. Neurobiol.* 26, 356–360.

Boztug K, Carson MJ, Pham-Mitchell N, Asensio VC, DeMartino J, Campbell IL **2002**, Leukocyte infiltration, but not neurodegeneration, in the CNS of transgenic mice with astrocyte production of the CXC chemokine ligand 10, *J. Immunol.* 169, 1505–1515.

Brown KA **2001**, Factors modifying the migration of lymphocytes across the blood-brain barrier, *Int. Immunopharmacol.* 1, 2043–2062.

Carson MJ, Reilly CR, Sutcliffe JG, Lo D **1998**, Mature microglia resemble immature antigen-presenting cells, *Glia* 22, 72–85.

Cinque P, Vago L, Mengozzi M, Torri V, Ceresa D, Vicenzi E, Transidico P, Vagani A, Sozzani S, Mantovani A, Lazzarin A, Poli G **1998**, Elevated cerebrospinal fluid levels of monocyte chemotactic protein-1 correlate with HIV-1 encephalitis and local viral replication, *AIDS*, 12, 1327–1332.

Dallasta LM, Pisarov LA, Esplen JE, Werley JV, Moses AV, Nelson JA, Achim CL **1999**, Blood-brain barrier tight junction disruption in human immunodeficiency virus-1 encephalitis, *Am. J. Pathol.* 155, 1915–1927.

del Rio-Hortega P **1932**, Microglia, in *Cytology and Cellular Pathology of the Nervous System*, ed. W Penfield, Hoeber, New York, pp. 481–534.

de Vos AF, van Meurs M, Brok HP, Boven LA, Hintzen RQ, van der Valk P, Ravid R, Rensing S, Boon L, 't Hart BA, Laman JD **2002**, Transfer of central nervous system autoantigens and presentation in secondary lymphoid organs, *J. Immunol.* 169, 5415–5423.

Ehrlich P **1885**, *Das Sauerstoffbedürfnis des Organismus,* In *Eine farbanalytische Studie*, Hirschwald, Berlin.

Fife BT, Huffnagle GB, Kuziel WA, Karpus WJ **2000**, CC chemokine receptor 2 is critical for induction of experimental autoimmune encephalomyelitis, *J. Exp. Med.* 192, 899–905.

Fife BT, Kennedy KJ, Paniagua MC, Lukacs NW, Kunkel SL, Luster AD, Karpus WJ **2001**, CXCL10 (IFN-gamma-inducible protein-10) control of encephalitogenic CD4+ T cell accumulation in the central nervous system during experimental autoimmune encephalomyelitis, *J. Immunol.* 166, 7617–7624.

Fischer HG, Reichmann G **2001**, Brain dendritic cells and macrophages/microglia in central nervous system inflammation, *J. Immunol.* 166, 2717–2726.

Flugel A, Bradl M, Kreutzberg GW, Graeber MB **2001**, Transformation of donor derived bone marrow precursors into host microglia during autoimmune CNS inflammation and during the retrograde response to axotomy, *J. Neurosci. Res.* 66, 74–82.

Ge S, Song L, Pachter JS **2005**, Where is the blood-brain barrier ... really? *J. Neurosci. Res.* 79, 421–427.

Glass WG, Liu MT, Kuziel WA, Lane TE **2001**, Reduced macrophage infiltration and demyelination in mice lacking the chemokine receptor CCR5 following infection with a neurotropic coronavirus, *Virology* 288, 8–17.

Gloor SM, Wachtel M, Bolliger MF, Ishihara H, Landmann R, Frei K **2001**, Molecular and cellular permeability control at the blood-brain barrier, *Brain Res. Brain Res. Rev.* 36, 258–264.

Gonzalez-Scarano F, Martin-Garcia J **2005**, The neuropathogenesis of AIDS, *Nat. Rev. Immunol.* 5, 69–81.

Graeber MB, Streit WJ **1990**, Perivascular microglia defined, *Trends Neurosci.* 13, 366.

Greter M, Heppner FL, Lemos MP, Odermatt BM, Goebels N, Laufer T, Noelle RJ, Becher B **2005**, Dendritic cells permit immune invasion of the CNS in an animal model of multiple sclerosis, *Nat. Med.* 11, 328–334.

Griffin DE **1997**, Cytokines in the brain during viral infection: clues to HIV-associated dementia, *J. Clin. Invest.* 100, 2948–2951.

Guillemin GJ, Brew BJ **2004**, Microglia, macrophages, perivascular macrophages, and pericytes: a review of function and identification, *J. Leukoc. Biol.* 75, 388–397.

Hailer NP, Heppner FL, Haas D, Nitsch R **1998**, Astrocytic factors deactivate antigen presenting cells that invade the central nervous system, *Brain Pathol.* 8, 459–474.

Harling-Berg CJ, Park TJ, Knopf PM **1999**, Role of the cervical lymphatics in the Th2-type hierarchy of CNS immune regulation, *J. Neuroimmunol.* 101, 111–127.

Hess DC, Abe T, Hill WD, Studdard AM, Carothers J, Masuya M, Fleming PA, Drake CJ, Ogawa M **2004**, Hematopoietic origin of microglial and perivascular cells in brain, *Exp. Neurol.* 186, 134–144.

Hickey WF **2001**, Basic principles of immunological surveillance of the normal central nervous system, *Glia* 36, 118–124.

Hickey WF, Kimura H **1988**, Perivascular microglial cells of the CNS are bone marrow-derived and present antigen in vivo, *Science* 239, 290–292.

Hofmann N, Lachnit N, Streppel M, Witter B, Neiss WF, Guntinas-Lichius O, Angelov DN **2002**, Increased expression of ICAM-1, VCAM-1, MCP-1, and MIP-1 alpha by spinal perivascular macrophages during experimental allergic encephalomyelitis in rats, *BMC Immunol.* 3, 11.

Huang DR, Wang J, Kivisakk P, Rollins BJ, Ransohoff RM **2001**, Absence of monocyte chemoattractant protein 1 in mice leads to decreased local macrophage recruitment and antigen-specific T helper cell type 1 immune response in experimental autoimmune encephalomyelitis, *J. Exp. Med.* 193, 713–726.

Izikson L, Klein RS, Charo IF, Weiner HL, Luster AD **2000**, Resistance to experimental autoimmune encephalomyelitis in mice lacking the CC chemokine receptor (CCR)2, *J. Exp. Med.* 192, 1075–1080.

Kamradt T, Mitchison A **2001**, Tolerance and autoimmunity, *N. Engl. J. Med.* 344, 655–664.

Kantor R, Bakhanashvili M, Achiron A **2003**, A mutated CCR5 gene may have favorable prognostic implications in MS, *Neurology* 61, 238–240.

Karpus WJ, Kennedy KJ **1997**, MIP-1alpha and MCP-1 differentially regulate acute and relapsing autoimmune encephalomyelitis as well as Th1/Th2 lymphocyte differentiation, *J. Leukoc. Biol.* 62, 681–687.

Karpus WJ, Ransohoff RM **1998**, Chemokine regulation of experimental autoimmune encephalomyelitis: temporal and spatial expression patterns govern disease pathogenesis, *J. Immunol.* 161, 2667–2671.

Karpus WJ, Lukacs NW, McRae BL, Strieter RM, Kunkel SL, Miller SD **1995**, An important role for the chemokine macrophage inflammatory protein-1 alpha in the pathogenesis of the T cell-mediated autoimmune disease, experimental autoimmune encephalomyelitis, *J. Immunol.* 155, 5003–5010.

Kaur C, Hao AJ, Wu CH, Ling EA **2001**, Origin of microglia, *Microsc. Res. Tech.* 54, 2–9.

Kieseier BC, Seifert T, Giovannoni G, Hartung HP **1999**, Matrix metalloproteinases in inflammatory demyelination: targets for treatment, *Neurology* 53, 20–25.

Klein RS, Izikson L, Means T, Gibson HD, Lin E, Sobel RA, Weiner HL, Luster AD **2004**, IFN-inducible protein 10/CXC chemokine ligand 10-independent induction of experimental autoimmune encephalomyelitis, *J. Immunol.* 172, 550–559.

Kosel S, Egensperger R, Bise K, Arbogast S, Mehraein P, Graeber MB **1997**, Long-lasting perivascular accumulation of major histocompatibility complex class II-positive lipophages in the spinal cord of stroke patients: possible relevance for the immune privilege of the brain, *Acta Neuropathol. (Berl.)* 94, 532–538.

Kremlev SG, Roberts RL, Palmer C **2004**, Differential expression of chemokines and chemokine receptors during microglial activation and inhibition, *J. Neuroimmunol.* 149, 1–9.

Kreutzberg GW **1996**, Microglia: a sensor for pathological events in the CNS, *Trends Neurosci.* 19, 312–318.

Kurz H, Korn J, Christ B **2004**, Morphogenesis of CNS vessels, *Cancer Treat. Res.* 11, 33–50.

Kwidzinski E, Mutlu LK, Kovac AD, Bunse J, Goldmann J, Mahlo J, Aktas O, Zipp F, Kamradt T, Nitsch R, Bechmann I **2003**, Self-tolerance in the immune privileged CNS: lessons from the entorhinal cortex lesion model, *J. Neural Transm. Suppl.* 65, 29–49.

Lassmann H, Zimprich F, Vass K, Hickey WF **1991**, Microglial cells are a component of the perivascular glia limitans, *J. Neurosci. Res.* 28, 236–243.

Lassmann H, Schmied M, Vass K, Hickey WF **1993**, Bone marrow derived elements and resident microglia in brain inflammation, *Glia* 7, 19–24.

Lehnardt S, Lachance C, Patrizi S, Lefebvre S, Follett PL, Jensen FE, Rosenberg PA, Volpe JJ, Vartanian T **2002**, The toll-like receptor TLR4 is necessary for lipopolysaccharide-induced oligodendrocyte injury in the CNS, *J. Neurosci.* 22, 2478–2486.

Lehnardt S, Massillon L, Follett P, Jensen FE, Ratan R, Rosenberg PA, Volpe JJ, Vartanian T **2003**, Activation of innate immunity in the CNS triggers neurodegeneration through a Toll-like receptor 4-dependent pathway, *Proc. Natl Acad. Sci. USA* 100, 8514–8519.

Letendre SL, Lanier ER, McCutchan JA **1999**, Cerebrospinal fluid beta chemokine concentrations in neurocognitively impaired individuals infected with human immunodeficiency virus type 1, *J. Infect. Dis.* 180, 310–319.

Lewandowski M **1900**, Zur Lehre von der Cerebrospinalflüssigkeit, *Z. Klein Med*, 40, 480–494.

Ling EA, Wong WC **1993**, The origin and nature of ramified and amoeboid microglia: a historical review and current concepts, *Glia* 7, 9–18.

Liu ZQ, Bohatschek M, Pfeffer K, Bluethmann H, Raivich G **2005**, Major histocompatibility complex (MHC2+) perivascular macrophages in the axotomized facial motor nucleus are regulated by receptors for interferon-gamma (IFN-gamma) and tumor necrosis factor (TNF), *Neuroscience* 131, 283–292.

Lokensgard JR, Hu S, Sheng W, van Oijen M, Cox D, Cheeran MC, Peterson PK **2001**, Robust expression of TNF-alpha, IL-1beta, RANTES, and IP-10 by human microglial cells during nonproductive infection with herpes simplex virus, *J. Neurovirol.* 7, 208–219.

Luabeya MK, Dallasta LM, Achim CL, Pauza CD, Hamilton RL **2000**, Blood-brain barrier disruption in simian immunodeficiency virus encephalitis, *Neuropathol. Appl. Neurobiol.* 26, 454–462.

Ludowyk PA, Willenborg DO, Parish CR **1992**, Selective localisation of neurospecific T lymphocytes in the central nervous system, *J. Neuroimmunol.* 37, 237–250.

Luster AD **1998**, Chemokines – chemotactic cytokines that mediate inflammation, *N. Engl. J. Med.* 338, 436–445.

Ma M, Wei T, Boring L, Charo IF, Ransohoff RM, Jakeman LB **2002**, Monocyte recruitment and myelin removal are delayed following spinal cord injury in mice with CCR2 chemokine receptor deletion, *J. Neurosci. Res.* 68, 691–702.

Mahad DJ, Ransohoff RM **2003**, The role of MCP-1 (CCL2) and CCR2 in multiple sclerosis and experimental autoimmune encephalomyelitis (EAE), *Semin. Immunol.* 15, 23–32.

Matyszak MK, Denis-Donini S, Citterio S, Longhi R, Granucci F, Ricciardi-Castagnoli P **1999**, Microglia induce myelin basic protein-specific T cell anergy or T cell activation, according to their state of activation, *Eur. J. Immunol.* 29, 3063–3076.

Mengozzi M, De Filippi C, Transidico P, Biswas P, Cota M, Ghezzi S, Vicenzi E, Mantovani A, Sozzani S, Poli G **1999**, Human immunodeficiency virus replication induces monocyte chemotactic protein-1 in human macrophages and U937 promonocytic cells, *Blood* 93, 1851–1857.

Middleton J, Patterson AM, Gardner L, Schmutz C, Ashton BA **2002**, Leukocyte extravasation: chemokine transport and presentation by the endothelium, *Blood* 100, 3853–3860.

Misu T, Onodera H, Fujihara K, Matsushima K, Yoshie O, Okita N, Takase S, Itoyama Y **2001**, Chemokine receptor expression on T cells in blood and cerebrospinal fluid at relapse and remission of multiple sclerosis: imbalance of Th1/Th2-associated chemokine signaling, *J. Neuroimmunol.* 114, 207–212.

Muller DM, Pender MP, Greer JM **2005**, Blood-brain barrier disruption and lesion localisation in experimental autoimmune encephalomyelitis with predominant cerebellar and brainstem involvement, *J. Neuroimmunol.* 160, 162–169.

Narumi S, Kaburaki T, Yoneyama H, Iwamura H, Kobayashi Y, Matsushima K **2002**, Neutralization of IFN-inducible protein 10/CXCL10 exacerbates experimental autoimmune encephalomyelitis, *Eur. J. Immunol.* 32, 1784–1791.

Nibbs R, Graham G, Rot A **2003**, Chemokines on the move: control by the chemokine "interceptors" Duffy blood group antigen and D6, *Semin. Immunol.* 15, 287–294.

Pachter JS, de Vries HE, Fabry Z **2003**, The blood-brain barrier and its role in immune privilege in the central nervous system, *J. Neuropathol. Exp. Neurol.* 62, 593–604.

Persidsky Y, Zheng J, Miller D, Gendelman HE **2000**, Mononuclear phagocytes mediate blood-brain barrier compromise and neuronal injury during HIV-1-associated dementia, *J. Leukoc. Biol.* 68, 413–422.

Platten M, Steinman L **2005**, Multiple sclerosis: trapped in deadly glue, *Nat. Med.* 11, 252–253.

Plumb J, McQuaid S, Mirakhur M, Kirk J **2002**, Abnormal endothelial tight junctions in active lesions and normal-appearing white matter in multiple sclerosis, *Brain Pathol.* 12, 154–169.

Poritz LS, Garver KI, Tilberg AF, Koltun WA **2004**, Tumor necrosis factor alpha disrupts tight junction assembly, *J. Surg. Res.* 116, 14–18.

Priller J, Flugel A, Wehner T, Boentert M, Haas CA, Prinz M, Fernandez Klett F, Prass K, Bechmann I, de Boer BA, Frotscher M, Kreutzberg GW, Persons DA, Dirnagl U **2001**, Targeting gene-modified hematopoietic cells to the central nervous system: use of green fluorescent protein uncovers microglial engraftment, *Nat. Med.* 7, 1356–1361.

Ransohoff RM, Hamilton TA, Tani M, Stoler MH, Shick HE, Major JA, Estes ML, Thomas DM, Tuohy VK **1993**, Astrocyte expression of mRNA encoding cytokines IP-10 and JE/MCP-1 in experimental autoimmune encephalomyelitis, *FASEB J.* 7, 592–600.

Ransohoff RM, Kivisakk P, Kidd G **2003**, Three or more routes for leukocyte migration into the central nervous system, *Nat. Rev. Immunol.* 3, 569–581.

Reese TS, Karnovsky MJ **1967**, Fine structural localization of a blood-brain barrier to exogenous peroxidase, *J. Cell Biol.* 34, 207–217.

Rock RB, Gekker G, Hu S, Sheng WS, Cheeran M, Lokensgard JR, Peterson PK **2004**, Role of microglia in central nervous system infections, *Clin. Microbiol. Rev.* 17, 942–964.

Rollins BJ **1996**, Monocyte chemoattractant protein 1: a potential regulator of monocyte recruitment in inflammatory disease, *Mol. Med. Today* 2, 198–204.

Rot A, von Andrian UH **2004**, Chemokines in innate and adaptive host defense: basic chemokinese grammar for immune cells, *Annu. Rev. Immunol.* 22, 891–928.

Santambrogio L, Belyanskaya SL, Fischer FR, Cipriani B, Brosnan CF, Ricciardi-Castagnoli P, Stern LJ, Strominger JL, Riese R **2001**, Developmental plasticity of CNS microglia, *Proc. Natl Acad. Sci. USA* 98, 6295–6300.

Schmidtmayerova H, Nottet HS, Nuovo G, Raabe T, Flanagan CR, Dubrovsky L, Gendelman HE, Cerami A, Bukrinsky M, Sherry B **1996**, Human immuno-deficiency virus type 1 infection alters chemokine beta peptide expression in human monocytes: implications for recruitment of leukocytes into brain and lymph nodes, *Proc. Natl Acad. Sci. USA* 93, 700–704.

Serafini B, Columba-Cabezas S, Di Rosa F, Aloisi F **2000**, Intracerebral recruitment and maturation of dendritic cells in the onset and progression of experimental autoimmune encephalomyelitis. *Am. J. Pathol.* 157, 1991–2002.

Sievers J, Parwaresch R, Wottge HU **1994**, Blood monocytes and spleen macrophages differentiate into microglia-like cells on monolayers of astrocytes: morphology, *Glia* 12, 245–258.

Simard AR, Rivest S **2004**, Bone marrow stem cells have the ability to populate the entire central nervous system into fully differentiated parenchymal microglia, *FASEB J.* 18, 998–1000.

Simpson JE, Newcombe J, Cuzner ML, Woodroofe MN **1998**, Expression of monocyte chemoattractant protein-1 and other beta-chemokines by resident glia and inflammatory cells in multiple sclerosis lesions, *J. Neuroimmunol.* 84, 238–249.

Song L, Pachter JS **2004**, Monocyte chemoattractant protein-1 alters expression of tight junction-associated proteins in brain microvascular endothelial cells, *Microvasc. Res.* 67, 78–89.

Sorensen TL, Tani M, Jensen J, Pierce V, Lucchinetti C, Folcik VA, Qin S, Rottman J, Sellebjerg F, Strieter RM, Frederiksen JL, Ransohoff RM **1999**, Expression of specific chemokines and chemokine receptors in the central nervous system of multiple sclerosis patients, *J. Clin. Invest.* 103, 807–815.

Sorensen TL, Trebst C, Kivisakk P, Klaege KL, Majmudar A, Ravid R, Lassmann H, Olsen DB, Strieter RM, Ransohoff RM, Sellebjerg F **2002**, Multiple sclerosis: a study of CXCL10 and CXCR3 co-localization in the inflamed central nervous system, *J. Neuroimmunol.* 127, 59–68.

Sorensen TL, Ransohoff RM, Strieter RM, Sellebjerg F **2004**, Chemokine CCL2 and chemokine receptor CCR2 in early active multiple sclerosis, *Eur. J. Neurol.* 11, 445–449.

Sozzani S, Locati M, Zhou D, Rieppi M, Luini W, Lamorte G, Bianchi G, Polentarutti N, Allavena P, Mantovani A **1995**, Receptors, signal transduction, and spectrum of action of monocyte chemotactic protein-1 and related chemokines, *J. Leukoc. Biol.* 57, 788–794.

Stoll G, Jander S, Schroeter M **2002**, Detrimental and beneficial effects of injury-induced inflammation and cytokine expression in the nervous system, *Adv. Exp. Med. Biol.* 513, 87–113.

Takahashi K, Yamamura F, Naito M **1989**, Differentiation, maturation, and proliferation of macrophages in the mouse yolk sac: a light-microscopic, enzyme-cytochemical, immunohistochemical, and ultrastructural study, *J. Leukoc. Biol.* 45, 87–96.

Tran EH, Kuziel WA, Owens T **2000**, Induction of experimental autoimmune encephalomyelitis in C57BL/6 mice deficient in either the chemokine macrophage inflammatory protein-1alpha or its CCR5 receptor, *Eur. J. Immunol.* 30, 1410–1415.

Trebst C, Ransohoff RM **2001**, Investigating chemokines and chemokine receptors in patients with multiple sclerosis: opportunities and challenges, *Arch. Neurol.* 58, 1975–1980.

Trifilo MJ, Lane TE **2003**, Adenovirus-mediated expression of CXCL10 in the central nervous system results in T-cell recruitment and limited neuropathology, *J. Neurovirol.* 9, 315–324.

Unger ER, Sung JH, Manivel JC, Chenggis ML, Blazar BR, Krivit W **1993**, Male donor-derived cells in the brains of female sex-mismatched bone marrow transplant recipients: a Y-chromosome specific in situ hybridization study, *J. Neuropathol. Exp. Neurol.* 52, 460–470.

Vallieres L, Sawchenko PE **2003**, Bone marrow-derived cells that populate the adult mouse brain preserve their hematopoietic identity, *J. Neurosci.* 23, 5197–5207.

Wang F, Graham WV, Wang Y, Witkowski ED, Schwarz BT, Turner JR **2005**, Interferon-gamma and tumor necrosis factor-alpha synergize to induce intestinal epithelial barrier dysfunction by up-regulating myosin light chain kinase expression, *Am. J. Pathol.* 166, 409–419.

Weiss JM, Downie SA, Lyman WD, Berman JW **1998**, Astrocyte-derived monocyte-chemoattractant protein-1 directs the transmigration of leukocytes across a model of the human blood-brain barrier, *J. Immunol.* 161, 6896–6903.

Weiss JM, Nath A, Major EO, Berman JW **1999**, HIV-1 Tat induces monocyte chemoattractant protein-1-mediated monocyte transmigration across a model of the human blood-brain barrier and up-regulates CCR5 expression on human monocytes, *J. Immunol.* 163, 2953–2959.

Wolburg H, Wolburg-Buchholz K, Engelhardt B **2005**, Diapedesis of mononuclear cells across cerebral venules during experimental autoimmune encephalomyelitis leaves tight junctions intact, *Acta Neuropathol. (Berl.)* 109, 181–190.

Zang YC, Samanta AK, Halder JB, Hong J, Tejada-Simon MV, Rivera VM, Zhang JZ **2000**, Aberrant T cell migration toward RANTES and MIP-1 alpha in patients with multiple sclerosis. Overexpression of chemokine receptor CCR5, *Brain* 123, 1874–1882.

8
The Bipolar Astrocyte: Polarized Features of Astrocytic Glia Underlying Physiology, with Particular Reference to the Blood-Brain Barrier

N. Joan Abbott

8.1
Introduction

During the development of the mammalian central nervous system (CNS) from the simple neural tube, astrocytes derived from multipotent precursor cells frequently show a transition from a bipolar to a multipolar form, i.e. from a columnar or spindle-shaped cell with a single process at either end of the cell body to a cell with multiple processes, including the highly branched phenotype that gave rise to the term "astrocyte" (star-shaped or stellate cells). The bipolar phenotype, with distinct apical and basal poles at opposite ends of the cell derives from the epithelial organization and function of the neural tube. In the mature CNS, astrocytes with a wide variety of morphologies can be observed, but interestingly, many of them preserve aspects of a polarized phenotype, even when the apical-basal (bipolar) distinction is less clear. The mammalian blood-brain barrier (BBB) formed by the brain endothelium is induced and maintained by chemical factors derived from the underlying neural tissue, and the polarized processes of astrocytes associated with the endothelium are responsible for the induction of a number of features of the BBB phenotype. Moreover, polarized signaling between astrocytes and endothelium is important in the modulation of BBB function. This chapter reviews the origin of polarity in the astrocyte population and ways in which this polarity is crucial to function, with particular emphasis on the BBB interface.

8.2
Formation of the Neural Tube

The CNS of vertebrates is formed from the ectoderm of the embryo. In the process of "neurulation", the dorsal ectoderm (skin) folds inwards to create the neural plate along the longitudinal axis, followed by fusion of the lips of the plate (neural folds) to form the neural tube (Fig. 8.1). The neural tube then separates

Blood-Brain Interfaces: From Ontogeny to Artificial Barriers.
Edited by R. Dermietzel, D.C. Spray, M. Nedergaard
Copyright © 2006 WILEY-VCH Verlag GmbH & Co. KGaA, Weinheim
ISBN: 3-527-31088-6

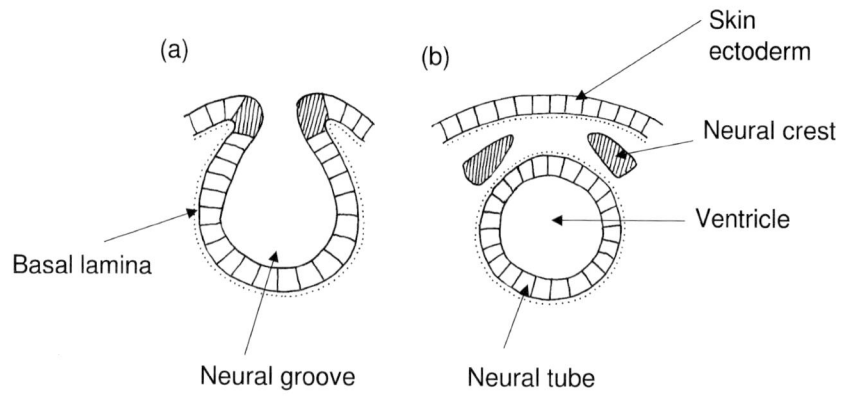

Fig. 8.1 Apical-basal polarity set up during neurulation. During development, the mammalian CNS forms by invagination of the surface ectoderm of the neural plate (A), which subsequently seals up and sinks below the surface to form the neural tube (B). The lips of the neural plate (neural folds) become the migratory population of neural crest cells, which contribute to the peripheral nervous system.

and sinks below the skin, with cells from the edges of the fold (neural crest) forming a migratory population and giving rise to many cells of the peripheral nervous system [1]. The cells of the skin ectoderm, like all epithelial layers forming an interface between tissue and a fluid-filled space (here the external environment) express a strong polarity, the apical cell membrane facing the fluid space, and the basal cell membrane secreting and apposed to a layer of specialized extracellular matrix, the basal lamina/basement membrane. This apical-basal polarity is retained in the neuroepithelial cells forming the single-layered neural tube (Fig. 8.1). As with many epithelia facing an internalized fluid cavity (e.g. small intestine, kidney tubule), the apical membrane surface area may be increased by the presence of fine processes (cilia, microvilli).

8.3
Origin of Neurons and Glia

The neuroepithelial cells are capable of forming both neurons and macroglia – the latter comprising oligodendrocytes, astrocytes, and ependymoglia [2]. Cell division among the earliest neuroepithelial cells forms the ventricular zone (VZ), a pseudostratified epithelium adjacent to the lumen of the neural tube [3]. Within the early VZ can be observed a prominent population of radially orientated cells spanning the thickness of the neural tube, which show many characteristics in common with mature astrocytes, including the expression of GLAST-type glutamate transporters, brain lipid-binding protein (BLBP), and the glial-fibrillary acidic protein GFAP [4]. This class of cells has been called the *radial glia* (RG); and they help guide the migration of immature neurons to the cortex. The RG successively replace the neuroepithelial cells and represent more fate-restricted progenitors, most RG giving rise to a single cell type, either astrocytes, oligodentrocytes or neurons [4]. Meanwhile, cells dividing within the VZ migrate into the adjacent subventricular zone (SVZ), which in turn becomes the major germinal site. Early neuroblasts in the SVZ first generate a population of neurons, before activation of a molecular "germinal switch" leads to the later production of oligodendrocytes and astrocytes [5, 6].

Recent evidence suggests that within the spinal cord astrocytes and oligodendrocytes are derived from different precursor populations, with oligodendrocytes arising from oligodendrocyte precursor cells (OLPs) in the ventral pMN domain, while astrocytes arise from more dorsal (p2) domains [6, 7]. Thus, the generation of astrocytes occurs in two distinct phases during embryonic and fetal stages of development, deriving initially from the radial glia, and later from migratory progenitors that emerge from the dorsolateral SVZ to colonize adjacent gray and white matter. The precise sequence of events and mechanisms responsible for the generation of neurons, oligodendrocytes, and astrocytes is not yet fully understood, but it is already known to depend on the location within the CNS [8].

The germinal switch from neurons to glia is determined by a combination of intrinsic (genetic) and extrinsic signals, with some reciprocal interactions, e.g. forced expression of the oligodendrocyte gene *Olig2* inhibits astrocyte differentiation, and forced expression of the activated Notch receptor *Notch1a* promotes formation of the astrocyte lineage, including RG and Müller cells in the retina [6, 9]. The sequence of events involved in the neuron-glial switch is better understood for oligodendrocytes than astrocytes, as more marker proteins for the stages of development and differentiation have been identified for oligodendrocytes.

The major phase of neurogenesis (birth of neurons) occurs before birth in most mammals, with neuronal precursors migrating along the scaffold formed by RG to their final positions, where synapse formation and connectivity are established. In the adult, the sites of neurogenesis from endogenous precursors are predominantly confined to two major sites – the SVZ giving rise to the rostral migratory stream (RMS) of neurons migrating to the olfactory bulbs, and the hilus of the dentate gyrus generating cells for adjacent regions of the hippocampus [3, 10].

The early-appearing radial glia have a clear bipolar morphology, with the basal membrane forming the *glia limitans* adjacent to the pia at the outer surface of the brain and the apical membrane facing the fluid-filled cavity of the neural tube that will become the adult ventricular system [2]. Once the major wave of neuronal migration is complete, the majority of RG retract one or both processes from the inner and outer brain surfaces, develop a more branching morphology, and begin to express other features of differentiated astrocytes [11], while processes from the basolateral parts of the cell can contact a number of cell types, from adjacent astrocytes to capillary endothelial cells to neurons. An idealized ependymoglial cell showing these contacts is illustrated in Fig. 8.2 [12]. Interestingly, it was recently found that the RG guiding radial neuronal migration later extend tangentially running basal processes that channel the mi-

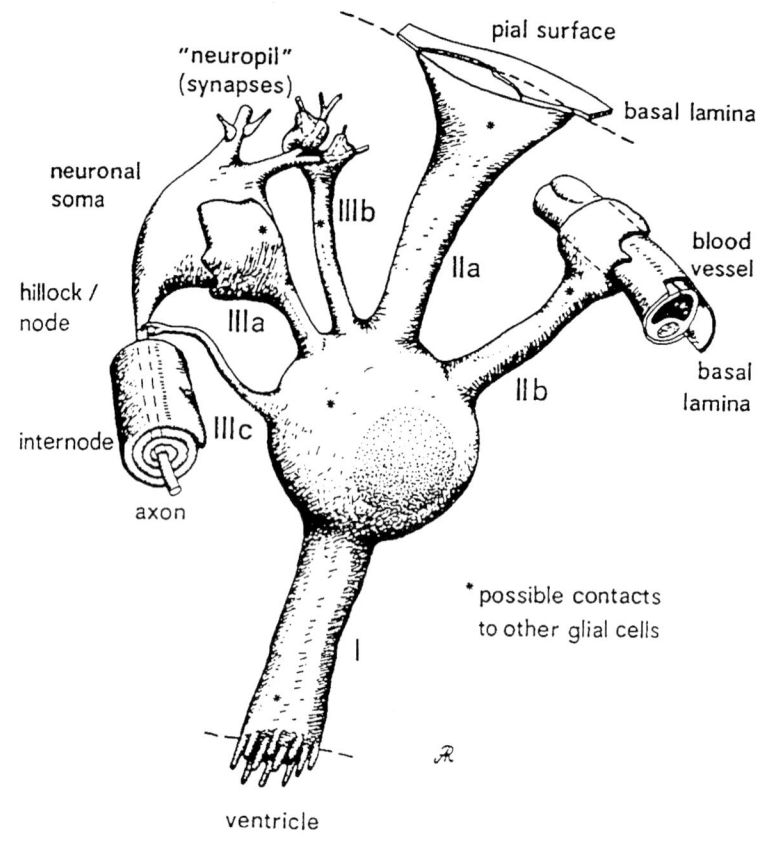

Fig. 8.2 Idealized ependymoglial cell: diagrammatic representation of apical and basal processes from a "generic" ependymoglial cell derived from radial glia, showing the potential for cell-cell interactions. The apical process (I) initially makes contact with the ventricle, but may later retract. Basal or baso-lateral processes may contact a number of other structures and cell types, including the basal lamina at ectoderm-mesoderm interfaces at the pial surface (IIa) and blood vessels (IIb), and neuronal somata (IIIa), synapses (IIIb), or axons, especially in the region of the axon hillock and nodes of Ranvier (IIIc). (Modified from [12] with permission).

gration of the RMS [13]. Even more intriguingly, the migrating neuroblasts that form a moving chain of cells destined for the olfactory bulb express a number of glial features [14]. Thus, the glial cells of these germinal zones may provide both the scaffold and the migratory population – suggesting that these cells retain a multipotent phenotype.

The cell layer left lining the ventricular cavity after formation of the SVZ is termed the ependyma, and the third class of macroglia, the ependymoglia, express some glial features and retain contact with this CSF-containing space. They include the tanycytes of the mature CNS and the Müller cells of the retina [2].

8.4
Morphology of Glial Polarity in Adult CNS

Within the adult central nervous system, a variety of different astrocyte morphologies can be observed (Fig. 8.3). Reichenbach [2] illustrated 11 different astrocytic-ependymoglial types, based on location and morphology; and, interestingly, at least eight of these make specialized contacts on cerebral microvessels. As immunocytochemical characterization of these glia becomes more complete, it will be possible to establish whether the differences in morphology and contacts relate to differences in expression of marker proteins, and more importantly, differences in function.

In a number of locations in the adult CNS, cells related to RG or ependyma retain a clear bipolar morphology. Thus, ependymal cells themselves generally

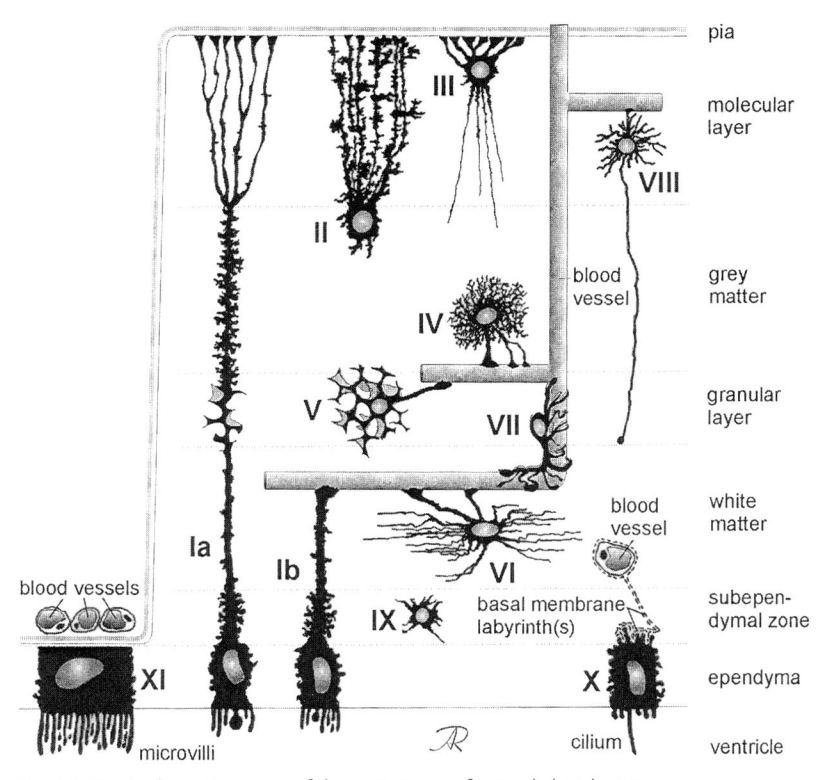

Fig. 8.3 Semi-schematic survey of the main types of astroglial and ependymoglial cells and their localization in different layers/specialized regions of the central nervous system. (I) Tanycyte (a, pial; b, vascular), (II) radial astrocyte (Bergmann glial cell), (III) marginal astrocyte, (IV) protoplasmic astrocyte, (V) velate astrocyte, (VI) fibrous astrocyte, (VII) perivascular astrocyte, (VIII) interlaminar astrocyte, (IX) immature astrocyte/glioblast, (X) ependymocyte, (XI) choroid plexus cell. (From [2] with permission).

have extensive apical microvilli contacting the ventricular CSF and a basal membrane associated with a prominent basal lamina, also called the "subependymal basement membrane" or basal membrane labyrinth (Fig. 8.3) [15, 16]. The specialized ependymal cells forming the choroid plexus (CP) epithelium show similar morphology and polarity, but the space between the basal laminae of epithelium and capillaries is filled with a "stromal" extracellular matrix. In the eye, both the retinal receptor layer and the retinal pigment epithelium (RPE) are ependymal derivatives, with their apical surfaces apposed to each other across the restricted remnant of the CSF-containing space between them [17].

Tanycytes are glia that retain an apical connection to the CSF, while the basal process makes contact either with other tanycytes or with capillaries. They are prominent components of neurosecretory zones (e.g. lining the floor of the third ventricle) [18]. Müller cells are effectively the RG of the retina. Depending on the species, the apical microvilli of Müller cells are more or less prominent and a well developed basal end foot faces the vitreous [2]. Where the apical connection to the ventricular or other fluid space is lost, the glial apical-basal polarity is less clear-cut (Fig. 8.2) but the basal specialisations, especially end feet on blood vessels and the *glia limitans* (Figs. 8.3 and 8.4) show the retained polarity.

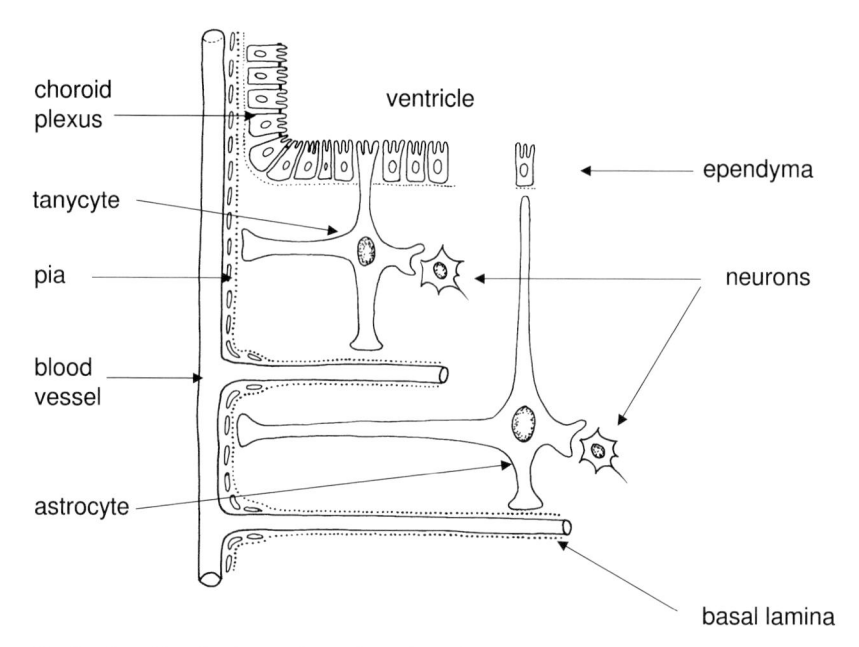

Fig. 8.4 Apical-basal polarity of ependymoglia and astrocytes in the adult CNS. In tanycytes, the apical process remains in contact with the CSF of the ventricles, while the basolateral processes may contact cells within the parenchyma, at the pial surface, and forming blood vessels. More commonly, the apical process is reduced, separating from the ependymal cell lining the ventricle, while the basolateral processes are retained.

There appears to be some plasticity between the bipolar and multipolar astrocytic forms. Moreover, some agents capable of triggering the change have been identified. Thus, transforming growth factor alpha (TGFα) is able to induce the transformation of polygonal astrocytes into a bipolar (RG-like) form in culture, with associated increase in the expression of GFAP and a reduction in motility [9]. In vivo, branched forms may revert to the bipolar form under some conditions; thus, damage and grafting may cause local stellate astrocytes to assume the bipolar shape [19], implying considerable plasticity. A subset of glial progenitor cells (and tumors) express gp130, important in glial differentiation, including conversion from multipolar to bipolar morphology [20]. Elevation of D-cyclin-associated kinases (cdk) is seen in some gliomas; and transfection to elevate cdk6 (but not cdk4) results in a change from multipolar to bipolar form [21]. There is remarkable sexual dimorphism in the arcuate nucleus of the hypothalamus, with complex stellate astrocytes in males and simpler bipolar cells in females – likely resulting from differences in gonadal steroid exposure during development and related to sexually dimorphic regulation of neuroendocrine secretions from the pituitary in the adult [22]. A population of bipolar astrocytes (GFAP-ve, vimentin +ve) is present in actively remyelinating MS lesions [23]. Taken together, these studies indicate considerable plasticity in morphology, even within the adult astrocyte population, and reversion to a less differentiated bipolar phenotype under some pathological conditions.

8.5
Astrocyte Spacing and Boundary Layers

Even in the earliest camera lucida drawings of brain sections by Lenhossek [24] and Golgi [25] (see [26]), the astrocytes appear to show a rather regular spacing, but given the fact that the Golgi impregnation method of silver staining fills only a proportion of cells present, the significance of this was unclear. In culture, it is observed that astrocytes go through a series of morphological changes, from bipolar (generally non-contacting) to stellate (contacting), including the extension and contraction of processes that result in regular spacing [27]. Astrocytes at laminar boundaries separating the association/commissural and perforant path afferents of the hippocampal dentate gyrus show a more polarized morphology than the stellate astrocytes typical of non-boundary zones, although there is no simple relation to the morphology of the laminar boundary [28]. Such studies suggest that cell morphology, polarity, and territory is influenced by interactions of astrocytes with each other and with extracellular molecules defining tissue boundaries. The result is not only morphological polarisation appropriate to function, but also an appropriate spacing that allows individual astrocytes to control a defined three-dimensional sector of parenchymal space [29, 30].

8.6
Origin and Molecular Basis of Cell Polarity

Cell polarity is traditionally seen as a segregation of protein and lipid repertoires, i.e. distinct sets of membrane proteins (receptors, ion channels, transporters, adhesion molecules) specific for a particular cellular domain [31]. In epithelia, the clear apical/basolateral polarity is established and maintained by a number of mechanisms [32–34]: (1) cell-cell contact is sufficient to trigger the segregation of marker proteins of the apical and basolateral membrane domains to distinct regions of the membrane, (2) cell-cell contact induces association of tight junction proteins, including ZO-1, to the apex of the lateral membrane and the transmembrane proteins act as a diffusional barrier ("fence"), restricting the migration of proteins between domains, (3) establishment of the epithelial axis requires the formation of a basal lamina and cell-substratum contact, and (4) reversal of cell polarity can occur if the apical side of the cell is presented with an appropriate extracellular matrix. Rho GTPases, including cdc42, are important in establishing cell polarity; and it was recently found that the Par3/Par6/aPKC complex acts as an evolutionarily conserved polarization signal; and a dynamic feedback loop at the apical junctional complex with Scrib/Dig/Lgl and Crumbs/Pals1/PATJ complexes organizes and defines the boundaries of polarized domains [35].

Within the adult mammalian CNS, ependymoglial derivatives make tight junction-coupled barrier layers at a number of locations, including the specialized ependyma overlying circumventricular organs (CVO), tanycytes enclosing neurosecretory zones, the choroid plexus epithelium, and the retinal pigment epithelium (Fig. 8.5). The strong apical-basal morphological polarity of these layers reflects their functional polarity (see below).

In cells which do not have tight junctions, polarity can be defined by stabilization of protein complexes at the membrane by scaffolding proteins and selective targeting of proteins along the secretory or endosomal pathways, e.g. involving localization of the Golgi apparatus to the apical region for polarized secretion [31]. Thus, in neurons, significant polarization between dendritic and axonal domains can be achieved, and even without tight junctions, specialization of the membrane and its protein complexes in the region of the axon initial segment can act as a barrier to diffusion within the membrane. Further specialization can occur even within one domain, e.g. subcompartments within the dendritic tree. In migrating astrocytes, mPar6 and PKCzeta are localized to the leading edge of the cell; and they reorient the microtubule organizing center (MTOC) to the side of the nucleus facing the direction of migration [36, 37]. Cdc42 acts through the Par3/Par6/aPKCzeta complex to phosphorylate and hence inhibit glycogen synthase kinase 3β (GSK-3β), allowing association of the tumor suppressor protein adenomatous polyposis coli (APC) with microtubules, stabilizing the leading edge [37]. This generates a new polarity independent of the original polarity of cells within the neural tube.

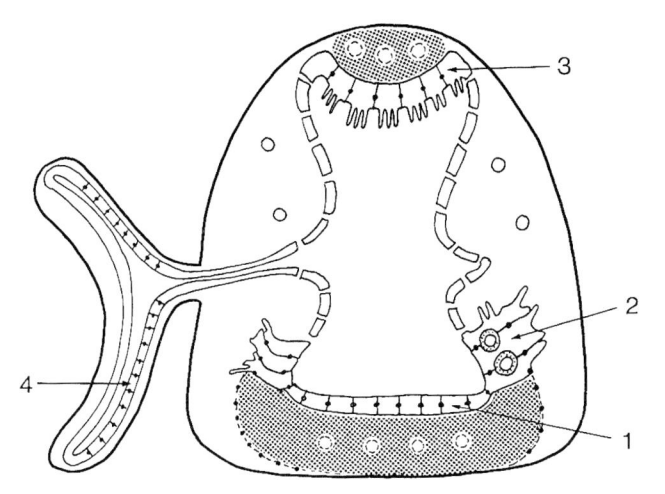

Fig. 8.5 Ependymoglial derivatives coupled by tight junctions and contributing to barrier layers in the adult mammalian CNS. (1) Ependyma overlying neurosecretory zones (circumventricular organs) such as the median eminence, (2) tanycytes surrounding neurosecretory zones, (3) choroid plexus epithelial cells, (4) retinal pigment epithelium. Capillaries with tight endothelium are shown as circles with solid line, leaky endothelium with dashed line. (Based on [137], with permission).

8.7
Functional Polarity of Astrocytes and Other Ependymoglial Derivatives

Epithelial cell layers characterized by occluding tight junctions and significant transepithelial fluid movement can show functional polarization of two main types: secretory, with a net movement of fluid and solutes from basal to apical side (e.g. salivary gland acinus), and absorptive, with net movement of solutes and fluid from apical to basal side (e.g. kidney proximal tubule) [38]. A number of mechanisms are available for generating the solute and fluid movements; and individual solutes may be transported against the fluid stream. Most epithelial cells use a basolateral Na,K,ATPase to generate a Na^+-K^+ ionic gradient across the cell membrane. The subsequent direction and composition of the secretory or absorptive flow is regulated by the differential distribution of apical and basal ion channels and transporters, and where large volume flows occur, by the presence of aquaporin (AQP) water channels. In the choroid plexus epithelium, as for other neuroepithelial cells, the Na,K,ATPase is on the apical side and drives a Na^+, Cl^- and HCO_3^- rich secretion of CSF into the ventricles, with water flows mediated by apical AQP1 channels [39]. Several mechanisms acting together give rise to a net absorption of K^+, from CSF to blood. In the retinal pigment epithelium (RPE), with a similarly polarized transporter distribution, the movement of water must be in the opposite direction to maintain the close physical and functional association of receptors and RPE and to avoid retinal detachment [40]. The high density of AQP4 on the (basal) end feet of astrocytes on capillaries and at the pial surface suggests the potential for significant water flux via astrocytes, both constitutively and following neural activity, from the parenchyma to the perivascular space and the CSF of the subarachnoid space [41, 42]. The brain endothelium has undetectable AQP1 and only very low levels of AQP4 [43], consistent with this being a large surface area membrane specialized for regulating molecular flux rather than sustaining large water movements [44]. Hence, under normal conditions, it is unlikely that transendothelial flux represents a significant mode of water clearance. Instead, the perivascular basal lamina at the capillary level, connecting with perivascular (Virchow-Robin) spaces around larger vessels may be a major route for water clearance from the parenchyma, with the CSF of the subarachnoid space being the sink. However, given that the total number of AQP4 channels on fine astrocytic processes within the neuropil is about equal to those on perivascular end feet [41], a significant water flux is also likely through the parenchyma, some of which may exit across the ependyma into the ventricular CSF. Within the neuropil, extracellular water movements are expected to be mainly by diffusion, intercellular clefts being too narrow to permit significant convective (bulk) flow. However, bulk flow has been documented along the wider (low resistance) pathways created by axon tracts and perivascular spaces [44].

Within the population of processes from highly branched stellate astrocytes, it is frequently possible to observe processes on both neurons and blood vessels. A number of proteins are clustered on the perivascular end feet, including AQP4, Kir 4.1, and rSlo K_{Ca} channels [45]. Astrocytes and Müller cells often express different classes of transmitter transporters compared with neurons, and although not specifically documented, these would be expected to be mainly on the neuron-facing processes of the glia – thus, while most retinal neurons express GLT1v and EAAC1 glutamate transporters, those on astrocytes and Müller cells are predominantly GLAST [46]. Astrocytes have also been shown to express GAT1 and 2/3 GABA transporters [47]. Limitrin is a new member of the transmembrane-type immunoglobulin superfamily specifically localized to the polarized astroglial end feet and *glia limitans* [48].

A number of anchoring proteins are responsible for the polarized clustering of proteins in the BBB (abluminal (basal) membrane of brain endothelial cells), the blood-CSF barrier, and the perivascular end feet of astrocytes. Recent studies indicate the particular importance of utrophin in the basolateral membrane of the choroid plexus epithelium and vascular endothelial cells, while the short C-terminal isoform of dystrophin (Dp71) is localized to the glial end feet [49]. Both proteins act as anchors for the dystrophin-associated protein complex (DPC) composed of isoforms of syntrophin, dystroglycan, and dystrobrevin, which in turn associates with AQP water channels. Studies showing disturbed ion and water regulation in transgenic animals [49] confirm the importance of the polarized function at these key blood-CNS tissue interfaces.

8.8
Secretory Functions of Astrocytes

Astrocytes have been shown to express and release a number of agents capable of acting on other cells within the nervous system, including other macroglia, neurons, endothelial cells, and microglia. Substances released include neurotransmitter and modulator substances, cytokines, chemokines, and growth factors (Table 8.1); and the agents may have inductive effects (influencing the phenotype of the recipient cell), or act as shorter-term signaling molecules (see below). The majority of these studies have been on cells in culture, as it is very difficult to do the equivalent experiments in vivo or in brain slices. From the observed differences in morphology and cell-cell associations in vivo, one would predict that the perivascular and perineuronal cell processes of astrocytes could release different agents, but studies have generally not been done to test this idea. There are likely to be interesting differences in the effective extracellular compartment size between perisynaptic release sites (narrow extracellular spaces and synaptic clefts) [50] and perivascular release sites (facing the extracellular matrix surrounding the blood vessel, frequently with an expanded perivascular space). These differences in volume will give different degrees of dilution of the released agents; and the likely channeling of ISF flow in low-resistance perivascular spaces will be an additional factor, affecting both the dilution and the sphere of influence of the agent [44, 51]. These are some of the complexities likely to result from differences in the behavior of different astrocyte processes.

8.9
Induction of BBB Properties in Brain Endothelium

It is clear from the above discussion that the polarized properties of astrocytes play an important role at the blood-brain interface, particularly in relation to the physiology of ion and water homeostasis. The way in which close communication between astrocytes and endothelium leads to induction of BBB properties is covered in detail by Galla's team [52] and Wolburg [53] (Chapters 16 and 4 in this volume). Here it is simply worth noting that these inductive influences depend on the polarized properties of both astrocytes and endothelium; and if either breaks down, for example by disruption of the anchoring proteins that stabilize the clustering of channels and transporters in particular membrane domains, then induction may be impaired [54–57].

Even before much was known about the mechanisms underlying the induction of barrier layers at specific sites within the CNS (in mammals: brain endothelium, choroid plexus epithelium, RPE, tanycytes, ependyma surrounding neurosecretory zones; in invertebrates: predominantly glial layers [17, 58]), it was proposed that induction could require the combined influence of factors from the blood side and from the brain side, with the final sites of barrier location being the cell layers giving the greatest separation of the two [17]. It was

Table 8.1 Substances secreted by astrocytes.

Category	Substance	Reference	Notes
(A) Peptides/growth factors, proteins	Angiotensinogen	77, 78	
	GDNF	79–84	
	BDNF	81–85	
	NGF	82–86	
	IGF1	87	Insulin-like growth factor
	ADNP	88	+NAP, eight amino acids
	FGF1 (aFGF)	89	See also ApoE
	FGF2 (bFGF)	90	
	FLRG	91	Follistatin-related gene
	Endozepines	92	Neuropeptide related to Octadecaneuropeptide (ODN)
	S100β	93–95	
	GM-CSF	96	Granulocyte-macrophage colony-stimulating factor
	Clusterin	97	Sulphated glycoprotein
	APP	98	Amyloid precursor protein
	>30 proteins (not all identified)	99	Proteases, protease inhibitors, carrier proteins, antioxidants
(B) Cytokines	TGF-α	87, 100	
	TGF-β1	87, 100–105	
	IL-1β, -6, -10, -12	1β: 106, 6: 107, 108, 10: 109, 12p40: 110	
	TNFα	111, 112	
(C) Chemokines	MCP-1	113	
	IL-8	114, 115	
	MIP3α/CCL20	116	
	Fractaline	117	
(D) Extracellular matrix proteins, modulators	TenascinC	118	
	MMP1, -2	119	Matrix metalloproteinases
	MMP2, -3, -9 +tPA	120	
	MMP9	121	
	TIMP-1	99, 119, 122	Tissue inhibitors of metalloproteinases

Table 8.1 (continued)

Category	Substance	Reference	Notes
(E) Sterols, lipidic substances	Cholesterol, desmosterol, labosterol	123	
	ApoA1	124	Cholesterol efflux to ApoA-1
	ApoE	89, 125–127	
	ApoJ	128	ApoE synthesis, HDL secretion
	PGD2	129	Prostaglandins
	PGE2	87, 129, 130	
	Hydroxypregnane one	87	3-α-hydroxy-5-α-pregnane-20-one
	Arachidonic acid	131	
(F) Neurotransmitters, neuroactive agents	GABA	132	
	Glutamate	133	
	ATP	134–136	
	Nitric oxide	108	

proposed that when the tight junctions within an epithelial or endothelial layer became sufficiently occluding to restrict the diffusion of potential inducing proteins, then this cell layer would experience the greatest difference between apical and basal microenvironments, and would be the one in which barrier induction would subsequently be most effective. Moreover, as this layer tightened during development, it would automatically limit the access of plasma factors to underlying cells, confining the barrier site to a single cell layer. This is still a useful working hypothesis, which has been able to accommodate more recent cellular and molecular advances. Many of the factors released by astrocytes (Table 8.1) are able to induce specific features of the BBB phenotype in brain endothelial cells [59, 60]. In some cases, close contact between the basolateral surfaces of the endothelial and glial cells is more effective in BBB induction than astrocyte-conditioned medium, which may reflect the need to preserve the correct cell polarity for optimal interaction [61].

It is now clear that, in addition to the important role of tight junctions in the induction and polarization of CNS barrier properties, the extracellular matrix is a critical anchoring site mediating the polarizing influences at the endothelial-astrocyte interface [53, 62] and in other polarized ependymoglial derivatives, such as the RPE [63]. New information is becoming available about regional differences in tight junctions, extracellular components, astrocyte polarity, and the inductive factors released by astrocytes and endothelium. This will aid our understanding of the more subtle aspects of induction, and disturbances in pathology, with possible insights leading to prospects for repair.

8.10
Astrocyte-Endothelial Signaling

In addition to the longer-term inductive influences between astrocytes and endothelium discussed above, evidence exists for significant short-term cell-cell signaling between these cells [59, 64–66], with the polarized end feet of astrocytes on the capillary endothelium being the most likely key site for this signaling in vivo. Of the range of substances released by astrocytes (Table 8.1), a number are classic neurotransmitters and modulators, capable of acting not only on neurons but also on endothelium, pericytes, smooth muscle, and other perivascular cells [58].

For many of these agents, receptor mechanisms have been reported in brain endothelium [59, 67]. Evidence comes largely from cell culture studies in which molecular and functional examinations were used to identify the endothelial receptors and signal transduction pathways involved [59, 68]. Where it has been possible to use in situ or brain slice preparations, broadly similar results are found, although in some cases it appears that the intact system expresses a more limited receptor profile than seen in vitro (e.g. it has not been possible to detect functional P2Y2 receptors on capillary endothelium in slices, although these are well documented in cultured cells [59, 69]).

The downstream consequences of activation of these receptors include modulation of a number of brain endothelial functions [70, 71], including permeability (tight junction modulation) [64, 67], transport [65, 66, 72], and enzymatic activity. In many of the cell culture studies, it has not been possible to determine whether the receptors are on the apical or basolateral membrane of the endothelium, although the latter would be expected for targets of astrocyte-released agents. However, in culture models preserving a relatively restricting paracellular (tight junctional) permeability, the efficacy of compounds applied to the apical and basal membranes can be compared [73–75]. In the in situ single pial microvessel preparation, agents are generally applied in the superfusate (basal side of endothelium), but can also be introduced via the lumen (apical side) [76]. Studies in brain slices offer the potential to examine whether signals released by astrocytes act only or directly on endothelial cells, or whether other cell types are involved [69].

8.11
Conclusion

This chapter has highlighted the way in which astrocyte polarity derives initially from polarity in the developing neural tube, but that this is modified thereafter by a combination of intrinsic and extrinsic influences that contribute to the bipolar and multipolar morphology of adult astrocytes and related ependymoglial derivatives in different brain regions. The importance of astrocyte polarity for function is emphasized, particularly in relation to induction of BBB features in the brain endothelium, and to secretion of agents capable of influencing neighboring cells, including other astrocytes, neurons, and endothelium. As the details of regional differences in receptors and transmitters on the fine processes of astrocytes become better understood, it will be easier to define the functional domains equivalent to the sphere of influence of individual astrocytes, and the different activities regulated by processes onto specific neighboring cells. In turn, these specific interactions are likely to depend on local signals that regulate the polarized properties of the apposed cell processes.

References

1 D. H. Sanes, T. A. Reh, W. A. Harris **2005**, *Development of the Nervous System*, 2nd edn. Academic Press, San Diego.
2 A. Reichenbach, H. Wolburg **2004**, Astrocytes and ependymal glia, in *Neuroglia*, 2nd edn, eds. H. Kettenmann, B. R. Ransom, Oxford University Press, New York.
3 L. Conti, T. Cataudella, E. Cattaneo **2003**, *Pharmacol. Res.* 47, 289–297.
4 M. Götz, W. B. Huttner **2005**, *Nature Rev. Mol. Cell Biol.* 6, 777–788.
5 M. Zerlin, S. W. Levison, J. E. Goldman **1995**, *J. Neurosci.* 15, 7238–7249.
6 D. H. Rowitch **2004**, *Nat. Rev. Neurosci.* 5, 409–419.
7 Q. Zhou, D. J. Anderson **2002**, *Cell* 109, 61–73.
8 J. E. Goldman **2004**, Lineages of astrocytes and oligodendrocytes, in *Neuroglia*, 2nd edn, eds. H. Kettenmann, B. R. Ransom, Oxford University Press, New York.
9 R. Zhou, X. Wu, O. Skalli **2001**, *Brain Res. Bull.* 56, 37–42.
10 D. Y. Zhu, S. H. Liu, H. S. Sun, Y. M. Lu **2003**, *J. Neurosci.* 23, 223–229.
11 N. Ulfig, F. Neudorfer, J. Bohl **1999**, *J. Anat.* 195, 87–100.
12 A. Reichenbach, S. R. Robinson **1995**, Ependymoglia and ependyma-like cells, in *Neuroglia*, eds. H. Kettenmann, B. R. Ransom, Oxford University Press, New York.
13 J. A. Alves, P. Barone, S. Engelender, M. N. M. Froes, J. R. Menezes **2002**, *J. Neurobiol.* 52, 251–265.
14 A. Alvarez-Buylla, J. M. Garcia-Verdugo **2002**, *J. Neurosci.* 22, 629–634.
15 H. Leonhardt **1972**, *Z. Zellforsch.* 127, 392–406.
16 H. F. Cserr, C. Patlak **1992**, Secretion and bulk flow of interstitial fluid, in *Physiology and Pharmacology of the Blood-Brain Barrier*, ed. M. W. B. Bradbury, Springer, Heidelberg.
17 N. J. Abbott **1992**, Comparative physiology of the blood-brain barrier, in *Physiology and Pharmacology of the Blood-Brain Barrier*, ed. M. W. B. Bradbury, Springer, Heidelberg.
18 L. M. Garcia-Segura, M. M. McCarthy **2004**, *Endocrinology* 145, 1082–1086.
19 B. R. Leavitt, C. S. Hernit-Grant, J. D. Macklis **1999**, *Exp. Neurol.* 157, 43–57.

73 L. L. Rubin, D. E. Hall, S. Porter, K. Barbu, V. Cannon, H. C. Horner, M. Janatpour, C. W. Liaw, K. Manning, J. Morales **1991**, *J. Cell Biol.* 115, 1725–1735.
74 M. A. Deli, L. Descamps, M. P. Dehouck, R. Cecchelli, F. Joó, C. S. Abraham, G. Torpier **1995**, *J. Neurosci. Res.* 41, 717–726.
75 B. Kis, A. M. Deli, H. Kobayashi, C. S. Abraham, T. Yanagita, H. Kaiya, T. Isse, R. Nishi, S. Gotoh, K. Kangawa, A. Wada, J. Greenwood, M. Niwa, H. Yamashita, Y. Ueta **2001**, *NeuroReport* 12, 4139–4142.
76 A. S. Easton, P. A. Fraser **1998**, *J. Physiol.* 507, 541–547.
77 M. J. McKinley, A. L. Albiston, A. M. Allen, M. L. Mathai, C. N. May, R. M. McAllen, B. J. Oldfield, F. A. Mendelsohn, S. Y. Chai **2003**, *Int. J. Biochem. Cell Biol.* 35, 901–918.
78 M. Sherrod, X. Liu, X. Zhang, C. D. Sigmund **2005**, *Am. J. Physiol. Integ. Comp. Physiol.* 288, R539–R546.
79 R. J. Wordinger, W. Lambert, R. Agarwal, X. Liu, A. F. Clark **2003**, *Mol. Vis.* 9, 249–256.
80 Z. Zhao, S. Alam, R. W. Oppenheim, D. M. Prevette, A. Evenson, A. Parsadanian **2004**, *Exp. Neurol.* 190, 356–372.
81 W. S. Lambert, A. F. Clark, R. J. Wordinger **2004**, *Mol. Vis.* 10, 289–296.
82 K. Ohta, A. Fujinami, S. Kuno, A. Sakakimoto, H. Matsui, Y. Kawahara, M. Ohta **2004**, *Pharmacology* 71, 162–168.
83 K. Ohta, M. Ohta, I. Mizuta, A. Fujinami, S. Shimazu, N. Sato, F. Yoneda, K. Hayashi, S. Kuno **2002**, *Neurosci. Lett.* 328, 205–208.
84 K. Ohta, S. Kuno, I. Mizuta, A. Fujinami, H. Matsui, M. Ohta **2003**, *Life Sci.* 73, 617–626.
85 M. Toyomoto, M. Ohta, K. Okumara, H. Yano, K. Matsumoto, S. Inoue, K. Hayashi, K. Ikeda **2004**, *FEBS Lett.* 562, 211–215.
86 M. Lipnik-Strangelj, M. Carman-Krzan **2004**, *Inflamm. Res.* 53, 245–252.
87 K. M. Dhandapani, V. B. Mahesh, D. W. Brann **2003**, *Exp. Biol. Med.* 228, 253–260.
88 I. Gozes, I. Divinski, I. Pilzer, M. Fridkin, D. E. Brenneman, A. D. Spier **2003**, *J. Mol. Neurosci.* 20, 315–322.
89 T. Tada, J. Ito, M. Asai, S. Yokoyama **2004**, *Neurochem. Int.* 45, 23–30.
90 A. G. Trentin, C. B. De Aguiar, R. C. Garcez, M. Alvarez-Silva **2003**, *Glia* 42, 359–369.
91 G. Zhang, Y. Ohsawa, S. Kametaka, M. Shibata, S. Waguri, Y. Uchiyama **2003**, *J. Neurosci. Res.* 72, 33–45.
92 O. Masmoudi, P. Gandolfo, J. Leprince, D. Vaudry, A. Fournier, C. Patte-Mensah, H. Vaudry, M. C. Tonon **2003**, *FASEB J.* 17, 17–27.
93 M. Leite, J. K. Frizzo, P. Nardin, L. M. de Almeida, F. Tramontina, C. Gottfried, C. A. Goncalves **2004**, *Brain Res. Bull.* 64, 139–143.
94 D. Goncalves, J. Karl, M. Leite, L. Rotta, C. Salbego, E. Rocha, S. Wofchuk, C. A. Goncalves **2002**, *NeuroReport* 13, 1533–1535.
95 F. Tramontina, S. Conte, D. Goncalves, C. Gottfried, L. V. Portela, L. Vinade, C. Salbego, C. A. Goncalves **2002**, *Cell. Mol. Neurobiol.* 22, 373–378.

96 A. Zaheer, S. N. Mathur, R. Lim **2002**, *Biochem. Biophys. Res. Comm.* 294, 238–244.
97 A. K. Wiggins, P. J. Shen, A. L. Gundlach **2003**, *Mol. Brain Res.* 114, 20–30.
98 C. Kim, C. H. Jang, J. H. Bang, M. W. Jung, I. Joo, S. U. Kim, O. Mook-Jung **2002**, *Neurosci. Lett.* 324, 185–188.
99 M. Lafon-Cazal, O. Adjali, N. Galeotti, J. Poncet, P. Jouin, V. Homburger, J. Bockaert, P. Marin **2003**, *J. Biol. Chem.* 278, 24438–24448.
100 H. Jung, S. R. Ojeda **2002**, *Hormone Res.* 57 [Suppl. 2], 31–34.
101 O. Sousa, V. de L. Romao, V. M. Neto, F. C. Gomes **2004**, *Eur. J. Neurosci.* 19, 1721–1730.
102 T. C. de Sampaio e Spohr, R. Martinez, E. F. da Silva, V. M. Neto, F. C. Gomes **2002**, *Eur. J. Neurosci.* 16, 2059–2069.
103 S. Bouret, S. De Serrano, J. C. Beauvillain, V. Prevot **2004**, *Endocrinology* 145, 101.
104 V. Prevot, A. Cornea, A. Mungenast, G. Smiley, S. R. Ojeda **2003**, *J. Neurosci.* 23, 10622–10632.
105 I. H. Zwain, A. Arroyo, P. Amato, S. S. Yen **2002**, *Neuroendocrinology* 75, 375–383.
106 N. Didier, I. A. Romero, C. Creminon, A. Wijkhuisen, J. Grassi, A. Mabondzo **2003**, *J. Neurochem.* 86, 246–254.
107 H. Takanaga, T. Yoshitake, S. Hara, C. Yamasaki, M. Kunimoto **2004**, *J. Biol. Chem.* 279, 15441–15447.
108 F. S. Shie, M. D. Neely, I. Maezawa, H. Wu, S. J. Olson, G. Jurgens, K. S. Montine, T. J. Montine **2004**, *Am. J. Pathol.* 164, 1173–1181.
109 A. Ledeboer, J. J. Breve, A. Wierinckx, S. van der Jagt, A. F. Bristow, J. E. Leysen, F. J. Tilders, A. M. Van Dam **2002**, *Eur. J. Neurosci.* 16, 1175–1185.
110 A. Rasley, K. L. Bost, I. Marriott **2004**, *J. Neurovirol.* 10, 171–180.
111 M. S. de Freitas, T. C. Spohr, A. B. Benedito, M. S. Caetano, B. Margulis, U. G. Lopes, V. Moura-Neto **2002**, *Brain Res.* 958, 359–370.
112 T. Magnus, A. Chan, R. A. Linker, K. V. Toyka, R. Gold **2002**, *J. Neuropathol. Exp. Neurol.* 61, 760–766.
113 M. P. Brenier-Pinchart, E. Blanc-Gonnet, P. N. Marche, F. Berger, F. Durand, P. Ambroise-Thomas, H. Pelloux **2004**, *Acta Neuropathol.* 107, 245–249.
114 A. C. Jauneau, A. Ischenko, P. Chan, M. Fontaine **2003**, *FEBS Lett.* 537, 17–22.
115 P. Saas, P. R. Walker, A. L. Quiquerez, D. E. Chalmers, J. F. Arrighiu, A. Lienard, J. Boucarut, P. Y. Dietrich **2002**, *NeuroReport* 13, 1921–1924.
116 E. Ambrosini, S. Columba-Cabezas, B. Serafini, A. Muscella, F. Aloisi **2003**, *Glia* 41, 290–300.
117 K. Hatori, A. Nagai, R. Heisel, J. K. Ryu, S. U. Kim **2002**, *J. Neurosci. Res.* 69, 418–426.
118 T. Nishio, S. Kawaguchi, T. Iseda, T. Kawasaki, T. Hase **2003**, *Brain Res.* 990, 129–140.

119 P. Giraudon, C. Malcus, A. Chalon, P. Vincent, S. Khuth, A. Bernard, M. F. Belin **2003**, *J. Soc. Biol.* 197, 103–112.
120 S. Deb, J. Wenjun Zhang, P. E. Gottschall **2003**, *Brain Res.* 970, 205–213.
121 K. Arai, S. R. Lee, E. H. Lo **2003**, *Glia* 43, 254–264.
122 R. Suryadevara, H. Holter, K. Borgmann, R. Persidsky, C. Labenz-Zink, Y. Persidsky, H. E. Gendelman, L. Wu, A. Ghorpade **2003**, *Glia* 44, 47–56.
123 A. L. Mutka, S. Lusa, M. D. Linder, E. Jokitalo, O. Kopra, M. Jauhiainen, E. Ikonen **2004**, *J. Biol. Chem.* 279, 48654–48662.
124 V. Hirsch-Reinshagen, S. Zhou, B. L. Burgess, L. Bernier, S. A. McIsaac, J. Y. Chan, G. H. Tansley, J. S. Cohn, M. R. Hayden, C. L. Wellington **2004**, *J. Biol. Chem.* 279, 41197–41207.
125 Y. Liang, S. Lin, T. P. Beyer, Y. Zhang, X. Wu, K. R. Bales, R. B. De Mattos, P. C. May, S. D. Li, X. C. Jiang, P. I. Eacho **2004**, *J. Neurochem.* 88, 623–634.
126 A. Naidu, Q. Xu, R. Catalano, B. Cordell **2002**, *Brain Res.* 958, 100–111.
127 S. Ueno, J. Ito, Y. Nagayasu, T. Furukawa, S. Yokoyama **2002**, *Biochim. Biophys. Acta* 1589, 261–272.
128 J. Saura, V. Petegnief, X. Wu, Y. Liang, S. M. Paul **2003**, *J. Neurochem.* 85, 1455–1467.
129 V. Prevot, C. Rio, G. J. Cho, A. Lomniczi, S. Heger, C. M. Neville, N. A. Rosenthal, S. R. Ojeda, G. Corfas **2003**, *J. Neurosci.* 23, 230–239.
130 C. Rozenfeld, R. Martinez, R. T. Figueiredo, M. T. Bozza, F. R. Lima, A. L. Pires, P. M. Silva, A. Bonomo, J. Lannes-Vieira, W. De Souza, V. Moura-Neto **2003**, *Infect. Immun.* 71, 2047–2057.
131 B. C. Kramer, J. A. Yabut, J. Cheong, R. Jno Baptiste, T. Robakis, C. W. Olanow, C. Mytilineou **2002**, *Neuroscience* 114, 361–372.
132 N. Echigo, Y. Moriyama **2004**, *Neurosci. Lett.* 367, 79–84.
133 D. S. Evanko, Q. Zhang, R. Zorec, P. G. Haydon **2003**, *Glia* 47, 233–240.
134 A. Abdipranoto, G. J. Liu, E. L. Werry, M. R. Bennett **2003**, *NeuroReport* 14, 2177–2181.
135 S. Coco, F. Calegari, E. Pravettoni, D. Pozzi, E. Taverna, P. Rosa, M. Matteoli, C. Verderio **2003**, *J. Biol. Chem.* 278, 1354–1362.
136 G. Arcuino, J. H. Lin, T. Takano, C. Liu, L. Jiang, Q. Gao, J. Kang, M. Nedergaard **2002**, *Proc. Natl Acad. Sci. USA* 99, 9840–9845.
137 N. J. Abbott, M. Bundgaard, H. F. Cserr **1986**, Comparative physiology of the blood-brain barrier, in *The Blood-Brain Barrier in Health and Disease*, eds. A. J. Suckling, M. G. Rumsby, M. W. B. Bradburg, Ellis Horwood-VCH, Chichester, pp. 52–72.

9
Responsive Astrocytic Endfeet: the Role of AQP4 in BBB Development and Functioning

Grazia P. Nicchia, Beatrice Nico, Laura M. A. Camassa, Maria G. Mola, Domenico Ribatti, David C. Spray, Alejandra Bosco, Maria Svelto, and Antonio Frigeri

9.1
Introduction

Maintenance of a constant internal osmotic environment is essential for normal cerebral activity. To this end, the blood-brain barrier (BBB) assumes an extremely important role in conjunction with astrocytes, which are involved in the regulation of water and ion homeostasis besides that of neurotransmitter metabolism and nutrient support of neurons.

Astrocytes are very complex cells, capable of responding to a variety of external stimuli. One of their main functions is to control brain water and ionic homeostasis in order to optimize the interstitial space for synaptic transmission. To this purpose, the anatomy of astrocytes is characterized by an irregular cell body with abundant leaflet-like processes which: (1) come into contact with most synapses, (2) form endfeet on the brain surface to form the external "glia limitans", and (3) form endfeet that completely envelop brain microvessels. Astrocytes are involved in the maintenance of the BBB, whose physiology is the product of collaboration between endothelial cells, pericytes, and astrocyte footprocesses. There are many transporters and channels at the level of perivascular astrocytic endfeet and they are responsible for solute transport in the CNS, since the BBB is impermeable to polar substances. These channels/transporters are responsible for important mechanisms, such as the transport of glucose, the main brain energy substrate, from the blood. In fact, astrocytic endfeet are enriched in glucose transporters and are capable of converting all or some of the glucose that enters their vascular endfeet to lactate before uptake by neurons. Moreover, astrocytes maintain the extracellular K^+ concentration at levels which optimize neuronal function, by actively taking up K^+ from the extracellular space and, at the level of astrocyte foot-processes, the excess intracellular K^+ is shunted into the vascular system. In addition, chloride, sodium, and hydrogen ions are extruded into the vascular system across the cell membrane of perivascular astrocytes.

9.2
Astrocyte Endfeet and BBB Maintenance

Astrocyte endfeet form part of the BBB, a complex glio-vascular system that controls the homeostasis of the central nervous system (CNS), preventing the non-specific passage of hydrophilic solutes between the blood and neuropil. Morphological and biochemical features of the brain capillaries, including endothelial tight junctions (TJs), the absence of endothelial fenestrations and vesicular transport, and the presence of membrane carriers, selectively regulate nutrient transport into the brain, thus ensuring neuronal protection (Betz and Goldstein 1986; Pardridge 1988; Reese and Karnowsky 1967). In addition to endothelial cells, the BBB is made up of pericytes, involved in angiogenetic processes and in the neuroimmune function (Balabanov et al. 1996; Hirschi and D'Amore 1997) embedded in the basement membrane.

Endothelial TJs are the major structure responsible for restricting the paracellular escape of blood solutes across the cerebral endothelium, as well as for its polarity and high electrical resistance in vivo (Crone 1986; Madara et al. 1985, 1992). Structurally, TJs are composed of a complex belt-like zonula occludens close to the capillary lumen and integral membrane proteins (such as claudin, occludin, JAM proteins) and cytoplasmic accessory proteins (such as ZO-1, ZO-2, ZO-3, cingulin, 7H6) are involved in its molecular composition (Citi et al. 1988; Furuse et al. 1993, 1998; Jesaitis and Goodenough 1994; Martin-Padura et al. 1998; Stevenson et al. 1986; Zhong et al. 1994).

Astrocytes form a continuous perivascular sheath closely apposed to the endothelial cells and pericytes, and become the main mediators between endothelial cells and neural tissue. Astrocytes are considered to be inducers of both the barrier and the permeability properties of the endothelium (Risau 1992). The earliest data on the inductive influence of astrocytes and the neural microenvironment on barrier features in the vessels came from studies performed on grafts of neural tissue in the coelomic cavity of different animals, which demonstrated that the newly formed vessels originating from the host displayed BBB characteristics (Stewart and Wiley 1981). Later, Janzer and Raff (1987), on injecting astrocytes into the anterior eye chamber of syngenic rats, demonstrated for the first time that a functional BBB was induced in non-neural endothelial cells, which vascularized the astrocytic aggregates. However, Krum and Rosenstein (1989) reported that astrocytes did not influence the barrier properties of the host endothelium in transplanted superior cervical ganglia and the same authors (1993) reported that no changes in endothelial permeability were detected after glial endfeet were damaged with a gliotoxin. These contrasting results might be due to different grafting techniques and graft vascularization analysis.

Important findings on the relationship between astrocytes and endothelial cells derive from the coculture of endothelial cells and astrocytes or the use of conditioned media (CM). Even if caution in the interpretation of in vitro results must be used as reported by Holash et al. (1993), due to dedifferentiation of endothelial cells, it is commonly accepted that astrocytes exert an inductive action on the morphological, biochemical, and functional barrier features of endothelial cells. Brain endothelial cells cocultured with glial cells display a significant increase in TJ formation, together with an increase in enzymatic systems such as γ-glutamyl transpeptidase (γ-GT), Na$^+$K$^+$ATPase, alkaline phosphatase, and transporters such as neutral amino acids glucose (Beck et al. 1984; Cancilla and De Bault 1983; Meresse et al. 1989; Raub et al. 1992; Rubin et al. 1991; Tao Cheng et al. 1987). In addition, an up-regulation of the low-density lipoprotein (LDL) receptors and P-glycoprotein was also reported (Dehouck et al. 1994; Gaillard et al. 2000). Furthermore, endothelial cells cultured with astrocyte CM manifest an increase in TJ formation and electrical resistance, as well as in the expression of γ-GT, ATPase, HT7, and neurothelin, suggesting the presence of soluble factors released by astrocytes, whose molecular composition has not yet been defined (Arthur et al. 1987; Lobrinus et al. 1992; Maxwell et al. 1987; Rubin et al. 1991).

An important clue to the chemical nature of astrocyte-released soluble factors was given by Lee et al. (2003) who demonstrated that, after treatment of the cultured endothelial cells with *src*-suppressed C-kinase substrate (SseCKs)-CM, a factor stimulating the expression of angiopoietin-1 in astrocytes, TJ proteins increased and permeability to ^{3}H-sucrose decreased. Moreover, various molecules such as transforming growth factor beta (TGF-β), fibroblast growth factor-2 (FGF-2) and interleukin-6 (Il-6) are able to induce some of the BBB properties induced by glial cells (Hoheisel et al. 1998; Sobue et al. 1999; Tran et al. 1999; Utsumi et al. 2000).

BBB induction seems to be more effective in coculture where the glial processes come into contact with the basal endothelial surface. Hayashi et al. (1997) showed that non-neural endothelial cells developed barrier features when closely apposed to cocultured astrocytes, suggesting a key role of cell-cell contacts. Although a BBB induction was also observed with a neuronal membrane fraction (Tontch and Bauer 1991), astrocytes played the most important role in maintaining BBB properties.

A particular role in the BBB is played by the perivascular astrocyte endfeet, which form a continuous perivascular layer and maintain cerebral homeostasis through a direct or indirect action on the endothelial cells, by regulating the ionic flux occurring during neuronal activity (Sun et al. 1995). The BBB also controls water transport, which is very important in CNS physiology, because it is involved with cerebro-spinal fluid (CSF) production, fluid transport across the endothelium and osmolality compensation in potassium siphoning (Nagelhus et al. 1998, 1999; Newman 1995; Walz and Hinks 1985). On the astrocytic endfeet membranes, a number of structures called orthogonal arrays of particles (OAPs) have been identified by means of a freeze-fracture technique (Dermietzel and Leibstein 1978; Neuhaus 1990). OAPs decrease on BBB damage (Wolburg et al. 1986), suggesting a role in the interaction between astrocytes and endothelial membrane. OAPs are associated with K$^+$ siphoning (Newman 1995) and it has recently been demonstrated that a specific water channel protein, called Aquaporin-4 (AQP4; Ma et al. 1995; Rash et al. 1998), is involved in their molecular composition. Astrocytes seem to be influenced by their interaction with endothelial cells. OAPs increase after cocultivation with brain endothelial cells (Tao-Cheng et al. 1990); and their developmental expression takes place in parallel with TJ development (Nico et al. 1994). Moreover, the expression of AQP4 and glial fibrillary acidic protein (GFAP) at glial endfeet occurs in parallel with zonula occludens-1 (ZO-1) at endothelial TJs (Nico et al. 2004) and, in dystrophic *mdx* mice, open TJs are coupled with perivascular endfeet alterations (Nico et al. 2003).

Astrocytes show a strong polarization in adult brains, where the BBB is structurally and functionally well developed, while they appear unpolarized in embryos when the BBB is absent or not completely differentiated (Bertossi et al. 1993; Wolburg 1995). In addition, the antioxidative activity of both astrocytes and endothelial cells in culture is higher than in monocultures of astrocytes (Schroeter et al. 1999) and it has been reported that endothelial cells release a factor which stimulates DNA synthesis in astrocytes (Estrada et al. 1990). Finally, it is noteworthy that astrocytes produce agrin, a heparan sulfate proteoglycan present in the subendothelial basal membrane of brain vessels and involved in BBB development (Barber and Lieth 1997; Warth et al. 2004). Thus, astrocytes may also contribute to BBB maintenance through the production of extracellular matrix components.

9.3
Astrocyte Endfeet and BBB Development

The development of CNS vasculature begins with perineural plexus vessels which penetrate the neuroectodermal layer and with new vessels originating from angiogenetic processes (Bar et al. 1995; Dermietzel and Krause 1991; Risau 1989, 1997; Risau and Wolburg 1990). These vessels radially penetrate the proliferative neuroectodermal tissue and connect with other vascular sprouts to give rise to a deep subventricular plexus in the brain anlage (Roncali et al. 1986). Brain angiogenesis is mediated by angiogenic cytokines such as fibroblast growth factor-2 (FGF-2) and vascular endothelial growth factors (VEGFs) produced by the embryonic brain, which, by linking with their endothelial receptors, are strongly mitogenic for endothelial cells (Risau et al. 1988 a, b). While FGF-2 and its receptor expression do not correlate with BBB differentiation, VEGF and VEGF receptor expression (VEGFR-1, VEGFR-2) is coupled to brain vessel growth and differentiation. VEGF, in fact, is more highly expressed during embryonic life; and the newly formed vessels penetrate into the developing brain on the basis of the VEGF gradient produced by neuroepithelial cells (Risau 1997).

Other angiogenic growth factors and their receptors are involved in brain vessel maturation, such as angiopoietin-1, -2 and their receptors Tie-1, -2 (Davis and Yancopoulos 1999), members of the transforming growth factor-β (TGFβ) signaling pathways (such as TGFβ1, ALK-1, SMAD5, SMAD6, endoglin), and the ephrin family (for a review, see Gale and Yancopoulos 1999), as well as specific endothelial adhesion molecules, such as vascular endothelial (VE)-cadherin (Carmeliet and Collen 2000).

The first embryonic brain vessels have no barrier properties. They are lined by a leaky endothelium with numerous vesicles and vacuoles, with no junction systems and are surrounded by neuroblast bodies and large perivascular spaces (Nico et al. 1999).

During embryonic development, brain vessels progressively acquire barrier properties through progressive decreases in permeability and by expressing specific endothelial transporters and antigens (Dermietzel and Krause 1991; Dermietzel et al. 1992; Risau and Wolburg 1990). One of the most debated questions concerning BBB differentiation is the time-course of its maturation. By means of intravascular injection of permeability markers such as horseradish peroxidase (HRP), it has been demonstrated that, in some avian species, precocious brain development is already associated with a structurally and functionally developed BBB in embryonic life (Risau et al. 1986a, b; Roncali et al. 1986; Wakai and Hirokawa 1978), whereas in rodents the BBB is completely differentiated and functioning only in post-natal life (Schulze and Firth 1992, 1993). The barrier vessel features are developmentally established by structural modifications involving both endothelial TJ and glial perivascular endfeet differentiation, as well as differential phenotypic antigen expression (Fig. 9.1 A, B).

Freeze-fracture techniques and ultramicroscopic analysis on chick embryo brain and isolated microvessels have demonstrated that the spatio-temporal development of the complex networks of TJ strands occur in parallel with a complete sheet of the glial perivascular endfeet and mature OAPs (Fig. 9.1 C, D), together with basal membrane and pericyte recruitment (Nico et al. 1994, 1997). Moreover, along with a reduction in vessel permeability, BBB morphofunctional maturation is coupled with the expression of TJ proteins such as ZO-1 and glial endfeet proteins such as AQP4 and GFAP (Nico et al. 1999, 2001, 2004), confirming a mutual relationship between endothelial and glial cells in the induction of their membrane specialization.

Recent studies demonstrated that the development of the glial endfeet expressing AQP4 is particularly important for controlling BBB differentiation and functioning. Immunogold electron microscopy (Fig. 9.1 D) demonstrated that this water channel protein is expressed in a polarized way on the glial membrane facing the vessels (Frigeri et al. 1995a, b; Nielsen et al. 1997) and that its developmental expression occurs in parallel with OAP and TJ formation, and reduction of the perivascular embryonic spaces, depending upon water flux regulation (Nico et al. 2001).

These results confirm the involvement of AQP4 in the molecular composition of OAPs and indicate a role of the glial endfeet in controlling the water flux balance at a BBB level, starting from embryonic life. Moreover, after BBB destruction obtained by treating embryos with LPS, glial swelling and brain edema occurs coupled with TJ opening and AQP4 reduction, showing a relation between glial endfeet polarity and BBB maintenance.

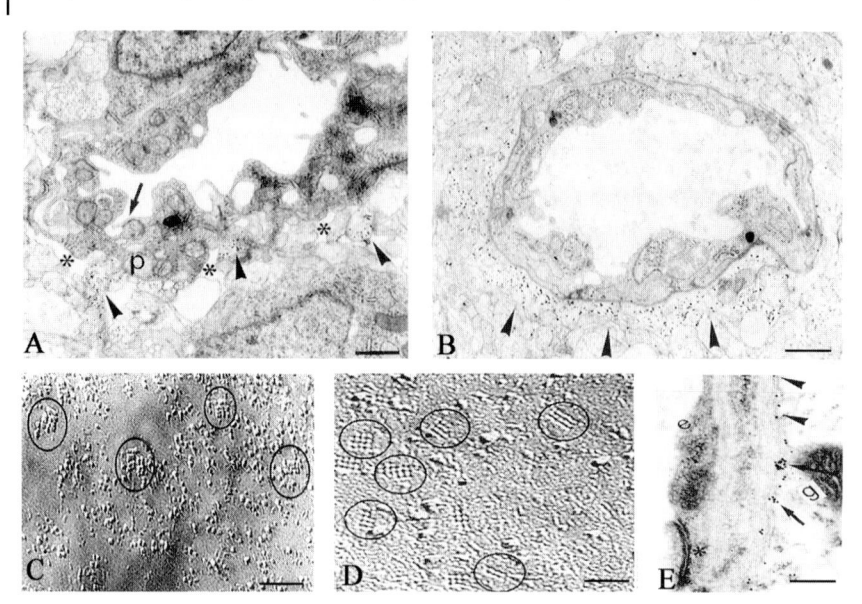

Fig. 9.1 Ultrastructural features of the glial endfeet during BBB development in 14-day chick embryos (A, C) and 20-day embryos (B, D, E).
(A) The microvessel wall is composed of endothelial cells with short TJs (arrow) and pericytes (p), which are discontinuously surrounded by isolated astroglial endfeet (arrowhead). Narrow perivascular spaces (asterisks) are recognizable.
(B) A continuous layer of glial endfeet filled with glycogen granules (arrowhead) surrounds the basal lamina of a microvessel lined by a thin endothelium (e) and pericytes.
(C) Replica from a fractured microvessel showing the P-face of the plasma membrane of an astrocytic endfoot with small quadrangular OAPs (circles).
(D) Replicas from fractured microvessels showing the E-faces of perivascular astroglial endfeet with a number of assembly pits with orthogonal symmetry (circles) corresponding to OAPs of the P-face.
(E) Ultrastructural immunodetection of AQP4: numerous immunogold particles, singularly (arrowhead) or in clusters (arrow), decorate the plasma membranes of the astroglial endfeet (g) facing the vessel endothelium. Note the unlabeled endothelial cells (e) joined by a TJ (asterisk).

The close developmental relationship between glial endfeet and endothelial cells has been furthermore demonstrated in dystrophic *mdx* mice, a model of Duchenne muscular dystrophy, as well as in *mdx* embryos (Nico et al. 2003, 2004). We found that BBB alterations were developmentally established in *mdx* mouse brains (Nico et al. 2004).

Biochemical maturation of the BBB takes place through changes in the expression of endothelial transporters and enzymatic systems (for reviews, see Betz and Goldstein 1986; Dermietzel and Krause 1991; Pardridge 1988; Risau and Wolburg 1990, 1991) which appear tightly coupled with astrocyte differentiation. In brain microvessels isolated from chick embryos, a high activity of alkaline phosphatase (ALKP) and neutral aminoacids (NAAs) and glucose transport levels was demonstrated before TJ formation (Nico et al. 1997). Moreover, in situ observations showed that endothelial junction differentiation was paralleled by a decrease in both endothelial pinocytosis and perivascular endfeet arrangement (Bertossi et al. 1992, 1993), further corroborating the notion that BBB structural and metabolic development is controlled by the glioneural environment. A number of other markers and proteins are also up- or down-regulated in their endothelial expression cells during BBB development (Risau et al. 1986a), such as the transferrin receptor (Kissel et al. 1998), the non-receptor tyrosine kinase Lyn (Achen et al. 1995), the HT7-antigen (Risau et al. 1986b), the caveolar protein PV-1 (Stan et al. 1999), or the MECA 32 antigen (Hallmann et al. 1995).

It is noteworthy that some endothelial transporters, such as P-glycoprotein, involved in the removal of lipophilic molecules and BBB differentiation (Qin and Sato 1995; Schinkel et al. 1994), or the glucose transporters (Glut-1), whose expression increases during development (Dermietzel et al. 1992), are also expressed in adults by perivascular glial endfeet (Morgello et al. 1995; Golden and Pardridge 2000), indicating a developmentally acquired astrocyte control in the pathways of glucose and other metabolites at BBB level.

9.4
Astrocyte Endfeet and BBB Damage

Disruption of the BBB is a consistent event occurring in the development of several CNS diseases, including demyelinating lesions in the course of relapsing multiple sclerosis, stroke, DMD, and *mdx* models, but also mechanical injuries (Ke et al. 2001, 2002), neurological insults (Vizuete et al. 1999), septic encephalopathy (Davies 2002), some brain tumors (Saadoun et al. 2002) where the BBB is poorly developed (Groothuis et al. 1991), and permanent ischemia or, more commonly, transient ischemia followed by reperfusion (Ke et al. 2001). In most cases, these pathological conditions are associated with an increase in microvascular permeability, vasogenic edema, swollen astrocyte endfeet, and BBB disruption.

Astrocytes are activated very early in and around focal ischemic brain regions (Norenberg 1998). Reactive astrocytes are hypertrophic and hyperplastic, and can be identified by their increased expression of GFAP. Moreover, astrocyte swelling may negatively affect the already impaired blood circulation by causing vascular constriction. Nevertheless, endothelial cell swelling, thrombi formation, or vasoconstriction may also explain localized reductions in blood flow.

Astrocytes play a significant role in host defense as well as in the pathogenesis of infectious and autoimmune diseases of the CNS. Astrocytes can be recruited to positively interact with T cells and seem to be involved in the down-regulation of immune responses. They produce mediators that suppress MHC induction and reduce the activity of T cells. In fact, cocultures of macrophages with glial cells induce lower expression of adhesion molecule and MHC-II complexes, thus down-regulating their capacity to present antigens (Hailer et al.

1998). Furthermore, astrocytes are capable of causing apoptosis in activated T cells that interact with locally presented (auto)antigen. Finally, astrocytes have been identified as one important source of chemokines for monocyte infiltration into the CNS (Weiss et al. 1998).

Interaction between astrocytes and the immune system leads to an altered production of neurotoxins and neurotrophins by these cells, which intervene in the pathogenesis of HIV-associated dementia, multiple sclerosis, and Alzheimer disease; and a common feature in these diseases is the interaction of macrophages and astrocytes (Minagar et al. 2002). Moreover, stromal-derived factor-1 (SDF-1)-induced activation of CXCR4 on astrocytes leads to TNF secretion, thereby contributing to neuronal apoptosis (Bezzi et al. 2001).

Astrocytomas are tumors composed predominantly of neoplastic astrocytes. In malignant astrocytomas, such as glioblastomas, the BBB is leaky, as visualized by the immunohistochemistry of plasma proteins using biopsy specimens (Seitz and Wechsler 1987), as well as by heterogeneous contrast enhancement upon neuroradiological examination (Roberts et al. 2001). Moreover, endothelial cells in tumor vessels show abnormal structural features, such as frequent fenestrations, prominent pinocytotic vesicles, lack of perivascular glial endfeet, as well as opening, loss, and/or abnormal morphology of TJs (Engelhard and Groothuis 1999).

Experimental data have demonstrated that some astrocytoma cells implanted into rat brain become vascularized by leaky vessels, suggesting either a deficit in production of inductive factors by proliferating glioma cells, or an enhanced production of permeability factors that counteract the inductive effects (Bauer and Bauer 2000).

9.5
The Role of Aquaporins in BBB Maintenance and Brain Edema

9.5.1
AQPs Expression and Functional Roles

All the movements of solute/ions occurring at perivascular glial endfeet are associated with the movement of water. In the brain, as in other organs, water passes through plasma membranes via three distinct mechanisms (Agre et al. 2002; Verkman et al. 1996). The first is by facilitated diffusion and occurs through particular water channels called "aquaporins", the second is by cotransport with organic or inorganic ions, and the third is by diffusion through the lipid bilayer. Brain neuropil, unlike epithelia in other organs, has a very complex three-dimensional structure, making it difficult to predict the direction and magnitude of water flux in physiological conditions (Nagelhus et al. 2004). Water transport has to be directly or indirectly coupled with homeostatic processes, such as glutamate uptake, potassium clearance, etc. Three aquaporins are expressed in brain: AQP1 is expressed in the choroid plexus epithelium (Nielsen et al. 1993), where it has been shown to

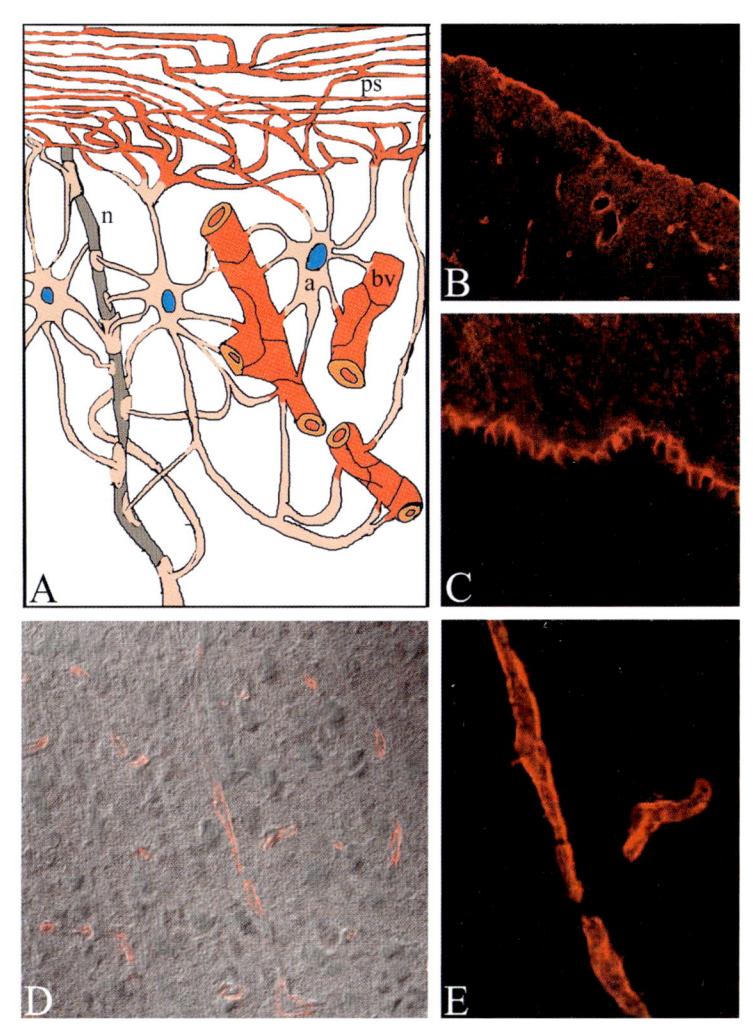

Fig. 9.2 AQP4 expression in brain glial cells. (A) Schematic diagram showing the relationship between astrocytes (a), pial surface (ps) and blood vessels (bv). Astrocytes are irregularly shaped cells with many radiating processes that surround neurons (n) and the surface of blood vessels and extend to the ependymal and pial surface forming the glial limiting membrane.
(B) AQP4 immunostaining (in red) is concentrated in astrocytic processes forming the glia limitans and surrounding intracerebral capillaries.
(C) High AQP4 expression is also found on the ependymal cells that are in contact with the CSF in the ventricular system.
(D) Differential interference contrast image of the brain cortex showing AQP4 staining (red) along the perivascular glial endfeet.
(E) Confocal analysis of AQP4 labeling at the perivascular glia membrane facing blood vessels.

facilitate cerebrospinal fluid secretion (Oshio et al. 2004); AQP9 is found in a subset of astrocytes and in catecholaminergic neurons (Badaut et al. 2004); and the most expressed water channel in the CNS is AQP4 (Frigeri et al. 1995 a, b; Nielsen et al. 1997), principally by the astrocytes of the brain cortex (Fig. 9.2). AQP4 seems to be of primary importance in the brain's handling of water (Agre et al. 2004; Amiry-Moghaddam et al. 2003 b; Manley et al. 2004; Papadopoulos et al. 2004 a b). Immunolocalization studies together with AQP4/GFAP double-staining experiments showed AQP4 expression strongly concentrated in astrocyte processes in direct contact with blood vessels (Frigeri et al. 1995 a, b; Nicchia et al. 2000). A confocal 3D reconstruction of a brain vessel stained with AQP4 antibodies shows strong AQP4 polarization where the glial endfeet are in contact with endothelial cells and demonstrates that the glial AQP4 positive endfeet completely envelop the vessel at the level of the BBB interface. High AQP4 expression has also been demonstrated at the subpial level, where the endfeet of numerous astrocytes form the glial limiting membrane (Frigeri et al. 1995 a, b), by the basolateral membrane of ependymal cells lining the intracerebral ventricles (Frigeri et al. 1995 a, b; Nielsen et al. 1997), and in spinal cord astrocytes (Fig. 9.3). Electron microscopy studies on ultrathin sections revealed that AQP4 gold particles appear to be numerous along the membranes of the perivascular glial endfeet facing the capillaries rather than those facing the neuropil (Fig. 9.4; Amiry-Moghaddam et al. 2003 b; Nicchia et al. 2000; Nico et al. 2001; Nielsen et al. 1997). AQP4 is also expressed in the glial

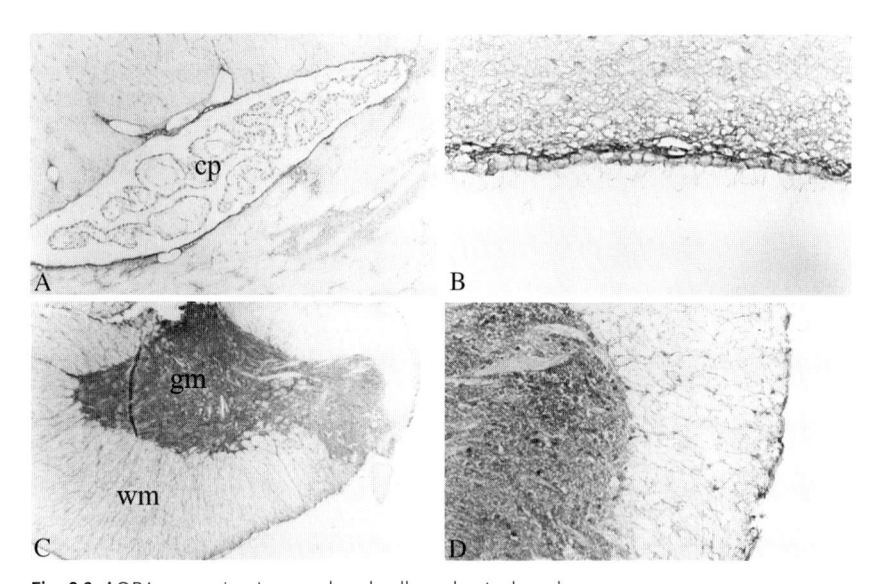

Fig. 9.3 AQP4 expression in ependymal cells and spinal cord.
(A) AQP4 immunoperoxidase staining in the ependymal cell layer.
Note that the choroid plexus (cp) cells are unstained.
(B) Ependymal cells lining the ventricle display strong basolateral staining.
(C, D) Dense AQP4 staining in the astrocytes of the spinal cord gray matter (gm). In white matter (wm), the staining is limited to fibrous astrocytes.

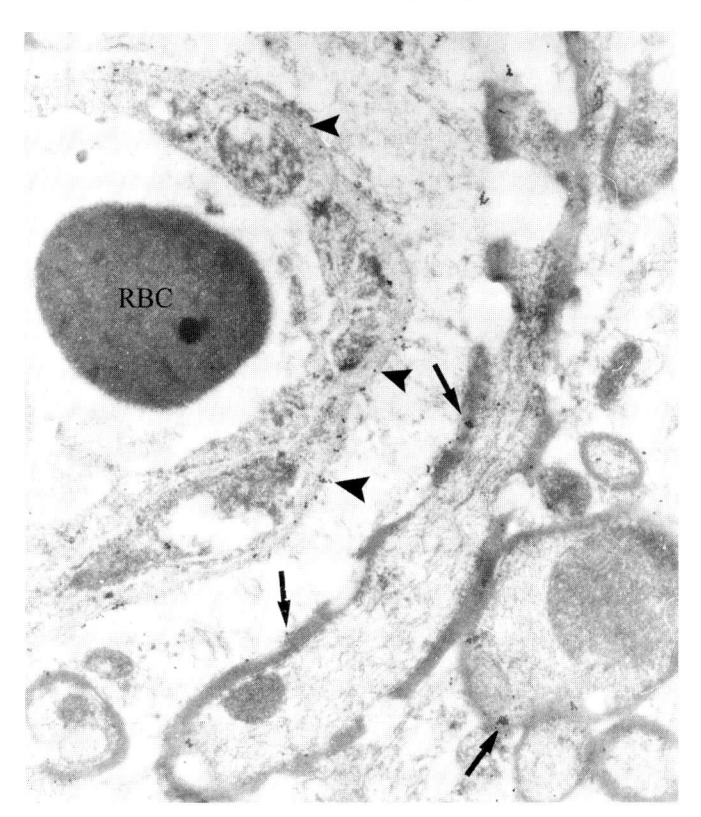

Fig. 9.4 Immunogold electron microscopy analysis of AQP4 expression in brain. AQP4 expression is mainly concentrated at the glial endfeet (arrowheads), while a few gold particles are detected on the neuronal side (arrows) of glial processes. RBC, red blood cell.

lamellae of the osmosensitive organs (Nielsen et al. 1997). Recently, it was reported that AQP4 is also expressed, at very low levels, by brain endothelial cells (Amiry-Moghaddam et al. 2004) and at non-endfeet membranes (Amiry-Moghaddam et al. 2003b; Nielsen et al. 1997). Although in situ hybridization studies reported AQP4 expression in neurons (Venero et al. 1999, 2001, 2004), double-staining with the neuronal marker NFH (Neurofilament H) and electron microscopy analysis showed no AQP4 immunoreactivity in the bodies and cell processes of neuronal cells (Nagelhus et al. 1998; Nicchia et al. 2000; Nielsen et al. 1997). Strong AQP4 enrichment at the perivascular and subpial endfeet membranes suggests its cell biological function in governing homeostatic processes at the synaptic level (Nagelhus et al. 2004; Nicchia et al. 2004). For example, it has been suggested that AQP4 works in concert with Kir 4.1 to effect potassium buffering (Nagelhus et al. 2004). Indeed, in retinal Muller cells, AQP4 subcellular distribution matches that of Kir 4.1 (Nagelhus et al. 1999). Moreover, recent experiments performed with a-syntrophin null (a-syn$^{-/-}$) mice support this hypothesis

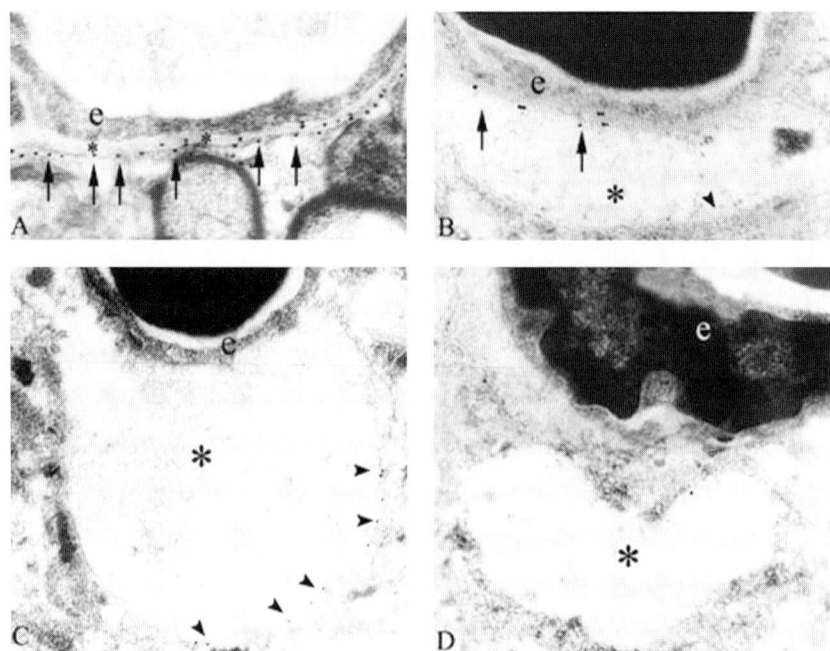

Fig. 9.6 Immunogold electron microscopy of AQP4 in brain cortex of control, *mdx* and *mdx*^{3cv} mice.
(A) Control brain. Numerous AQP4 gold particles (arrows) decorate the glial membranes, separated by a basal lamina (asterisk) from the blood vessel. Magnification: ×20000.
(B) *mdx* brain from 1.5-year *mdx* mouse. A small number of gold particles are attached to the swollen glial endfoot (asterisk) facing the blood vessels. Note the unlabeled glial membrane facing the neuropil (arrowheads). Magnification: ×20000.
(C, D) *mdx*^{3cv} brain. Swollen glial endfeet (asterisk) enveloping two vessels are labeled by rare gold particles at the membrane facing the neuropil (C; ×12 000) or are devoid of gold particles (D; ×20000) at the membranes facing the vessel and the neuropil. e, endothelial cell. (Reproduced from Nicchia et al. (2004), with permission).

to the same extent in DMD human muscle biopsies (Frigeri et al. 2002), suggesting that an altered osmotic balance may be seriously involved in the alteration of dystrophic muscles. Interestingly, the analysis of AQP4 expression in brain astrocyte foot processes of *mdx* mice revealed an age-related reduction of this water channel associated with swollen astrocyte processes (Frigeri et al. 2001) and BBB breakdown (Nico et al. 2003, 2004), suggesting a close relationship between BBB integrity and control of the water flux by astroglial cells (Nico et al. 2001). However, AQP4 reduction in *mdx* mice was significant only in older mice (Frigeri et al. 2001). To investigate whether the presence of other dystrophin isoforms in the *mdx* mouse, in particular DP71, may delay the reduction in AQP4 expression, we studied *mdx*^{3cv} animals. The immunolabeling of perivascular astroglial endfeet membranes of a large number of *mdx*^{3cv} vessels in several brain cortex areas displayed very reduced AQP4 staining (Fig. 9.6). Immunogold electron microscopy revealed that, in young mdx^{3cv} mice, few AQP4-gold particles were found at the glial endfeet surrounding the vessels, which appeared swollen (Fig. 9.6 C, D). Furthermore we noted that, in some vessels, a small number of gold particles were found at the level of glial membranes facing the neuropil (Fig. 9.6 D). This result indicates that AQP4 expression is more acutely affected in the mdx^{3cv} than in the mdx mouse and that AQP4 reduction starts much earlier in the mdx^{3cv} than in the mdx mouse. At 2 months of age, the mdx^{3cv} mouse displayed a strong reduction in AQP4 immunolabeling, in conjunction with dilated glial perivascular spaces.

The presence of DP71 in the brain seems to play a crucial role in the correct anchoring and stability of AQP4 at the glial perivascular endfeet. In normal mice, AQP4 expression at the perivascular glial endfeet determines rapid water movement, driven by the osmotic gradient in order to maintain brain water homeostasis (Amiry-Moghaddam et al. 2003b, 2004; Frigeri et al. 2001; Manley et al. 2000; Nico et al. 2003; Vajda et al. 2002). The reduction in AQP4 associated with swollen glial processes found in mdx^{3cv} indicates that water homeostasis is severely altered. In conclusion, our results suggest that DP71 plays a critical role in AQP4 expression at the astrocyte endfeet membrane adjacent to blood vessels in the cerebral cortex; and they strongly suggest that, as in mdx^{3cv}, AQP4 reduction is involved in the brain modification occurring in DMD patients in which dystrophin mutation involves the DP71 isoform.

9.5.2.2 The α-Syntrophin Null Mice

The α-syntrophin knockout mouse is the latest model available to study the role of AQP4 in brain physiology since, like mdx mice, it lacks AQP4 expression at perivascular level. α-Syntrophin is an adaptor protein, one of the dystrophin-associated proteins (DAPs), whose PSD95-disc large-ZO-1 (PDZ) domain is proposed to link the C-terminus of AQP4. In favor of this hypothesis, chemical cross-linking and coimmunoprecipitations from brain have demonstrated AQP4 in association with a complex, including dystrophin, beta-dystroglycan, and syntrophin (Neely et al. 2001). Moreover, studies on AQP4 expression in a-Syn$^{-/-}$ mice have demonstrated that AQP4 is absent at the level of skeletal muscle sarcolemma and that the polarized subcellular localization of AQP4 is reversed, being markedly reduced in astrocyte perivascular and subpial endfeet membranes, but present at higher than normal levels in membranes facing the neuropil (Neely et al. 2001). However, the need for a chemical crosslinker to obtain AQP4 and α-syntrophin co-immunoprecipitation and the demonstration that the expression and stability of AQP4 in the sarcolemma does not always decrease when α-syntrophin is strongly reduced (Frigeri et al. 2002) raises concerns about a direct interaction between AQP4 and α-syntrophin. Further biochemical studies are required to demonstrate the interaction between these two proteins.

Likewise, in mdx mice, perivascular and subpial astroglial endfeet have been shown to be swollen in brains of a-Syn$^{-/-}$ mice compared with wild-type mice (Amiry-Moghaddam et al. 2003b), suggesting that the clearance of water generated by brain metabolism is reduced when AQP4 is absent or strongly reduced. In contrast, in pathological conditions, brain edema was attenuated after water intoxication in a-Syn$^{-/-}$ mice (Amiry-Moghaddam et al. 2004) as well as in AQP4 knock-out mice (Manley et al. 2000) and in mdx and mdx^{3cv} mice (Frigeri et al. unpublished data) and also after transient cerebral ischemia in a-Syn$^{-/-}$ mice (Amiry-Moghaddam et al. 2003b).

However, the greater contribution of a-Syn$^{-/-}$ mice is related to studying the role of AQP4 in potassium buffering. In fact, since the expression of the inwardly rectifying K$^+$ channel, Kir4.1, is grossly unmodified in these transgenics, they have been used to demonstrate that AQP4 and K$^+$ channels work together to obtain an isosmotic clearance of K$^+$ after neuronal activation. In fact, K$^+$ clearance was prolonged up to 2-fold in a-Syn$^{-/-}$ mice compared with wild-type mice (Amiry-Moghaddam et al. 2003a).

9.5.2.3 The Effect of Lipopolysaccharide on AQP4 Expression

In order to clarify the role of AQP4 during conditions of BBB damage, we induced BBB disruption by LPS treatment in the optic tectum of 20-day chicken embryos in which the BBB is morphofunctionally well developed (Nico et al. 2001; Roncali et al. 1986). The BBB alterations induced by LPS during meningoencephalitis are characterized by a loss of endothelial tightness and an increase in vesicular content (Quagliarello et al. 1986). LPS-treated brains showed the presence of severe edema with swollen perivascular astrocytes, in agreement with the commonly accepted idea that glial swelling is a consequence of BBB damage. Immunogold and immunoperoxidase analysis revealed only a faint labeling of AQP4 protein on the astrocytic processes in the neuropil and around microvessels. The reduction of AQP4 protein expression on swollen glial endfoot associated with the damaged BBB further supports the close relationship between BBB function and water flux mediated by AQP4 perivascular expression and suggests that this water channel plays a key role in the genesis of brain edema.

9.5.2.4 AQP4 in Astrocytomas

AQP4 is massively up-regulated in astrocytomas and this correlates with BBB opening assessed by contrast-enhanced computed tomographs (Saadoun et al. 2002). Warth et al. (2004) demonstrated that the redistribution of AQP4 in human glioblastomas correlates with a loss of agrin immunoreactivity. These authors postulated that agrin determines the polarity of astrocytes by binding alpha dystroglycan; and this polarity might be the precondition for the astrocytes' ability to induce or maintain the BBB properties of brain endothelial cells.

9.6
AQP4 Expression in Astrocyte-Endothelial Cocultures

As discussed earlier, the specific structure of the BBB is based on the partnership of brain endothelial cells and astrocytes. Numerous in vivo studies have described interactive influences between astrocytes and vascular cells. In particular, several studies aimed to characterize the influence of astrocytes on endothelial cells (Arthur et al. 1987; Beuckmann and Galla 1998; Hurwitz et al. 1993; Janzer and Raff 1987; Stewart and Wiley 1981; Tao-Cheng et al. 1987; Tout et al. 1993), but the inductive influence of brain endothelium on astrocytes was also demonstrated (Beck et al. 1984; Estrada et al. 1990; Sperri et al. 1997; Wagner and Gardner 2000). Therefore, the maintenance of the BBB appears to depend on the continuous exchange of signals between astrocytes and endothelial cells; and disturbance of this exchange may be involved in several pathologies involving BBB dysfunction. In recent years, cocultures of these two cell types have been developed and used extensively as in vitro models to study aspects of barrier induction and modulation (Abbott 2002; Bauer and Bauer 2000; Hayashi et al. 1997; Krämer et al. 2001; Reinhart and Gloor 1997; Rubin et al. 1991).

Astrocyte primary cultures constitutively express large amounts of AQP4 protein (Nicchia et al. 2000, 2003). Immunofluorescence analysis revealed that, although the expression of AQP4 appears to be clearly detectable on the plasma membrane with staining of the cell periphery, a significant amount of AQP4

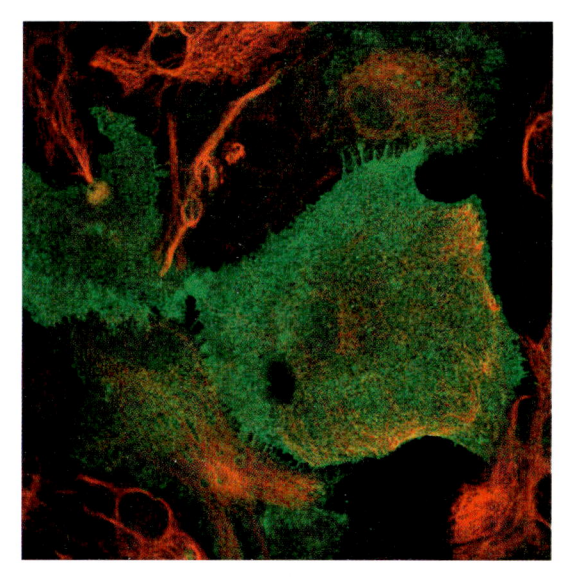

Fig. 9.7 AQP4 expression in primary culture astrocytes. Double immunofluorescent confocal analysis of AQP4 (green) and GFAP (red) in purified astrocyte cultures grown on glass coverslips. Note the intense AQP4 staining at the cell periphery.

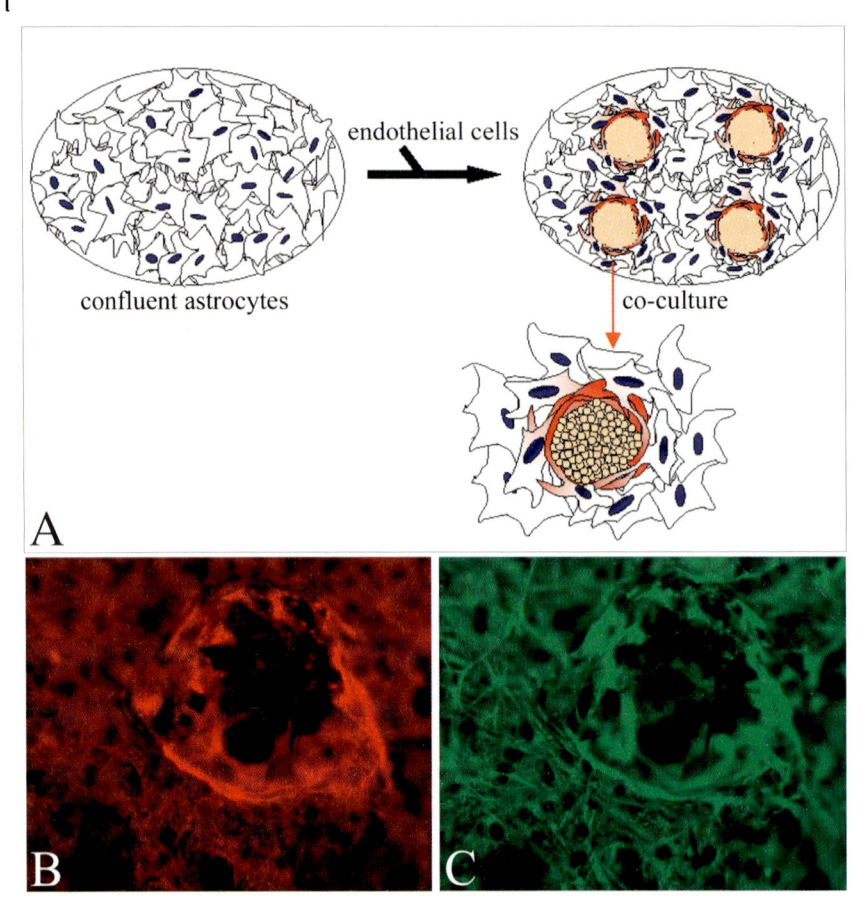

Fig. 9.8 AQP4 expression analysis in rat astrocytes cocultured with endothelial cells (bEnd3) on glass coverslips.
(A) Schematic diagram illustrating the coculture system.
(B) AQP4 antibodies strongly stained the astrocyte processes in close contact with endothelial cells.
(C) GFAP immunostaining of astrocyte endfeet surrounding a group of bEnd3 cells.

staining is observed in the cell cytoplasm (Fig. 9.7), indicating that the polarized expression of AQP4 observed in vivo is not maintained in purified astrocyte cultures. Thus, in brain the localization of AQP4 in the astrocytic processes close to the abluminal side of endothelial cells suggests that its position is chiefly determined by the BBB composition.

To determine whether AQP4 polarized expression observed in brain could be reestablished in vitro and thus to dissect the mechanism that determines the anchoring of AQP4 in the astrocytic perivascular domain, we first analyzed the role of endothelial cells. To this purpose, astrocytes were cocultured with endothelial

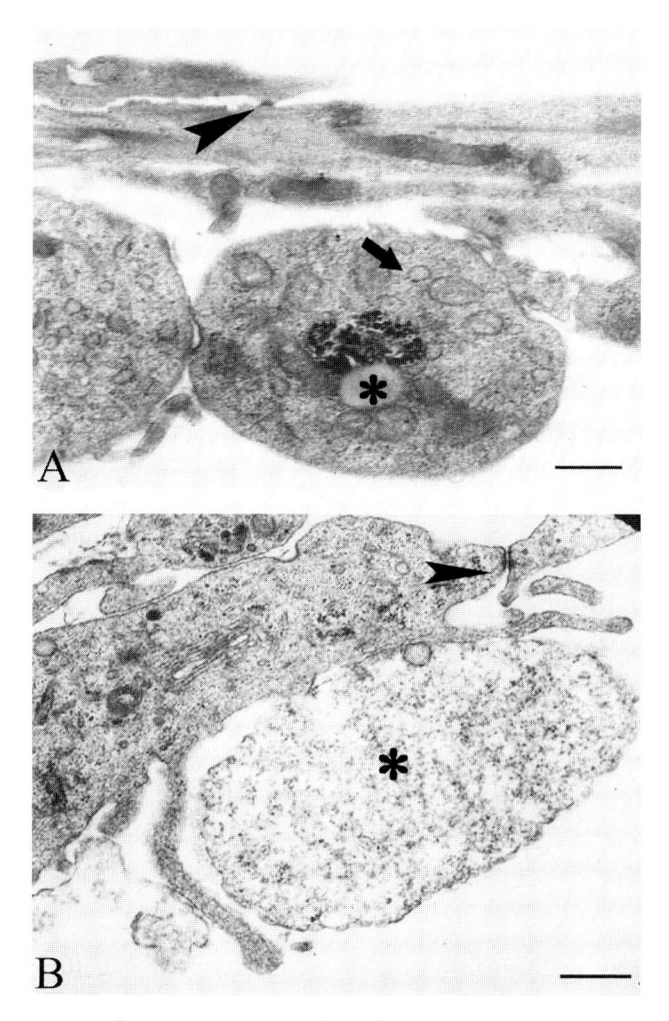

Fig. 9.9 Electron microscopy analysis of astrocytes and endothelial cells in coculture.
(A, B) Ultrathin sections of CE cocultured with astrocytes show thin layers of endothelial cells (asterisks) sealed by small TJs (arrowheads) and subtended by rounded processes of glial cells, containing glycogen granules and vesicles (A, arrow). Scale bars: A=0.5 µm; B=1.05 µm.

cells. For these studies, we used immortalized endothelial cells, bEnd3, originally derived from mouse brain capillaries (Williams et al. 1988, 1989) and freshly prepared rat astrocytes. Cells were cocultured in two different models: (1) by plating bEnd3 cells onto a confluent layer of astrocytes using the same medium as for the astrocyte cultures, or (2) by cultivating astrocytes on one surface of a porous membrane and endothelial cells on the opposite surface. In the first model, the two cell types are in direct contact with each other and after 7–14 days of coculture, the

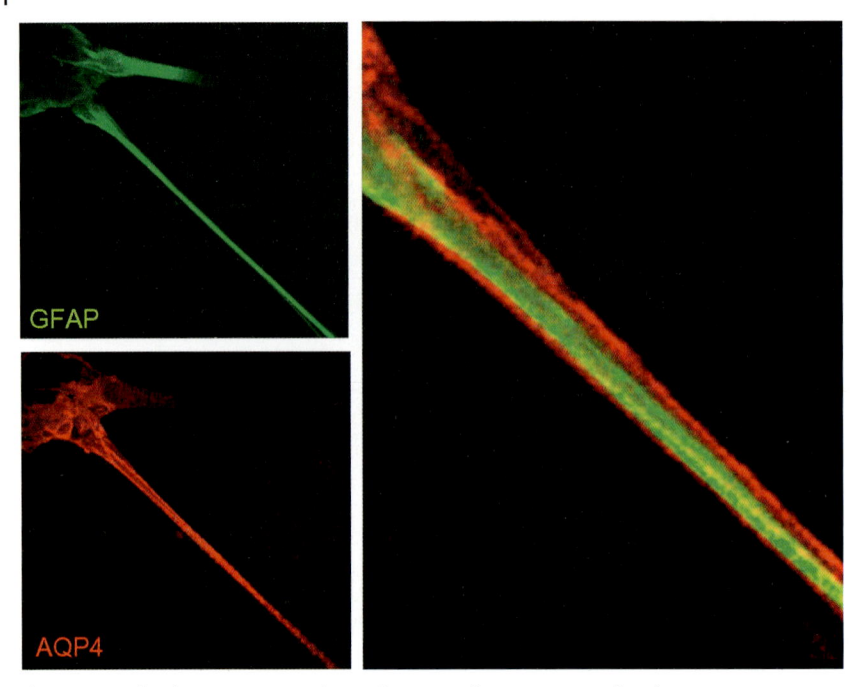

Fig. 9.10 Confocal microscopy analysis of AQP4 and GFAP protein distribution in astrocyte processes. Double immunostaining with AQP4 (red) and GFAP (green) antibodies. Note that the AQP4 staining is mainly distributed over the surface membrane.

presence of endothelial cells determined dramatic changes in the morphology of astrocytes (Yoder 2002), transforming them from a confluent flat monolayer into islands that were interconnected by elongated multicellular columns and thick processes (Fig. 9.8). Immunostaining with GFAP antibody confirmed that astrocytes were the cells forming those structures (Fig. 9.8C). In many areas of the coculture preparation, endothelial cells were arranged to form capillary-like structures (Fig. 9.8B) that were not seen when endothelial cells were grown alone or with other cell lines. Thus, a specific mutual induction occurs in cocultures. Indeed, other immortalized cell lines (glioma C6, RG2, neuronal GT1) were not able to induce similar morphological changes (Yoder 2002).

Electron microscopy analysis revealed the presence of TJs between endothelial cells and thin astrocyte processes surrounding them (Fig. 9.9). The immunofluorescence analysis of AQP4 expression in the coculture system revealed an extraordinary redistribution of AQP4 in the astrocytes. In cocultured astrocytes, AQP4 staining appeared increased and was distinctly localized to those astrocytic processes in close contact to endothelial cells (Fig. 9.10). The staining was strongly associated with the plasma membrane, with little or no intracellular staining. Furthermore, the astrocyte somata also displayed membrane staining, but this was much fainter compared with that seen in the processes and was

9.6 AQP4 Expression in Astrocyte-Endothelial Cocultures

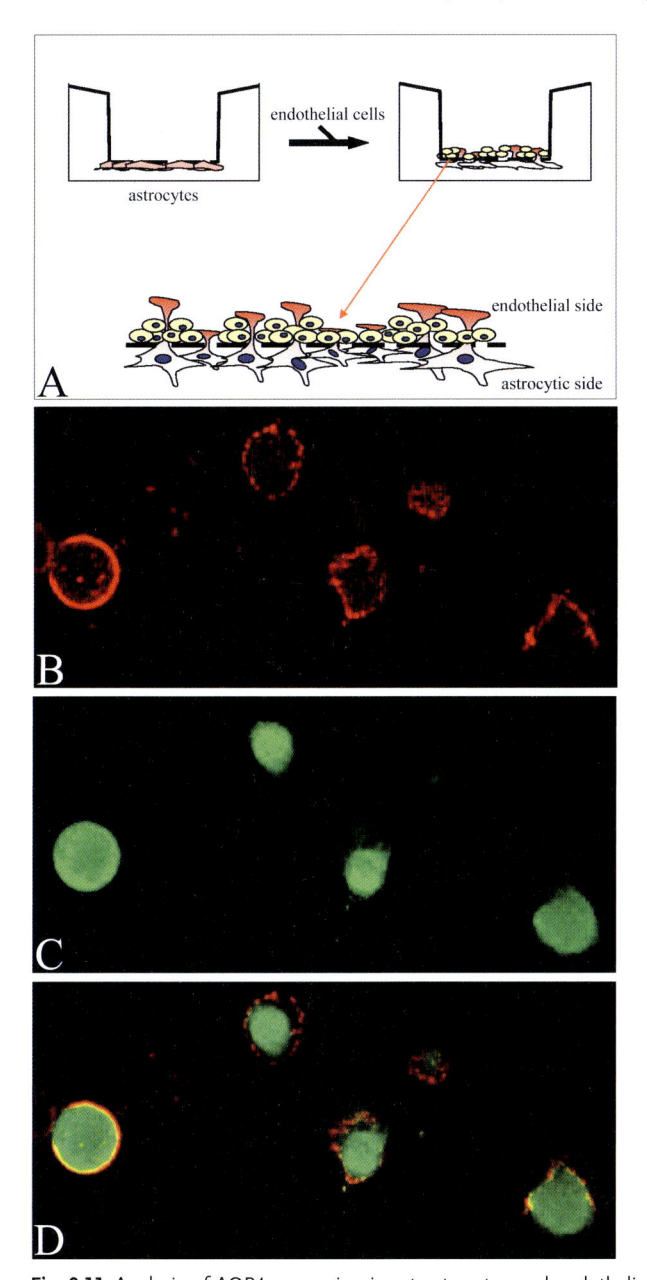

Fig. 9.11 Analysis of AQP4 expression in rat astrocytes and endothelial cells (bEnd3) cocultured on 1-μm porous membranes.
(A) Schematic diagram illustrating the coculture system.
(B–D) Confocal microscopy analysis showing the presence of astrocyte processes on the endothelial side of the membrane. AQP4 antibodies specifically stained astrocyte processes (B), as demonstrated by GFAP immunostaining (C), in close proximity to bEnd3 cells. (D) Merged image of B, C.

mainly comparable with levels seen in astrocytes alone. These results suggest that the presence of endothelial cells is necessary to induce the formation of astrocyte processes, which in turn are required for the localization of AQP4.

In the second model (Fig. 9.11), endothelial cells were seeded on an insert membrane having 1.0-μm pores, on the opposite surface of which astrocytes had been directly seeded and grown. This system spatially separates the two cell types and thereby permits astrocytes to make contact with endothelial cells through their processes. Immunofluorescence analysis revealed strong membrane AQP4 expression in these GFAP-positive processes that traversed the pores and were in close contact with endothelial cells. No GFAP/AQP4 staining was detected on the opposite side of astrocytes when these cells were grown alone on the membrane insert, indicating that no cell processes passed through the pores. Furthermore, neuroblastoma cells (N2A) were not able to induce the formation of astrocyte processes. To our knowledge, this is the first evidence of a membrane protein whose polarized expression in vivo can be efficiently re-established in vitro. This induction of AQP4 polarization in astrocytic processes by close apposition to cocultured endothelial cells could be due either to a cellular differentiation process induced by endothelial cells (the result of a physical interaction between the two cell types), or to a diffusion of soluble signals. Preliminary coculture experiments performed using inserts with smaller pore sizes (0.2 μm) or astrocytes treated with conditioned medium did not reveal any phenotypic change, indicating that soluble molecules seem not to be involved in this phenomenon, whereas endothelial cells are required for the formation of cell processes. Future studies aim to analyze in detail the molecular aspect of this inductive effect on AQP4 polarity in coculture.

References

Abbott, N. J. **2002**, *J. Anat.* 200, 629–638.
Achen, M. G., Clauss, M., Schnürch, H., Risau, W. **1995**, *Differentiation* 59, 15–24.
Agre, P., King, L. S., Yasui, M., Guggino, W. B., Ottersen, O. P., Fujiyoshi, Y., Engel, A., Nielsen, S. **2002**, *J. Physiol.* 542, 3–16.
Agre, P., Nielsen, S., Ottersen, O. P. **2004**, *Neuroscience* 129, 849–850.
Amiry-Moghaddam, M., Williamson, A., Palomba, M., Eid, T., de Lanerolle, N. C., Nagelhus, E. A., Adams, M. E., Froehner, S. C., Agre, P., Ottersen, O. P. **2003 a**, *Proc. Natl Acad. Sci. USA*, 100, 13615–13620.
Amiry-Moghaddam, M., Otsuka, T., Hurn, P. D., Traystman, R. J., Haug, F. M., Froehner, S. C., Adams, M. E., Neely, J. D., Agre, P., Ottersen, O. P., Bhardwaj, A. **2003 b**, *Proc. Natl Acad. Sci. USA* 100, 2106–2111.
Amiry-Moghaddam, M., Xue, R., Haug, F. M., Neely, J. D., Bhardwaj, A., Agre, P., Adams, M. E., Froehner, S. C., Mori, S., Ottersen, O. P. **2004**, *FASEB J.* 18, 542–544.
Arthur, F. E., Shivers, R. R., Bowman, P. D. **1987**, *Brain Res.* 433, 155–159.

Badaut, J., Petit, J. M., Brunet, J. F., Magistretti, P. J., Charriaut-Marlangue, C., Regli, L. **2004**, *Neuroscience* 128, 27–38.

Balabanov, R., Washington, R., Wagnerova, J., Dore-Duffy, P. **1996**, *Microvasc. Res.* 52, 2127–2142.

Bar, R. S., Peacock, M. L., Spanheimer, R. G., Veenstra, R., Hoak, J. C. **1995**, *Diabetes* 29, 478–481.

Barber, A. J., Lieth, E. **1997**, *Dev. Dyn.* 208, 62–74.

Bauer, H. C., Bauer, H. **2000**, *Cell. Mol. Neurobiol.* 20, 13–28.

Beck, D. W., Vinters, H. V., Hart, M. N., Cancilla, P. A. **1984**, *J. Neuropathol. Exp. Neurol.* 43, 219–224.

Bertossi, M., Mancini, L., Favia, A., Nico, B., Ribatti, D., Virgintino, D., Roncali, L. **1992**, *Biol. Struct. Morphol.* 4, 144–152.

Bertossi, M., Roncali, L., Nico, B., Ribatti, D., Mancini, L., Virgintino, D., Fabiani, G., Guidazzoli, A. **1993**, *Anat. Embryol.* 188, 21–29.

Betz, A. L., Goldstein, G. W. **1986**, *Annu. Rev. Physiol.* 48, 241–250.

Beuckmann, C. T., Galla, H. J. **1998**, Tissue culture of brain endothelial cells: induction of blood-brain barrier properties by brain factors, in *Introduction to the Blood-Brain Barrier: Methodology, Biology, and Pathology*, ed. W. M. Pardridge, Cambridge University Press, Cambridge, pp. 79–85.

Bezzi, P., Domercq, M., Brambilla, L., Galli, R., Schols, D., De Clercq, E., Vescovi, A., Bagetta, G., Kollias, G., Meldolesi, J., Volterra, A. **2001**, *Nat. Neurosci.* 4, 702–710.

Binder, D. K., Oshio, K., Ma, T., Verkman, A. S., Manley, G. T. **2004**, *NeuroReport* 15, 259–262.

Blake, D. J., Kroger, S. **2000**, *Trends Neurosci.* 23, 92–99.

Blake, D. J., Hawkes, R., Benson, M. A., Beesley, P. W. **1999**, *J. Cell Biol.* 147, 645–658.

Bulfield, G., Siller, W. G., Wight, P. A., Moore, K. J. **1984**, *Proc. Natl Acad. Sci. USA* 81, 1189–1192.

Cancilla, P. A., De Bault, L. E. **1983**, *J. Neuropathol. Exp. Neurol.* 42, 191–199.

Carmeliet, P., Collen, D. **2000**, *J. Physiol.* 429, 47–62.

Citi, S., Sabanay, H., Jakes, R., Geiger, B., Kendrik-Jones, J. **1988**, *Nature* 333, 272–276.

Cox, G. A., Phelps, S. F., Chamberlain, J. S. **1993**, *Nat. Genet.* 4, 87–93.

Crone, C. **1986**, *Ann. N.Y. Acad. Sci.* 481, 174–185.

Davies, D. C. **2002**, *J. Anat.* 200, 639–646.

Davis, S., Yancopoulos, G. D. **1999**, *Curr. Top. Microbiol. Immunol.* 237, 173–185.

Dehouck, B., Dehouck, M. P., Fruchart, J. C., Cecchelli, R. **1994**, *J. Cell Biol.* 126, 465–473.

Dermietzel, R., Krause, D. **1991**, *Int. Rev. Cytol.* 127, 57–109.

Dermietzel, R., Leibstein, A. G. **1978**, *Cell Tissue Res.* 1861, 97–110.

Dermietzel, R., Krause, D., Kremer, M., Wang, C., Stevenson, B. **1992**, *Dev. Dyn.* 193, 152–163.

Engelhard, H. H., Groothuis, D. G. **1999**, in *The Blood-Brain Barrier: Structure, Function and Response to Neoplasia*, eds. M. D. Berger, C. B. Wilson, Gliomas, Saunders, Pa., pp. 115–121

Estrada, C., Bready, J. V., Berliner, J. A., Pardridge, W. M., Cancilla, P. A. **1990**, *J. Neuropathol. Exp. Neurol.* 49, 539–549.

Frigeri, A., Gropper, M. A., Umenishi, F., Kawashima, M., Brown, M., Verkman, A. S. **1995 a**, *J. Cell Sci.* 108, 2993–3002.

Frigeri, A., Gropper, M. A., Turck, C. W., Verkman, A. S. **1995 b**, *Proc. Natl Acad. Sci. USA* 92, 4328–4331.

Frigeri, A., Nicchia, G. P., Verbavatz, J. M., Valenti, G., Svelto, M. J. **1998**, *Clin. Inv.* 102, 695–703.

Frigeri, A., Nicchia, G. P., Nico, B., Quondamatteo, F., Herken, R., Roncali, L., Svelto, M. **2001**, *FASEB J.* 15, 90–97.

Frigeri, A., Nicchia, G. P., Repetto, S., Bado, M., Minetti, C., Svelto, M. **2002**, *FASEB J.* 16, 1120–1122.

Furuse, M., Hirase, T., Itoh, M., Nagafuchi, A., Yonemura, S., Tsukita, S., Tsukita, S. *J. Cell Biol.* **1993**, 123, 1777–1788.

Furuse, M., Fujita, K., Hiiragi, T., Fujimoto, K., Tsukita, S. **1998**, *J. Cell Biol.* 141, 1539–1550.

Gaillard, P. J., van der Sandr, J. C., Voorwinden, L. H., Vu d, Nielsen, J. L., de Boer, A. G., Breimer, D. D. **2000**, *Pharm. Res.* 17, 1198–1205.

Gale, N. W., Yancopoulos, G. D. **1999**, *Genes Dev.* 13, 1055–1066.

Golden, P. L., Pardridge, W. M. **2000**, *Cell. Mol. Neurobiol.* 20, 165–181.

Gorecki, D. C., Monaco, A. P., Derry, M. J., Walker, A. P., Barnard, E. A., Barnard, P. J. **1992**, *Hum. Mol. Genet.* 1, 505–510.

Groothuis, D. R., Vriesendorp, F. J., Kupfer, B., Warnke, P. C., Lapin, G. D., Kuruvilla, A., Vick, N. A., Mikhael, M. A., Patlak, C. S. **1991**, *Ann. Neurol.* 30, 581–588.

Hailer, N. P., Heppner, F. L., Haas, D., Nitsch, R. **1998**, *Brain Pathol.* 8, 459–474.

Hallmann, R., Mayer, D. N., Berg, E. L., Broermann, R., Butcher, E. C. **1995**, *Dev. Dyn.* 202, 325–332.

Hayashi, Y., Nomura, M., Yamagishi, S., Harada, S., Yamashita, J., Yamamoto, H. **1997**, *Glia* 19, 13–26.

Hirschi, K. K., D'Amore, P. A. **1997**, *Exs* 79, 419–427.

Hoheisel, D., Nitz, T., Franke, H., Wegener, J., Hakvoort, A., Tilling, T., Galla, H. J. **1998**, *Biochem. Biophys. Res. Commun.* 247, 312–315.

Holasch, J. A., Noden, D. M., Stewart, P. A. **1993**, *Dev. Dyn.* 197, 14–25.

Hosaka, Y., Yokota, T., Miyagoe-Suzuki, Y., Yuasa, K., Imamura, M., Matsuda, R., Ikemoto, T., Kameya, S., Takeda, S. **2002**, *J. Cell Biol.* 158, 1097–1107.

Hurwitz, A. A., Berman, J. W., Rashbaum, W. K., Lyman, W. D. **1993**, *Brain Res.* 625, 238–243.

Imamura, M., Ozawa, E. **1998**, *Proc. Natl Acad. Sci. USA* 95, 6139–6144.

Jancsik, V., Hajos, F. **1999**, *Brain Res.* 831, 200–205.

Janzer, R. C., Raff, M. C. **1987**, *Nature* 325, 253–257.

Jesaitis, L. A., Goodenough, D. A. **1994**, *J. Cell Biol.* 124, 949–962.

Ke, C., Poon, W. S., Ng, H. K., Pang, J. C., Chan, Y. **2001**, *Neurosci. Lett.* 301, 21–24.

Ke, C., Poon, W. S., Ng, H. K., Lai, F. M., Tang, N. L., Pang, J. C. **2002**, *Exp. Neurol.* 178, 194–206.

Kimelberg, H. K., Ransom, B. R. **1986**, Physiological and pathological aspects of astrocytic swelling, in *Astrocytes. Cell Biology and Pathology of Astrocytes*, eds. S. Federoff, A. Vernadakis, Academic Press, San Diego, pp. 77–127.

Kissel, K., Hamm, S., Schulz, M., Vecchi, A., Garlanda, C., Engelhardt, B. **1998**, *Histochem. Cell Biol.* 110, 63–72.

Krämer, S. D., Abbott, N. J., Begley, D. J. **2001**, Biological models to study blood-brain barrier permeation, in *Pharmacokinetic Astrocytes and BBB permeability, Optimization in Drug Research: Biological, Physicochemical and Computational Strategies*, ed. B. Testa, H. van de Waterbeemd, G. Folkers, R. Guy, Wiley-VCH, Weinheim, pp. 127–153.

Krum, J. M., Rosenstein, J. M. **1989**, *Exp. Neurol.* 103, 203–212.

Krum, J. M., Rosenstein, J. M. **1993**, *Dev. Brain Res.* 74, 41–50.

Lee, S. W., Kim, W. J., Choi, Y. K., Song, H. S., Son, M. J., Gelman, I. H., Kim, Y. J., Kim, K. W. **2003**, *Nat. Med.* 9, 900–906.

Lobrinus, J. A., Juillerat-Jeanneret, L., Darekar, P., Schlosshauer, B., Janzer, R. C. **1992**, *Dev. Brain Res.* 70, 207–211.

Ma, Y., Miyano, K. E., Cowan, P. L., Aglitzkiy, Y., Karlin, B. A. **1995**, *Phys. Rev. Lett.* 74, 478–481.

Madara, J. L., Dharmsathaphorn, K. **1985**, *J. Cell Biol.* 101, 2124–2133.

Madara, J. L., Parkos, C., Colgan, S., Nusrat, A., Atisook, K., Kaoutzani, P. **1992**, *Ann. N.Y. Acad. Sci.* 664, 47–60.

Manley, G. T., Fujimura, M., Ma, T., Noshita, N., Filiz, F., Bollen, A. W., Chan, P., Verkman, A. S. **2000**, *Nat. Med.* 6, 159–163.

Manley, G. T., Binder, D. K., Papadopoulos, M. C., Verkman, A. S. **2004**, *Neuroscience* 129, 983–991.

Martin-Padura, I., Lostaglio, S., Schneemann, M., Williams, L., Romano, M., Fruscella, P., Panzeri, C., Stoppacciaro, A., Ruco, L., Villa, A., Simmons, D., Dejana E. **1998**, *J. Cell Biol.* 142, 117–127.

Maxwell, K., Berliner, J. A., Cancilla P. A. **1987**, *Brain Res.* 410, 309–314.

Meinild, A. K., Klaerke, D. A., Zeuthen, T. J. **1998**, *Biol. Chem.* 273, 32446–32451.

Meresse, S., Dehouck, M. P., Delorme, P., Bensaid, M., Tauber, J. P., Delbart, C., Fruchart, J. C., Cecchelli, R. **1989**, *J. Neurochem.* 53, 1363–1371.

Minagar, A., Shapshak, P., Fujimura, R., Ownby, R., Heyes, M., Eisdorfer, H. J. **2002**, *Neurol. Sci.* 202, 13–23.

Morgello, S., Uson, R. R., Schwartz, E. J., Haber, R. S. **1995**, *Glia* 14, 43–54.

Nagelhus, E. A., Veruki, L. M., Torp, R., Haugh, F. M., Laake, J. H., Nielsen, S. **1998**, *J. Neurosci.* 18, 2506–2519.

Nagelhus, E. A., Horio, Y., Inanobe, A., Fujita, A., Haugh, F. M., Nielsen, S., Kurachi, Y., Ottersen, P. **1999**, *Glia* 26, 47–54.

Nagelhus, E. A., Mathiisen, T. M., Ottersen, O. P. **2004**, *Neuroscience* 129, 905–913.

Neely, J. D., Amiry-Moghaddam, M., Ottersen, O. P., Froehner, S. C., Agre, P., Adams, M. E. **2001**, *Proc. Natl Acad. Sci. USA* 98, 14108–14113.

Neuhaus, J. **1990**, *Glia* 3, 241–245.

Newman, E. A. **1995**, Glial regulation of extracellular potassium, in *Neuroglia*, eds. H. Kettenman, B. R. Ramson, Oxford University Press, New York, pp. 717–731.

Nicchia, G. P., Frigeri, A., Liuzzi, G. M., Santacroce, M. P., Nico, B., Procino, G., Quondamatteo, F., Herken, R., Roncali, L., Svelto, M. **2000**, *Glia* 31, 29–38.

Nicchia, G. P., Frigeri, A., Liuzzi M. G., Svelto, M. **2003**, *FASEB J.* 17, 1508–1510.

Nicchia, G. P., Nico, B., Camassa, L. M. A., Mola, M. G., Loh, N., Dermietzel, R., Spray, D. C., Svelto, M., Frigeri, A. **2004**, *Neuroscience* 129, 935–945.

Nico, B., Cantino, D., Sassoè-Pognetto, M., Bertossi, M., Roncali, L. **1994**, *J. Submicrosc. Cytol. Pathol.* 26, 193–209.

Nico, B., Cardelli, P., Fiori, A., Riccetelli, L., Giglio, R. M., Strom, R., Sassoè-Pognetto, M., Cantino, D., Bertossi, M., Ribatti, D., Roncali, L. **1997**, *Microvasc. Res.* 53, 79–91.

Nico, B., Quondamatteo, F., Herken, R., Marzullo, A., Corsi, P., Bertossi, M., Russo, G., Ribatti, D., Roncali, L. **1999**, *Dev. Brain Res.* 114, 161–169.

Nico, B., Frigeri, A., Nicchia, G. P., Quondamatteo, F., Herken, R., Erede, M., Ribatti, D., Svelto, M., Roncali, L. **2001**, *J. Cell Sci.* 114, 1297–1307.

Nico, B., Frigeri, A., Nicchia, G. P., Corsi, P., Ribatti, D., Quondamatteo, F., Herken, R., Girolamo, F., Marzullo, A., Svelto, M., Roncali, L. **2003**, *Glia* 42, 235–251.

Nico, B., Nicchia, G. P., Frigeri, A., Corsi, P., Mangieri, D., Ribatti, D., Svelto, M., Roncali, L. **2004**, *Neuroscience* 125, 921–935.

Nielsen, S., Smith, B. L., Christensen, E. I., Agre, P. **1993**, *Proc. Natl Acad. Sci. USA* 90, 7275–7279.

Nielsen, S., Nagelhus, E. A., Amiry-Moghaddam, M., Bourque, C., Agre, P. **1997**, *J. Neurosci.* 17, 171–180.

Norenberg, M. D. **1998**, *Blackwell Sci.* 1998, 113–128.

Oshio, K., Binder, D. K., Yang, B., Schecter, S., Verkman, A. S., Manley, G. T. **2004**, *Neuroscience* 127, 685–693.

Papadopoulos, M. C., Manley, G. T., Krishna, S., Verkman, A. S. **2004a**, *FASEB J.* 18, 1291–1293.

Papadopoulos, M. C., Saadoun, S., Binder, D. K., Manley, G. T., Krishna, S., Verkman, A. S. **2004b**, *Neuroscience* 129, 1011–1020.

Pardridge, W. M. **1988**, *Annu. Rev. Pharmacol. Toxicol.* 28, 25–39.

Qin, Y., Sato, T. N. **1995**, *Dev. Dyn.* 202, 172–180.

Quagliarello, V. J., Long, W. J., Scheld, W. M. **1986**, *J. Clin. Invest.* 77, 1084–1095.

Rash, J. E., Yasumura, T., Hudson, C. S., Agre, P., Nielsen, S. **1998**, *Proc. Natl Acad. Sci. USA* 95, 11981–11986.

Raub, T. J., Kluentzel, S. L., Sawada, G. A. **1992**, *Exp. Cell Res.* 199, 330–340.

Reese, T. S., Karnowsky, M. J. **1967**, *J. Cell Biol.* 34, 207–217.

Reinhart, C. A., Gloor, S. M. **1997**, *Toxicol. In Vitro* 11, 513–518.

Risau, W. **1989**, *News Physiol. Sci.* 4, 151–153.
Risau, W. **1992**, *Ann. N.Y. Acad. Sci.* 633, 405–419.
Risau, W. **1997**, *Nature* 386, 671–674.
Risau, W., Wolburg, H. **1990**, *Trends Neurosci.* 13, 174–178.
Risau, W., Wolburg, H. **1991**, *Trends Neurosci.* 14, 15.
Risau, W., Hallmann, R., Albrecht, U. **1986a**, *Dev. Biol.* 117, 537–545.
Risau, W., Hallmann, R., Albrecht, U., Henke-Fahle, S. **1986b**, *EMBO J.* 5, 3179–3183.
Risau, W., Gautschi-Sova, P., Bohlen, P. **1988**, *EMBO J.* 7, 959–962.
Roberts, P., Chumas, P.D., Picton, S., Bridges, L., Livingstone, J.H., Sheridan, E. **2001**, *Cancer Genet. Cytogenet.* 131, 1–12.
Roncali, L., Nico, B., Ribatti, D., Pertossi, M., Mancini, L. **1986**, *Acta Neuropathol.* 70, 193–201.
Rubin, L.L., Hall, D.E., Porter, S., Barbu, K., Cannon, C., Horner, H.C., Janatpour, M., Liaw, C.W., Manning, K., Morales, J., Tanner, L.I., Tomaselli, K.J. **1991**, *J. Cell Biol.* 115, 1725–1735.
Saadoun, S., Papadopoulos, M.C., Davies, D.C., Krishna, S., Bell, B.A. **2002**, *J. Neurol. Neurosurg. Psychiatry* 72, 262–265.
Schinkel, A.H., Smit, J.J., van Tellingen, O., Beijnen, J.H., Wagenaar, E., Deemter, L., Mol, C.A., van der Valk, M.A., Robanus- Maandag, E.C., te Riele, H.P., Berns, A.J.M., Borst, P. **1994**, *Cell* 77, 491–502.
Schroeter, M.L., Mertsch, K., Giese, H., Muller, S., Sporbert, A., Hickel, B., Blasig, I.E. **1999**, *FEBS Lett.* 449, 241–244.
Schulze, C., Firth, J.A. **1992**, *J. Cell Sci.* 101, 647–655.
Schulze, C., Firth, J.A. **1993**, *J. Cell Sci.* 104, 773–782.
Seitz, R.J., Wechsler, W. **1987**, *Acta Neuropathol.* 73, 145–152.
Sobue, K., Yamamoto, N., Yoneda, K., Hodgson, M.E., Yamashiro, K., Tsuruoka, N., Tsuda, T., Katsuya, H., Miura, Y., Asai, K., Kato, T. **1999**, *Neurosci. Res.* 35, 155–164.
Sperri, P.E., Grant, M.B., Gomez, J., Vernadakis, A. **1997**, *Dev. Brain. Res.* 104, 205–208.
Stan, R.V., Ghitescu, L., Jacobson, B.S., Palade, G.E. **1999**, *J. Cell Biol.* 145, 1189–1198.
Stevenson, B.R., Siciliano, D., Mooseker, M.S., Goodenough, D.A. **1986**, *J. Cell Biol.* 103, 755–766.
Stewart, P.A., Wiley, M.J. **1981**, *Dev. Biol.* 84, 183–192.
Sun, D., Lytle, C., O'Donnell, M. **1995**, *Am. J. Physiol.* 269, C1506–C1512.
Tao-Cheng, J.H., Nagy, Z., Brightman, M.W. **1987**, *J. Neurosci.* 7, 3293–3299.
Tao-Cheng, J.H., Nagy, Z., Brightman, M.W. **1990**, *J. Neurocytol.* 19, 143–153.
Tontch, M.U., Bauer, H.C. **1991**, *Brain Res.* 539, 247–253.
Tout, S., Chan-Ling, T., Hollander, H., Stone, S. **1993**, *J. Neurosci.* 55, 291–301.
Tran, N.D., Correale, J., Schreiber, S.S., Fisher, M. **1999**, *Stroke* 30, 1671–1678.
Utsumi, H., Chiba, H., Kamimura, Y., Osanai, M., Igarashi, Y., Tobioka, H., Mori, M., Sawada, N. **2000**, *Am. J. Physiol. Cell Physiol.* 279, C361–C368.

Vajda, Z., Pedersen, M., Fuchtbauer, E. M., Wertz, K., Stodkilde-Jorgensen, H., Sulyok, E., Doczi, T., Neely, J. D., Agre, P., Frokiaer, J., Nielsen, S. **2002**, *Proc. Natl Acad. Sci. USA* 99, 13131–13136.

Venero, J. L. , Vizuete, M. L., Ilundain, A. A., Machado, A., Echevarria, M., Cano, J. **1999**, *Neuroscience* 94, 239–250.

Venero, J. L., Vizuete, M. L., Machado, A., Cano, J. **2001**, *Prog. Neurobiol.* 63, 321–336.

Venero, J. L., Machado, A., Cano, J. **2004**, *Curr. Pharm. Des.* 10, 2153–2161.

Verkman, A. S., van Hoek, A. N., Ma, T., Frigeri, A., Skach, W. R., Mitra, A., Tamarappoo, B. K., Farinas, J. **1996**, *Am. J. Physiol.* 270, C12–C30.

Vizuete, M. L., Vizuete, M. L., Venero, J. L., Vargas, C., Ilundain, A. A., Echevarria, M., Machado, A., Cano, J. **1999**, *Neurobiol. Dis.* 6, 245–258.

Wagner, S., Gardner, H. **2000**, *Neurosci. Lett.* 284, 105–108.

Wakai, S., Hirokawa, N. **1978**, *Cell. Tissue Res.* 195, 195–203.

Walz, W., Hinks, E. C. **1985**, *Brain Res.* 343, 44–51.

Warth, A., Kroger, S., Wolburg, H. **2004**, *Acta Neuropathol.* 107, 311–318.

Weiss, J. M., Downie, S. A., Lyman, W. D., Berman, J. W. **1998**, *J. Immunol.* 161, 6896–6903.

Williams, R. L., Courtneidge, S. A., Wagner, E. F. **1988**, *Cell* 52, 121–131.

Williams, R. L., Risau, W., Zerwes, H. G., Drexler, H., Aguzzi, A., Wagner, E. F. **1989**, *Cell* 57, 1053–1063.

Wolburg, H. **1995**, *J. Hirnforsch.* 36, 239–258.

Wolburg, H., Neuhaus, J., Pettmann, B., Labourdette, G., Sensenbrenner, M. **1986**, *Neurosci. Lett.* 72, 25–30.

Yoder, E. J. **2002**, *Glia* 38, 137–145.

Zhong, Y., Enomoto, K., Isomura, H., Sawada, N., Minase, T., Oyamada, M., Konishi, Y., Mori, M. **1994**, *Exp. Cell Res.* 214, 614–620.

Part III
Hormonal and Enzymatic Control of Brain Vessels

10
The Role of Fibroblast Growth Factor 2 in the Establishment and Maintenance of the Blood-Brain Barrier

Bernhard Reuss

10.1
Introduction

The establishment and maintenance of a functional blood-brain barrier (BBB) is an important prerequisite for the special milieu of the central nervous system (CNS) and thus of proper brain function. In accordance with this, during alterations of the BBB during pathological injury and its restoration afterwards, it is important for the brain to cope with the special requirements for brain reorganization during wound healing. Fibroblast growth factors (FGFs) are amongst the earliest identified growth factors and have been shown to influence the growth and differentiation of brain microvascular endothelial cells, and thus to be involved in the establishment, maintenance, and restoration of the BBB. Especially recent findings on the role of FGF-2 (also known as basic FGF) for BBB formation and maintenance in vivo have brought this growth factor back onto the agenda of scientific interest. This chapter is intended to give an overview on past and present findings on the role of FGF-2 for BBB maintenance and function in the intact and lesioned CNS.

10.2
Role of FGF-2 in the Regulation of BBB Formation

10.2.1
Expression of FGF-2 in Astrocytes and Endothelial Cells of the Rodent Brain

Astrocytes and endothelial cells are both structural and regulatory partners for the formation and maintenance of the BBB; and thus an important question with regard to a role for FGF-2 in BBB-formation is whether FGF-2 and its appropriate receptors are expressed in either astrocytes and/or endothelial cells.

In general during development of the CNS, a switch from a predominantly neuronal expression of FGF-2 during prenatal phases towards a primarily astrocytic expression during postnatal phases has been described [1]. According to this at embryonic day (E)13 of rat development, FGF-2 is expressed in various neuronal precursor cell types. On E18, the cerebral cortex shows strong FGF-2 immunoreactivity and at the first postnatal day (PND1), FGF-2 is primarily located in neurons of the hippocampal formation. During PND4–PND6, astroglial expression of FGF-2 begins, with the adult pattern of FGF-2 immunoreactivity in astrocytes from all brain regions being established at PND20. At that time neuronal FGF-2 expression is restricted to particular brain regions, such as the cingulate cortex and the hippocampal formation (CA2 field) [1]) This distribution seems to persist in the adult brain, where expression of FGF-2 has been demonstrated mainly in astrocytes [2], but also in special neuronal subpopulations [3, 4].

Besides such a predominant neural localization, at least during development, FGF-2 is also expressed in brain microvascular endothelial cells and therefore seems to be directly involved in endothelial cell differentiation during pre- and postnatal brain development. Support for this comes from a study by Schechter et al. [5], who demonstrated that, at the onset of vascularization during E15–E18 in the pituitary gland of the rat, the cytoplasm of the invading endothelial cells from immature capillaries shows intense immunoreactivity for FGF-2. Later (E19–E20), numbers of FGF-2-positive capillaries greatly decrease, with foci of released FGF-2 remaining still evident within the presumptive pericapillary spaces throughout gestation. In adult animals, however, capillary endothelial cells did not contain immunostainable FGF-2 in their cytoplasm [5]. Likewise a study by Marin [6] in the pineal gland demonstrated immunoreactivity for FGF-2 in endothelial cells and perivascular spaces from PND20 onwards. This expression reached a maximum at the age of PND30–PND45, but unlike the pituitary gland, FGF-2 remains to be expressed also in later life [6]. This suggests that in several brain regions FGF-2 is only expressed during pre- and postnatal development, whereas in others FGF-2 expression continues to be present in the adult brain.

With respect to expression of fibroblast growth factor receptors (FGFRs), in the adult brain expression of FGFR-1 mRNA and protein seems to be restricted to neurons, whereas FGFR-2 and -3 mRNAs are located preferentially in glial cells such as oligodendrocytes and astrocytes [7–9]. Again, FGFRs are also expressed in brain microvascular endothelial cells, as has been demonstrated for FGFR-1 in the median eminence of the pituitary gland of the rat [10]. Further evidence provided an in vitro study by Bastaki et al. [11] where FGFR-1 protein and mRNA were detected in a brain microvascular endothelial cell line.

10.2.2
Induction of BBB Properties in Endothelial Cells by Soluble Factors

Despite the above-mentioned localization of immunoreactive FGF-2 and FGFR-1, brain endothelial cells are not capable to induce BBB properties on their own, but need to receive signals from the surrounding tissues (i.e. from astrocytes). Such factors have indeed been found in astrocyte-conditioned medium and have been identified as hormones and protein growth factors [12, 13]. Thus, transforming growth factor(TGF)-β_1 [14], glia-derived neurotrophic factor (GDNF [15, 16], interleukin (IL)-6, hydrocortisone [17], and FGF-2 [13, 18] have all been shown to induce at least some of the specific barrier properties of brain microvessels.

With respect to FGF-2, Roux et al. [18] demonstrated in an immortalized cell line of rat brain endothelial cells (RBE4) that FGF-2 is able to induce RBE4 monolayers cultured on collagen-coated dishes to reorganize into three-dimensional tube-like structures. In addition, these endothelial tubes show increased activity of the BBB-associated enzymes γ-glutamyl transpeptidase (γ-GTP) and alkaline phosphatase, as compared with cell cultures growing only in two dimensions [18]. Likewise, as demonstrated by Sobue et al. [13], FGF-2 induces BBB properties in an immortalized endothelial cell line, as revealed by increased L-glucose permeability and alkaline phosphatase activity [13]. This effect could be blocked by the application of FGF-2 specific antiserum. However, the previous finding of Boado et al. [19], who demonstrated induction of the BBB-specific glucose transporter GLUT1 by FGF-2 in brain endothelial cells, could not be reproduced. There was also no observable induction of the expression of multidrug resistance genes which are normally present in brain endothelial cells [13].

10.2.3
Indirect Astrocyte Mediated Effects Seem to Play a Role in FGF-2-Dependent Changes in Endothelial Cell Differentiation

Although direct actions of FGF-2 on brain endothelial cell differentiation are of great importance, astroglial FGF-2 is also able to influence endothelial cell differentiation by indirect mechanisms. Thus, FGF-2 is able to elicit changes in astrocyte differentiation in an autocrine manner, which then indirectly affect endothelial cell differentiation and BBB properties. An important mediator for such indirect effects of FGF-2 on endothelial cell differentiation seems to be TGF-β_1, which can be induced in astrocytes by FGF-2 and, upon its induction, modulates BBB-like differentiation of brain endothelial cells. This has been shown by Garcia et al. [20], who used a coculture model to demonstrate that local activation of TGF-β_1 release from astrocytes is responsible for establishing at least some of the barrier properties in brain microvascular endothelial cells. Likewise Dohgu et al. [21] could show that TGF-β_1 lowers the permeability of endothelial cells in culture and thus is involved in keeping the BBB function.

Another feature of astrocytes that seems to have a major impact on the differentiation of brain microvascular endothelial cells and on BBB properties seems to be the intermediary filament proteins such as the glial fibrillary acidic protein (GFAP). In support of this, Pekny et al. [22] could demonstrate that astrocytes from wild-type mice are able (whereas astrocytes from GFAP-deficient mice are not able) to induce BBB properties in aortic endothelial cells in an in vitro coculture model. Since FGF-2 is well known to influence the synthesis and phosphorylation of GFAP in astrocytes [23–25], this could then indirectly lead to changes in differentiation of brain microvascular endothelial cells by the above-mentioned GFAP-dependent mechanism. This has indeed been demonstrated in a study by our group [26], where mice with a single or double deficiency in genes for FGF-2 and/or FGF-5 revealed reduced levels of GFAP in different brain regions. In addition, FGF-deficient animals showed also a leaky BBB as demonstrated by albumin extravasation as well as reduced levels of the tight junction proteins occludin and ZO-1 [26]. Whether this regulation also involves changes in astroglial expression and release of TGF-β_1 will have to be clarified by additional experiments.

10.2.4
Involvement of FGF-2 in the Regulation of BBB Properties in the Pathologically Altered Brain

Another important aspect of FGF-2 function in the regulation of endothelial BBB properties might be also its implications for understanding the regulation of BBB permeability under pathological circumstances. Disturbed BBB permeability is a common feature of several disease conditions in the CNS, including ischemia, tumor growth, and demyelinating disorders [27–29]. Likewise, changes in FGF-2 expression under these conditions also suggest a role for FGF-2 in the regulation of BBB permeability under pathological circumstances.

For example, after focal brain ischemia by middle cerebral artery occlusion (MCAO), expression of FGFR-1 mRNA is induced in capillary endothelium in the corpus callosum and internal capsule [30]. Together with increased astroglial FGF-2 expression found in the brains of rats suffering from MCAO dependent ischemia [31], this suggests that stimulation of FGFR-positive capillary endothelial cells by astroglia-derived FGF-2 could play a role in the restoration of blood flow and the reorganization of brain vascularization after ischemic damage. This view is further supported by the fact that intracerebral xenografts of mouse bone marrow cells in adult rats facilitate the restoration of cerebral blood flow and BBB by the release of FGF-2 [32].

Similar conclusions are drawn from an in vitro study of Brown et al. [33], which demonstrated that conditioned medium from the C6 glioma cell line can induce BBB properties in brain microvessel endothelial cells and protect them against hypoxia-induced BBB breakdown. This protective effect is accompanied by significantly higher levels of FGF-2 in C6-conditioned medium and increased expression of the tight junction protein claudin-1 in endothelial cells [33]. Likewise, inhibition of vascularization during tumor growth of C6 and 9L glioma cell lines by angiostatin is accompanied by decreased expression of vascular endothelial growth factor (VEGF) and increased expression of FGF-2 mRNA, suggesting that in this case proliferation of endothelial cells and thus vasculariza-

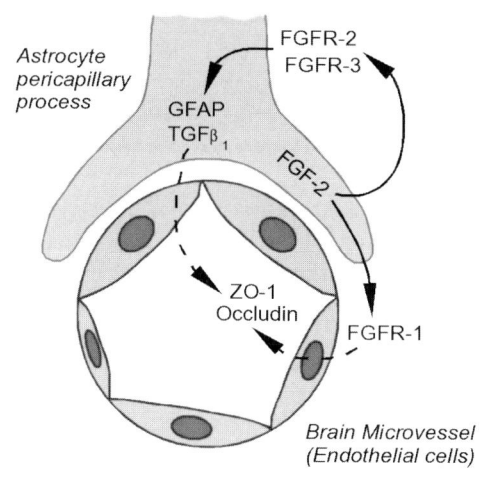

Fig. 10.1 Direct and indirect modes of action are thought to play a role in FGF-2-dependent regulation of microvascular endothelial cell differentiation.

tion might be reduced in parallel with a higher rate of terminal differentiation of endothelial cells [34].

A group of disorders where FGF-2-dependent regeneration of BBB properties in endothelial cells could be involved are the demyelinating diseases, an important feature of which is the breakdown of the BBB (for a review, see [35]). In accordance with this, in an experimental in vivo demyelination/remyelination model, levels of FGF-2 and of its receptors are greatly increased, with their peak expression being at the initial stage of remyelination [36].

10.3
Future Perspectives

In conclusion, FGF-2 is one of the key regulators for the formation of the BBB during brain development and for BBB maintenance in the adult brain (for an overview, see Fig. 10.1). In addition, it seems to be even more important for the restoration of proper BBB functions after brain injury and a lack in FGF-2 seems to be the cause of BBB leakiness in solid brain tumors. Of special significance therefore is the indirect impact of FGF-2 on endothelial cell functions by regulating the interactions of the latter with astroglial perivascular endfeet. Together these properties of FGF-2 make it an interesting candidate for pharmacological intervention in the regulation of blood flow and revascularization after brain injury or BBB permeability during the treatment of solid brain tumors.

References

1 Gomez-Pinilla F, Lee JW, Cotman CW **1994**, Distribution of basic fibroblast growth factor in the developing rat brain, *Neuroscience* 61, 911–923.
2 Kuzis K, Reed S, Cherry NJ, Woodward WR, Eckenstein FP **1995**, Developmental time course of acidic and basic fibroblast growth factors' expression in distinct cellular populations of the rat central nervous system, *J. Comp. Neurol.* 358, 142–153.
3 Eckenstein FP, Shipley GD, Nishi R **1991a**, Acidic and basic fibroblast growth factors in the nervous system: distribution and differential alteration of levels after injury of central versus peripheral nerve, *J. Neurosci.* 11, 412–419.
4 Eckenstein F, Woodward WR, Nishi R **1991b**, Differential localization and possible functions of aFGF and bFGF in the central and peripheral nervous systems, *Ann. N.Y. Acad. Sci.* 638, 348–360.
5 Schechter J, Pattison A, Pattison T **1996**, Basic fibroblast growth factor within endothelial cells during vascularization of the anterior pituitary, *Anat. Rec.* 245, 46–52.
6 Marin F, Boya J, Calvo JL, Lopez-Munoz F, Garcia-Maurino JE **1994**, Immunocytochemical localization of basic fibroblast growth factor in the rat pineal gland, *J. Pineal Res.* 16, 44–49.
7 Asai T, Wanaka A, Kato H, Masana Y, Seo M, Tohyama M **1993**, Differential expression of two members of FGF receptor gene family, FGFR-1 and FGFR-2 mRNA, in the adult rat central nervous system, *Brain Res. Mol. Brain Res.* 17, 174–178.
8 Yazaki N, Hosoi Y, Kawabata K, Miyake A, Minami M, Satoh M, Ohta M, Kawasaki T, Itoh N **1994**, Differential expression patterns of mRNAs for members of the fibroblast growth factor receptor family, FGFR-1-FGFR-4, in rat brain, *J. Neurosci. Res.* 37, 445–452.
9 Reuss B, Hertel M, Werner S, Unsicker K **2000**, Fibroblast growth factors-5 and -9 distinctly regulate expression and function of the gap junction protein connexin43 in cultured astroglial cells from different brain regions, *Glia* 30, 231–241.
10 Gonzalez AM, Logan A, Ying W, Lappi DA, Berry M, Baird A **1994**, Fibroblast growth factor in the hypothalamic-pituitary axis: differential expression of fibroblast growth factor-2 and a high affinity receptor, *Endocrinology* 134, 2289–2297.
11 Bastaki M, Nelli EE, Dell'Era P, Rusnati M, Molinari-Tosatti MP, Parolini S, Auerbach R, Ruco LP, Possati L, Presta M **1997**, Basic fibroblast growth factor-induced angiogenic phenotype in mouse endothelium. A study of aortic and microvascular endothelial cell lines, *Arterioscler. Thromb. Vasc. Biol.* 17, 454–464.
12 Hurst RD, Fritz IB **1996**, Properties of an immortalised vascular endothelial/glioma cell co-culture model of the blood-brain barrier, *J. Cell Physiol.* 167, 81–88.

13 Sobue K, Yamamoto N, Yoneda K, Hodgson ME, Yamashiro K, Tsuruoka N, Tsuda T, Katsuya H, Miura Y, Asai K, Kato T **1999**, Induction of blood-brain barrier properties in immortalized bovine brain endothelial cells by astrocytic factors, *Neurosci. Res.* 35, 155–164.

14 Tran ND, Correale J, Schreiber SS, Fisher M **1999**, Transforming growth factor-beta mediates astrocyte-specific regulation of brain endothelial anticoagulant factors, *Stroke* 30, 1671–1678.

15 Igarashi Y, Utsumi H, Chiba H, Yamada-Sasamori Y, Tobioka H, Kamimura Y, Furuuchi K, Kokai Y, Nakagawa T, Mori M, Sawada N **1999**, Glial cell line-derived neurotrophic factor induces barrier function of endothelial cells forming the blood-brain barrier, *Biochem. Biophys. Res. Commun.* 261, 108–112.

16 Utsumi H, Chiba H, Kamimura Y, Osanai M, Igarashi Y, Tobioka H, Mori M, Sawada N **2000**, Expression of GFRα-1, receptor for GDNF, in rat brain capillary during postnatal development of the BBB, *Am. J. Physiol. Cell Physiol.* 279, C361–C368.

17 Hoheisel D, Nitz T, Franke H, Wegener J, Hakvoort A, Tilling T, Galla HJ **1998**, Hydrocortisone reinforces the blood-brain barrier properties in a serum free cell culture system, *Biochem. Biophys. Res. Commun.* 247, 312–315.

18 Roux F, Durieu-Trautmann O, Chaverot N, Claire M, Mailly P, Bourre JM, Strosberg AD, Couraud PO **1994**, Regulation of gamma-glutamyl transpeptidase and alkaline phosphatase activities in immortalized rat brain microvessel endothelial cells, *J. Cell Physiol.* 159, 101–113.

19 Boado RJ, Wang L, Pardridge WM **1994**, Enhanced expression of the blood-brain barrier GLUT1 glucose transporter gene by brain-derived factors, *Brain Res. Mol. Brain Res.* 22, 259–267.

20 Garcia CM, Darland DC, Massingham LJ, D'Amore PA **2004**, Endothelial cell-astrocyte interactions and TGF-β are required for induction of blood-neural barrier properties, *Brain Res. Dev. Brain Res.* 152, 25–38.

21 Dohgu S, Yamauchi A, Takata F, Naito M, Tsuruo T, Higuchi S, Sawada Y, Kataoka Y **2004**, Transforming growth factor-β1 upregulates the tight junction and P-glycoprotein of brain microvascular endothelial cells, *Cell Mol. Neurobiol.* 24, 491–497.

22 Pekny M, Stanness KA, Eliasson C, Betsholtz C, Janigro D **1998**, Impaired induction of blood-brain barrier properties in aortic endothelial cells by astrocytes from GFAP-deficient mice, *Glia* 22, 390–400.

23 Eclancher F, Perraud F, Faltin J, Labourdette G, Sensenbrenner M **1990**, Reactive astrogliosis after basic fibroblast growth factor (bFGF) injection in injured neonatal rat brain, *Glia* 3, 502–509.

24 Eclancher F, Kehrli P, Labourdette G, Sensenbrenner M **1996**, Basic fibroblast growth factor (bFGF) injection activates the glial reaction in the injured adult rat brain, *Brain Res.* 737, 201–214.

25 Perraud F, Labourdette G, Eclancher F, Sensenbrenner M **1990**, Primary cultures of astrocytes from different brain areas of newborn rats and effects of basic fibroblast growth factor, *Dev. Neurosci.* 12, 11–21.

26 Reuss B, Dono R, Unsicker K **2003**, Functions of fibroblast growth factor (FGF)-2 and FGF-5 in astroglial differentiation and blood-brain barrier permeability: evidence from mouse mutants, *J. Neurosci.* 23, 6404–6412.

27 Pollay M, Stevens FA **1980**, Blood-brain barrier restoration following cold injury, *Neurol. Res.* 1, 239–245.

28 Saburina IN **1989**, Embryonic nervous tissue transplantation accelerates restoration of hypoxia-damaged blood-brain barrier in rats, *J. Hirnforsch.* 30, 737–745.

29 Broom KA, Anthony DC, Blamire AM, Waters S, Styles P, Perry VH, Sibson NR **2005**, MRI reveals that early changes in cerebral blood volume precede blood-brain barrier breakdown and overt pathology in MS-like lesions in rat brain, *J. Cereb. Blood Flow Metab.* 25, 204–216.

30 Yamada K, Sakaguchi T, Yuguchi T, Kohmura E, Otsuki H, Koyama T, Hayakawa T **2000**, Blood-borne macromolecule induces FGF receptor gene expression after focal ischemia, *Acta Neurochir. Suppl. (Wien)* 60, 261–264.

31 Wei OY, Huang YL, Da CD, Cheng JS **2000**, Alteration of basic fibroblast growth factor expression in rat during cerebral ischemia, *Acta Pharmacol. Sin.* 21, 296–300.

32 Borlongan CV, Lind JG, Dillon-Carter O, Yu G, Hadman M, Cheng C, Carroll J, Hess DC **2004**, Intracerebral xenografts of mouse bone marrow cells in adult rats facilitate restoration of cerebral blood flow and blood-brain barrier, *Brain Res.* 1009, 26–33.

33 Brown RC, Mark KS, Egleton RD, Huber JD, Burroughs AR, Davis TP **2003**, Protection against hypoxia-induced increase in blood-brain barrier permeability: role of tight junction proteins and NFkappaB, *J. Cell Sci.* 116, 693–700.

34 Kirsch M, Strasser J, Allende R, Bello L, Zhang J, Black PM **1998**, Angiostatin suppresses malignant glioma growth in vivo, *Cancer Res.* 58, 4654–4659.

35 Minagar A, Alexander JS **2003**, Blood-brain barrier disruption in multiple sclerosis, *Mult. Scler.* 9, 540–549.

36 Messersmith DJ, Murtie JC, Le TQ, Frost EE, Armstrong RC **2000**, Fibroblast growth factor 2 (FGF-2) and FGF receptor expression in an experimental demyelinating disease with extensive remyelination, *J. Neurosci. Res.* 62, 241–256.

11
Cytokines Interact with the Blood-Brain Barrier

Weihong Pan, Shulin Xiang, Hong Tu, and Abba J. Kastin

11.1
Introduction

After peripheral production, cytokines exert diverse effects on CNS functions. These CNS effects can occur in several ways. Cytokines can cross the blood-brain barrier (BBB) directly, reach the CNS by retrograde axonal transport, act on circumventricular organs, or activate secondary mediators without themselves entering the brain and spinal cord. This chapter will focus on the current knowledge from pharmacokinetic studies of cytokine permeation across the BBB and on the intracellular events of cytokine trafficking in cerebral microvessel endothelial cells. Representative cytokines discussed include the interleukins, tumor necrosis factor α, leukemia inhibitory factor, epidermal growth factor, basic fibroblast growth factor, transforming growth factor, and some neurotrophins. These cytokines have pronounced neuroendocrine and trophic effects. We will also discuss some cytokines and chemokines that alter the morphology and intracellular signaling pathways of the cerebral microvessel endothelial cells. They are implicated in CNS inflammation, tumor metastasis, cerebral amyloid angiopathy, and various etiologies of stroke. As shown in disease models such as spinal cord injury, experimental autoimmune encephalomyelitis, and stroke, the regulation of the interactions between the cytokines and the BBB play important roles in CNS physiology and pathology.

The BBB interfaces the parenchyma of brain and spinal cord and its supplying capillary vessels. The structural components are microvessel endothelial cells, pericytes, astrocytic endfeet, and extracellular matrix. The endothelial cells are joined by tight junctions, lined by a continuous basement membrane, and have reduced pinocytic vesicles and increased metabolic and enzymatic activity. These structural features are involved in the relative impermeability of the BBB to large proteins in the blood circulation. However, there are many instances in which cytokines produced in the periphery have CNS effects, and most of such actions are mediated by the BBB. In this review, we will discuss two principal ways by which cytokines interact with the BBB: (a) their transport or transcyto-

Blood-Brain Interfaces: From Ontogeny to Artificial Barriers.
Edited by R. Dermietzel, D.C. Spray, M. Nedergaard
Copyright © 2006 WILEY-VCH Verlag GmbH & Co. KGaA, Weinheim
ISBN: 3-527-31088-6

sis across the endothelial cells that are the structural backbone of the BBB and (b) their actions on these endothelial cells which result in altered endothelial function, cytotoxicity, or cell proliferation.

11.2
Identification of the Phenomena

11.2.1
Cytokines That Cross the BBB by Specific Transport Systems

Interleukins Interleukin-1α and -β are among the proinflammatory cytokines that have a saturable influx transport system [8, 12]. In their studies, Banks et al. [8, 12] reported an influx transfer constant of 0.25–0.43 µl g^{-1} min^{-1} for interleukin-1α and 0.47 µl g^{-1} min^{-1} for interleukin-1β. The initial volume of distribution was 20.1 µl g^{-1} and 16.5 µl g^{-1}, respectively. The relative stability of the interleukins in the circulating blood and in the brain has been shown by high performance liquid chromatography (HPLC) and acid precipitation. Although there is influx transport, shown by self-inhibition, there is no saturability of brain-blood efflux. This vectorial passage indicates that interleukin-1 is an important mediator in the communication between the CNS and the periphery.

Regional differences in the rate of transport illustrates that the transport system for interleukin-1α is physiologically relevant to CNS function. Interleukin-1α gets into the hypothalamus more rapidly than other parts and there is selective uptake in the posterior division of the septum [8, 9]. The region-specific uptake is probably related to the suppression of feeding behavior and catabolic metabolism induced by interleukin-1α. The circumventricular organs (CVOs) account for less than 5% of the total brain uptake of interleukin-1α, and permeation into the CVOs can still be saturable [74]. Since interleukin-1 is pyrogenic, the reduced brain permeation of interleukin-1β in aged animals may partially explain the diminished fever response in the elderly [47].

Interleukin-6 also has a saturable influx transport system, shared by human and mouse interleukin-6 and flagged mouse interleukin-6, but not by interleukin-1 or tumor necrosis factor α (TNFα) [10]. After intravenous delivery by a bolus injection, intact interleukin-6 is recovered from the cerebrospinal fluid (CSF) at 10 min and 30 min. The lack of saturable efflux indicates that this transport system is unidirectional.

Tumor Necrosis Factor α Although large doses of TNFα might disrupt the BBB and increase paracellular permeability of the BBB, TNFα can cross the BBB by a saturable influx transport system. Trace amounts of ^{125}I-TNFα can be detected by HPLC in blood, brain homogenate, and CSF 30 min after intravenous delivery, without increasing the permeability of co-administered ^{99m}Tc-albumin [31]. Most of the injected TNFα enters brain and spinal cord parenchyma. The regional differences in uptake are such that the spinal cord has higher permeation than the brain [53]. In the spinal cord, the cervical and lumbar segments have higher permeability reflected by a greater volume of distribution and faster influx rate. In the brain, the hypothalamus and occipital cortex appear to take up TNFa significantly faster than the rest of the brain [11]. The influx of TNFa is blocked by pre-incubation with a soluble receptor against the p75 receptor [13], and it is absent in double TNF receptor knockout mice [65]. The regulation of such receptor-mediated transport of TNFa will be further discussed later in this review.

Leukemia Inhibitory Factor and Cilliary Neurotrophic Factor Cilliary neurotrophic factor (CNTF) and leukemia inhibitory factor (LIF) cross the BBB by independent saturable transport systems [68, 69]. Both have moderately fast influx rates (0.46 µl g^{-1} min^{-1} for CNTF, 0.41 µl g^{-1} min^{-1} for LIF). Although the two cytokines have one shared receptor subunit, gp130, there is no known cross-inhibition. The high affinity receptor gp190 is apparently involved in the transport of LIF, since a blocking antibody specifically reduces the influx transfer constant of LIF in both mouse studies and cultured brain endothelial cells [56].

Insulin-Like Growth Factor 1 The availability of insulin-like growth factor 1 (IGF-1) in the blood circulation for crossing the BBB is significantly influenced by IGF binding proteins (IGFBPs). To deliver sufficient amounts of IGF-1 to the CNS compartment, one has to first saturate the binding sites of IGFBPs. Regardless, there is a saturable influx transport system for IGF-1 at the BBB [61], which is at least partially shared with that for insulin [71]. The cross-inhibition of transport between IGF-1 and insulin indicates a receptor-mediated mechanism. The possible beneficial effects of IGF-1 include reduction of neurodegeneration and amelioration of autoimmune damage to the CNS [40, 43]. Thus, manipulation of the transport system and design of IGF-1 variants that can cross the BBB easily should be important goals for future research.

Epidermal Growth Factor Like IGF-1, epidermal growth factor (EGF) is a trophic factor both in the periphery and the CNS. EGF has a saturable transport system from blood to brain that does not seem to involve its receptor, and there is no efflux transport [60]. A potential problem for the delivery of EGF and IGF-1 as therapeutic agents for CNS pathology is that both may promote the growth of tumors such as prostate cancer.

11.2.2
Cytokines That Permeate the BBB by Simple Diffusion

Glial Cell-Derived Neurotrophic Factor Glial cell-derived neurotrophic factor (GDNF) has relatively fast degradation in the blood circulation; such instability may be explained by its endogenous production in the CNS. There is no saturable transport for GDNF, but there is simple diffusion across the BBB to some extent. Unlike GDNF, many other neurotrophins, such as nerve growth factor and neurotrophin-3, can be transported from blood into the CNS [54, 75].

Interleukin 10 Interleukin 10 is more stable in blood than many cytokines, with no apparent degradation at 30 min after intravenous injection. It also has relatively high lipid solubility compared with other proteins of similar size (160 amino acids), having an octanol buffer partition coefficient of 0.183 ± 0.004 and hydrogen bonding of 0.965. Regardless, there is very limited simple diffusion of interleukin 10 from blood to brain in the mouse, and its apical-basolateral flux is similar to that of the paracellular permeability marker albumin in cultured endothelial cells [38].

Chemokines Of the CXCR family of chemokines, cytokine-induced neutrophil chemoattractant-1 (CINC1) has limited passage cross the BBB by simple diffusion but shows significant binding to the apical surface of cerebral microvessel endothelial cells [62]. Interleukin 8 is relatively more stable but also crosses the BBB only by simple diffusion, although its influx transfer constant is significantly lower than that of CINC1 [66]. The interactions between chemokines and the CXCR family of receptors are important in leukocyte migration and tumor metastasis to CNS parenchyma.

The macrophage inflammatory proteins MIP-1α and MIP-1β are very stable in blood because of their polymerization. They have a high apparent volume of distribution in brain; however, their high binding to the luminal surface of cerebral blood vessels did not lead to endocytosis and penetration across the BBB during the period of study [7].

11.2.3
Cytokines That Have Known Effects on Endothelial Cells

TNFα affects cytoskeletal arrangement and tight junction protein distribution, which in turn increases paracellular permeability at high doses [72]. Platelet-activating factor is involved in the re-arrangement of cytoskeletal organization induced by TNFα [19]. Depending on the concentration and species of endothelial cell, TNFα may stimulate nitric oxide production or activation of the phospholipase A2 pathway [27]. More importantly, TNFα induces the expression of intracellular adhesion molecule-1 [44, 46], vascular cell adhesion molecule-1, and E-selectin [14, 24]. The interactions of TNFα and the endothelial cells could promote both the transport and paracellular permeation of TNFα and other molecules, including the 4.7-kDa peptide urocortin [52]. The interactions are also important in mediating inflammatory responses and brain metastasis of cancer cells.

Endothelial cells express both p55 and p75 receptors for TNFα. In the human endothelial cell line ECV304 and other cells, p55 is predominantly located in the trans-Golgi network [37]. This subcellular localization is related to a C-terminal sequence of 23 amino acids that contains an acid patch and a dileucine mo-

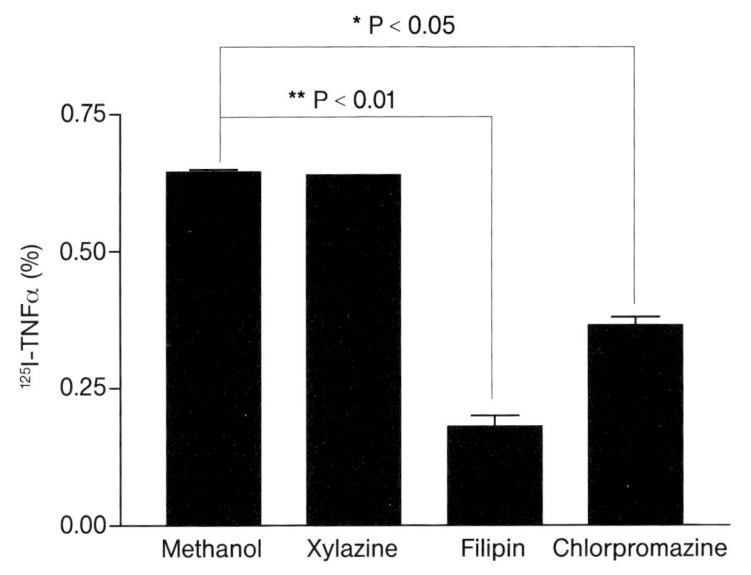

Fig. 11.1 Effects of filipin and chlorpromazine on the internalization of TNFα in TM-BBB4 cells. TM-BBB4 cells were incubated with filipin (5 µg ml^{-1}) or chlorpromazine (25 µg ml^{-1}) for 30 min at 37°C. The cells were incubated with filipin or chlorpromazine in the presence of ^{125}I-TNFα at 37°C for an additional 30 min. Surface binding was determined by an acid wash procedure. The percent of ^{125}I-TNFα internalized was determined after cell lysis. There was a significant decrease in the percent of ^{125}I-TNFα internalized in the presence of either filipin ($P<0.05$) or chlorpromazine ($P<0.01$).

tif which interact with membrane traffic adaptor proteins [82]. Upon phosphorylation, such as by the p42 mitogen-activated protein kinase (MAPK), p55 translocates to intracellular tubular structures associated with the endoplasmic reticulum [21]. The trafficking of the p75 receptor is less well known. The differential locations of the two subtypes of receptors may not only be important in cell signaling but also in receptor-mediated transport of TNFα across the BBB.

How does intracellular signal transduction then affect the endocytosis and intracellular trafficking of TNFα? In an immortalized mouse brain microvessel endothelial cell line, TM-BBB4, endocytosis of ^{125}I-TNFα is mediated both by clathrin-coated pits and caveolin. As shown in Fig. 11.1, both chlorpromazine (an inhibitor of the clathrin-mediated pathway) and filipin (an inhibitor of caveolin-mediated endocytosis) significantly reduced the internalization of TNFα at 30 min. Although it is known that p42 MAPK is involved in the translocation of the p55 receptor, inhibitors for MAPK, including PD98059 and SB203850, had no effect on endocytosis, the initial and rate-limiting step for transport of ^{125}I-TNFα (Fig. 11.2).

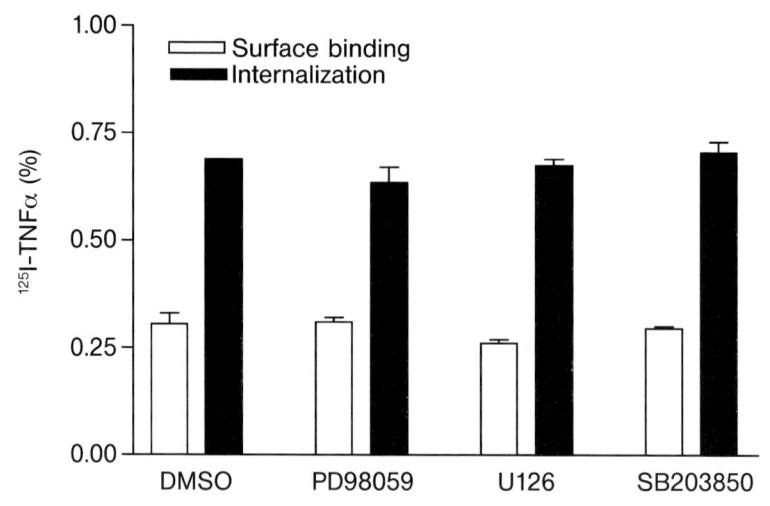

Fig. 11.2 Effect of MAPK inhibitors on the internalization of TNFα in TMBBB4 cells. TM-BBB4 cells were cultured with PD98059 (25 μM) U126 (1 μM), and SB203850 (1 μM) for 1 h at 37°C. ^{125}I-TNFα was added and the cells were incubated at 37°C for an additional 30 min. The surface binding was determined by an acid wash step; and the percent of internalized ^{125}I-TNFα was not significantly different among the groups.

Transforming growth factor α (TGFα) is a potent angiogenic factor [25]. TGFα can promote the interactions of tumor cells with the endothelial cells composing the BBB, facilitate cancer cell colonization and invasion across the BBB, and stimulate tumor growth inside the CNS [49]. Thus, TGFα plays a significant role in tumor metastasis to the brain. It may also induce proliferation of endothelial cells after mild injury and thereby contribute to the angiopathy of intracranial vessels [79]. Therefore, although TGFα does not have a saturable transport system for crossing the BBB [70], its interactions with the BBB are implicated in various CNS pathologies.

Chemokines and their receptors on the apical surface of endothelial cells are involved in inflammation, tumor metastasis, and endothelial proliferation. In particular, monocyte chemoattractant protein-1 (MCP-1) acts on the CCR2 receptor on endothelial cells, causes redistribution and down-regulation of tight junction proteins, and increases the permeability of the BBB, both in vivo and in vitro [80]. The expression of MCP-1 is up-regulated by proinflammatory cytokines including TNFα, IL-1β, and interferon γ [33].

11.3
Mechanisms of Cytokine Interactions with the BBB

11.3.1
Endocytosis of Cytokines by the Apical Surface of Endothelial Cells

The membrane events that lead to transport of a cytokine from blood to the CNS could involve carrier-mediated transport, receptor-mediated endocytosis, and adsorptive endocytosis. There are many factors that determine the nature of the endocytic pathways, such as the size of the endocytic vesicle, the cargo (ligands, receptors, lipids), and the mechanisms of vesicle formation. Clathrin, caveolin, or non-clathrin, non-caveolin-mediated endocytosis has been reviewed in detail by experts in cell biology [20].

Transport of TNFa across the BBB involves both p55 and p75 receptors. This was first shown in studies with TNF receptor knockout mice [65]. The absence of either the p55 or the p75 receptor reduces the influx transfer constant of ^{125}I-TNFa, an effect significant in the brain but less pronounced in the spinal cord. When both receptors are absent, however, transport of TNFa is completely abolished. The involvement of both receptors is also evident in transwell studies on primary mouse brain microvessel endothelial cells from double receptor knockout mice [59]. In addition to TNFa, compounds such as LIF, IGF-1, and the adipokine leptin also can be transported by receptor-mediated endocytosis.

Adsorptive endocytosis has been shown for ebiratide, an analog of adrenocorticotropic hormone. Ebiratide is a basic peptide with an isoelectric point of 10.0. Its blood-brain influx is saturable, energy-dependent, and inhibitable by other polycationic peptides and endocytosis inhibitors such as protamine and dansylcadaverine [83]. The occurrence of adsorptive endocytosis implies that the positively charged ebiratide moieties interact with the negatively charged cell surface in a specific manner. The binding affinity is lower than what would be expected for receptor-mediated endocytosis, but the transport capacity is substantially higher.

However, another basic polypeptide, basic fibroblast growth factor (bFGF), crosses the BBB with an intermediate capacity by specific binding to heparin sulfate proteoglycan; and its endocytosis is neither receptor- nor adsorption-mediated [22].

11.3.2
Intracellular Trafficking Pathways

In many types of polarized cells, the binding of a protein or peptide ligand to the cell surface receptor is followed by alterations in membrane dynamics. The ligand-receptor complex is concentrated in microdomains that are rich in either clathrin or caveolae. After a series of events to recruit adaptor molecules, membrane invagination leads to vesicular transport of the ligand-receptor complex to early endosomes. There are two likely fates. In one situation, the complex can be guided to lysosomes where increased acidity (lower pH) causes dissociation of the ligand from the receptor with recycling of the receptor back to the cell surface and degradation of the ligand. Alternatively, the complex can be trafficked to common endosomes and then elsewhere for a secretory pathway. It can also be directed to the Golgi complex and eventually to the secretory pathway. Regardless, the apical to basolateral migration of a cytokine may follow a tortuous path. Not all cytokine ligands can be transported across the endothelial cells, and probably even fewer pass the four-dimensional structure of the BBB.

It has been shown in recent years that monoubiquination plays an important role in determining the fate of an endocytosed protein [36]. Ubiquitin is an abundant and highly conserved peptide of 76 amino acids. It is added post-translationally to lysine residues on proteins and is recognized by the cellular machinery that targets them for proteolysis. Cytokine receptors, which are membrane proteins, might be susceptible to ubiquination, generating signals for regulated internalization and intracellular sorting.

11.3.3
Signal Transduction in Endothelial Cells

The fate of a cytokine inside the endothelial cells is largely dependent on its transport cargo. For instance, specific amino acid sequences in the cytoplasmic domain of a receptor can be the sorting signal to determine the site of degradation for a cytokine. How signal transduction triggered by a cytokine then affects its trafficking pattern is not clear.

11.3.4
Involvement of Other Cells Composing the BBB

In addition to endothelial cells, the BBB also contains pericytes, astrocytic endfeet, and neuronal components, and it is surrounded by extracellular matrix. The regulatory roles of elements of the BBB other than endothelial cells have been reviewed elsewhere [6].

11.4
Regulation of the Interactions of Cytokines with the BBB

Circadian Rhythm TNFα is a cytokine that can affect circadian rhythms in both physiological and pathological conditions. The uptake of TNFα by the spinal cord, but not by the brain and peripheral muscle, shows a multiphasic response that is not simply sinusoidal [57]. Circadian changes in BBB permeation are also seen for OSIP, an enkephalin analog [77], interleukin-1α, TNFα [57], and leptin [63].

Feeding Status, Obesity, and Cachexia Our preliminary study shows that the serum concentrations of TNFα and interleukin-6 are in the pg ml^{-1} range, whereas that for interleukin-1β is not detectable. Fasting and obesity do not seem to significantly affect the concentrations of these cytokines, whereas cachexia is known to be related to increased blood concentrations of proinflammatory cytokines [2]. Whether altered feeding status affects BBB permeation of cytokines is not clear.

Spinal Cord Injury and Brain Trauma Traumatic injury to the CNS induces cytokine production in a time- and region-dependent manner in situ; and there are increased levels of certain cytokines in the blood circulation. We have shown that both spinal cord injury and mild traumatic brain injury increase the specific transport of TNFα independently of barrier disruption [64, 67, 73]. Cytokines can also alter the expression of adhesion molecules such as E-selectin after trauma [24].

Primary and Metastatic Tumors, Inflammation, and Autoimmune Diseases Intracranial neoplasms, inflammation, and autoimmune disorders are related to a partially disrupted BBB and altered expression of cytokines and their receptors in the brain. Regarding direct permeation of cytokines across the BBB in such situations, we have shown that transport of TNFα is upregulated after experimental autoimmune encephalomyelitis [55].

11.5
Stroke and Other Vasculopathy

We have been focusing on the role of TNFα in stroke. Real-time PCR analysis of gene expression profiles in rats after transient middle cerebral artery occlusion (MCAO) shows that TNFα expression is increased 3 h after stroke and lasts for 24 h in the hemisphere ipsilateral to occlusion. This is accompanied by changes in the mRNA for interleukin-1, interleukin-6, E-selectin, and intercellular adhesion molecule-1 (ICAM-1), in the immunoreactivity for activated nuclear factor (NF)-6B, and in the infiltration of inflammatory cells [15]. After permanent MCAO, TNFα expression is increased not only in neurons, but also in astrocytes, microglia, choroid plexus, endothelial cells, infiltrating polymorphonuclear leukocytes, and probably pericytes. Upregulation of the p55 receptor for TNFα is present at 6 h and precedes that of the p75 receptor, which occurs at 24 h [18].

Both injurious and beneficial roles have been proposed for TNFα in the pathogenesis of cerebral ischemia [48]. On the one hand, blockade of TNFα actions by dimeric type I soluble TNF receptor (a TNFα binding protein) in BALB/C mice [50], anti-TNFα antibody Pl14, and a synthesis inhibitor CNI-1493 in Lewis rats [48] all reduce infarct volume after permanent MCAO. On the other hand, TNFα pretreatment induces neuroprotection against permanent MCAO in BALB/C mice with reduction of infarct size and CD11b-positive neutrophils and macrophages [51], whereas mice lacking the p55 TNFα receptor (R1) have a greater infarct volume than wild-type and p75 receptor knockout mice after MCAO and sustain more damage of CA3 hippocampal neurons by kainic acid excitotoxicity [29]. One possible explanation of these paradoxical results is that the TNFα blocking reagents were administered peripherally, whereas TNFα pre-treatment was given by the intracisternal route. The time of administration probably accounts for the different actions of TNFα, as the two types of studies most likely involve activation of different cellular targets. Regardless, TNFα is crucially involved in the initiation, progression, and regeneration processes after stroke, and in the development of tolerance to ischemia [32, 45].

As stroke alters BBB function and dynamically changes communication of the CNS with the periphery, transport of vascular TNFα across the BBB after stroke probably contributes to the effects of TNF in the CNS. We have shown that transport of TNFα is significantly increased 1 week after tMCAO, and this increased function is mediated by upregulation of the p55 and p75 receptors in endothelial cells [58].

Interactions of TNFα with cerebral endothelial cells affect not only its own transport but also the paracellular permeability of the BBB. TNFα changes cytoskeleton organization, tight junction protein expression, and production of serine proteases, including tissue plasminogen activator (tPA), urokinase plasminogen activator (uPA), and matrix metalloproteinases (MMPs) [23, 78, 81]. These serine proteases are involved in BBB disruption, tissue remodeling, and neural plasticity [1, 3, 17, 86]. Exogenous tPA as a thrombolytic agent binds to the LDL-receptor related protein (LRP) at the BBB and can induce opening of the BBB [84]. TNFα also induces cerebral endothelial cell expression of genes involved in cell adhesion, chemotaxis, apoptosis, neuroprotection, transcriptional regulation, etc. [28]. Thus, TNFα affects stroke progression and recovery at the BBB level.

11.6
Neurodegenerative Disorders

Amyotrophic lateral sclerosis (ALS) is a progressive and devastating disorder affecting motor neurons. There has been evidence that the general permeability of the BBB is increased in ALS patients, as indicated by increased serum and CSF levels of albumin, and increased CSF total protein and immunoglobulin [41, 42].

TNFα might exacerbate neurodegeneration by potentiating the release of the excitatory neurotransmitter glutamate [16]. In ALS patients, TNFα concentrations in blood are increased [76]. The linkage of TNFα and neuroinflammation is evident in an animal model of familial ALS by gain-of-function mutations in Cu, Zn superoxide dismutase (SOD1). TNFα mRNA expression appears in the spinal cord of young SOD1 mice before the development of symptoms of motor weakness or significant motor neuron loss and correlates with the severity of the disease [26]. The SOD1 mice have upregulation of the p55 TNFα receptor in the presymptomatic stage, supporting the concept that inflammation in the CNS occurs relatively early and precedes the onset of frank paralysis and accumulation of bulk protein oxidative damage [35]. Microarray analysis of cDNA from the spinal cord of SOD1 mice also shows that the mRNA expression of TNFα increases eight-fold and TNFα immunoreactivity is increased in microglia and motor neurons of the spinal cord. Along with other factors, it seems that the inflammatory process mediated by TNFα directly participates in neuronal death [85]. Further study in the spinal cord of SOD1 mice shows that TNFα, transforming growth factor β, interleukin 10, some chemokines, and other proinflammatory cytokines like interferon γ (IFNγ) and interleukin-1, -2, and -6 are also increased. However, nitrite efflux assays in cultured Walker EOC-20 microglia cells show that TNFα causes the maximal cellular response. The results suggest that TNFα is the principal driver for neuroinflammation, with its effects potentiated by co-stimulating cytokines and chemokines [34]. It is yet to be determined whether TNFα transport across the BBB is altered in ALS, and whether modulation of such transport can modify the course of neurodegeneration.

LIF is a neurotrophic cytokine to motor neurons, although the systemic side-effect of cachexia is seen when it is administered in large doses. In a pilot study, LIF appeared to reduce the loss of motor neurons in the lumbar spinal cord of SOD1 mice after daily intraperitoneal injection for 6 weeks [4]. Although a follow-up study from the same research group failed to reproduce the results, LIF seemed to have a positive effect on locomotor behavioral tests by delaying the onset and reducing the severity of symptoms [5]. The lack of a substantial and sustained benefit in this study could be related to multiple factors including sex difference, generation variability, sensitivity of the tests, as well as the dose, duration, and route of administration. In addition, a transport defect of LIF crossing the BBB may explain why the effects of LIF are not sustained after 3 months of age. In normal mice, a saturable transport system at the BBB is responsible for blood-brain and blood-spinal cord entry of LIF [68]. In the presence of dysregulation of the BBB, peripherally administered LIF may not reach its CNS targets effectively. This could also be part of the explanation for the inconsistent results with LIF in clinical trials of ALS patients [30, 39].

11.7 Summary

The BBB is a target of action for many cytokines; and it also serves as a dynamic regulatory interface in communications between the CNS and the rest of the body. A limited number of cytokines can permeate the BBB by way of saturable transport systems. The details of intracellular trafficking and the regulation of transport in pathophysiological states are the subjects of ongoing studies. The BBB provides a potential site for modulation of the effects of cytokines on the CNS.

Acknowledgment

This work was supported by the NIH (NS45751, NS46528, DK54880, AA12865).

References

1 Aoki, T., Sumii, T., Mori, T., Wang, X., Lo, E. H. **2002**, Blood-brain barrier disruption and matrix metalloproteinase-9 expression during reperfusion injury, *Stroke* 33, 2711–2717.
2 Argiles, J. M., Busquets, S., Lopez-Soriano, F. J. **2003**, Cytokines in the pathogenesis of cancer cachexia, *Curr. Opin. Clin. Nutr. Metab. Care* 6, 401–406.
3 Asahi, M., Wang, X., Mori, T., Sumii, T., Jung, J.-C., Moskowitz, M. A., Fini, M. E., Lo, E. H. **2001**, Effects of matrix metalloproteinase-9 gene knock-out on the proteolysis of blood-brain barrier and white matter components after cerebral ischemia, *J. Neurosci.* 21, 7724–7732.
4 Azari, M. F., Galle, A., Lopes, E. C., Kurek, J., Cheema, S. S. **2001**, Leukemia inhibitory factor by systemic administration rescues spinal motor neurons in the SOD1 G93A murine model of familial amyotrophic lateral sclerosis, *Brain Res.* 922, 144–147.
5 Azari, M. F., Lopes, E. C., Stubna, C., Turner, B. J., Zang, D., Nicola, N. A., Kurek, J. B., Cheema, S. S. **2003**, Behavioural and anatomical effects of systemically administered leukemia inhibitory factor in the SOD1$^{G93A\ G1H}$ mouse model of familial amyotrophic lateral sclerosis, *Brain Res.* 982, 92–97.
6 Balabanov, R., Dore-Duffy, P. **1998**, Role of the CNS microvascular pericyte in the blood-brain barrier, *J. Neurosci. Res.* 53, 637–644.
7 Banks, W. A., Kastin, A. J. **1996**, Reversible association of the cytokines MIP-1a and MIP-1b with the endothelia of the blood-brain barrier, *Neurosci. Lett.* 205, 202–206.
8 Banks, W. A., Kastin, A. J., Durham, D. A. **1989**, Bidirectional transport of interleukin-1 alpha across the blood-brain barrier, *Brain Res. Bull.* 23, 433–437.
9 Banks, W. A., Kastin, A. J., Gutierrez, E. G. **1993**, Interleukin-1α in blood has direct access to cortical brain cells, *Neurosci. Lett.* 163, 41–44.
10 Banks, W. A., Kastin, A. J., Gutierrez, E. G. **1994**, Penetration of interleukin-6 across the murine blood-brain barrier, *Neurosci. Lett.* 179, 53–56.
11 Banks, W. A., Moinuddin, A., Morley, J. E. **2001**, Regional transport of TNF-α across the blood-brain barrier in young ICR and young and aged SAMP8 mice, *Neurobiol. Aging* 22, 671–676.
12 Banks, W. A., Ortiz, L., Plotkin, S. R., Kastin, A. J. **1991**, Human interleukin (IL)1α, murine IL-1α and murine IL-1β are transported from blood to brain in the mouse by a shared saturable mechanism, *J. Pharmacol. Exp. Ther.* 259, 988–996.

13 Banks, W.A., Plotkin, S.R., Kastin, A.J. **1995**, Permeability of the blood-brain barrier to soluble cytokine receptors Neuroimmunomodulation 2: 161–165.
14 Barten, D.M., Ruddle, N.H. **1994**, Vascular cell adhesion molecule-1 modulation by tumor necrosis factor in experimental allergic encephalomyelitis, *J. Neuroimmunol.* 51, 123–133.
15 Berti, R., Williams, A.J., Moffett, J.R., Hale, S.L., Velarde, L.C., Elliott, P.J., Yao, C., Dave, J.R., Tortella, F.C. **2002**, Quantitative real-time RT-PCR analysis of inflammatory gene expression associated with ischemia-reperfusion brain injury, *J. Cereb. Blood Flow Metab.* 22, 1068–1079.
16 Bezzi, P., Domercq, M., Brambilla, L., Galli, R., Schols, D., DeClercq, E., Vescovi, A., Bagetta, G., Kollias, G., Meldolesi, J., Volterra, A. **2001**, CXCR4-activated astrocyte glutamate release via TNF-α: amplification by microglia triggers neurotoxicity, *Nat. Neurosci.* 4, 702–710.
17 Bhattacharjee, A.K., Kondoh, T., Ikeda, M., Kohmura, E. **2002**, MMP-9 and EBA immunoreactivity after papaverine mediated opening of the blood-brain barrier, *NeuroReport* 13, 2217–2221.
18 Botchkina, G.I., Meistrell, M.E.I., Botchkina, I.L., Tracey, K.J. **1997**, Expression of TNF and TNF receptors (p55 and p75) in the rat brain after focal cerebral ischemia, *Mol. Med.* 3, 765–781.
19 Camussi, G., Turello, E., Bussolino, F., Baglioni, C. **1991**, Tumor necrosis factor alters cytoskeletal organization and barrier function of endothelial cells, *Int. Arch. Allergy Appl. Immunol.* 96, 84–91.
20 Conner, S.D., Schmid, S.L. **2003**, Regulated portals of entry into the cell, *Nature* 422, 37–44.
21 Cottin, V., Van Linden, A., Riches, D.W.H. **1999**, Phosphorylation of tumor necrosis factor receptor CD120a (p55) by p42$^{mapk/erk2}$ induces changes in its subcellular localization, *J. Biol. Chem.* 274, 32975–32987.
22 Deguchi, Y., Okutsu, H., Okura, T., Yamada, S., Kimura, R., Yuge, T., Furukawa, A., Morimoto, K., Tachikawa, M., Ohtsuki, S., Hosoya, K., Terasaki, T. **2002**, Internalization of basic fibroblast growth factor at the mouse blood-brain barrier involves perlecan, a heparan sulfate proteoglycan, *J. Neurochem.* 83, 381–389.
23 Dobbie, M.S., Hurst, R.D., Klein, N.J., Surtees, R.A. **1999**, Upregulation of intercellular adhesion molecule-1 expression on human endothelial cells by tumour necrosis factor-alpha in an in vitro model of the blood-brain barrier, *Brain Res.* 830, 330–336.
24 Dore-Duffy, P., Washington, R.A., Balabanov, R. **1994**, Cytokine-mediated activation of cultured CNS microvessels: a system for examining antigenic modulation of CNS endothelial cells, and evidence for long-term expression of the adhesion protein E-selectin, *J. Cereb. Blood Flow Metab.* 14, 837–844.
25 Dvorak, H.F., Brown, L.F., Detmar, M., Dvorak, A.M. **1995**, Vascular permeability factor/vascular endothelial growth factor, microvascular hypermeability, and angiogenesis, *Am. J. Pathol.* 146, 1029–1039.

26 Elliott, J. L. **2001**, Cytokine upregulation in a murine model of familial amyotrophic lateral sclerosis, *Mol. Brain Res.* 95, 172–178.
27 Estrada, C., Gomez, C., Martin, C. **1995**, Effects of TNFa on the production of vasoactive substances by cerebral endothelial and smooth muscle cells in culture, *J. Cereb. Blood Flow Metab.* 15, 920–928.
28 Franzén, B., Duvefelt, K., Jonsson, C., Engelhardt, B., Ottervald, J., Wickman, M., Yang, Y., Schuppe-Koistinen, I. **2003**, Gene and protein expression profiling of human cerebral endothelial cells activated with tumor necrosis factor-a, *Mol. Brain Res.* 115, 130–146.
29 Gary, D. S., Bruce-Keller, A. J., Kindy, M. S., Mattson, M. P. **1998**, Ischemic and excitotoxic brain injury is enhanced in mice lacking the p55 tumor necrosis factor receptor, *J. Cereb. Blood Flow Metab.* 18, 1283–1287.
30 Giess, R., Beck, M., Goetz, R., Nitsch, R. M., Toyka, K. V., Sendtner, M. **2000**, Potential role of LIF as a modifier gene in the pathogenesis of amyotrophic lateral sclerosis, *Neurology* 54, 1003–1005.
31 Gutierrez, E. G., Banks, W. A., Kastin, A. J. **1993**, Murine tumor necrosis factor alpha is transported from blood to brain in the mouse, *J. Neuroimmunol.* 47, 169–176.
32 Hallenbeck, J. M. **2002**, The many faces of tumor necrosis factor in stroke, *Nat. Med.* 8, 1363–1368.
33 Harkness, K. A., Sussman, J. D., Davies-Jones, G. A., Greenwood, J., Woodroofe, M. N. **2003**, Cytokine regulation of MCP-1 expression in brain and retinal microvascular endothelial cells, *J. Neuroimmunol.* 142, 1–9.
34 Hensley, K., Fedynyshyn, J., Ferrell, S., Floyd, R. A., Gordon, B., Grammas, P., Hamdheydari, L., Mhatre, M., Mou, S., Pye, Q. N., Stewart, C., West, M., West, S., Williamson, K. S. **2003**, Message and protein-level elevation of tumor necrosis factor *a* (TNFa) and TNFa-modulating cytokines in spinal cords of the G93A-SOD1 mouse model for amyotrophic lateral sclerosis, *Neurobiology* 14, 74–80.
35 Hensley, K., Floyd, R. A., Gordon, B., Mou, S., Pye, Q. N., Stewart, C., West, M., Williamson, K. **2002**, Temporal patterns of cytokine and apoptosis-related gene expression in spinal cords of the G94A-SOD1 mouse model of amyotrophic lateral sclerosis, *J. Neurochem.* 82, 365–374.
36 Hicke, L., Dunn, R. **2003**, Regulation of membrane protein transport by ubiquitin and ubiquitin-binding proteins, *Annu. Rev. Cell Dev. Biol.* 19, 141–172.
37 Jones, S. J., Ledgerwood, E. C., Prins, J. B., Galbraith, J., Johnson, D. R., Pober, J. S., Bradley, J. R. **1999**, TNF recruits TRADD to the plasma membrane but not the *trans*-Golgi network, the principal subcellular location of TNF-R1, *J. Immunol.* 162, 1042–1048.
38 Kastin, A. J., Akerstrom, V., Pan, W. **2003**, Interleukin-10 as a CNS therapeutic: the obstacle of the blood-brain/blood-spinal cord barrier, *Mol. Brain Res.* 114, 168–171.
39 Kurek, J. B., Radford, A. J., Crump, D. E., Bower, J. J., Feeney, S. J., Austin, L., Byrne, E. **1998**, LIF (AM424), a promising growth factor for the treatment of ALS, *J. Neurol. Sci.* 160, S106–S113.

40 Lai, E.C., Felice, K.J., Festoff, B.W., Gawel, M.J., Gelinas, D.F., Murphy, M.F., Natter, H.M., Norris, F.H., Rudnicki, S.A. **1997**, Effect of recombinant human insulin-like growth factor-I on progression of ALS. A placebo-controlled study. The North America ALS/IGF-I study group, *Neurology* 49, 1621–1630.

41 Leonardi, A., Abbruzzese, G., Arata, L., Cocito, L., Vische, M. **1984**, Cerebrospinal fluid (CSF) findings in amyotrophic lateral sclerosis, *J. Neurol.* 231, 75–78.

42 Leonardi, A., Abbruzzese, G., Arata, L., Cocito, L., Vische, M. **2004**, Cerebrospinal fluid (CSF) findings in amyotrophic lateral sclerosis, *J. Neurol.* 231, 75–78.

43 Liu, X., Yao, D.L., Webster, H. **1995**, Insulin-like growth factor 1 treatment reduces clinical deficits and lesion severity in acute demyelinating experimental autoimmune encephalomyelitis, *Mult. Scler.* 1, 2–9.

44 Male, D., Rahman, J., Pryce, G., Tamatani, T., Miyasaka, M. **1994**, Lymphocyte migration into the CNS modelled in vitro: roles of LFA-1, ICAM-1 and VLA-4, *Immunology* 81, 366–372.

45 Masada, T., Hua, Y., Xi, G., Ennis, S.R., Keep, R.F. **2001**, Attenuation of ischemic brain edema and cerebrovascular injury after ischemic preconditioning in the rat, *J. Cereb. Blood Flow Metab.* 21, 22–33.

46 McCarron, R.M., Wang, L., Racke, M.K., McFarlin, D.E., Spatz, M. **1993**, Cytokine-regulated adhesion between encephalitogenic T lymphocytes and cerebrovascular endothelial cells, *J. Neuroimmunol.* 43, 23–30.

47 McLay, R.N., Kastin, A.J., Zadina, J.E. **2000**, Passage of interleukin-1-beta across the blood-brain barrier is reduced in aged mice: a possible mechanism for diminished fever in aging, *Neuroimmunomodulation* 8, 148–153.

48 Meistrell, M.E.I., Botchkina, G.I., Wang, H., Di Santo, E., Cockroft, K.M., Vishnubhakat, J.M., Ghezzi, P., Traces, K.J. **1997**, Tumor necrosis factor is a brain damaging cytokine in cerebral ischemia, *Shock* 8, 341–348.

49 Menter, D.G., Herrmann, J.L., Nicolson, G.L. **1995**, The role of trophic factors and autocrine/paracrine growth factors in brain metastasis, *Clin. Exp. Metastasis* 13, 67–88.

50 Nawashiro, H., Martin, D., Hallenbeck, J.M. **1997**, Inhibition of tumor necrosis factor and amelioration of brain infarction in mice, *J. Cereb. Blood Flow Metab.* 17, 229–232.

51 Nawashiro, H., Tasaki, K., Ruetzler, C.A., Hallenbeck, J.M. **1997**, TNF-α pretreatment induces protective effects against focal cerebral ischemia in mice, *J. Cereb. Blood Flow Metab.* 17, 483–490.

52 Pan, W., Akerstrom, V., Zhang, J., Pejovic, V., Kastin, A.J. **2004**, Modulation of feeding-related peptide/protein signals by the blood-brain barrier, *J. Neurochem.* 90, 455–461.

53 Pan, W., Banks, W.A., Kastin, A.J. **1997**, Permeability of the blood-brain and blood-spinal cord barriers to interferons, *J. Neuroimmunol.* 76, 105–111.

54 Pan, W., Banks, W. A., Kastin, A. J. **1998**, Permeability of the blood-brain barrier to neurotrophins, *Brain Res.* 788, 87–94.

55 Pan, W., Banks, W. A., Kennedy, M. K., Gutierrez, E. G., Kastin, A. J. **1996**, Differential permeability of the BBB in acute EAE: enhanced transport of TNF-α, *Am. J. Physiol.* 271, E636–E642.

56 Pan, W., Cain, C., Tu, H., Yu, Y., Zhang, L., Kastin, A. J. **2005**, Receptor-mediated transport of LIF across blood-spinal cord barrier is upregulated after SCI (submitted).

57 Pan, W., Cornelissen, G., Halberg, F., Kastin, A J. **2002**, Selected contribution: circadian rhythm of tumor necrosis factor-alpha uptake into mouse spinal cord, *J. Appl. Physiol.* 92, 1357–1362.

58 Pan, W., Ding, Y., Zhang, J., Kastin, A. J. **2005**, Stroke increases transport of TNFα across the blood-brain barrier that is partially mediated by the receptors (submitted).

59 Pan, W., Csernus, B., Kastin, A. J. **2003**, Upregulation of p55 and p75 receptors mediating TNF-alpha transport across the injured blood-spinal cord barrier, *J. Mol. Neurosci.* 21, 173–184.

60 Pan, W., Kastin, A. J. **1999**, Entry of EGF into brain is rapid and saturable, *Peptides* 20, 1091–1098.

61 Pan, W., Kastin, A. J. **2000**, Interactions of IGF-1 with the blood-brain barrier in vivo and in situ, *Neuroendocrinology* 72, 171–178.

62 Pan, W., Kastin, A. J. **2001 a**, Changing the chemokine gradient: CINC1 crosses the blood-brain barrier, *J. Neuroimmunol.* 115, 64–70.

63 Pan, W., Kastin, A. J. **2001 b**, Diurnal variation of leptin entry from blood to brain involving partial saturation of the transport system, *Life Sci.* 68, 2705–2714.

64 Pan, W., Kastin, A. J. **2001 c**, Increase in TNFα transport after SCI is specific for time, region, and type of lesion, *Exp. Neurol.* 170, 357–363.

65 Pan, W., Kastin, A. J. **2002**, TNFα transport across the blood-brain barrier is abolished in receptor knockout mice, *Exp. Neurol.* 174, 193–200.

66 Pan, W., Kastin, A. J. **2003**, Interactions of cytokines with the blood-brain barrier: implications for feeding, *Curr. Pharm. Des.* 9, 827–831.

67 Pan, W., Kastin, A. J., Bell, R. L., Olson, R. D. **1999**, Upregulation of tumor necrosis factor α transport across the blood-brain barrier after acute compressive spinal cord injury, *J. Neurosci.* 19, 3649–3655.

68 Pan, W., Kastin, A. J., Brennan, J. M. **2000**, Saturable entry of leukemia inhibitory factor from blood to the central nervous system, *J. Neuroimmunol.* 106, 172–180.

69 Pan, W., Kastin, A. J., Maness, L. M., Brennan, J. M. **1999**, Saturable entry of ciliary neurotrophic factor into brain, *Neurosci. Lett.* 263, 69–71.

70 Pan, W., Vallance, K., Kastin, A. J. **1999**, TGFα and the blood-brain barrier: accumulation in cerebral vasculature, *Exp. Neurol.* 160, 454–459.

71 Pan, W., Yu, Y., Nyberg, F., Kastin, A. J. **2005**, Direct and indirect interactions of growth hormone with the blood-brain barrier, *Peptides* **2005** (in press).

72 Pan, W., Zadina, J. E., Harlan, R. E., Weber, J. T., Banks, W. A., Kastin, A. J. **1997**, Tumor necrosis factor α: a neuromodulator in the CNS, *Neurosci. Biobehav. Rev.* 21, 603–613.

73 Pan, W., Zhang, L., Liao, J., Csernus, B., Kastin, A. J. **2003**, Selective increase in TNFα permeation across the blood-spinal cord barrier after SCI, *J. Neuroimmunol.* 134, 111–117.

74 Plotkin, S. R., Banks, W. A., Kastin, A. J. **1996**, Comparison of saturable transport and extracellular pathways in the passage of interleukin-1α across the blood-brain barrier, *J. Neuroimmunol.* 67, 41–47.

75 Poduslo, J. F., Curran, G. L. **1996**, Permeability at the blood-brain and blood-nerve barriers of the neurotrophic factors: NGF, CNTF, NT-3, BDNF, *Mol. Brain Res.* 36, 280–286.

76 Poloni, M., Facchetti, D., Mai, R., Micheli, A., Agnoletti, L., Francolini, G., Mora, G., Camana, C., Mazzini, L., Bachetti, T. **2000**, Circulating levels of tumor necrosis factor-alpha and its soluble receptors are increased in the blood of patients with amyotrophic lateral sclerosis, *Neurosci. Lett.* 287, 214.

77 Ramge, P., Kreuter, J., Lemmer, B. **1999**, Circadian phase-dependent antinociceptive reaction in mice determined by the hot-plate test and the tail-flick test after intravenous injection of dalargin-loaded nanoparticles, *Chronobiol. Int.* 16, 767–777.

78 Rosenberg, G. A. **2002**, Matrix metalloproteinases in neuroinflammation, *Glia* 39, 279–291.

79 Sidawy, A. N., Mitchell, M. E., Neville, R. F. **1998**, Peptide growth factors and signal transduction, *Semin. Vasc. Surg.* 11, 149–155.

80 Stamatovic, S. M., Shakui, P., Keep, R. F., Moore, B. B., Kunkel, S. L., Van Rooijen, N., Andjelkovic, A. V. **2005**, Monocyte chemoattractant protein-1 regulation of blood-brain barrier permeability, *J. Cereb. Blood Flow Metab.* (in press).

81 Stolphen, A. H., Guinan, E. C., Fiers, W., Pober, J. S. **1986**, Recombinant tumor necrosis factor and immune interferon act singly and in combination to reorganize human vascular endothelial cell monolayers, *Am. J. Pathol.* 123, 16–24.

82 Storey, H., Stewart, A., Vandenabeele, P., Luzio, J. P. **2002**, The p55 tumor necrosis factor receptor TNFR1 contains a *trans*-Golgi network localization signal in the C-terminal region of its cytoplasmic tail, *Biochem. J.* 366, 15–22.

83 Terasaki, T., Takakuwa, S., Saheki, A., Moritani, S., Shimura, T., Tabata, S., Tsuji, A. **1992**, Absorptive-mediated endocytosis of an adrenocorticotropic hormone (ACTH) analogue, ebiratide, into the blood-brain barrier: studies with monolayers of primary cultured bovine brain capillary endothelial cells, *Pharm. Res.* 9, 529–534.

84 Yepes, M., Sandkvist, M., Moore, E. G., Bugge, T. H., Strickland, D. K., Lawrence, D. A. **2003**, Tissue-type plasminogen activator induces opening of the blood-brain barrier via the LDL receptor-related protein, *J. Clin. Invest.* 112, 1533–1540.

Measurements using the rat adipose tissue bioassay also did not detect significant amounts of insulin in the CSF.

The first of these studies was by Elgee et al. and was conducted in 1954. Their objective was to investigate the biodistribution of insulin after intravenous (iv) injection; and brain was only one of many tissues studied. The study carefully considered many potential pitfalls. For example, the study used the power of radioactively labeled insulin to track biodistribution, it showed that radioactive insulin retained biological activity, it used acid precipitation to distinguish radioactive degradation products from intact insulin, it used fasted animals, and it used a ratio method (radioactivity in tissue vs radioactivity in whole body) to decrease statistical variation. The study was also comparative in that, while most of the studies were conducted in rats, a parallel arm of the study injected radioactive insulin into four "agonal" patients at 25 min to 8 h before their death. The study also suffered from a common problem of biodistribution studies which happen to include brain. Peripheral tissues, especially kidney and liver, took up the majority of radioactive insulin. In comparison, brain uptake seemed negligible. In rats, only erythrocytes took up less insulin than brain. In humans, little or no insulin was found in the CSF or brain. This finding fit well with the idea that the brain was not an insulin-sensitive tissue.

Haugaard et al. in the same year published a complementary study. They injected insulin labeled with either radioactive sulfur or radioactive iodine into rats or dogs. Again, these were pioneering studies addressing the fate of intravenous insulin and its distribution among the major tissues of the body. In one set of studies which included brain, they co-injected radioactive albumin as a measure of vascular space. Brain was the only tissue which did not obviously take up insulin. They concluded that little or no insulin entered the brain and, like Elgee et al., noted this was consistent with the view that "metabolism of the brain is... independent of the presence of insulin."

The above findings were reinforced a few years later by the study of Mahon et al. They assayed human CSF for insulin-like activity, using the rat adipose tissue technique. They estimated that they should be able to determine insulin levels as low as 3% of the concentration in serum with this method. However, they found no evidence for insulin in the CSF. They also injected iv radioactive insulin into humans and CSF obtained from the lumbar region. Radioactivity in blood and CSF was acid-precipitated to distinguish the degradation products. They conducted their study to 90 min, by which time they felt so much insulin had been degraded that the study should not be extended. Acid-precipitable radioactivity was only found in CSF in the 90-min sample. The counts were extremely low and in all cases were much less than 1% of the serum counts. Despite these negative results, the authors made the very conservative conclusion that the results did "not exclude the possibility that insulin is active in the central nervous system."

In retrospect, the study of Mahon et al. is the easiest to critique. The bioassay available to them was crude by the standards which would be set in the next few years by the work of Berson and Yalow. Taking CSF from the lumbar region was a distinct, but almost always necessary, disadvantage in humans. That same year, Grundy [1] would show that the CSF of the spine and the brain do not mix well. As such, the levels of peptides in brain CSF often correlate with peptide levels in blood, whereas levels in spinal CSF and blood usually do not [2].

At this point, it might have seemed that any insulin which was in the CNS was negligible. But the idea that insulin might affect the CNS was being reconsidered. Furthermore, studies had found that insulin injected into the brain produced hypoglycemia. This raised the possibility that insulin might act within the CNS. The invention of the radioimmunoassay by Berson and Yalow introduced a much more sensitive way to measure insulin. In 1967, Margolis and Altszuler used the radioimmunoassay to revisit the question of whether insulin could cross the BBB [3]. They found a low amount of insulin (3 $\mu U\ ml^{-1}$) in the CSF of dogs, or about 27% of blood levels. They then infused large amounts of insulin and glucose intravenously for up to 6 h. CSF levels of insulin also increased, although with a lag before reaching a steady state. The higher infusion rates produced higher levels of insulin in the CSF, but the increase was not proportionate. Approximations made from their figure 2 indicate that the CSF/serum ratio was about 0.27 (i.e. 27% of serum values) at the baseline serum value of 11 $\mu U\ ml^{-1}$, 0.07 at serum levels of 275 $\mu U\ ml^{-1}$, and 0.03 at serum levels of 1600 $\mu U\ ml^{-1}$. Thus, Margolis and Alszuler [3] concluded that insulin crossed the blood-CSF barrier by a saturable transport system rather than by passive diffusion.

Margolis and Altszuler followed up their study to show that insulin injected into the CNS could also enter the blood. Although they argued from their results that the efflux of insulin was not saturable but by the mechanism of CSF reabsorption (bulk flow), they also concluded that insulin was able to cross the

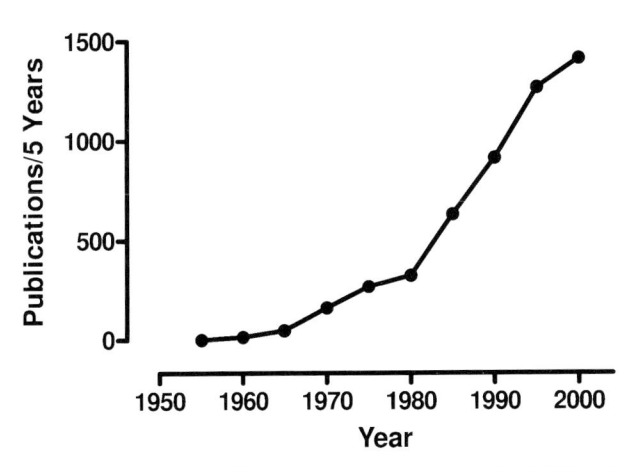

Fig. 12.1 Increase in publications for a search combining (CNS or brain) and insulin, using the Library of Congress. Intervals are for 5 years (for example, 1960 refers to the interval 1956–1960). For the first interval (1951–1960), there were no publications. For the last full 5-year interval (1996–2000), 1413 publications were located.

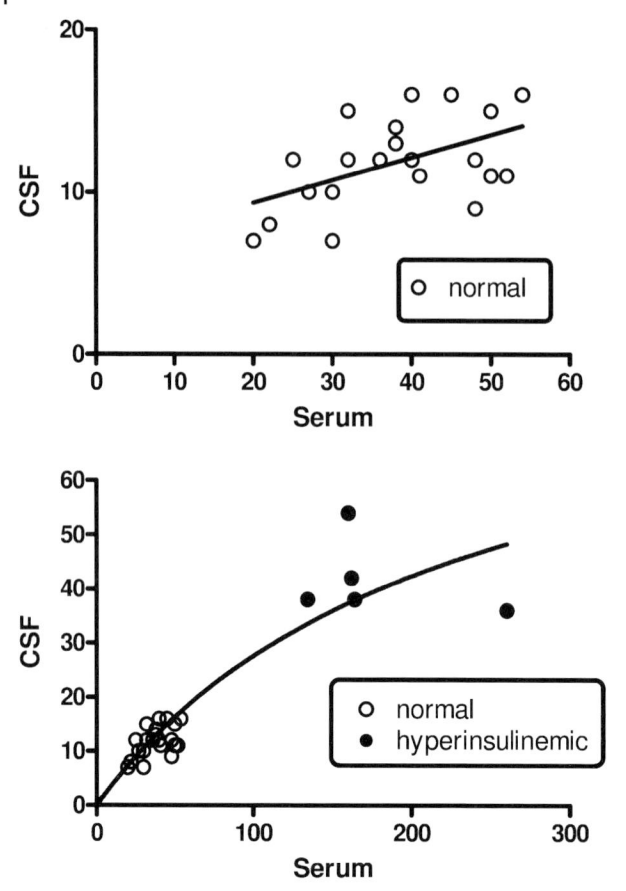

Fig. 12.2 Reanalysis of the data of Greco et al. (1970) by plotting serum and CSF values for baseline condition (open circles "normal", in upper and lower panels) and for hyperinsulinemic condition (closed circles, lower panel). Results show a significant relation for the baseline condition ($n=21$, $r=0.510$, $P<0.05$). Combined results for baseline condition and hyperinsulinemic condition show a nonlinear relation consistent with saturable transport of insulin across the BBB.

BBB in both directions, that is, from blood to CNS and from CSF to blood. Thirty years later, the implications of such bidirectional movement would be debated not only for insulin, but also for peptides, cytokines, feeding hormones, and many other substances. Ironically, the efflux of insulin from CSF to blood provided an explanation for how CNS insulin induced hypoglycemia. Rather than being mediated through central mechanisms, the insulin was merely entering blood to work through peripheral insulin receptors. However, within a few years, new studies would begin to redefine the role of insulin in the CNS. Figure 12.1 shows the increase in CNS insulin publications over the past 50 years from 1950 until 2000.

The idea that insulin could cross the BBB was reinforced by the study of Greco et al., who revisited the question of insulin in the CSF of man, this time using a radioimmunoassay. They examined diabetic and nondiabetic patients and split their analysis into two groups: those with normal levels of serum insulin and those with high levels of serum insulin. They found insulin in the CSF of all patients. In those patients with high serum insulin, the insulin in the CSF was also elevated, although not proportionately so. On this basis, they suggested that insulin crossed the BBB by an active mechanism as opposed to passive diffusion. Figure 12.2 reanalyzes the data from their 1970 paper by plotting serum and CSF values. Interestingly, a statistically significant linear correlation exists between the serum and CSF insulin values for patients with normal insulin levels and a nonlinear model fits all data. This is consistent with their conclusion of a saturable transport system for insulin from blood to CSF.

With these three studies, it might seem that the question of insulin passage across the BBB was once again settled, this time in favor of insulin crossing the BBB. But studies would occur throughout the 1970s and 1980s at prominent laboratories and published in prestigious journals that concluded that insulin was not able to cross the BBB [4, 5]. It fell to the next generation of studies to further define the relation between insulin and the BBB.

12.1.2
Debates Related to the Question of Permeability of the BBB to Insulin

In retrospect, three prominent issues seem to have impacted on the question of whether insulin could cross the BBB. Although these issues did not directly deal with the physical transfer of insulin across the BBB, they did frame the context in which the BBB studies were interpreted. These issues were:

1. Does the CNS make insulin? If the CNS could make its own insulin, then why would it need to import insulin from the outside? If the CNS did not make insulin, then did not the mere presence of insulin in the CSF and brain tissue indicate passage across the blood-brain and blood-CSF barriers?

2. Can proteins/peptides cross the BBB? The BBB was defined in the late 19th century by its ability to restrict dyes which bound tightly to albumin. This concept was rigorously generalized by many to all substances with peptide bonds, even the small, biologically active peptides. But the 1970s and 1980s found increasing proof that small peptides could cross in both the CNS-blood and blood-CNS directions. This suggested a form of communication between the CNS and peripheral tissues mediated through the BBB, an idea which resonated well with the ideas of Margolis and Altszuler in 1967. How much the ongoing work with small peptides influenced the insulin field is unclear, as insulin with its double-stranded structure does not meet the usual definitions of peptides.

3. Does the brain need insulin? Whether insulin was needed for glucose uptake by the CNS was a hotly debated topic. If insulin was not needed to either drive insulin across the BBB or into neurons, what useful role could it serve within the CNS? A slowly emerging counter-argument, foreshadowed by Margolis and Altszuler in their second paper, was that insulin had actions in addition to glucose uptake. Could insulin have roles within the CNS other than glucose regulation?

Each of these questions have aspects or derivative questions which are very much a topic of investigation. Full answers to them and a review of what aspects are currently being investigated are beyond the scope of this chapter. However, a summary of current thinking as it affects this review are given here:

1. Does the CNS make insulin? It seems generally agreed that the vast majority of insulin in the CNS of the healthy adult originates from the pancreas. If the CNS does make insulin, it seems to be localized regionally or to specific circumstances. However, as reviewed by Plata-Salaman [6], several studies have found a paradoxic increase in brain insulin levels in animals whose pancreatic insulin output had been greatly diminished by streptozotocin.

2. Can peptides/proteins cross the BBB? Many peptides and proteins have been shown to cross from CNS to blood or from blood to CNS. Such passage has been shown in some cases to provide the mechanism by which a peptide or protein originating in one compartment (e.g. the periphery) influences the function in the other compartment (e.g. the CNS).

3. Does the brain need insulin? The general consensus is currently that, whereas the CNS is largely "insulin-independent", defined as not requiring insulin for utilization of glucose, it is insulin-sensitive, i.e. insulin has many effects on brain function. Insulin receptors were found to occur throughout the CNS, including the cells which comprise the BBB. These receptors mediate several actions, some of which are discussed further below.

12.1.3
Does Insulin Cross the BBB? The Middle Years

The first 20 years of investigation of whether insulin crossed the BBB had produced mixed results. The earliest studies had failed to find insulin in the CSF or to show that radioactive insulin injected intravenously could be recovered from the CNS. Furthermore, it seemed likely that the brain did not require insulin for use of glucose. A second set of papers, however, used the newly invented, highly sensitive radioimmunoassay to show that insulin was in the CSF. Furthermore, it seemed that a correlation existed between the level of insulin in the CSF and the level in blood. Such a correlation was interpreted as evidence that CSF insulin was derived from serum.

The studies of the next 20 years (1970s–1980s) largely confirmed these suspicions. Two major lines of investigation in particular reinforced the idea that insulin could cross the BBB: (a) evidence of insulin in the CNS, (b) the demonstration of insulin receptors and binding sites within the CNS and at the BBB.

A seminal paper in the debate about insulin transport was published in 1977 [7]. This paper clearly showed that a correlation existed between the CSF and serum levels of insulin in dogs. A bolus injection of 0.2 µU of insulin increased serum levels by over 150-fold. CSF levels of insulin were also increased, but only by 2- to 3-fold. The same amount of insulin delivered as an infusion that elevated serum insulin levels by 4-fold also increased CSF levels of insulin. However, the increase in CSF was only about 2-fold and it was delayed 15–30 min after the start of the infusion. Bolus injections and infusions of glucose increased the serum but not CSF levels of insulin. However, the increases in serum insulin with glucose administration were much smaller and more transient than the increases seen with insulin administration. When the CSF levels were regressed against serum levels for all data (basal levels and steady-state levels after glucose and insulin infusion), a correlation existed.

However, this paper also pointed out difficulties that were complicating not only studies of insulin transport but also studies of other peptides and regulatory proteins. The results showed that CSF increases were small compared with serum levels, especially when studied by bolus injection. An apparent delay occurred between rises in serum and CSF levels, raising the question of which values of serum should be compared to which values of CSF. Comparison of the areas under the curve offered some degree of solution but also presented its own set of difficulties. These and related issues of how to analyze results for peptides and regulatory proteins were not fully resolved until the application of the methods of Patlak, Blasberg, and Fenstermacher [8, 9] several years later.

Subsequent studies refined the experimental approach and demonstrated correlations between CSF and serum insulin levels in rats, man, and baboons [10–12]. Additionally, insulin was clearly demonstrated to be present in brain tissue [13]. The delay in the increase in CSF levels was explained as the need for the insulin to move through a compartment between the CSF and serum, possibly the brain interstitial fluid [10]. Finally, the nonlinear association between CSF and serum levels of insulin was clearly demonstrated [14]. This nonlinearity could be readily explained if serum insulin entered the CNS by way of a saturable transporter.

However, not all studies agreed that insulin crossed the BBB or even that there was a correlation between CNS and serum levels of insulin [4, 5, 15, 16]. Instead, the presence of insulin in the CSF and brain was interpreted as representing CNS insulin production. Especially difficult to ignore were a series of experiments previously reviewed [6], in which animals with streptozotocin-induced insulinopenia had unaltered levels of insulin in the brain. Whether insulin was produced within the CNS in quantities sufficient to explain the levels found in CSF and CNS continued to complicate the question of insulin transport into the 1990s [6]. If no insulin was produced within the CNS, then CNS insulin must be derived from the periphery and so had crossed the BBB. To the degree that the CNS could produce insulin, the role of BBB transport was unclear.

At first it may seem that a strong correlation between CSF and serum levels demonstrates that the substance is crossing the BBB. However, correlations between CSF and serum constitute a weak or presumptive proof of BBB transport. This is because it could be that release from the CNS and peripheral sources are responding independently, but to the same stimulus. Such has been shown to be the case for arginine vasopressin [17].

An important parallel set of studies involved the demonstration of insulin binding sites at both the choroid plexus and the brain endothelium [18–21]. Although the presence of such binding sites was not controversial, what those binding sites represented was. Alternate interpretations, sometimes by the same laboratory, were that these sites represented g-coupled receptors, transporters, or enzymes. Alternate positions held, sometimes by the same laboratory, were that insulin did not cross the BBB at all, crossed the endothelial barrier but not the choroid plexus, crossed at the circumventricular organs, crossed the choroid plexus but not the endothelial barrier, or crossed at both the endothelial barrier and the choroid plexus [20, 22, 23]. In retrospect, the BBB has binding sites which fulfill all three of these functions. Brain endothelial cells, for example, can degrade insulin [21, 24], contains g protein-coupled receptors which affect brain endothelial cell function [25–28] and can transport insulin across the BBB. These binding sites are not static, but vary during development and with other conditions [29, 30], consistent with the pathophysiological alterations in insulin transport noted below.

12.1.4
Insulin, the BBB, and Pathophysiology: The Past Decade

As it became increasingly clear that the CNS production of insulin was minimal, at least in healthy adults [11, 31], and increasingly accepted that a portion of the binding sites at the BBB represented transporters, work began to focus less on whether insulin crossed the BBB and more on what the pathophysiological implications of such transport represented. Nevertheless, these latter studies revisited the issue of insulin transport and largely confirmed with a different set of techniques that insulin clearly crosses the BBB. Before examining the pathophysiological work, it may be worth examining the new work showing that insulin crosses the BBB.

Two variations on the older work of examining transport confirmed that insulin can cross the BBB. The first variation used radioactively labeled insulin in pharmacokinetic studies. The second variation used species-specific immunoassays to distinguish between exogenous and endogenous insulin. Both of these approaches solved the old problem of whether CNS insulin was made by the brain or transported from the periphery, using the same logic. Since the rodent brain could not make radioactive insulin, any radioactive insulin recovered from brain after its peripheral injection must have crossed the BBB. Similarly, any human insulin appearing in rodent brain after its peripheral injection must have crossed the BBB.

Poduslo et al. [32] in 1994 examined the ability of radioactively labeled insulin to cross the BBB and also the blood-nerve barrier. They compared the transport rate of insulin with that of several other proteins, including albumin. Insulin was transported across the BBB at a rate of 1.36–0.94 µl g^{-1} min^{-1}. This rate was well over 100-fold faster than the rate at which albumin leaked into the brain. These rates matched almost exactly the rates found by us several years later [33].

Human insulin injected intravenously appeared in the mouse brain [34]. Species-specific ELISAs capable of distinguishing between human and rodent insulin confirmed that the CNS insulin was of peripheral, exogenous origin. These studies showed a nonlinear relation between serum and brain levels of human insulin, consistent with saturable transport across the BBB. Fitting the data to a one-site hyperbola model showed that the transporter had reached half of its maximal capacity at a blood level of about 3.5 ng ml^{-1} and was about 30% saturated at levels as low as 1.0 ng ml^{-1}. Blood glucose was not affected by serum insulin levels below 1 ng ml^{-1}. This shows that substantial amounts of insulin are being transported into the brain at serum levels of insulin too low to cause hypoglycemia. This, in turn, suggests that insulin transport and CNS effects of insulin are important at physiologic, nonhypoglycemic levels of serum insulin.

A study conducted with radioactive insulin also showed that insulin transport occurs at physiological levels [33]. This study found insulin transport rates to range over 0.87–1.70 µl g^{-1} min^{-1}, rates which agreed very well with those found previously by Poduslo et al. [32]. A dose of 0.1 µg per mouse of unlabeled insulin inhibited transport across the BBB of radioactive insulin by about 50% without producing hypoglycemia. Again, this showed that substantial amounts of insulin are transported across the BBB at the physiologic, euglycemic levels of insulin.

These studies clearly showed, regardless of the ability of the CNS to produce insulin, that insulin crossed the BBB by a saturable mechanism. They showed that insulin transport was substantial, far exceeding the amount which could be accounted for by the residual leakiness of the BBB [32]. In addition, they unvieled that substantial amounts of insulin crossed the BBB at serum levels which did not cause hypoglycemia. Thus, insulin transport and the signaling between the peripheral tissues and the brain is likely important at physiological levels of insulin as well as at higher, hypoglycemic levels.

As an aside, these studies shed light on the great difficulties that the early studies faced in examining the questions of whether insulin crossed the BBB. As those studies had to elevate serum levels above the physiologic range, they were studying the transporter at its less efficient levels. The studies with radioactive insulin showed no delay in entry into brain tissue, suggesting that the delay in elevation of CSF levels may indeed be caused by the brain interstitial fluid compartment [10].

12.2
Pathophysiology of Insulin Transport

A hallmark of BBB transport systems across the BBB is that they are not static. Instead, they are modulated by various agents and events. As such, they change during development and maturation, with the physiological demands of the CNS, and are altered in disease states. This hallmark of BBB transporters extends to those systems for peptides and regulatory proteins in general and for insulin in particular. For example, insulin transport is increased in iron-deficient rats and decreased by aluminum or dexamethasone treatment [33, 35, 36]. Insulin binding is enhanced in newborns and the brain-blood efflux of insulin is retarded by starvation and tumor necrosis factor-alpha [30, 37], whereas insulin protein and receptor levels in brain decrease with aging [38]. Most of the other effects on insulin transport can be broadly related to obesity, diabetes, insulin resistance, and Alzheimer's disease. These are reviewed below.

12.2.1
Insulin, Obesity, and Diabetes

CNS insulin is important in brain maturation [8, 39], increases brain glucose utilization [40, 41], promotes synthesis of acetylcholine [41], alters norepinephrine and dopamine levels and turnover in selective brain regions [42, 43] in part by enhancing transport across the BBB of tryptophan and tyrosine [26], increases efferent sympathetic nerve activity [44], modulates neuronal responses in the olfactory bulb and amygdala to stimuli [45], affects pituitary sex hormone secretion [46], and alters auditory evoked potentials [47]. Many of these effects of CNS insulin are opposite to those of peripheral insulin. For example, CNS insulin induces hyperglycemia, hypoinsulinemia, and anorexia, decreases neuropeptide Y expression in the hypothalamus, and ultimately decreases body weight [48–53].

These effects are likely occurring at physiological levels of insulin, as antibodies to insulin given directly into the brain enhance feeding [51, 54] and increase bodyweight [51]. Mice with a selective knockout of insulin neuronal receptors have an increased food intake, increased serum insulin levels, insulin resistance, mild diet-induced obesity, increased serum leptin levels, hypogonadal hypogonadism, and dyslipidemia [46]. Taken together, these studies suggest that, at least for energy homeostasis, CNS insulin has a complex temporal action [55, 56]. Specifically, acute CNS insulin opposes peripheral insulin by increasing serum glucose, decreasing serum insulin, inducing insulin resistance, and reducing food intake. As CNS insulin is derivative of serum insulin, these CNS actions would be a counterbalance to the insulin released into serum in response to a meal and so would protect from hypoglycemia. Of these acute actions, decreased feeding ultimately exerts the most important chronic effect. By preventing obesity and its accompanying proinflammatory state, CNS insulin ultimately prevents the chronic insulin-resistant state that can lead to type 2 diabetes [55, 56].

Insulin transport across the BBB has been found to be reduced in obese dogs and rats [57, 58]. After insulin infusions, for example, obese *fa/fa* rats had both lower CSF insulin levels and lower CSF/serum ratios than did lean *fa/fa* rats. Brain levels of insulin also decreased in the obese rats. In comparison, transport of insulin into CSF and binding of radioactive insulin to brain capillaries was not different with diet-induced obesity in the *Osborne-Mendel* rat [59]. However, the obese rats weighed only 20% more than the controls, a difference that may be too small to detect obesity-related differences. These findings with the BBB reinforced other work that indicated that insulin was playing an important role in energy homeostasis. Work with marmots showed that CSF insulin levels decreased early during hibernation despite maintenance of serum insulin levels [58]. This strongly suggested that insulin transport across the BBB was turned off during hibernation, a period when feeding signals were not needed. CSF insulin also decreases with fasting, strongly suggesting that insulin transport is also impaired [12].

Overall, these findings suggest that CNS insulin is important in energy homeostasis. Impaired BBB transport of insulin and the resulting decrease in CNS insulin could lead to derangements in energy homeostasis which would favor diet-induced obesity and, ultimately, insulin resistance in peripheral tissues.

12.2.2
Insulin Resistance and Inflammatory States

We have shown that treatment with lipopolysaccharide (LPS) stimulates insulin transport across the BBB [60]. LPS, also termed endotoxin, is derived from the coat of gram negative bacteria and is a powerful immune stimulant. Since CNS insulin opposes many of the actions of peripheral insulin, LPS-induced enhancement of insulin transport into the brain could underlie the mechanisms by which proinflammatory states induce insulin resistance. Indeed, CNS insulin and peripheral LPS have many similar effects. Thus, enhanced insulin transport across the BBB is one mechanism by which LPS could induce anorexia, glycemic dysregulation, insulin resistance, dyslipidemia, learning and memory deficits, impaired sexual and reproductive functions, and impairment of motivation.

The actions of LPS on the BBB transport of insulin are likely mediated through the proinflammatory cytokines IL-1, IL-6, and TNF. These cytokines are known to be released by LPS and many of the effects of LPS are mediated through them. The literature supports these three cytokines in particular as playing a central role in mediating the effects of LPS on insulin transport. This is based on three lines of evidence: (1) These three cytokines have themselves been shown to mimic the effects of CNS insulin, including inducing insulin resistance, anorexia, body weight loss, dyslipidemia, cognitive effects, decreased motivation, and sexual/reproduction effects; (2) they are associated in clinical studies with insulin resistance in non-LPS-mediated proinflammatory states such as obesity, post-myocardial infarction, and lipodystrophic AIDS, and (3) they are known to alter other aspects of BBB function.

IL-6 and TNF induce insulin resistance [61–63], cause anorexia [64, 65], decrease bodyweight [66–69], affect cognitive function and motivation [66–68, 70], and affect sexual/reproductive function [67, 68], all actions ascribed to CNS insulin. IL-1 also has effects on cognition, motivation, and anorexia and is the main mediator of sickness behavior [67, 68, 71]. Based on studies with TNF and IL-6 knockout mice, Keller et al. [62] proposed that TNF-induced insulin resistance is mediated through IL-6, suggesting a cytokine cascade. IL-6 knockout mice are protected from the development of age-related obesity, hyperglycemia, and dyslipidemia [72]. In humans, the degree of anorexia induced by LPS correlates with blood levels of IL-6 and TNF [73]. In contrast, evidence suggests that IL-1 is an important regulator of physiological insulin regulation [74] and is more associated with insulin sensitivity. IL-1 induces the release of insulin [75, 76] and mediates LPS-induced hypoglycemia in mice [77]. These studies show that insulin resistance in inflammatory states is mediated in large part through TNF and IL-6, perhaps with opposition by IL-1. Although some of these actions are likely mediated directly on various tissues, effects of insulin transport into the CNS could be another mechanism.

The mentioned cytokines are associated in clinical studies with the insulin resistance in non-LPS mediated proinflammatory states such as obesity, post-myocardial infarction, and lipodystrophic AIDS. In obesity, adipose tissue is a source of TNF and interleukins [78, 79]. Insulin resistance develops in 40% of nondiabetic patients after myocardial infarction and the degree of insulin resistance correlates with serum levels of IL-6 [80]. In the *Nurses Health Study* of 32 826 women, serum levels of IL-6 and TNF significantly predicted the risk of developing diabetes, even after controlling for age, race, lifestyle factors, and BMI [81]. In the nondiabetic elderly, IL-1ra, IL-6 and TNF correlate with the degree of insulin resistance [82]. The finding that IL-1ra, the endogenous antagonist to IL-1, correlated with insulin resistance again suggests that IL-1 may be acting in opposition to the other two proinflammatory cytokines. Evidence suggests that the thiazolidinediones, a powerful class of antidiabetic drugs which work by enhancing insulin sensitivity, work in part by decreasing the production of proinflammtory cytokines from adipose tissue [83]. These studies show that insulin resistance occurs in proinflammatory states other than sepsis and that this insulin resistance is likely mediated by the proinflammatory cytokines.

Taken together, the literature shows that IL-6 and TNF recapitulate those effects associated with CNS insulin, that insulin resistance in various proinflammatory states is affected by these cytokines, and that these cytokines have numerous effects on BBB function. IL-1 may represent the nitric oxide-mediated opposition to the LPS-simulated transport of insulin. These points make it highly likely that the ability of LPS to alter insulin transport across the BBB is mediated through one or more of these cytokines.

These mechanisms likely mediate the insulin resistance of states other than just gram negative sepsis. A number of conditions associated with insulin resistance are now recognized as proinflammatory states. Obesity is the most studied example. Adipose tissue is a source of interleukins and TNF; and levels of proinflammatory cytokines are elevated in obese persons. This proinflammatory state as mediated through the proinflammatory cytokines is currently considered the mechanism by which obesity induces insulin-resistant diabetes [55]. Moreover, a wide variety of other conditions with insulin resistance are closely linked with proinflammatory status and range from post-myocardial infarction to AIDS lipodystrophy.

The American Academy of Clinical Endocrinologists and the American College of Endocrinology, based on task force recommendations, have recognized a new syndrome, referred to as the insulin resistance syndrome (IRS; in Sept/Oct 2003, Supplement 2 of volume 9 of *Endocrine Practice* was devoted to this topic). IRS is characterized as that compensatory phase of tissue resistance to insulin when serum levels of insulin are elevated but serum glucose levels are still in the normal range (as opposed to type 2 diabetes mellitus, in which hyperglycemia results from the inability of the compensatory hyperinsulinemia to overcome insulin resistance). IRS predisposes to or is associated with many of the same conditions [84] that form the metabolic X syndrome (which includes as one of its characteristics insulin-resistant diabetes mellitus [85]): hypertension, stroke, nonalcoholic fatty liver disease, polycystic ovary syndrome, and coronary artery disease. Furthermore, risk factors for IRS include obesity and a proinflammatory state as measured by acute phase reactants.

Taken together, these findings strongly suggest that insulin resistance induced by proinflammatory conditions forms a spectrum from severe hyperglycemia to IRS. This may not be coincidental, but based on evolutionary pressures. Lazar, in a recent viewpoint in *Science*, pointed out that insulin resistance has survival value in stresses such as infection, possibly by making glucose less available to insulin-sensitive tissues such as muscle and therefore more available to noninsulin-sensitive tissues such as immune cells and brain [55]. This reinforces the idea that those mechanisms induced by our model of LPS administration may also be operational at lower levels of inflammation, such as obesity. Besides insulin resistance, our hypothesis also links many other conditions. This is because an increase in CNS insulin mimics many of the other findings of proinflammatory states: anorexia, cognitive effects, effects on sexual activity and reproduction, and motivation. By studying the ability of LPS to mediate transport of insulin into the CNS, one can study mechanisms that recapitulate the classic glucose intolerance of sepsis. However, these same mechanisms in attenuated form likely underlie the insulin resistance seen in other proinflammatory states including IRS and obesity.

12.2.3
Insulin and Alzheimer's Disease

Insulin transport across the BBB may also be involved in Alzheimer's disease (AD). Islet amyloid deposits were first described in autopsy studies [86], even before Dr Alois Alzheimer first reported the presence of "senile plaques" in 1906.

These deposits have since been elucidated to consist of islet amyloid polypeptide (IAPP) [87], which bears a 90% structural similarity to amyloid A precursor protein [88]. This observation is particularly relevant in the light of recent epidemiological evidence, suggesting that AD may predispose to a state of insulin resistance [89], insulin hypersecretion [89], and type 2 diabetes mellitus [90]. Conversely, individuals with type 2 diabetes and hyperinsulinemia show an association with increased prevalence of AD [91, 92], as well as hippocampal and amygdalar atrophy on magnetic resonance imaging that is independent of vascular pathology [93]. A recent study also reported that diet-induced insulin resistance promoted brain β-amyloid formation in a transgenic mouse model of AD [94].

Further insight into this complex but fascinating link between AD and diabetes may be accrued from an understanding of the different mechanisms of insulin action in the CNS. On the one hand, insulin can facilitate memory via various modalities: acting in concert with glucose to increase cholinergic activity, which is greatly decreased in AD [95], facilitating long-term potentiation of memory by hippocampal synaptic plasticity, mediated by glutamate receptors [96] and insulin-sensitive glucose transporters such as GLUT4 and GLUT8 [97], and promoting neurogenesis [98]. At the same time, insulin plays a pivotal role as a neuromodulator involved in the regulation of vital homeostatic processes essential for cerebral function. Insulin dysregulation may therefore contribute to AD pathology through several mechanisms, including: oxidative stress through the formation of advanced glycation end-products [99], increased tau phosphorylation and neurofibrillary tangle formation [100], and increased β-amyloid aggregation through competitive inhibition of insulin-degrading enzyme (IDE), a metalloprotease present on brain microvessels that degrades both insulin and β-amyloid [101].

This field has considered whether the effects of insulin on the brain should be regarded as mainly an extension of its peripheral action (in other words, as a regulator of systemic energy homeostasis), or whether insulin exerts CNS effects that are clearly independent of its peripheral metabolic effects. Currently, the weight of evidence favors the latter viewpoint. Thus, findings in this field resonate with current viewpoints in the area of energy homeostasis.

Consistent with the notion of regional variation in insulin receptor density within the CNS [102], studies have indicated the prevalence of insulin receptors in the limbic and hippocampal regions [103]. A further extension of the concept of regional specificity is seen in the selective function of insulin receptors in these regions, where they have been shown to be involved in synaptic plasticity and long-term memory consolidation [104].

Accruing evidence also corroborates the contrasting effects of acute compared with chronic elevation of glucose and insulin in the brain [95]. Acutely raising plasma insulin levels, while maintaining euglycemia, can improve memory in Alzheimer's patients [105]. However, raising glucose levels while suppressing endogenous insulin secretion abolished the memory enhancing effects of glucose, suggesting the pivotal role of insulin-mediated memory facilitation in AD [106]. At first sight, this may seem surprising, as both chronic hyperinsulinemia and chronic hyperglycemia are associated with accelerated cognitive decline in the elderly [107]. However, this apparent discrepancy should be seen in the light of the mechanisms underlying spontaneous chronic hyperinsulinemia and chronic hyperglycemia, of which insulin resistance is an important denominator.

Consistent with these roles are two studies which have examined the ability of insulin to cross the BBB in animal models of Alzheimer's disease. One model used the SAMP8 mouse, a natural mutation which has an age-dependent overexpression of amyloid precursor peptide, age-dependent increases in amyloid beta protein, and age-dependent learning and memory deficits which are reversed by antibodies or anti-sense directed against amyloid beta protein [108, 109]. Uptake was higher in aged SAMP8 mice for the cerebellum and the thalamus [110]. A second model used double-transgenic mice which overexpressed both amyloid precursor protein and presenilin [111]. This model found higher uptake of insulin in four of six brain regions.

This brings us to the last related point of the influence of disease-dependent factors on CNS insulin action. Converging evidence suggests that impairment at the level of the BBB in the insulin signal transduction cascade, and possibly insulin growth factor-1 signaling, may be an early and pivotal event in the pathogenesis of AD [112]. In this instance, although neuronal insulin receptors are upregulated [113], they exhibit defective signal transduction, analogous to the peripheral insulin resistant state seen in type 2 diabetes mellitus. This has led to the hypothesis that AD may be the "insulin-resistant brain state" equivalent of noninsulin-dependent diabetes mellitus [114]. This is an attractive hypothesis, which may also account for the apparent differential effects of acute and chronic hyperinsulinemia in AD [115]. Acute administration of exogenous insulin may serve to temporarily overcome cerebral insulin resistance, thereby improving memory. In contrast, chronic "endogenous" hyperinsulinemia reflects a compensatory response to insulin resistance of peripheral tissues. The seemingly elevated insulin levels belie a state of relative insulin deficiency that prevails when this compensatory response becomes inadequate.

In support of this hypothesis is the observation that peripheral insulin infusion in humans increases the levels of β-amyloid in cerebrospinal fluid, indirectly supporting the putative role of hyperinsulinemia in the pathogenesis of AD [116]. In addition, different mechanisms appear to be involved within the nosology of AD [117]. In early onset familial AD (ca. 5% of all AD cases), defective signal transduction is due to an insulin-binding deficit at the α-subunit of the insulin receptor that results from the competitive binding of β-amyloid [118]. In late-onset sporadic AD (95% or more of cases), though, the age-related increase in CNS noradrenaline and cortisol levels inhibits the β-subunit of the neuronal insulin receptor at different sites, leading to receptor desensitization [117]. There is also evidence that the apolipoprotein E (APOE) genotype, a prominent genetic factor in sporadic AD, can modulate the risk for insulin resistance [119] and affect insulin degrading enzyme levels [120].

Novel strategies based on this common pathologic link between insulin and cerebrovascular Aβ amyloidoses are being explored for the treatment of AD.

These include the enhancement of Aβ-degrading enzymes, such as insulin-degrading enzyme [121], and insulin sensitizers such as rosiglitazone, a peroxisome proliferator-activated receptor-gamma (PPAR-gamma) agonist [122]. Although intranasal insulin has been shown to improve short-term memory and mood in healthy subjects without systemic effects [123], it is uncertain whether chronic administration in AD patients would aggravate cerebral insulin resistance and thereby impede cognition in the long term.

References

1 Grundy, H. F. **1962**, *J. Physiol.* 163, 457–465.
2 Banks, W. A., Kastin, A. J. **1985**, *Psychoneuroendocrinology* 10, 385–399.
3 Margolis, R. U., Altszuler, N. **1967**, *Nature* 215, 1375–1376.
4 Havrankova, J., Roth, J., Brownstein, M. J. **1979**, *J. Clin. Invest.* 64, 636–642.
5 Reiser, M., Lenz, E., Bernstein, H. G., Dorn, A. **1985**, *Hum. Neurobiol.* 4, 53–55.
6 Plata-Salaman, C. R. **1991**, *Neurosci. Biobehav. Rev.* 15, 243–258.
7 Woods, S. C., Porte, D. Jr. **1977**, *Am. J. Physiol.* 233, E331–E334.
8 Blasberg, R. G., Fenstermacher, J. D., Patlak, C. S. **1983**, *J. Cereb. Blood Flow Metab.* 3, 8–32.
9 Patlak, C. S., Blasberg, R. G., Fenstermacher, J. D. **1983**, *J. Cereb. Blood Flow Metab.* 3, 1–7.
10 Schwartz, M. W., Bergman, R. N., Kahn, S. E., Taborsky, G. L. Jr., Fisher, L. D., Sipols, A. J., Woods, S. C., Steil, G. M., Porte, D. Jr. **1991**, *J. Clin. Invest.* 88, 1272–1281.
11 Schwartz, M. W., Figlewicz, D. P., Baskin, D. G., Woods, S. C., Porte, D. Jr. **1992**, *Endocrine Rev.* 13, 387–414.
12 Strubbe, J. H., Porte, D. Jr., Woods, S. C. **1988**, *Physiol. Behav.* 44, 205–208.
13 Baskin, D. G., Porte, D. Jr., Guest, K., Dorsa, D. M. **1983**, *Endocrinology* 112, 898–903.
14 Baura, G. D., Foster, D. M., Porte, D. Jr., Kahn, S. E., Bergman, R. N., Cobelli, C., Schwartz, M. W. **1993**, *J. Clin. Invest.* 92, 1824–1830.
15 Ono, T., Steffens, A. B., Sasaki, K. **1983**, *Physiol. Behav.* 30, 301–306.
16 Havrankova, J., Brownstein, M., Roth, J. **1981**, *Diabetologia* 20 [Suppl], 268–273.
17 Szczepanska-Sadowska, E., Gray, D., Simon-Oppermann, C. **1983**, *Am. J. Physiol.* 245, R549–R555.
18 Werther, G. A., Hogg, A., Oldfield, B. J., McKinley, M. J., Figdor, R., Allen, A. M., Mendelsohn, F. A. **1987**, *Endocrinology* 121, 1562–1570.
19 Baskin, D. G., Brewitt, B., Davidson, D. A., Corp, E., Paquette, T., Figlewicz, D. P., Lewellen, T. K., Graham, M. K., Woods, S. C., Dorsa, D. M. **1986**, *Diabetes* 35, 246–249.
20 Frank, H. J. L., Pardridge, W. M. **1981**, *Diabetes* 30, 757–761.
21 Miller, D. W., Borchardt, R. T. **1991**, *J. Cell Biol.* 115, 261a.

22 Pardridge, W. M. **1983**, *Annu. Rev. Physiol.* 45, 73–82.
23 Pardridge, W. M., Frank, H. J. L., Cornford, E. M., Braun, L. D., Crane, P. D., Oldendorf, W. H. **1981**, Neuropeptides and the blood-brain barrier, in *Neurosecretion and Brain Peptides*, eds. Martin, J. B., Reichlin, S., Bick, K. L., Raven Press, New York, pp. 321–328.
24 Keller, B. T., Borchardt, R. T. **1987**, *Fed. Proc.* 46, 416.
25 Catalan, R. E., Martinez, A. M., Aragones, M. D., Miguel, B. G., Robles, A. **1988**, *Biochem. Biophys. Res. Comm.* 150, 583–590.
26 Tagliamonte, A., DeMontis, M. G., Olianas, M., Onali, P. L., Gessa, G. L. **1976**, *Adv. Exp. Med. Biol.* 69, 89–94.
27 Ayre, S. G., Skaletski, B., Mosnaim, A. D. **1989**, *Res. Comm. Chem. Pathol. Pharmacol.* 63, 45–52.
28 Cangiano, C., Cardelli-Cangiano, P., Cascino, A., Patrizi, M. A., Barberini, F., Rossi, F., Capocaccia, L., Strom, R. **1983**, *Biochem. Int.* 7, 617–627.
29 Duffy, K. R., Pardridge, W. M. **1987**, *Brain Res.* 420, 32–38.
30 Frank, H. J. L., Jankovic-Vokes, T., Pardridge, W. M., Morris, W. L. **1985**, *Diabetes* 34, 728–733.
31 Woods, S. C., Seeley, R. J., Baskin, D. G., Schwartz, M. W. **2003**, *Curr. Pharm. Des.* 9, 795–800.
32 Poduslo, J. F., Curran, G. L., Berg, C. T. **1994**, *Proc. Natl Acad. Sci. USA* 91, 5705–5709.
33 Banks, W. A., Jaspan, J. B., Huang, W., Kastin, A. J. **1997**, *Peptides* 18, 1423–1429.
34 Banks, W. A., Jaspan, J. B., Kastin, A. J. **1997**, *Peptides* 18, 1257–1262.
35 Ben-Shachar, D., Yehuda, S., Finberg, J. P., Spanier, I., Youdim, M. B. **1988**, *J. Neurochem.* 50, 1434–1437.
36 Baura, G. D., Foster, D. M., Kaiyala, K., Porte, D. Jr., Kahn, S. E., Schwartz, M. W. **1996**, *Diabetes* 45, 86–90.
37 Cashion, M. F., Banks, W. A., Kastin, A. J. **1996**, *Hormones Behav.* 30, 280–286.
38 Frolich, L., Blum-Degen, D., Bernstein, H. G., Engelsberger, S., Humrich, J., Laufer, S., Muschner, D., Thalheimer, A., Turk, A., Hoyer, S., Zochling, R., Boissl, K. W., Jellinger, K., Riederer, P. **1998**, *J. Neural Transm.* 105, 423–428.
39 Schubert, M., Brazil, D. P., Burks, D. J., Kushner, J. A., Ye, J., Flint, C. L., Farhang-Fallah, J., Dikkes, P., Warot, X. M., Rio, C. , Corfas, G., White, M. F. **2003**, *J. Neurosci.* 23, 7084–7092.
40 Henneberg, N., Hoyer, S. **1994**, *Neurosci. Lett.* 175, 153–156.
41 Hoyer, S. **2003**, *Pharmacopsychiatry* 36 [Suppl. 1], S62–S67.
42 Kwok, R. P., Juorio, A. V. **1988**, *Neurochem. Res.* 13, 887–892.
43 Montefusco, O., Assini, M. C., Missale, C. **1983**, *Acta Diabetol. Lat.* 20, 71–77.
44 Muntzel, M. S., Morgan, D. A., Mark, A. L., Johnson, A. K. **1994**, *Am. J. Physiol.* 267, R1350–R1355.
45 Cain, D. P. **1975**, *Brain Res.* 99, 69–83.
46 Bruning, J. C., Gautam, D., Burks, D. J., Gillette, J., Schubert, M., Orban, P. C., Klein, R., Krone, W., Muller-Wieland, D., Kahn, C. R. **2000**, *Science* 289, 2122–2125.

47 Kern, W., Born, J., Schreiber, H., Fehm, H. L. **1999**, *Diabetes* 48, 557–563.
48 Ajaya, B., Haranath, P. S. **1982**, *Indian J. Med. Res.* 75, 607–615.
49 Brief, D. J., Davis, J. D. **1984**, *Brain Res. Bull.* 12, 571–575.
50 Hatfield, J. S., Millard, W. J., Smith, C. J. V. **1974**, *Pharmacol. Biochem. Behav.* 2, 223–226.
51 McGowan, M. K., Andrews, K. M., Grossman, S. P. **1992**, *Physiol. Behav.* 51, 753–766.
52 Florant, G. L., Singer, L., Scheurink, A. J. W., Park, C. R., Richardson, R. D., Woods, S. C. **1991**, *Physiol. Behav.* 49, 335–338.
53 Schwartz, M. W., Sipols, A. J., Marks, J. L., Sanacora, G., White, J. D., Scheurink, A., Kahn, S. E., Baskin, D. G., Woods, S. C., Figlewicz, D. P., Porte, D. Jr. **1992**, *Endocrinology* 130, 3608–3615.
54 Strubbe, J. H., Mein, C. G. **1977**, *Physiol. Behav.* 19, 309–313.
55 Lazar, M. A. **2005**, *Science* 307, 373–375.
56 Schwartz, M. W., Porte, D. Jr. **2005**, *Science* 307, 375–379.
57 Kaiyala, K. J., Prigeon, R. L., Kahn, S. E., Woods, S. C., Schwartz, M. W. **2000**, *Diabetes* 49, 1525–1533.
58 Stein, L. J., Dorsa, D. M., Baskin, D. G., Figlewicz, D. P., Porte, D. Jr., Woods, S. C. **1987**, *Endocrinology* 121, 1611–1615.
59 Israel, P. A., Park, C. R., Schwartz, M. W., Green, P. K., Sipols, A. J., Woods, S. C., Porte, D. Jr., Figlewicz, D. P. **1993**, *Brain Res. Bull.* 30, 571–575.
60 Xaio, H., Banks, W. A., Niehoff, M. L., Morley, J. E. **2001**, *Brain Res.* 896, 36–42.
61 Lang, C. H., Dobrescu, C., Bagby, G. J. **1992**, *Endocrinology* 130, 43–52.
62 Keller, C., Keller, P., Giralt, M., Hidalgo, J., Pedersen, B. K. **2004**, *Biochem. Biophys. Res. Comm.* 321, 179–182.
63 Kim, H. J., Higashimori, T., Park, S. Y., Choi, H., Dong, J., Kim, Y. J., Noh, H. L., Cho, Y. R., Cline, G., Kim, Y. B., Kim, J. K. **2004**, *Diabetes* 53, 1060–1067.
64 Langhans, W., Hrupka, B. **1999**, *Neuropeptides* 33, 415–424.
65 Bodnar, R. J., Pasternak, G. W., Mann, P. E., Paul, D., Warren, R., Donner, D. B. **1989**, *Cancer Research* 15, 6280–6284.
66 Bluthe, R. M., Parnet, P., Dantzer, R., Kelley, K. W. **1991**, *Neurosci. Res. Comm.* 15, 151–158.
67 Kent, S., Bret-Dibat, J. L., Kelley, K. W., Dantzer, R. **1996**, *Neurosci. Biobehav. Rev.* 20, 171–175.
68 Larson, S. J., Dunn, A. J. **2001**, *Brain Behav. Immun.* 15, 371–387.
69 Hotamisligil, G. S., Spiegelman, B. M. **1994**, *Diabetes* 43, 1271–1278.
70 Pugh, C. R., Fleshner, M., Watkins, L. R., Maier, S. F., Rudy, J. W. **2001**, *Neurosci. Biobehav. Rev.* 25, 29–41.
71 Kelley, K. W., Bluthe, R. M., Dantzer, R., Zhou, J.-H., Shen, W.-H., Johnson, R. W., Broussard, S. R. **2003**, *Brain Behav. Immun.* 17, S112–S118.
72 Di Gregorio, G. B., Hensley, L., Lu, T., Ranganathan, G., Kern, P. A. **2004**, *Am. J. Physiol. Endocrinol. Metabol.* 287, E182–E187.

73 Reichenberg, A., Kraus, T., Haack, M., Schuld, A., Pollmacher, T., Yirmiya, R. **2002**, *Psychoneuroendocrinology* 27, 945–956.
74 Matsuki, T., Horai, R., Sudo, K., Iwakura, Y. **2003**, *J. Exp. Med.* 198, 877–888.
75 Green, I. C., Delaney, C. A., Cunningham, J. M., Karmiris, V., Southern, C. **1993**, *Diabetologia* 36, 9–16.
76 Eizirik, D. L., Sandler, S., Welsh, N., Juntti-Berggren, L., Berggren, P. O. **1995**, *Mol. Cell. Endocrinol.* 111, 159–165.
77 Ogure, S., Motegi, K., Iwakura, Y., Endo, Y. **2002**, *Clin. Diagn. Lab. Immunol.* 9, 1307–1312.
78 Heilbronn, L. K., Smith, S. R., Ravussin, E. **2003**, *Curr. Pharm. Des.* 9, 1411–1418.
79 Chaldakov, G. N., Stankulov, I. S., Hristova, M., Ghenev, P. I. **2003**, *Curr. Pharm. Des.* 9, 1023–1031.
80 Choi, K. M., Lee, K. W., Kim, S. G., Kim, N. H., Park, C. G., Seo, H. S., Oh, D. J., Choi, D. S., Baik, S. H. **2005**, *J. Clin. Endocrinol. Metabol.* 90, 175–180.
81 Hu, F. B., Meigs, J. B., Li, T. Y., Rifai, N., Manson, J. E. **2004**, *Diabetes* 53, 693–700.
82 Abbatecola, A. M., Ferrucci, L., Grella, R., Bandinelli, S., Bonafe, M., Barbieri, M., Corsi, A. M., Lauretani, F., Franceschi, C., Paolisso, G. **2004**, *J. Am. Geriatr. Soc.* 52, 399–404.
83 Fasshauer, M., Paschke, R. **2003**, *Diabetologia* 46, 1594–1603.
84 Task Force, A. **2003**, *Endocrine Pract.* 9, 9–21.
85 Hansen, B. C., Saye, J., Wennogle, L. P. **1999**, *The Metabolic Syndrome X: Convergence of Insulin Resistance, Glucose Intolerance, Hypertension, Obesity, and Dyslipidemia – Searching for the Underlying Defects*, New York Academy of Sciences, New York, N.Y.
86 Opie, E. L. **1901**, *J. Exp. Med.* 5, 527–540.
87 Mosselman, S., Hoppener, J. W., Zandberg, J., van Mansfeld, A. D., Guerts van Kessel, A. H., Jansz, H. S. **1988**, *FEBS Lett.* 239, 227–232.
88 Janson, J., Laedtke, T., Parisi, J. E., O'Brien, O., Petersen, R. C., Butler, P. C. **2004**, *Diabetes* 53, 474–481.
89 Razay, G., Wilcock, G. K. **1994**, *Age and Ageing* 23, 396–399.
90 Carantoni, M., Zuliani, G., Munari, M. R., d'Elia, K., Palmieri, E., Fellin, R. **2000**, *Dementia Geriatric Cognitive Disorders* 11, 176–180.
91 Ott, A., Stolk, R. P., van Harskamp, F., Pols, H. A., Hofman, A., Breteler, M. M. **1999**, *Neurology* 53, 1907–1909.
92 Luschsinger, J. A., Tang, M. X., Shea, S., Mayeux, R. **2004**, *Neurology* 63, 1187–1192.
93 den Heijer, T., Vermeer, S. E., van Dijk, E. J., Prins, N. D., Koudstaal, P. J., Hofman, A., Breteler, M. M. **2003**, *Diabetologia* 46, 1604–1610.
94 Ho, L., Qin, W., Pompl, P. N., Xiang, Z., Wang, J., Zhao, Z., Peng, Y., Cambareri, G., Rocher, A., Mobbs, C. V., Hof, P. R., Pasinetti, G. M. **2004**, *FASEB J.* 18, 902–904.
95 Watson, G. S., Craft, S. **2004**, *Eur. J. Pharmacol.* 490, 115–125.

96 Trudeau, F., Gagnon, S., Massicotte, G. **2004**, *Eur. J. Pharmacol.* 490, 177–186.
97 McEwen, B. S., Reagan, L. P. **2004**, *Eur. J. Pharmacol.* 490, 13–24.
98 Jackson-Guilford, J., Leander, J. D., Nisenbaum, L. K. **2000**, *Neurosci. Lett.* 293, 91–94.
99 Grossman, H. **2003**, *CNS Spectr.* 8, 815–823.
100 Park, C. R. **2001**, *Neurosci. Biobehav. Rev.* 25, 311–323.
101 Qui, W. Q., Walsh, D. M., Ye, Z., Vekrellis, K., Zhang, J., Podlisny, M. B., Rosner, M. R., Safavi, A., Hersh, L. B., Selkoe, D. J. **1998**, *J. Biol. Chem.* 273, 32730–32738.
102 Havrankova, J., Roth, J., Brownstein, M. **1978**, *Nature* 72, 827–829.
103 Schulingkamp, R. J., Pagano, T. C., Hung, D., Raffa, R. B. **2000**, *Neurosci. Biobehav. Rev.* 24, 855–872.
104 Zhao, W. Q., Chen, H., Quon, M. J., Alkon, D. L. **2004**, *Eur. J. Pharmacol.* 490, 71–81.
105 Craft, S., Newcomer, J., Kanne, S., Dagogo-Jack, S., Cryer, P., Sheline, Y., Luby, J., Dagogo-Jack, A., Alderson, A. **1996**, *Neurobiol. Aging* 17, 123–130.
106 Craft, S., Asthana, S., Newcomer, J. W., Wilkinson, C. W., Matos, I. T., Baker, L. D., Cherrier, M., Lofgreen, C., Latendresse, S., Petrova, A., Plymate, S., Raskind, M., Grimwood, K., Veith, R. C. **1999**, *Arch. Gen. Psychiatry* 56, 1135–1140.
107 Gispen, W. H., Biessels, G. J. **2000**, *Trends Neurosci.* 23, 542–549.
108 Morley, J. E., Farr, S. A., Kumar, V. B., Banks, W. A. **2002**, *Peptides* 23, 589–599.
109 Banks, W. A., Morley, J. E. **2003**, *J. Gerontol.*
110 Banks, W. A., Farr, S. A., Morley, J. E. **2000**, *J. Gerontol. Biol. Sci.* 55A, B601–B606.
111 Poduslo, J. F., Curran, G. L., Wengenack, T. M., Malester, B., Duff, K. **2001**, *Neurobiol. Aging* 8, 555–567.
112 Carro, E., Torres-Aleman, I. **2004**, *Eur. J. Pharmacol.* 490, 127–133.
113 Frolich, L., Blum-Degen, D., Bernstein, H. G., Engelsberger, S., Humrich, J., Laufer, S., Muschner, D., Thalheimer, A., Turk, A., Hoyer, S., Zochling, R., Boissl, K. W., Jellinger, K., Riederer, P. **1998**, *J. Neural Transm.* 105, 423–438.
114 Hoyer, S. **1998**, *J. Neural Transm.* 105, 415–422.
115 Biessels, G. J., Bravenboer, B., Gispen, H. W. **2004**, *Eur. J. Pharmacol.* 490, 1–4.
116 Watson, G. S., Peskind, E. R., Asthana, S., Purganon, K., Wait, C., Chapman, D., Schwartz, M. W., Plymate, S., Craft, S. **2003**, *Neurology* 60, 1899–1903.
117 Hoyer, S. **2004**, *Eur. J. Pharmacol.* 490, 115–125.
118 Xie, L., Helmerhorst, E., Taddel, K., Plewright, B., van Bronswijk, W., Martins, R. **2002**, *Neuroscience* 22, 1–5.
119 Craft, S., Asthana, S., Schelenberg, G., Cherrier, M., Baker, L. D., Newcomer, J., Plymate, S., Latendresse, S., Petrova, A., Raskind, M., Peskind, E., Lofgreen, C., Grimwood, K. **1999**, *Neuroendocrinology* 70, 146–152.

120 Anon. **2003**, *Am. J. Physiol.* 162, 313–319.
121 Turner, A.J., Fisk, L., Nalivaeva, N.N. **2004**, *Ann. N.Y. Acad. Sci.* 1035, 1–20.
122 Watson, G.S., Craft, S. **2004**, *CNS Drugs* 17, 27–45.
123 Benedict, C., Hallschmid, M., Hatke, A., Schultes, B., Fehm, H.L., Born, J., Kern, W. **2004**, *Psychoneuroendocrinology* 29, 1326–1334.

13
Glucocorticoid Hormones and Estrogens: Their Interaction with the Endothelial Cells of the Blood-Brain Barrier

Jean-Bernard Dietrich

13.1
Introduction

The blood-brain barrier (BBB) isolates the central nervous system from the blood. The BBB is mainly composed of capillary endothelial cells, which normally block nonadherent circulating leukocytes in the bloodstream, but permit their transmigration into neighboring tissues following an inflammatory stimulus. Many adhesion molecules are known to be necessary for this recruitment of leukocytes. The expression of these adhesion molecules in brain-derived microvascular endothelial cells can be regulated by glucocorticoid hormones. This important class of hormones has proven particularly useful in the treatment of multiple sclerosis, a disease resulting from the chronic inflammation of the central nervous system and mediated by the infiltration of inflammatory cells of the immune system.

Estrogens can also modulate the expression of these endothelial adhesion molecules, thus regulating the leukocyte-endothelial cell interactions. In addition, estrogens are useful for the treatment of experimental autoimmune encephalomyelitis, an animal model with similarities to multiple sclerosis (MS).

In this chapter, the effects of glucocorticoids (GC) and estrogens on brain endothelial cells and their use as therapeutic tools for the treatment of autoimmune diseases are discussed.

There is no doubt that GC hormones probably affect every organ in mammals, but many of their effects are specific for particular cells or tissues. GC mediate their action through a cytoplasmic receptor which has several functional domains and activation of this receptor directly or indirectly regulates the transcription of target genes [1, 2].

The brain is a major target tissue for GC and these hormones interact with endothelial cells, especially those of the BBB. These BBB endothelial cells can be distinguished from others in the body by the absence of fenestrations and more extensive tight junctions, which actually limit the flux of hydrophilic molecules across the BBB [3, 4].

demonstrated. Acute inflammation is a phenomenon where rapid infiltration of polymorphonuclear neutrophils is followed by monocytes and eventually lymphocytes [21]. Leukocyte recruitment into extravascular tissue is indeed regulated by specific combinations of adhesion and signaling molecules [22–24].

GC are able to exert their anti-inflammatory effect by inhibiting the expression of the genes coding cytokines and adhesion molecules as well as by redirecting the traffic of lymphocytes [25, 26]. Expression of inducible nitric oxide synthase, an enzyme required for nitric oxide (NO) synthesis, is repressed by GC [27]. NO synthase is known to contribute to the inflammation process [28]. GC also have a direct effect on the expression of lipid mediators of inflammation, because they can inhibit the activity of enzymes involved in their biosynthesis. A good example is the regulation of cytokine-induced cyclo-oxygenase 2 by DEX [29].

13.2.3
Regulation of Adhesion Molecules Expression by GC in Endothelial Cells

The adhesion molecules involved in leukocyte recruitment are well documented: they include selectins, integrins, and members of the immunoglobulin superfamily (ICAM-1,-2, VCAM-1, PECAM-1) and CD 99 (Table 13.1). Leukocytes emigrate from the bloodstream into inflamed tissues by sequential events. These intercellular adhesion events with endothelial cells lining the vascular wall are now well characterized and described in many papers [24, 30–32].

A general paradigm for this process is now accepted. The first steps, tethering and rolling, are mediated by the members of the selectin family: L-selectins expressed mainly by the leukocytes, and P- and E-selectins expressed by platelets and endothelial cells respectively. The counter-ligands of the selectins are sialylated, fucosylated receptors. However, some leukocytes use the interaction between alpha 4-beta 1 integrin (VL4, CD49d-CD29) and vascular cell adhesion molecule 1 (VCAM-1) for both these steps. The second action, rolling, brings the leukocyte into close contact with the endothelium, which can then be activated by chemokines [33], platelet activating factor [24], or other soluble or membrane-bound molecules. These signals activate the beta 2 integrins (CD11–CD18 family) specific for leukocytes and allow the firm adhesion of these cells on the endothelial luminal surface. The counter-receptors on endothelial cells are members of the immunoglobulin gene superfamily. Intercellular adhesion molecules (ICAM-1, ICAM-2) are involved as well as VCAM-1. Leukocytes then squeeze between the opposing endothelial cells and this last step is known as transmigration or diapedesis. Homophilic interactions between the same molecules expressed on both endothelial cells and leukocytes mediate this process. PECAM-1 (CD31) plays an important role in the transmigration of monocytes, neutrophils, and natural killer cells, both in vitro [34, 35] and in vivo [36]. Diapedesis and subsequent migration across the basal lamina and interstitial tissues are stimulated by soluble and membrane-bound chemoattractants from the inflammatory environment (bacterial cell products, chemokines synthesized by

Table 13.1 Adhesion molecules involved in inflammation: leukocyte adhesion molecules and their ligands on endothelial cells, with CD and integrin nomenclatures in brackets.

Leukocyte adhesion molecules	L-Selectin (CD62L)	PSGL-1 (CD162)	E-Selectin ligand-1, CLA bearing sialyl-Lewis$_x$	VLA-4 (CD49d/CD18; $\alpha_4\beta_1$)	LFA-1 (CD11a/CD18; $\alpha_L\beta_2$)	Mac-1, CR3 (CD11b/CD18; $\alpha_M\beta_2$)	PECAM-1 (CD31)	(CD99)
Endothelial cell counter-ligand	Sialyl-Lewis$_x$ on appropriate ligand	P-selectin (CD62P)	E-selectin (CD62P)	VCAM-1 (CD106)	ICAM-1 (CD54), ICAM-2 (CD102)	ICAM-1 (CD54)	PECAM-1 (CD31)	(CD99)
Actions	Tethering (T), rolling (R)	T, R	T, R	T, R tight adhesion (TA)	TA	TA	Diapedesis (D)	D
Leukocytes involved	Neutrophils (N), monocytes (M), lymphocytes B&T (B&T), natural killer cells (NK)	N, M, B&T, NK	N, M, B&T, NK	N, M, B&T, NK	M, B&T, NK	N, M, NK	N, M, NK, subsets of T	N, M, B&T, NK

host cells in response to inflammatory stimuli). The role of a chemoattractant gradient was recently documented [37]. CD99 also plays a major role in the migration of monocytes through endothelial junctions [38]. In addition, the molecule JAM-1 (junctional adhesion molecule 1) has been identified and shown to be involved in tight adhesion or transmigration of leukocytes, depending on its apical or junctional localization on endothelial cells. JAM-1 is involved in such interactions via JAM-1/LFA-1 contacts [39, 40].

How do GC act on the expression of these adhesion molecules at the level of endothelial cells, especially the endothelial cells of the BBB? This is an important question, not only on a fundamental level but also in relation to the anti-inflammatory properties of GC. Among GC, DEX is a synthetic glucocorticoid used as a potent anti-inflammatory product and also for studies of gene expression in many cell lines or tissues. The effects of DEX and another GC, methylprednisolone (MP), on the expression of the genes coding these adhesion molecules are summarized in Table 13.2. Expression of E-selectin is upregulated by cytokines and lipopolysaccharides (LPS) in human brain microvessel endothelial cells (HBECs) [15] and by TNF-alpha and LPS in aortic endothelial cells and in HUVECs. This effect is markedly reduced in the presence of DEX, both in aortic endothelial cells [41] and in HUVECs [14]. Inflammatory cytokines augment synthesis of P-selectin in human tissues affected by allergic or chronic inflammation [42, 43]. In human endothelial cells, the interleukin-4-induced expression of P-selectin, which is constitutively synthesized by endothelial cells, is not affected by DEX but is decreased by the proteasome inhibitor N-acetyl-leucinyl-leucinyl-norleucinal-H, the antioxidant pyrrolidine dithiocarbamate, or sodium salicylate [44]. DEX can also induce regulated by GC on circulating leukocytes [69]. ANXA-1 is involved in the regulation of all steps of the adhesion cascade [67, 70]. ANXA-1 does not affect CAM gene expression on either resting or activated lymphocytes [71]: this protein influences leukocyte adhesion events by modulating CAM functions but not by regulating CAM gene transactivation. ANXA-1 in vivo effects on leukocyte adhesion and transmigration cause conformational/functional changes of surface CAM. Thus, these effects can be considered as nongenomic [72].

Finally, induction of CAM gene expression may be regulated in different cell types by various cytokines and chemokines. This highlights the inhibitory effect of GC on pro-inflammatory mediators. For example, GC have been shown to inhibit lymphocyte CAM by interfering with a set of IL-2-gamma-dependent cytokines [73].

13.2.5
Glucocorticoids, Cerebral Endothelium and Multiple Sclerosis

GC effects on brain vascular endothelial cells are not only observed at the level of the regulation of adhesion molecules expression. Other genes are under GC control. In human brain-derived microvascular endothelial cells (HBECs), DEX (1 µM) treatment for 24 h decreased the number of high affinity endothelin-1 binding sites. In this way, GC may counteract some endothelin-induced events in the cerebral endothelium (permeability changes, adhesion molecule expression) which are involved in the development of inflammatory and/or cerebrovascular brain disorders [74]. Statins induce apoptosis in many cell types. The effect of fluvastatin (FS) at a micromolar concentration was tested on the endothelial cell line EA.hy926. DEX (1 µM) blocked FS-induced apoptosis as well as apoptosis induced by serum deprivation, TNF-alpha, oxidation, DNA damage, and mitochondria disruption. This study suggests that GC play a role in preventing vascular injury and explains why statins are not toxic to vascular endothelial cells in vivo [75]. Histamine H_1 and H_2 receptors in rat brain endothelial cells (RBE4 cells) are downregulated by DEX at the mRNA level. This mechanism may be involved in GC-mediated effects on cerebrovascular permeability and brain edema [76]. DEX decreases transmonolayer permeability in cultured rat brain endothelial cells: this can be demonstrated using sucrose, fluorescein, and dextrans of up to 20 kDa in GPNT rat brain endothelial cells treated with 1 µM DEX. The continuity of the tight junctional protein zonulae occludens-1 (ZO-1) is characteristic of the BBB at inter-endothelial cell-cell contacts. A more regular and continuous cortical ZO-1 distribution was observed following DEX treatment. Thus, these results suggest that GC can induce a more differentiated BBB phenotype in these cells by affecting tight junction (TJ) structure [77].

HBECs under basal conditions produce interleukin-8 (IL-8) and monocyte-chemoattractant protein-1 (MCP-1), as shown by the expression of their mRNA transcripts [78]. Beta-chemokines play an important role in the recruitment of mononuclear cells into the brain during a variety of neuroinflammatory conditions, such as experimental allergic encephalomyelitis (EAE) and MS [79, 80]. Monocyte migration could be significantly inhibited by the addition of blocking antibodies to MCP-1 using HBECs in an in vitro model of the migration of cells of systemic immune origin across the BBB during the initiation of an inflammatory response within the central nervous system [81]. High-affinity saturable binding sites for MCP-1 are present along human brain microvessels [82]. This suggested that the CC chemokine receptor (CCR2) is expressed by the brain endothelium [83]. Both occludin and ZO-1 are part of TJs and disruption of the TJs could foster leukocyte extravasation. In addition to its chemotactic activity, MCP-1 has been shown to alter BBB integrity during CNS inflammation. Exposure to MCP-1 caused a loss in ZO-1 immunoreactivity at the inter-endothelial junctional region in both isolated brain microvessels and cultured brain microvessel endothelial cells (BMECs). A similar effect on occludin was observed in cultured BMECs. Finally, expression of caveolin-1, a major structural component of membrane microdomains thought to be functionally complexed with TJs, was also altered when MCP-1 was added to BMECs and microvessels [84]. Considering the pivotal role of MCP-1 in the transendothelial migration of lymphocytes, it is worth noting that DEX treatment reduces the upregulation of secreted MCP-1 after activation with TNF-alpha, IL1-beta, and IFN-gamma. This effect of DEX has been observed in vitro, using rat brain (GP8/3.9) and a retinal (JG/2) vascular endothelial cell line expressing MCP-1. In this study, DEX treatment (1 µM) significantly inhibited MCP-1 secretion in GP8/3.9 endothelial cells stimulated by TNF-alpha, IL1-beta, and IFN-gamma, by 22.7±2.4%, 24.5±6.4%, and 31.2±4.7% respectively. These in vitro findings support the hypothesis that CNS endothelial cells play an active role in the recruitment of inflammatory cells from the circulation. This could be mediated by enhanced expression of chemokines like MCP-1. Downregulation of endothelial derived MCP-1 by DEX found in vitro may account for the beneficial effects of GC observed in vivo in the treatment of autoimmune neuroinflammatory diseases [85].

Matrix metalloproteases are also present in the endothelial cells of the CNS, where their altered expression may contribute to the disruption of the BBB. DEX can partially inhibit MMP-9 cytokine-upregulated expression in rat brain cell lines like GP8/3.9. The effect of GC on MMP production may be one of the possible mechanisms by which steroids affect BBB permeability [86]. MMP production is elevated in MS patients and has a role in BBB disruption in MS. In this case, GC also inhibit the activity of these MMP [87].

GC are currently used as immunosuppressive and anti-inflammatory drugs (with quite variable doses ranging from very low to extremely high, depending on the disease). GC pulse therapy, which allows a strong induction of leukocyte apoptosis, is now the standard approach used for the treatment of neuroinflammatory diseases [88]. Apoptosis in this case contributes to the downregulation of T cell activity and terminates the inflammatory response. MP augments apoptosis of T cells in situ and a clear dose-response effect is observed in treatment of adaptative transfer EAE [89]. In vivo, unstimulated peripheral leukocytes shows enhanced apoptosis in the three groups of MS patients [90].

The use of interferon, particularly IFN-beta, for the treatment of MS is now well documented [91–94]. The effects of GC and type I interferons (IFN-alpha, IFN-beta) used simultaneously were investigated in vivo. Both drugs additively decreased BBB permeability, but did not prevent the increase of permeability induced by a pro-inflammatory stimulus (LPS). Thus, the beneficial effect of this treatment for MS did not seem to be mediated through a direct action at the level of BBB. A more general sensitivity is involved, because pretreatment by type I IFNs potentiated the effect of GC by two orders of magnitude. Type I IFNs may possibly restore the dysfunctional T helper I (Th1/Th2) balance associated with MS [95].

IFN beta has been approved for the treatment of MS and has been demonstrated to stabilize BBB integrity in vivo. Using a coculture model (brain endothelial cells with astrocytes), Kraus and coworkers showed that recombinant IFN-beta inhibited an increased paracellular permeability for small tracers. Permeability changes were accompanied by minor changes in the staining of TJ-associated proteins (occludin, claudin-3, -5, ZO-1, -2) in brain endothelial cell monolayers. Under conditions with low paracellular permeabilities, the monolayer resembled a typical cobblestone-like pattern and the TJ-associated proteins at the cellular borders appeared continuous and homogeneous. In culture conditions leading to increased permeabilities, the staining appeared to be dispersed or concentrated at the cell-cell contact zones of endothelial cells [94].

High-dose GC therapy is currently used in MS relapses and one mode of action of GC could be their inhibition of the cytokine-induced expression of CAM, which mediates leukocyte-BBB interactions and finally chronic recruitment of leukocytes across the BBB. The severe side-effects of long-term GC therapies are well known [95, 96]. To avoid these side-effects, the development of new steroidal drugs has been undertaken with the hope that these drugs could be beneficial for MS patients. These molecules would specifically transrepress NF-kappaB-mediated expression of CAM [58, 97]. For cell-specific drug delivery, endothelial cells were considered as attractive targets, especially due to their direct contact with the bloodstream. To deliver GC into activated endothelial cells at the inflammatory site, DEX was conjugated to a monoclonal antibody recognizing E-selectin. It was demonstrated that this immunoconjugate was internalized by activated, not resting, endothelium via the lysosomal pathway. After intracellular degradation, pharmacologically active DEX was released, which was able to downregulate IL-8 gene [98]. In a recent study using HUVECs, the conjugate (10 µg ml^{-1}) was found to decrease the levels of ICAM-1 and VCAM-1 after 24 h stimulation with TNF-alpha. However, the expression of 24 genes was downregulated (by 50% or more) by free DEX and the immunoconjugate after 6 h stimulation with TNF-alpha, showing that intracellularly delivered DEX is able to exert its pleiotropic anti-inflammatory effect [99].

13.3
Estrogens and the Endothelial Cells of the BBB

13.3.1
Mechanisms of Action of Estrogens

Many recent reviews dealing with the mechanisms of action of estrogens (ES) are available [100–103]. Because the cerebral vasculature is an important target tissue for these hormones, a precise understanding of their effects at the molecular level is required. ES can act by either direct or indirect genomic pathways as well as by nongenomic effects, allowing multiple actions in a single cell. The direct genomic mechanism involves two nuclear estrogen receptors (ER): ER-alpha and ER-beta. The formation of the ES-ER complex and its dimerization leads to activation and permits association with an estrogen response element (ERE) or with an ER-dependent AP1 response element. In the latter case, the complex binds to fos-jun heterodimers which then bind to the AP-1 response element [104].

This complex thus acts as a transcription factor at the genomic level. Indirect genomic effects include activation of ER-linked second-messenger systems such as adenylyl cyclase, protein kinases A and C and MAP kinase (AC/PKC, cAMP/PKA, MAPK/ERK). Consequently, many cell substrates are phosphorylated. Among them are transcriptional regulators like CREB which in turn interact with the DNA regulatory CRE (cAMP-responsive element). This can then allow indirect regulation of the expression of genes (without ERE).

Nongenomic effects can occur at low or high estrogen concentrations, either at nanomolar or lower levels or at micromolar concentrations. These nongenomic effects are characterized by rapid changes in the cell physiology that occur on the order of minutes. At high concentrations, nongenomic effects involve antioxidant effects not mediated by known ERs [105]. The question remains whether these changes are mediated by a modified form of nuclear ER that is associated with the plasma membrane (mER) or a completely novel membrane receptor. Recently, a membrane-associated receptor was characterized in a rat hypothalamic endothelial cell line (D12 cells). These brain-derived endothelial cells express a membrane-associated protein (67 kDa molecular mass, similar to that of ER-alpha), colocalized with caveolae-enriched membranes. However, the data suggest that this mER, related to nuclear ER-alpha in D12 cells, is nevertheless biochemically and immunologically distinct [106]. At the lower estrogen concentrations, nongenomic effects have been described in neurons, and at high estrogen concentrations, neuroprotective effects have been reported for a number of cell culture models, but not endothelial cells [105]. A good example of nongenomic action of ES is vasodilation, induced in a matter of seconds to minutes. This rapid effect is a result of the release of vasoactive molecules which regulate ionic fluxes on smooth muscle and vascular endothelial cells [103, 107].

13.3.2
Endothelial Cells as Targets of Estrogens

Indeed, the variety of ES effects on the cell, and the fact that ES can be ferried through the bloodstream to specific addresses, highlight endothelial cells as an important target for these hormones. Several ES actions are now described on cerebral vessels or endothelial cells. 17beta-estradiol (E2) is known to be a neuroprotective hormone as it protects cultured neurons against a wide range of insults. This effect is also observed in vivo [108]. For example, E2 and a low concentration of tamoxifen promoted cytoprotection of cultured rat cerebral endothelial cells treated with 3-nitropropionic acid, a mycotoxin inducing brain damage [109]. Maintenance of endothelial nitric oxide synthase (eNOS) activity is critical to vascular function, as diminished NO availability contributes to endothelial dysfunction [110]. Along these lines, it is of interest to note that chronic treatment with E2 increased eNOS mRNA and protein levels in vivo [111]. This activation may occur via the phosphoinositolkinase-Akt pathway in human endothelial cells [112, 113].

In addition, E2 modulates glucose transporter-1 protein and mRNA in the BBB endothelium. E2 treatment causes a dose- and time-dependent increase in GLUT-1 protein expression in microvessels, thus facilitating the transport of glucose, the primary fuel of the mammalian brain, across the BBB [114]. Interestingly, E2 inhibits IFN-gamma-induced class II major histocompatibility complex (MHC) expression by a novel mechanism, involving the modification of the histone acetylation status of the class II MHC promoter. One has to keep in mind that aberrant expression of class II MHC is suspected to be one of the factors leading to some autoimmune diseases, such as multiple sclerosis [115].

The presence of ER-alpha has been detected in female rat cerebral vessels and the localization of this receptor in endothelial cells allowed the detection of multiple forms of ER-alpha. Chronic exposure to E2 significantly increased the expression of these forms at the protein level [116]. In addition, HUVECs and human coronary artery cells also contain low levels of ER; and E2 also upregulates the expression of these ER [117]. Recently, using differential display analysis, three other genes were found to be upregulated by E2 in female aortic endothelial cells. Significant increases in mRNA expression (more than 2-fold) were measured for aldose reductase, caspase homologue-alpha protein, and plasminogen activator inhibitor-1 intron e. These genes may be of potential importance for vascular function in human endothelial cells [118]. These few examples demonstrate the importance of endothelial cells as targets for ES action.

13.3.3
Adhesion Molecules are Regulated by Estrogens in Endothelial Cells

One of the most important events in inflammation is clearly the adhesion of leukocytes to the endothelium, followed by their emigration through the BBB. These events are regulated by the precise interactions of complementary adhesion molecules on leukocytes and endothelial cells [23, 30]. For these reasons, understanding the effects of ES on the expression of these adhesion molecules is crucial, to determine whether ES affects the immune response in a suppressive or facilitative way. There are conflicting reports dealing with the effects of ES on the cytokine-induced expression of adhesion molecules in endothelial cells. For example, E-selectin expression is enhanced by E2 at a physiological concentration (10 nM) during the first hours of exposure to TNF-alpha in TNF-alpha-stimulated HUVECs. VCAM-1 expression is also enhanced by E2 at 24 h [119]. However, after longer exposure periods (48 h) and at pharmacological doses, ES decreases the expression of cytokine-induced CAM genes. E2 strongly inhibits (60–80%) E-selectin, ICAM-1, and VCAM-1 induction in HUVECs activated by interleukin-1 [120]. E2 downregulates the E-selectin promoter through either ER-alpha or ER-beta, requiring the NF-kappaB site at positions –94 to –85 within the promoter [121]. These two examples clearly show that the dual effects of ES depend on the time and dose of exposure.

E2 is also able to increase the expression of beta 1, alpha 5 and alpha 6 integrin mRNA in HUVECs, an increase followed by a later enhancement in the surface expression of these integrins [122].

E2, but not the alpha enantiomer, inhibits basal and interleukin-1 beta-stimulated expression of ICAM-1 as well as NF-kappaB activation in immortalized rat brain endothelial cells [123].

VCAM-1 is an important molecule, mediating mononuclear cell adhesion to endothelial cells. In ECV 304 cells transfected with an ER alpha expression plasmid to overexpress this receptor, E2 inhibited the activating effects of TNF-alpha on NF-kappaB activation, VCAM-1 expression, and adhesion of monocytes. These findings suggested that E2 can suppress inflammatory cell adhesion to vascular endothelial cells that possess functional ES receptors [124]. E2 also decreases VCAM-1 expression by an inhibitory effect at the level of the transcription factors NF-kappaB, AP-1, and GATA. This was observed in LPS-induced VCAM-1 expression in human vascular endothelial cells [125]. IL1-beta also induced VCAM-1 expression in HUVECs and this effect could be suppressed by preincubation with E2 (250 pg ml^{-1} or 500 pg ml^{-1}) [126]. 17-Epiestrol, an estrogen metabolite, is more potent than 17-beta E2 in suppressing TNF-alpha-induced VCAM-1 expression, an action modulated at least in part through nitric oxide [127]. These results are summarized in Table 13.3.

13.3.4
Estrogens and Experimental Autoimmune Encephalomyelitis

ES treatment has been shown to protect against EAE, a disease considered as an animal model of MS [128]. Such an effect was first observed after a long-term treatment of EAE in mice with high doses of hormone [129]. More recent data suggest an effect of low-dose therapy: diestrus levels (<100 pg ml^{-1} in serum) of E2 significantly reduce the clinical manifestation of active EAE in both

Table 13.3 Adhesion molecules expressed in endothelial cells (ECV 304: human endothelial-like cells; HBECs: human brain endothelial cells; HUVECs: human umbilical vein endothelial cells; HSVECs: human saphenous vein endothelial cells) and involved in leukocyte recruitment at the BBB. Effects of ES.

Name of cell or cell line	Name of gene	Description	Inducers of gene expression	ES used (effects on gene expression)	Ref.
HBECs, HUVECs	E-selectin, ELAM-1	Selectin, endothelial leukocyte adhesion molecule	TNF-alpha, interleukin-1beta	E2, 17beta-estradiol (downregulation)	109, 117
HUVECs	Integrins ($a5$, $a6$, $\beta1$)	Integrins	no induction	E2 (upregulation)	122
HUVECs	ICAM-1 (CD54)	Intercellular adhesion molecule	Interleukin-1beta	E2 (downregulation)	109, 117, 124
HUVECs	VCAM-1 (CD106)	Vascular cell adhesion molecule	TNF-alpha (24 hours exposure)	E2 (upregulation)	109
ECV 304	VCAM-1 (CD106)	Vascular cell adhesion molecule	TNF-alpha	E2 (downregulation)	124
HUVECs	VCAM-1 (CD106)	Vascular cell adhesion molecule	TNF-alpha	17-epiestrol (downregulation)	127
HUVECs	VCAM-1 (CD106)	Vascular cell adhesion molecule	Interleukin-1beta	E2 (downregulation)	117, 123
HSVECs	VCAM-1 (CD106)	Vascular cell adhesion molecule	LPS	E2 (downregulation)	120

male and female mice [130]. In addition, the mechanism of E2 protection appears to involve both systemic inhibition of TNF-alpha expression and local recruitment of inflammatory cells into the CNS of mice developing EAE [131]. A recent study demonstrated that, although ES is generally assumed to act by modulating immune functions, E2 treatment can inhibit EAE without affecting autoantigen-specific T cell responsiveness and type 1 cytokine production. In fact, the beneficial effect of E2 does not involve ER-alpha signaling in blood-derived cells. These authors suggest a role for ER-alpha expressed in CNS-resident microglia or endothelial cells in the mediation of this protective effect [132]. Nonlymphocytic cells, such as macrophages, dendritic cells, or other nonlymphocytic cells, were suggested to be primarily responsive to E2 treatment in EAE [133].

DNA microarrays have been used to evaluate the effects of E2 on gene expression in EAE and MS. In the case of EAE, E2 treatment affected about 10% of the genes tested, and only 18 cytokine, chemokine/receptor, adhesion molecule, or activation genes were up- or downregulated more than 2.4-fold by this treatment. Thus, it was clearly shown by this study that the ES effect is restricted to several specific genes. In mice splenocytes, CTLA-4 (known to inhibit T cell activation), two interferon gamma-induced genes, TGF-beta3, IL-18, chemokines, VCAM-1, and a disintegrin metalloprotease (thought to regulate TNF-alpha production) were found to be upregulated. In contrast, TNF-alpha, an important proinflammatory cytokine in EAE and MS, RANTES (known to be increased in EAE and MS), and NCAM (increased in cerebrospinal fluid from patients with active MS) were downregulated. In conclusion, this set of unexpected E2-sensitive genes may be of interest for the development of novel strategies for the treatment of EAE and possibly MS [134]. The hormone estriol, which is elevated during pregnancy, has been shown to alleviate EAE when administered in vivo at these levels [135]. Recently, a pilot clinical trial using oral estriol was completed and relapsing-remitting MS patients demonstrated significantly decreased gadolinium enhancing lesions [136]. Accordingly, the immunomodulatory effects of oral estriol therapy were assessed and significant increases in the levels of IL-5 and IL-10 and a decrease in TNF-alpha were observed in stimulated peripheral blood mononuclear cells [137].

New genes involved in MS were also found: increased transcripts of genes encoding inflammatory cytokines, particularly IL-6 and IL-17, interferon-gamma, and associated downstream pathways, were observed by microarray analysis of MS lesions [138]. Some products of these genes have been chosen as targets for EAE therapy in mice. For example, when granulocyte colony-stimulating factor (G-CSF) was given before the onset of EAE, its influence is essentially exerted on the acute rather than the chronic stage of the disease. The reversal of EAE with G-CSF has been reported [139]. The EAE result is corroborated with the transcriptional analysis of active and silent MS lesions: G-CSF is upregulated in acute, but not in chronic MS lesions and the effect on EAE is more pronounced in the acute phase of the disease [138].

13.4
Conclusions and Perspectives

The activated cerebrovascular endothelium is an important target for both GC and ES. By regulating the expression of genes involved in leukocyte-endothelial cell interactions, these hormones can play a crucial role in the modulation of recruitment of activated leukocytes during chronic inflammatory diseases. In addition, this activated endothelium is an attractive target for pharmacological intervention in order to inhibit endothelial cell activation as well as the consequent recruitment of activated leukocytes in order to improve the therapy of such diseases. GC are commonly used in anti-inflammatory therapy. In particular, acute relapses in MS are treated with GC. GC-Dependent suppression of the expression of the genes coding CAM is now an accepted mechanism of action of these hormones. In the case of ICAM-1 and ICAM-2, which have been identified as essential regulators of transendothelial migration of autoaggressive T cells, GC can modulate CAM expression by the repression of NF-kappaB signaling [58, 61]. Elevated levels of ICAM-1 have been described in MS during relapses [140], and conversely, so have ICAM-1 decreases in MS during remission or following treatment with high-dose GC therapy [141]. However, long-term therapies using

GC are usually accompanied by many severe side-effects: atrophy of the skin, myopathy, osteoporosis, and psychosis. To improve this therapy, new GR ligands have been synthesized with highly potent immunosuppressive and anti-inflammatory properties, but reduced side-effects. These new molecules (dissociated GC) have been shown to possess anti-inflammatory and immunosuppressive activity as potent as the glucocorticoid prednisolone in two in vivo models. These dissociated GC strongly inhibit AP-1, but have little or no transactivating activity. These results have provided a novel concept for drug discovery [142].

Another possibility for the selective inhibition of endothelial cell activation is the development of selective immunoconjugates like the DEX-anti-E-selectin immunoconjugate, which is successfully internalized by activated endothelial cells [98, 99]. The intercellularly delivered DEX is then fully able to exert its pleiotropic anti-inflammatory activity [143]. In the future, analogous strategies may also prove useful for delivering dissociated GCs to inhibit endothelial cell activation.

Acknowledgment

I thank Dr. Nancy Grant for critical reading of the manuscript.

References

1 Danielsen, M. **1995**, Structure and function of the glucocorticoid receptor, in *Nuclear Hormone Receptors*, ed. Parker M.G., Academic Press, London, pp. 39–78.
2 Schaaf, M.J.M., Cidlowski, J.A. **2003**, Molecular mechanisms of glucocorticoid action and resistance, *J. Steroid Biochem. Mol. Biol.* 83, 37–48.
3 Sawada, N., Murata, M., Kikuchi, K., Osanai, M., Tobioka, H., Kojima, T., Chiba, H. **2003**, Tight junctions and human diseases, *Med. Electron. Microsc.* 36, 147–156.
4 Ballabh, P., Braun, A., Nedergaard, M. **2004**, The blood-brain barrier: an overview. Structure, regulation, and clinical implications, *Neurobiol. Dis.* 16, 1–13.
5 Green, S., Chambon, P. **1991**, The oestrogen receptor: from perception to mechanism, in *Nuclear Hormone Receptors*, ed. Parker M.G., Academic Press, London, pp. 15–38.
6 Truss, M., Beato, M. **1993**, Steroid hormone receptor: interaction with deoxyribonucleic acid and transcription factors, *Endocr. Rev.* 14, 459–479.
7 De Bosscher, K., Vanden Bergh, W., Haegeman, G. **2000**, Mechanisms of anti-inflammatory action and of immunosuppression by glucocorticoids: negative interference of activated glucocorticoid receptor with transcription factors, *J. Neuroimmunol.* 109, 16–22.
8 Hayashi, R., Wada, H., Ito, K., Adcock, I.M. **2004**, Effects of glucocorticoids on gene transcription, *Eur. J. Pharmacol.* 500, 51–62.

9 Barnes, P. J. **1998**, Anti-inflammatory actions of glucocorticoids: molecular mechanisms, *Clin. Sci.* 94, 557–572.

10 Pitzalis, C., Pipitone, N., Peretti, M. **2002**, Regulation of leukocyte-endothelial interactions by glucocorticoids, *Ann. N.Y. Acad. Sci.* 966, 101–118.

11 Almawi, W. Y., Melemedjian, O. K. **2002**, Negative regulation of nuclear factor-kappaB activation and function by glucocorticoids, *J. Mol. Endocrinol.* 28, 69–78.

12 Baldwin, A. S. **1996**, The NF-κB and IκB proteins: new discoveries and insights, *Annu. Rev. Immunol.* 14, 74–80.

13 Catron, K. M., Brickwood, J. R., Shang, C., Li, Y., Shannon, M. F., Parks, T. P. **1995**, Cooperative binding and synergistic activation by RelA and C/EBPbeta on the intercellular adhesion molecule-1 promoter, *Cell Growth Differ.* 9, 949–959.

14 Ray, K. P., Farrow, S., Daly, M., Talabot, F., Searle, N. **1997**, Induction of the E-selectin promoter by interleukin 1 and tumor necrosis factor alpha, and inhibition by glucocorticoids, *Biochem. J.* 328, 707–715.

15 Wong, D., Dorovini-Zis, K. **1996**, Regulation by cytokines and lipopolysaccharide of E-selectin expression by human brain microvessel endothelial cells in primary culture, *J. Neuropathol. Exp. Med.* 55, 225–235.

16 Tronche, F., Kellendonk, C., Reichardt, M., Schütz, G. **1998**, Genetic dissection of glucocorticoid receptor function in mice, *Curr. Opin. Gen. Dev.* 8, 532–538.

17 Norman, A., Mizwicki, M. T., Norman, D. P. **2004**, Steroid-hormone rapid actions, membrane receptors and a conformation ensemble model, *Nat. Rev. Drug Discov.* 3, 27–41.

18 Kapajos, J. J., van den Berg, A., Borghuis, T., Banas, B., Huitema, S., Poelstra, K., Bakker, W. W. **2004**, Enhanced ecto-pyrase activity of stimulated endothelial or mesangial cells is downregulated by glucocorticoids in vitro, *Eur. J. Pharmacol.* 501, 191–198.

19 Solito, E., Mulla, A., Morris, J. F., Christian, H. C., Flower, R. J., Buckingham J. C. **2003**, Dexamethasone induces rapid serine-phosphorylation and membrane translocation of annexin 1 in a human folliculostellate cell line via a novel nongenomic mechanism involving the glucocorticoid receptor, protein kinase C, phosphatidylinositol 3-kinase, and mitogen-activated protein kinase, *Endocrinology* 144, 1164–1174.

20 Metchnikoff, E. **1892**, *Leçons sur la Pathologie Comparée de l'Inflammation*, Masson, Paris, reissued as: Metchnikoff, E. **1968**, *Lectures on Comparative Pathology of Inflammation*, Dover, New York.

21 Walzog, B., Gaehtgens, P. **2000**, Adhesion molecules: the path to the new understanding of acute inflammation, *News Physiol. Sci.* 15, 107–113.

22 Springer, T. A. **1995**, Traffic signals on endothelium for lymphocyte recirculation and leukocyte emigration, *Annu. Rev. Physiol.* 57, 827–872.

23 Zimmerman, G. A., McIntyre, T. M., Mehra, M., Prescott, S. M. **1996**, Endothelial cell-associated platelet-activating factor: a novel mechanism for signalling intercellular adhesion, *J. Cell. Biol.* 110, 529–540.

24 Ulbrich, H., Eriksson, E. E., Lindbom, L. **2003**, Leukocyte and endothelial cell adhesion molecules as targets for therapeutic interventions in inflammatory disease, *Trends Pharmacol. Sci.* 24, 640–647.

25 Cato, A. C. B., Wade, E. **1996**, Molecular mechanisms of anti-inflammatory response, *Bioessays* 18, 371–378.

26 Wikström, A. C. **2003**, Glucocorticoid action and novel mechanisms of steroid resistance: role of glucocorticoid receptor-interacting proteins for glucocorticoid responsiveness, *J. Endocrinol.* 178, 331–337.

27 Radomski, M. W., Palmer, R. M. J., Moncada, S. **1990**, Glucocorticoids inhibit the expression of an inducible, but not the constitutive, nitric oxide synthase in vascular endothelial cells, *Proc. Natl Acad. Sci. USA* 87, 10043–10047.

28 MacMicking, J. D., Nathan, C., Hom, G., Chartrain, N., Fletcher, D. S., Trumbauer, M., Stevens, K., Xie, Q. W., Sokol, K., Hutchison, N. **1995**, Altered responses to bacterial infection and endotoxic shock in mice lacking inducible nitric oxide synthase, *Cell* 81, 641–650.

29 Mitchell, J. A., Belvisi, M. G., Akarasereenont, P., Robbins, R. A., Known, O. J., Croxtall, J., Barnes, P. J., Vane, J. R. **1994**, Induction of cyclo-oxygenase-2 by cytokines in human pulmonary epithelial cells: regulation by dexamethasone, *Br. J. Pharmacol.* 113, 1008–1014.

30 Muller, W. A. **2002**, Leukocyte-endothelial cell interactions in the inflammatory response, *Lab. Invest.* 82, 521–533.

31 Muller, W. A. **2003**, Leukocyte-endothelial cell interactions in leukocyte transmigration and the inflammatory response, *Trends Immunol.* 24, 326–333.

32 Harlan, J. M., Winn, R. K. **2002**, Leukocyte-endothelial interactions: clinical trials of anti-adhesion therapy, *Crit. Care Med.* 30, S214–S219.

33 Campbell, J. J., Butcher, E. C. **1998**, Chemokines in tissue-specific and microenvironment specific lymphocyte homing, *Curr. Opin. Immunol.* 12, 336–341.

34 Muller, W. A., Wigl, S. A., Deng, X., Phillips, D. M. **1993**, PECAM-1 is required for transendothelial migration of leukocytes, *J. Exp. Med.* 178, 449–460.

35 Berman, M. E., Xie, Y., Muller, W. A. **1996**, Roles of platelet/endothelial cell adhesion molecule-1 (PECAM-1) in natural killer cell transendothelial migration and $\beta 2$ integrin activation, *J. Immunol.* 156, 1515–1524.

36 Liao, F., Huynh, H. K., Eiroa, A., Greene, T., Polizzi, E., Muller, W. **1997**, Migration of monocytes across endothelium and passage through extracellular matrix involve separate molecular domains of PECAM-1, *J. Exp. Med.* 182, 1337–1343.

37 Cinamon, G., Shinder, V., Alon, R. **2001**, Shear forces promote lymphocyte migration across vascular endothelium bearing apical chemokines, *Nat. Immunol.* 2, 515–522.

38 Schenkel, A. R., Mahmdouh, Z., Chen, X., Liebman, R. M., Muller, W. M. **2002**, CD99 plays a major role in the migration of monocytes through endothelial junctions, *Nat. Immunol.* 3, 143–150.

39 Ostermann, G., Weber, K. S. G., Zernecke, A., Schroder, A, Weber, C. **2002**, JAM-1 is a ligand of the beta(2) integrin LFA-1 involved in transendothelial migration of leukocytes, *Nat. Immunol.* 3, 151–158.

40 Aurrand-Lions, M., Johnson-Leger, C., Imhof, B. A. **2002**, The last molecular fortress in leukocyte trans-endothelial migration, *Nat. Immunol.* 3, 116–118.

41 Brostjan, C., Anrather, J., Csizmadia, V., Natarajan, G., Winkler, H. **1997**, Glucocorticoids inhibit E-selectin expression by targeting NF-kappaB and not ATF/c-jun, *J. Immunol.* 158, 3836–3844.

42 Grober, J. S., Bowden, B. L., Ebling, H., Athey, B., Thompson, C. B., Fox, D. A., Stoolman, L. M. **1993**, Monocyte-endothelial adhesion in chronic rheumatoid arthritis: in situ detection of selectin and integrin-dependent interactions, *J. Clin. Invest.* 91, 2609–2619.

43 Johnson-Tidey, R. R., McGregor, J. L., Taylor, P. R., Poston, R. N. **1994**, Increase in the adhesion molecule P-selectin in the endothelium overlying atherosclerotic plaques, *Am. J. Pathol.* 144, 952–961.

44 Xia, L., Pan, J., Yao, L., McEver, R. P. **1998**, A proteasome inhibitor, an antioxydant, or a salicylate, but not a glucocorticoid, blocks constitutive and cytokine-induced expression of P-selectin in human endothelial cells, *Blood* 91, 162–163.

45 De Coupade, C., Solito, E., Levine, J. D. **2003**, Dexamethasone enhances interaction of endogenous annexin 1 with L-selectin and triggers shedding of L-selectin in the monocytic cell line U-397, *Br. J. Pharmacol.* 140, 133–145.

46 Cronstein, B. N., Kimmel, S. C., Levin, R. I., Martiniuk, F., Weissmann, G. **1992**, A mechanism for the antiinflammatory effects of corticosteroids: the glucocorticoid receptor regulates leukocyte adhesion to endothelial cells and expression of endothelial-leukocyte adhesion molecule-1 and intercellular adhesion molecule-1, *Proc. Natl Acad. Sci. USA* 89, 9991–9995.

47 Aziz, K. E., Wakefield, D. **1996**, Modulation of endothelial cell expression of ICAM-1, E-selectin, and VCAM-1 by beta-estradiol, progesterone and dexamethasone, *Cell. Immunol.* 167, 79–85.

48 Burke-Gaffney, A., Hellewell, P. **1996**, Regulation of ICAM-1 by dexamethasone in a human vascular endothelial cell line EAhy926, *Am. J. Physiol.* 270, C552–C561.

49 Tan, K. H., Dobbie, M. S., Felix, R. A., Barrand, M. A., Hurst, R. D. **2001**, A comparison of the induction of immortalized endothelial cell impermeability by astrocytes, *NeuroReport* 12, 1329–1334.

50 Dietrich, J. B., Zaepfel, M., Kuchler-Bopp, S. **1999**, Dexamethasone represses 3,5,5′-triiodothyronine-stimulated expression of intercellular adhesion molecule-1 in the human cell line ECV 304, *Cell. Biol. Toxicol.* 15, 269–277.

51 Dufour, A., Corsini, E., Gelati, M., Ciusani, E., Zaffaroni, M., Giombini, S., Massa, G., Salmaggi, A. **1998**, Modulation of ICAM-1, VCAM-1 and HLA-DR by cytokines and steroids on HUVECs and human brain endothelial cells, *J. Neurol. Sci.* 157, 117–121.

52 Gelati, M., Corsini, E., Dufour, A., Massa, G., Giombini, S., Solero, C. L., Salmaggi, A. **2000**, High-dose methylprednisolone reduces cytokine-induced

adhesion molecules on human brain endothelial cells, *Can. J. Neurol.* 27, 241–244.

53 Droogan, A. G., Crockard, A. D., McMillan, S. A., Hawkins, S. A. **1998**, Effects of intravenous methylprednisolone therapy on leukocyte and soluble adhesion molecule expression, *Neurology* 50, 224–229.

54 Radi, Z. A., Kehrli Jr, M. R., Ackermann, M. R. **2001**, Cell adhesion molecules, leukocyte trafficking, and strategies to reduce leukocyte infiltration, *J. Vet. Intern. Med.* 15, 516–529.

55 Persidsky, Y. **1999**, Model systems for the studies of leukocyte migration across the blood-brain barrier, *J. Neurovirol.* 5, 579–590.

56 Gaillard, P. J., Voorwinden, L. H., Nielsen, J. L., Ivanov, A., Atsumi, R., Engman, H., Ringbom, C., de Boer, A. G., Breimer, D. D. **2001**, Establishment and functional characterization of an in vitro model of the blood-brain barrier, comprising a co-culture of brain capillary endothelial cells and astrocytes, *Eur. J. Pharmacol.* 12, 215–222.

57 Marshall, D., Haskard, D. O. **2002**, Clinical overview of leukocyte adhesion and migration: where are we now? *Semin. Immunol.* 14, 133–140.

58 Engelhardt, B. **2000**, Role of glucocorticoids on T cell recruitment across the blood-brain barrier, *Z. Rheumatol.* 59 [Suppl. 2], II/18–II/21.

59 Kraus, J., Ochsmann, P., Engelhardt, B., Bauer, R., Kern, A., Traupe, H., Dorndorf, W. **1998**, Soluble and cell surface ICAM-3 in blood and cerebrospinal fluid of patients with multiple sclerosis: influence of methylprednisolone treatment and relevance as markers for disease activity, *Acta Neurol. Scand.* 101, 135–139.

60 Kraus, J., Ochsmann, P., Engelhardt, B., Schiel, C., Hornig, C., Bauer, R., Kern, A., Traupe, H., Dorndorf, W. **2000**, Soluble and cell surface ICAM-1 in blood and cerebrospinal fluid as markers for disease activity in multiple sclerosis, *Acta Neurol. Scand.* 98, 102–109.

61 Reiss, Y., Hoch, G., Deutsch, U., Engelhardt, B. **1998**, T cell interaction with ICAM-1 deficient endothelium in vitro: requisite role for ICAM-1 and VCAM-1 in transendothelial migration of T cells, *Eur. J. Immunol.* 28, 3086–3099.

62 Carpen, O., Pallai, P., Staunton, D. E., Springer T. A. **1992**, Association of intercellular adhesion molecule-1 (ICAM-1) with actin-containing cytoskeleton and alpha-actinin, *J. Cell. Biol.* 118, 1223–1234.

63 Vaheri, A., Carpen, O., Heiska, L., Helander, T. S., Jaaskelainen, J., Majander-Nordenswan, P., Sainio, M., Timonen, T., Turunen, O. **1997**, The ezrin protein family: membrane-cytoskeleton interactions and diseases association, *Curr. Opin. Cell. Biol.* 9, 659–666.

64 Helander, T. S., Carpen, O., Turunen, O., Kovanen, P., Vaheri, A., Timonen, T. **1996**, ICAM-2 redistributed by ezrin as a target for killer cells, *Nature* 382, 265–268.

65 Smith, M. D., Ahern, M. J., Brooks, P. M., Roberts-Thomson, P. J. **1988**, The clinical and immunological effects of pulse methylprednisolone therapy in

rheumatoid arthritis. III. Effects on immune and inflammatory indices in synovial fluid, *J. Rheumatol.* 15, 238–241.

66 Gelati, M., Corsini, E., Dufour, A., Ciusani, E., Massa, G., Frigiero, S., Milanes, C., Nespolo, A., Salmaggi, A. **1997**, Reduced adhesion of PBMNCs to endothelium in methylprednisolone-treated MS patients: preliminary results, *Acta Neurol. Scand.* 96, 283–292.

67 Pitzalis, C., Pipitone, N., Peretti, M. **2002**, Regulation of leukocyte-endothelial interactions by glucocorticoids, *Ann. N.Y. Acad. Sci.* 966, 101–118.

68 Pitzalis, C., Pipitone, N., Bajocchi, G., Hall, N., Goulding, N., Lee, G., Kingsley G., Lanchbury, J., Panayi, G. **1997**, Corticosteroids inhibit lymphocyte binding to endothelium and intercellular adhesion: an additional mechanism for their anti-inflammatory and immunosuppressive effect, *J. Immunol.* 158, 5007–5016.

69 Euzger, E., Flower, R.J., Goulding, N.J., Perretti, M. **1999**, Differential modulation of annexin I binding sites on monocytes and neutrophils, *Med. Inflamm.* 8, 53–62.

70 Pitzalis, C., Pipitone, N., Peretti, M. **2001**, Glucocorticoids and leukocyte adhesion, in *Glucocorticoids*, vol. 1, eds. N.J. Goulding, R.J. Flower, Birkhäuser, Basel, pp. 105–118.

71 Goulding, N.J., Ougsbourn, S., Pipitone, N., Biagini, P., Gerli, R., Pitzalis, C. **1999**, The inhibitory effect of dexamethasone on lymphocyte adhesion molecule expression and intercellular aggregation is not mediated by lipocortin 1, *Clin. Exp. Immunol.* 118, 376–383.

72 Goulding, N.J., Flower, R.J. **2001**, Glucocorticoid biology – a molecular maze and clinical challenge, in *Glucocorticoids*, vol. 1, eds. N.J. Goulding, R.J. Flower, Birkhäuser, Basel, pp. 119–127.

73 Pipitone, N., Sinha, M., Theodoridis, N., Goulding, N., Hall, M., Lanchbury, J., Corrigall, V., Panayi, G., Pitzalis, C. **2001**, The glucocorticoid inhibition of LFA-1 and CD2 expression in human mononuclear cells is reversed by IL-1, IL-7 and IL-15, *Eur. J. Immunol.* 31, 2135–2142.

74 Stanimirovic, D.B., McCarron, R.M., Spatz, M. **1994**, Dexamethasone downregulates endothelin receptors in human cerebromicrovascular endothelial cells, *Neuropeptides* 26, 145–152.

75 Newton, C.J., Ran, G., Xie, Y.X., Bilko, D., Burgoyne, C.H., Adams, I., Abidia, A., McCollum, P.T., Atkins, S.L. **2002**, Statin-induced apoptosis of vascular endothelial cells is blocked by dexamethasone, *J. Endocrinol.* 174, 7–16.

76 Karlstedt, K., Sallmen, T., Eriksson, K.S., Lintunen, M., Couraud, P.O., Joo, F., Panula, P. **1999**, Lack of histamine synthesis and down-regulation of H1 and H2 receptor mRNA levels by dexamethasone in cerebral endothelial cells, *J. Cereb. Blood Flow Metab.* 9, 321–330.

77 Romero, I.A., Radewicz, K., Jubin, E., Michel, C.C., Greenwood, J., Couraud, P.O., Adamson, P. **2003**, Changes in cytoskeletal and tight junctional proteins correlate with decreased permeability induced by dexamethasone in cultured rat brain endothelial cells, *Neurosci. Lett.* 344, 112–116.

78 Prat, A., Biernacki, K., Lavoie, J.-F., Poirier, J., Duquette, P., Antel, J. P. **2002**, Migration of multiple sclerosis lymphocytes through brain endothelium, *Arch. Neurol.* 59, 391–397.

79 Simpson, J. E., Newcombe, J., Cuzner, M. L., Woodroofe, M. N. **1998**, Expression of beta-chemokines RANTES and MIP-beta by human brain microvessel endothelial cells in primary culture, *J. Neuroimmunol.* 84, 238–249.

80 Ransohoff, R. M. **1999**, Mechanisms of inflammation in MS tissue: adhesion molecules and chemokines, *J. Neuroimmunol.* 134, 57–68.

81 Seguin, R., Biernacki, K., Rotondo, R. L., Antel, J. P. **2003**, Regulation and functional effects of monocyte migration across human brain-derived endothelial cells, *J. Neuropathol. Exp. Neurol.* 62, 412–419.

82 Andjelkovic, A. V., Pachter, J. S. **2000**, Central nervous system endothelium in neuroinflammatory, neuroinfectious and neurodegenerative disease, *J. Neurosci. Res.* 51, 423–430.

83 Dzenko, K. A., Andjelkovic, A. V., Kuziel, W. A., Pachter, J. S. **2001**, The chemokine receptor CCR2 mediates the binding and internalization of monocyte chemoattractant protein-1 along brain microvessels, *J. Neurosci.* 21, 9214–9223.

84 Song, L., Pachter, J. S. **2004**, Monocyte chemoattractant protein-1 alters expression of tight junction-associated proteins in brain microvascular endothelial cells, *Microvasc. Res.* 67, 78–89.

85 Harkness, K. A., Sussman, J. D., Davies-Jones, G. A., Greenwood, J., Woodroofe, M. N. **2003**, Cytokine regulation of MCP-1 expression in brain and retinal microvascular endothelial cells, *J. Neuroimmunol.* 142, 1–9.

86 Harkness, K. A., Adamson, P., Sussman, J. D., Davies-Jones, G. A., Greenwood, J., Woodroofe, M.N. **2000**, Dexamethasone regulation of matrix metalloprotease expression in CNS vascular endothelium, *Brain* 123, 698–709.

87 Rosenberg, G. A., Dencoff, B. S., Correa, N., Reiners, M., Ford, C. C. **1996**, Effects of steroids on CSF matrix metalloproteinases in multiple sclerosis: relation to blood-brain barrier injury, *Neurology* 46, 1626–1632.

88 Gold, R., Buttgereit, F., Toyka, K. V. **2001**, Mechanism of glucocorticosteroid hormones: possible implications for therapy of neuroimmunological disorders, *J. Neuroimmunol.* 117, 1–8.

89 Schmidt, J., Gold, R., Schönrock, L., Zettl, U. K., Hartung, H. P., Toyka, K. V. **2000**, T-cell apoptosis in situ in experimental autoimmune encephalomyelitis following methylprednisolone pulse therapy, *Brain* 123, 1431–1441.

90 Leussink, V., Jung, S., Merschdorf, U., Toyka, K. V., Gold, R. **2001**, High-dose methylprednisolone therapy in multiple sclerosis induces apoptosis in peripheral blood leukocytes, *Arch. Neurol.* 58, 91–97.

91 Corsini, E., Gelati, M., Dufour, A., Massa, G., Nespolo, A., Ciusani, E., Milanese, C., La Mantia, L., Salmaggi, A. **1997**, Effects of beta-IFN-1b treatment in MS patients on adhesion between PBMNCs, HUVECs and MS-HBECs: an in vivo and in vitro study, *J. Neuroimmunol.* 79, 76–83.

92 Schluep, M., Bogousslavsky, J. **1997**, Emerging treatments in multiple sclerosis, *Eur. Neurol.* 38, 216–221.

93 Minagar, A., Long, A., Ma, T., et al. **2003**, Interferon (IFN)-beta 1a and IFN-beta 1b block-IFN-gamma-induced disintegration of endothelial junction integrity and barrier. *Endothelium* 10, 299–307.

94 Kraus, J., Ling, A. K., Hamm, S., Voigt, K., Ochsmann, P., Engelhardt, B. **2004**, Interferon-beta stabilizes barrier characteristics of brain endothelial cells in vitro, *Ann. Neurol.* 56, 192–205.

95 Gaillard, P. J., van der Meide, P. H., de Boer, A. G., Breiner, D. D. **2001**, Glucocorticoid and type I interferon interactions at the blood-brain barrier: relevance for drug therapies for multiple sclerosis, *NeuroReport* 12, 2189–2193.

96 Miller, W. L., Blake Tyrrel, J. **1995**, The adrenal cortex, in *Endocrinology and Metabolism*, eds. Felig, P., Baxter J. D., Frohmann, L. A., McGraw-Hill, New York, pp. 555–711.

97 Reichardt, H. M., Tuckermann, J. P., Bauer, A., Schütz, G. **2000**, Molecular genetic dissection of glucocorticoid receptor function in vivo, *Z. Rheumatol.* 59 [Suppl. 2], II/1–II/5.

98 Everts, M., Kok, J., Asgeirsdottir, S. A., Melgert, B. N., Moolenaar, T. J., Koning, G. A., van Luyn, M. J., Meijer, D. K., Molema, G. **2002**, Selective intracellular delivery of dexamethasone into activated endothelial cells using an E-selectin-directed immunoconjugate, *J. Immunol.* 168, 883–889.

99 Asgeirsdottir, S. A., Kok, R. J., Everts, M., Meijer, D. K. F., Molema, G. **2003**, Delivery of pharmacologically active dexamethasone into activated endothelial cells by dexamethasone-anti-E-selectin immunoconjugate, *Biochem. Pharmacol.* 65, 1729–1739.

100 Hall, J. M., Couse, J. F., Korach, K. M. **2001**, The multifaced mechanisms of estradiol and estrogen receptor signaling, *J. Biol. Chem.* 276, 36869–36872.

101 Nadal, A., Diaz, M., Valverde, M. A. **2001**, The estrogen trinity: membrane, cytosolic and nuclear effects, *News Physiol. Sci.* 16, 251–255.

102 Mendelsohn, M. E. **2002**, Genomic and nongenomic effects of estrogen in the vasculature, *Am. J. Cardiol.* 90 [suppl.], 3F–6F.

103 Simoncini, T., Mannella, P., Fornari, L., Caruso, A., Varone, G., Genazzani, A. R. **2004**, Genomic and non-genomic effects of estrogens on endothelial cells, *Steroids* 69, 537–542.

104 Paech, K., Webb, P., Kuiper, G. G. J. M., Nilsson, S., Gustafsson, J.-A., Kushner, P. J., Scanlan, T. S. **1997**, Differential ligand activation of estrogen receptors ER alpha and ER beta at AP1 sites, *Science* 277, 1508–1510.

105 Lee, S. J., McEwen, B. S. **2001**, Neurotrophic and neuroprotective actions of estrogens and their therapeutic implications, *Annu. Rev. Pharmacol. Toxicol.* 41, 569–591.

106 Deecher, D. C., Swiggard, P., Frail, D. E., O'Connor, L. T. **2003**, Characterization of a membrane-associated estrogen in a rat hypothalamic cell line (D12), *Endocrine* 22, 211–223.

107 Gilligan, D. M., Quyyumi, A. A., Cannon III, R. O. **1994**, Effects of physiological levels of estrogen on coronary vasomotor function in postmenopausal women, *Circulation* 89, 2545–2551.

108 Behl, C. **2002**, Oestrogen as neuroprotective hormone, *Nat. Rev. Neurosci.* 3, 433–442.

109 Mogami, M., Hida, H., Hayashi, Y., Kohri, K., Kodama, Y., Jung, C.G., Nishino, H. **2002**, Estrogen blocks 3-nitropropionic acid-induced [Ca^{2+}] increase and cell damage in cultured rat cerebral endothelial cells, *Brain Res.* 956, 116–125.

110 Shaul, P.W. **2002**, Regulation of endothelial nitric oxide synthase: location, location, location, *Annu. Rev. Physiol.* 64, 749–774.

111 Stirone, C., Chu, Y., Sunday, L., Duckles, S.P., Krause, D.N. **2003**, 17-*β*-estradiol increases endothelial nitric oxide synthase mRNA copy number in cerebral blood vessels: quantification by real-time polymerase chain reaction, *Eur. J. Pharmacol.* 478, 35–38.

112 Haynes, M.P., Sinha, D., Russell, K.S., Collinge, M., Fulton, D., Morales-Ruiz M., Sessa, W.C., Bender, J.R. **2000**, Membrane estrogen receptor engagement activates endothelial nitric oxyde synthase via the PI3-Akt pathway in human endothelial cells, *Circ. Res.* 87, 677–682.

113 Xu, R., Sowers, J.R., Skafar, D.F., Ram, J.L. **2001**, Hydrocortisone modulates the effect of estradiol on endothelial nitric oxide synthase expression in human endothelial cells, *Life Sci.* 69, 2811–2817.

114 Shi, J., Simpkins, J.W. **1997**, 17beta-estradiol modulation of glucose transporter expression in blood-brain barrier, *Am. J. Physiol.* 272, E1016–E1022.

115 Adamski, J., Ma, Z., Nozell S., Benveniste, E.M. **2004**, 17-*β*-estradiol inhibits class II major histocompatibility complex (MHC) expression: influence on histone modifications and CBP recruitment to the class II MHC promoter, *Mol. Endocrinol.* 18, 1963–1974.

116 Stirone, C., Duckles, S.P., Krause, D.N. **2003**, Multiple forms of estrogen receptor-alpha in cerebral blood vessels: regulation by estrogen, *Am. J. Physiol. Endocrinol. Metab.* 284, E184–E192.

117 Kim-Schulze, S., McGowan, K.A., Hubchak, S.C., Cid, M.B., Kleinman, G.L., Greene, G.L., Schnaper, H.W. **1996**, Expression of an estrogen receptor by human coronary artery and umbilical vein endothelial cells, *Circulation* 94, 1402–1407.

118 Villablanca, A.C., Lewis, K.A., Rutledge, J.C. **2002**, Time- and dose-dependent differential upregulation of three genes by 17beta-estradiol in endothelial cells, *J. Appl. Physiol.* 92, 1064–1073.

119 Cid, M.C., Kleinman, H.K., Grant, D.S., Schnaper, H.W., Fauci, A.S., Hoffman, G.S. **1994**, Estrogens and the vascular endothelium, *J. Clin. Invest.* 93, 17–25.

120 Caulin-Glaser, T., Watson, C.A., Pardi, R., Bender, J.R. **1996**, Effects of 17beta-estradiol on cytokine-induced endothelial cell adhesion molecule expression, *J. Clin. Invest.* 98, 36–42.

121 Tyree, C.M., Zou, A., Allegretto, E.A. **2002**, 17beta-estradiol inhibits cytokine induction of the human E-selectin promoter, *J. Steroid Biochem. Mol. Biol.* 80, 291–297.

122 Cid, M.C., Esparza, J., Schnaper, W.H., Juan, M., Yague, J., Grant, D.S., Urbano-Marquez, A., Hoffman, G.S., Kleinman, H.K. **1999**, Estradiol enhances endothelial cell interactions with extracellular matrix proteins via an increase in integrin expression and function, *Angiogenesis* 3, 271–280.

123 Galea, E., Santizo, R., Feinstein, D.L., Adamson, P., Greenwood, J., Koenig, H.M., Pelligrino, D.A. **2002**, Estrogen inhibits NFkappaB-dependent inflammation in brain endothelium without interfering with IkappaB degradation, *NeuroReport* 13, 1469–1472.

124 Mori, M., Tsukahara, F., Yoshioka, T., Irie, K., Ohta, H. **2004**, Suppression by 17beta-estradiol of monocyte adhesion to vascular endothelial cells is mediated by estrogen receptors, *Life Sci.* 75, 599–609.

125 Simoncini, T., Maffei, S., Basta, G., Barsacchi, G., Genazzani, A.R., Liao, J.K., De Caterina, R. **2000**, Estrogens and glucocorticods inhibit endothelial vascular cell adhesion molecule-1 expression by different transcriptional mechanisms, *Circ. Res.* 87, 19–25.

126 Nakai, K., Itoh, C., Hotta, K., Yoshizumi, M., Hiramori, K. **1994**, Estradiol-17beta regulates the induction of VCAM-1 mRNA expression by interleukin-1beta in human umbilical vein endothelial cells, *Life Sci.* 54, PL221–PL227.

127 Mukherjee, T.K., Nathan, L., Dinh, H., Reddy, S.T., Chaudhuri, G. **2003**, 17-Epiestrol, an estrogen metabolite, is more potent than estradiol in inhibiting vascular cell adhesion molecule 1 (VCAM-1) mRNA expression, *J. Biol. Chem.* 278, 11746–11752.

128 Offner, H. **2004**, Neuroimmunoprotective effects of estrogen and derivatives in experimental autoimmune encephalomyelitis: therapeutic implications for multiple sclerosis, *J. Neurosci. Res.* 78, 603–624.

129 Jansson, L., Olson, T., Holmdahl, R. **1994**, Estrogen induces a potent suppression of experimental autoimmune encephalomyelitis and collagen-induced arthritis in mice, *J. Neuroimmunol.* 53, 203–207.

130 Bebo, B.F., Fyfe-Johnson, A., Adlard, K., Beam, A.G., Vandenbark, A.A., Offner, H. **2001**, Low-dose estrogen therapy ameliorates experimental autoimmune encephalomyelitis in two different inbred mouse strains, *J. Immunol.* 166, 2080–2089.

131 Ito, A., Buenafe, A.C., Matejuk, A., Zamora, M., Silverman, M., Dwyer, J., Vandenbark, A.A., Offner, H. **2002**, Estrogen treatment down-regulates TNF-alpha production and reduces the severity of experimental autoimmune encephalomyelitis in cytokine knockout mice, *J. Immunol.* 167, 542–552.

132 Garridou, L., Laffont, S., Douin-Echinard, V., Coureau, C., Krust, A., Chambon, P., Guery, C. **2004**, Estrogen receptor alpha signaling in inflammatory leukocytes is dispensable for 17beta-estradiol-mediated inhibition of experimental autoimmune encephalomyelitis, *J. Immunol.* 173, 2345–2342.

133 Polanczyk, M.J., Jones, R.E., Subramanian, S., Afentoulis, M., Rich, C., Zakroczymski, M., Cooke, P., Vandenbark, A.A., Offner, H. **2004**, T lymphocytes do not directly mediate the protective effects of estrogen on experimental autoimmune encephalomyelitis, *Am. J. Pathol.* 165, 2069–2077.

14.2
Metalloproteinases in Brain Microvessels: Types and Functions

Metalloproteinases (MPs), often referred to as metalloproteases or metallopeptidases, are proteolytic enzymes which use a metal ion for their catalytic activities. Extracellular MPs are crucial for some cell functions entailing complex roles at the cell surface. MPs located at the BBB comprise aminopeptidase A (EC 3.4.11.7) [3], aminopeptidase N (EC 3.4.11.2, also known as aminopeptidase M) [4], carboxypeptidase N [5], angiotensin-converting enzyme (ACE, EC 3.4.15.1) [6], low levels of the neutral endopeptidase 24.11 (enkephalinase, NEP, EC 3.4.24.11) [7], the serine peptidases dipeptidyl peptidase II and IV (DPP) [2, 8], and matrix metalloproteinases [9] (see Table 14.1). Most of them have a catalytic domain with a zinc-binding motif, others contain a cobalt ion.

The aminopeptidases are ectoenzymes, which cleave either the first or the second peptide bond within a protein to release the N-terminal amino acid which consists of a dipeptide. Generally they are Zn^{2+}-dependent, although aminopeptidase A is activated by Ca^{2+}. Aminopeptidase A, N(M), and W have been detected in isolated brain microvessels. Aminopeptidase A is suggested to be involved in cerebrovascular metabolism of angiotensins whereas aminopeptidase N(M) has been particularly implicated in the inactivation of enkephalins [3]. Bestatin has been shown to be a slow-binding, competitive inhibitor for most aminopeptidases [10]. The neutral endopeptidase 24.11 (enkephalinase) also cleaves enkephalins.

Table 14.1 Localization of metalloproteinases at the cerebral microvasculature.

Type	Synonym	Location at the BBB
Aminopeptidase A	Glutamyl-aminopeptidase (EC 3.4.11.7)	Pericytes
Aminopeptidase N(M)	Alanin aminopeptidase (EC 3.4.11.2)	Pericytes, microglia, (astrocytes?, endothelium?)
Aminopeptidase W		Cerebral microvasculature
Angiotensin-converting enzyme (ACE)	Peptidyl dipeptidase A (EC 3.4.15.1)	Endothelium
Carboxypeptidase N		Endothelium
Endopeptidase 24.11	Enkephalinase	Cerebral microvasculature, pericytes
Dipeptidylpeptidase IV (DPP IV)	Serine peptidase IV	Endothelium
Dipeptidylpeptidase II (DPP II)	Serine peptidase II	Cerebral microvasculature
Matrix metalloproteinases (MMPs; see Table 14.2)	Matrixins	Endothelial cells, pericytes, astrocytes, microglia, neurons

The angiotensin-converting enzyme (ACE) is a Zn^{2+}-dependent metalloproteinase activated by Cl^- ions. It hydrolyzes dipeptides from the C-terminus and converts angiotensin I to the active vasoconstrictor angiotensin II. It also inactivates the vasodilatator bradykinin. Both angiotensin II and bradykinin not only constitute vasoactive substances, but also act as neurotransmitters. ACE may therefore have a vital role in the regulation of cerebral blood pressure and in peptide metabolism. Similar to ACE, the metalloproteinase carboxypeptidase N was found to cleave the kinins kallidin and bradykinin and to degrade vasopressin into fragments.

The dipeptidylpeptidases (DPP) II and IV are serine peptidases that degrade small peptides and proteins containing proline residues (e.g. substance P) which is resistant to degradation by aminopeptidase N(M) [2].

One larger family of metalloproteinases identified in cerebral microvessels with zinc-dependent proteolytic activity are the matrix metalloproteinases (MMPs), which under normal and pathological conditions possess complex functions at the cell surface and within the extracellular matrix of the central nervous system. MMPs are able to degrade virtually all types of extracellular matrix components and are known to be responsible for the maintenance, turnover, and integrity of the extracellular matrix, suggesting they play a central role in several biological processes such as embryogenesis, organogenesis, wound healing, angiogenesis, and in multiple steps of inflammation.

Under normal physiological conditions, MMP expression and activity are precisely balanced at different levels including: (a) the level of transcription and translation, (b) activation of latent pro-enzymes by other proteinases and free radicals, and (c) inhibition by endogenous tissue inhibitors of metalloproteinases (TIMPs) and α-macroglobulins. Several synthetic peptides have also been shown to be inhibitors of MMP activities [11, 12].

TIMPs are specific MMP inhibitors that participate in controlling the local activities of MMPs in tissues by binding to the catalytic site of MMPs. Four TIMPs (TIMP-1, -2, -3, -4) have been identified in vertebrates. Their expression is differentially regulated during ontogenic development and tissue remodeling. Changes of TIMP levels are considered to be clinically important because they directly affect the level of local MMP activities; and a loss of MMP control may result in pathological conditions.

When the delicate balance between the production and proteolytic activity of MMPs on the one hand and their inhibition by TIMPs on the other hand is disturbed towards increased MMP activity, uncontrolled breakdown and widespread destruction of the extracellular matrix occurs. Conversely, excessive expression of TIMPs leads to the restriction of physiological proteolysis and a net buildup of extracellular matrix proteins, resulting in fibrosis. In the process of neovascularization, the key role of MMPs and TIMPs is the control of dysregulated vascular growth which is characteristic for a number of angiogenic diseases.

Twenty-eight MMP genes have been identified so far, of which 23 are found in humans (see Table 14.2). Most of them are multidomain proteins which are either secreted by the cell into the extracellular space or anchored to the plasma membrane [13].

Table 14.2 Matrix metalloproteinase family members.

Matrix metalloproteinase	Function
Collagenases	Collagenolytic activity, digestion of ECM and non-ECM molecules
MMP-1 (collagenase-1)	
MMP-8 (collagenase-2)	
MMP-13 (collagenase-3)	
MMP-18 (collagenase-4, identified in *Xenopus*)	
Stromelysins	
MMP-3 (stromelysin-1)	Activation of pro-MMP-9, expressed in endothelial cells
MMP-10 (stromelysin-2)	
MMP-11 (stromelysin-3)	
Matrilysins	
MMP-7 (matrilysin-1)	
MMP-26 (matrilysin-2, endometase)	
Gelatinases	Cleave gelatine, fibronectin, elastin and different collagens
MMP-2 (gelatinase-A)	Expressed in astrocytes, Schwann cells, and peripheral nerves
MMP-9 (gelatinase-B)	Expressed in brain endothelial cells, astrocytes, pericytes, and microglia; markedly upregulated in inflammation
Membrane-type MMPs	
MMP-14 (MT1-MMP)	Activation of pro-MMP-2
MMP-15 (MT2-MMP)	
MMP-16 (MT3-MMP)	
MMP-17 (MT4-MMP)	
MMP-24 (MT5-MMP)	
MMP-25 (MT6-MMP)	
Other MMPs	
MMP-12 (metalloelastase)	Macrophage migration; mainly expressed by alveolar macrophages
MMP-19	Autoantigen from patients with rheumatoid arthritis
MMP-20 (enamelysin)	Digestion of amelogenin; exclusively expressed by odontoblasts
MMP-22 = MMP-21	
MMP-23	
MMP-28 (epilysin)	Wound repair

MMPs, collectively called matrixins, can be divided into six subgroups, depending on their substrate specificity, sequence similarity and domain organisation [14, 15].

Transcription of many MMPs is promoted by inflammatory cytokines, growth factors, chemokines, oncogenes, and cell-cell or cell-matrix interactions. Further, it is known that MMPs are intracellularly synthesized as latent enzymes (pro-MMPs) and most pro-MMPs are activated extracellularly. The activation process is a stepwise mechanism which can be triggered either by proteinases in vivo or by chemical agents, low pH, and heat treatment, as shown in vitro. Activation factors include parts of the plasminogen-plasmin cascade [16], as well as other MMPs which proteolytically cleave the propeptide region of effectors which disrupt the interaction between cysteine and zinc (the so-called "cysteine switch" mechanism) to trigger activation [17]. A few MMPs are processed at the cell membrane into fully active enzymes [18].

Active MMPs are differentially inhibited by the binding of TIMPs. TIMP-1 inhibits MMP-1, MMP-3 and MMP-9 more effectively than TIMP-2. TIMP-3 inhibits MMP-2 and MMP-9 [19], whereas TIMP-4 is a good inhibitor of all classes of MMPs without remarkable preference [20]. Interestingly, TIMPs are also required for the activation of some MMPs [21]. These complex mechanisms are described exemplarily for the cell surface activation of MMP-2 through a membrane-type MMP (MT-MMP)-mediated cascade. MT-MMPs are membrane proteins which are either glycosyl-phosphatidylinositol-anchored or transmembrane proteins. They frequently function as activating proteins, e.g. MT1-MMP (MMP-14) has been found to bind to a complex formed by TIMP-2 and pro-MMP-2 to facilitate the activation of MMP-2 [22]. TIMP-2 mediates pro-MMP-2 activation in a dose-dependent manner. Pro-MMP-2 is processed at low TIMP-2 concentrations, but inhibited at higher TIMP-2 concentrations [23]. MT2-MMP may activate pro-MMP-2 through a TIMP-2-independent route [24]. MMP-2 in concert with MT1-MMP can activate pro-MMP-13.

Two recently discovered new families of MMPs also appear to play an important role in brain function, the ADAMs (*a* *d*isintegrin *a*nd *m*etalloproteinase) [25] and the ADAMTs (*a* *d*isintegrin *a*nd *m*etalloproteinase with a *t*hrombospondin motif) [26]. ADAMs are with few exceptions transmembrane proteins, while the ADAMTs are secreted molecules, some of which bind to the extracellular matrix [27].

These enzymes cleave a number of extracellular matrix molecules and remove ectodomain molecules from the cell surface, such as TNF-α receptor, interleukin-6, L-selectin and syndecans [28]. Literature on the role of ADAMs and ADAMTs in the BBB is still sparse and little is known about the regulation of their activity.

14.3
Cerebral Endothelial Cells and Metalloproteinases

MMPs are able to cleave the basal lamina macromolecules that line the BBB microvessels, leading to disruption of the BBB and modulation of capillary permeability [9]. However, it is to date poorly understood if and how MMPs are involved in the regulation of BBB function. Regulation of paracellular BBB permeability is a complex process involving the intracellular signaling and rearrangement of tight junction proteins in cerebral endothelial cells. Several studies have reported markedly increased BBB permeability after exposure to TNF-α, interleukin-1β, interferon-γ, histamine, and growth factors.

It is known from in vivo studies that intracerebral MMP injection leads to extensive leakage of the BBB and T-cell recruitment to the lesion site and demyelination. This lesion could be reduced by MMP inhibitors [29]. In vitro investigations by Lohmann et al. [30] show that MMP-2, present in serum, decreases the transendothelial electrical resistance of cultured porcine brain endothelial cells. In the same model, the glucocorticoid hydrocortisone suppresses the secretion of MMPs by brain endothelial cells and simultaneously results in improved barrier function. Moreover, proinflammatory cytokines and chemokines, which markedly decrease the barrier function of cerebral endothelial cells, have been shown to regulate the secretion and activation of MMPs [31], suggesting MMPs to be the missing link in BBB permeability regulation by cytokines.

Thus, clear evidence is given for the role of MMPs in endothelial tight junction regulation. However, the molecular mechanism of action of MMPs on the tight junction complex remains to be elucidated. Lohmann et al. [30] investigated the regulatory mechanisms of tight junction permeability and MMP interaction in a well characterized in vitro model of the BBB, based on primary cultured pig brain capillary endothelial cells of a polarized phenotype which displays a tight and intact cellular barrier [32]. The tightness of the cellular BBB in vitro can be directly measured as transendothelial electrical resistance (TEER, see Chapter 16), reflecting the permeability of the tight junctions for small ions and solutes [33].

Zymographic analysis revealed that pig brain endothelial cells secrete MMP-9 apically (to the blood side in vivo). Lohmann et al. [30] show that secretion of MMP-9 is markedly reduced after the addition of hydrocortisone (550 nM) to the culture medium, which at the same time results in improved barrier function. Fluorimetric analysis of enzymatic activity revealed that hydrocortisone concomitantly reduces MMP activity at the BBB in vitro.

A basolateral (to the brain side in vivo) secretion of MMP-9 by cerebral endothelial cells in vitro was also demonstrated by immunocytochemistry, where MMP-9 was found to be attached to the extracellular matrix. In addition, endogenous MMP-inhibitors in form of TIMP-1 and TIMP-2 were detected in an in vitro BBB model by immunocytochemistry and Western blot analysis. The studies of Herron et al. [34, 35] describe the production of pro-MMP-1 and pro-MMP-3 in rabbit cerebral endothelial cell cultures. These authors observed that the absence of enzymatic activity is due to the simultaneous production of inhibitors. The inhibitors are secreted by endothelial cells into the medium and are at least partially bound to MMPs in the extracellular matrix. In this way cerebral endothelial cells in vitro seem to balance their proteolytic side under these conditions.

MMP activity at the BBB in vitro can be significantly increased by addition of the tyrosine phosphatase inhibitor phenylarsine oxide (PAO) in a time- and dose-dependent fashion [30]. Increased MMP activity was paralleled by a significant decrease in TEER, indicating impaired barrier function, and morphological changes, implying the disruption of cell-cell contacts and the formation of intercellular gaps. Moreover, Western blot analysis of tight junction proteins revealed PAO-induced proteolysis of occludin, with a generated fragment of approximately 51 kDa [30, 36], but the tight junction proteins zonula occludens-1 (ZO-1) and claudin-5 remained intact. Tyrosine phosphatase inhibition by pervanadate without concomitant MMP activation did not induce occludin proteolysis. Occludin has previously been shown to be a substrate for serine proteases [37]. As occludin proteolysis at the BBB in vitro could specifically be prevented by MMP inhibitors, GM 6001, zinc ions and 1,10-phenanthroline [30, 36], there is now direct evidence for increased MMP activity upon exposing the cells to PAO, leading to occludin proteolysis and loss of barrier integrity. The described data give direct evidence of an involvement of MMPs in the regulation of BBB function in vitro and suggest that cleavage of the tight junction protein occludin by MMPs is a possible molecular mechanism of BBB regulation. Figure 14.1 schematically depicts the described activation pathways of MMPs in cerebral endothelial cells.

It appears that increased MMP activity at the BBB leads to MMP-dependent cleavage of the tight junction protein occludin and a profound disruption of cell-cell contacts, resulting in markedly compromised barrier integrity and function. Additionally, Asahi et al. [38] reported that MMP-9 can proteolytically cleave ZO-1. Thus, extracellular activity of MMPs could be a crucial and site-restricted determinant of the unique properties of BBB interendothelial tight junctions.

Comparable results were shown recently in diabetic retinopathy, which involves the breakdown of endothelial tight junctions of the blood-retinal barrier (see Chapter 26). The retinas of diabetic animals demonstrated elevated levels of MMP-2, MMP-9, and MMP-14 mRNA, accompanied by an increase in vascular permeability and a proteolytic degradation of the tight junction protein occludin [39]. Behzadian et al. [40] demonstrated that retinal endothelial cells express MMP-9 when treated with TGF-β or when cocultured with glial cells (astrocytes or Muller glial cells) and that both TGF-β and MMP-9 increase retinal endothelial cell permeability. This mechanism may also contribute to the breakdown of the blood-retinal barrier.

Miyamori et al. [41] could identify the tight junction protein claudin 5 as one factor which promotes the activation of pro-MMP-2 mediated by MT1-MMP in 293 T cell cultures. These results suggest a more dynamic function for claudins than simply being structural constituents of tight junctions; and they also indi-

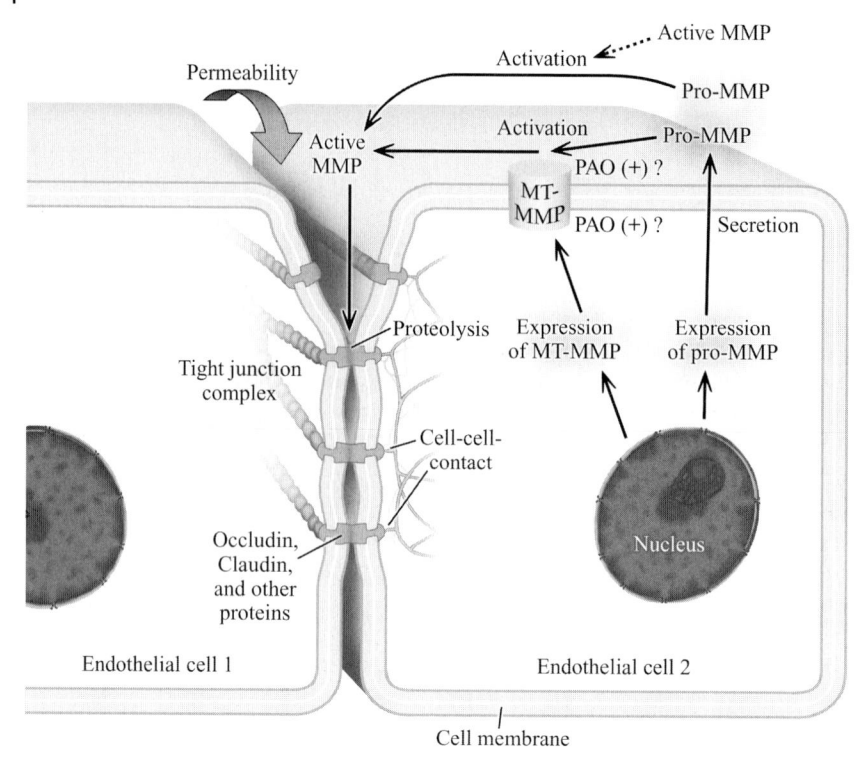

Fig. 14.1 Possible mechanism of BBB permeability regulation by MMPs. Cerebral endothelial cells secrete pro-MMPs. Pro-MMPs are proteolytically activated by other active MMPs or by cell membrane-bound active membrane-type MT-MMPs. In a next step, active MMPs proteolytically cleave the tight junction protein occludin within the first extracellular loop near the N-terminus [30]. This has a destabilizing effect on cell-cell contacts and results in increased transendothelial permeability. In this context, phenylarsenoid (PAO) has a feasible effect on extracellular MMP activity, either by increasing the expression of membrane-bound MT-MMP or by binding pro-MMPs to MT-MMPs.

cate a precisely located effect of MMPs at the tight junction complex. These regulatory processes are supplemented by novel findings on the roles of the low density lipoprotein (LDL) receptor family which show that LDL receptor-related protein (LRP) interacts as a regulator of vascular tone, permeability of the BBB, and expression of MMPs [42]. Besides the described endothelial expression of MMP-9, it is known that astrocytes, oligodendrocytes, microglia, and neurons can also secrete MMP-9 [43].

14.4
Perivascular Cells and Metalloproteinases

14.4.1
Pericytes

A predominant role within the BBB is played by the cells closest to the cerebral endothelium, the perivascular cells ("pericytes", see Chapter 5). Pericytes are positioned along the vessel axis, forming numerous cytoplasmic processes that encircle endothelial cells. However, they are separated from the cerebral endothelium by a common basement membrane. They take up blood-borne substances through endo- and phagocytosis and they are recognized as immunocompetent antigen-presenting cells, which participate in angiogenesis. Furthermore, they are suggested to play a regulatory function in the control of capillary growth and tube formation [44]. These findings provide evidence that cerebral pericytes act as a "second line of defense" to maintain metabolic homeostasis between blood and brain [45]. Balabanov and Dore-Duffy [46] postulated that pericytes constitute an important component of the BBB, with an involvement in virtually all processes at the BBB. However, the involvement of pericytes in BBB regulation is still a matter of discussion (see Chapter 5).

Immunocytochemical studies have shown the existence of BBB-specific enzymes in pericytes, e.g. γ-glutamyltranspeptidase (γ-GT) [47], alkaline phosphatase, and the metalloproteinases aminopeptidase A [48, 49], and aminopeptidase N(M) (APN) [45, 50].

Aminopeptidase A (APA) is the only enzyme so far discovered in the cerebral microvasculature that degrades angiotensin II to angiotensin III [3]. APA shows highest expression in brain regions that lack a tight BBB endothelium, such as the median eminence and choroid plexuses [48], whereas aminopeptidase N is confined to sites of tight BBB microvessels. The abundance of aminopeptidase A at the circumventricular organs correlates well with the fact that these brain regions are highly enriched in angiotensin II receptors [51].

To date, only APN has been clearly characterized and its ultrastructural localization as an extracytoplasmic membrane-bound enzyme of cerebral pericytes (pericytic aminopeptidase N; "pAPN") has been described [45, 50]. APN was first reported by Solhonne et al. [52] as a membrane-bound ectopeptidase with a broad activity for almost all unsubstituted oligopeptides and is involved in the extracellular degradation and inactivation of endogenous opioid peptides, e.g. enkephalines and endorphines. pAPN is not expressed at the endothelial side of the BBB frontier. Microvasculature of brain regions that lack BBB properties, such as the area postrema, median eminence, and choroid plexuses, are devoid of pAPN expression. pAPN is asymmetrically distributed at the pericytic plasma membrane with preference at the abendothelial sites. At the pericytic cell processes, this asymmetric distribution is less evident. The predominant localization of pAPN at the parenchymal front rather than at the vascular site indicates a primarily parenchymal control function of pAPN. Thus, endogenous unsubstituted peptides can be degraded easily before entering the periendothelial space. In contrast, the lack of pAPN in the circumventricular organs allows for the exchange of endogenous peptides from blood to brain and vice versa.

The onset of pAPN in embryonic rat brain is known to occur on day E18. A steady-state level is reached at postnatal day 6–8 [50].

The exclusive expression of pAPN at cerebral pericytes indicates an important metabolic function within the BBB control mechanisms. In this context, primary cultured endothelial cells as well as cocultures with astrocytes and pericytes are useful in vitro alternatives to investigate blood vessels in vivo for studying the BBB function.

Pericytes are also involved in specific microvascular diseases and in angiogenesis. In cocultures with endothelial cells, pericytes secrete an active form of TGF-β that controls endothelial growth and appears to mediate pericyte-endothelium interactions, like guiding the migrating endothelial cells and forming connections between newly formed sprouts [44, 53, 54].

Angiotensin II can contribute to the regulation of retinal neovascularization by stimulating the migration of pericytes in retinal microvessels, augmenting latent MMP-2 activity, and doubling the TIMPs activity [55]. Pericytic angiotensin II-induced chemotaxis is mediated by antibodies against platelet-derived growth factor (PDGF) and involves TGF-β.

Chantrain et al. [56] demonstrate that MMP-9 in neuroblastoma contributes to angiogenesis by promoting blood vessel morphogenesis with pericyte recruitment.

Girolamo et al. [57], using human developing brain microvascular cultures, show the concerted interplay between endothelial cells and pericytes in angiogenesis regulation by expression of MMP-2. Endothelial cells express pro-MMP-2 and pericytes express active MMP-2 [57]. The active MMP-2-containing pericytes may be considered as cells cooperating with the endothelial cells to bring about basal lamina degradation and play a role as "pioneers" that dissect the brain parenchyma and allow endothelial sprouting during angiogenesis.

Together, these expression patterns highlight the enzymatic and metabolic aspect of the pericytic barrier function.

14.4.2
Astrocytes and Microglia

Astrocyte endfeet cover much of the capillary's basal surface and they are known to induce BBB formation [58]. They participate in a variety of homeostatic functions and are involved in mechanisms of neural injury and repair (see Chapters 8 and 9). Furthermore, astrocytes induce the differentiation of cerebral capillary endothelial cells into cells with BBB characteristics; and this induction is mediated by soluble products released by the astrocytes [59]. Metalloproteinase expression is also described in astrocytes, but most of the investigations of MMP expression in astrocytes have been done under cell culture conditions. For example, Apodaca et al. [60] showed the presence of MMP-9 and MMP-2 as well as the inhibitors of these MMPs in conditioned cell culture medium from fetal astrocytes stimulated with phorbol esters. Similar observations were done by Gottschall and Yu [61] when astrocyte cultures were stimulated with LPS, IL-1β, or TNF-α. Under these conditions, astrocytes secreted MMP-1, MMP-2, and pro-MMP-9. MMP-1 has been shown to be toxic to human neurons in culture [62]. LPS and IL-1β also stimulate the production of MMP-3 in astrocytes [43]. However, astrocytes and microglia in coculture produce an active form of MMP-9 when stimulated by LPS [63].

Leveque et al. [64] reported on the effect of HIV infection and TNF-α on the expression of MMP-2 and MMP-9 in astrocytes. HIV infection increases the production of pro-MMP-2 and pro-MMP-9, suggesting that astrocyte-HIV contact may lead to extracellular matrix activation.

Primary mouse astrocytic cultures stimulated with various cytokines and cellular growth factors (IL1-β, TNF-α, EGF) display increased levels of MMP-3 mRNA. Interferon-γ inhibits this response. The mRNA accumulation was preceeded by activation of the transcription factors NFκB and AP-1, but this effect might not solely be responsible for the cytokine-induced expression of MMP-3 mRNA in astrocytes [65].

Further factors influencing the expression of MMPs in astrocytes include bradykinin and the heat-shock protein HSP 70. Bradykinin, an inflammatory mediator, has been shown to induce the expression of MMP-9 in rat astrocytes; and activation of p42/p44 mitogen-activated protein kinase (MAPK) mediated through NFκB pathways is essential for bradykinin-induced MMP-9 gene expression [66]. In contrast, HSP 70, which is synthesized in response to a variety of stress factors including ischemia, suppresses MMP-2 and MMP-9 production in astrocytes [67]. These findings suggest that HSP 70 may play a protective role by down-regulating MMPs in stress situations.

Another key homeostatic mechanism of astrocytes in tissue repair is maintained through their production of TIMP-1. Gardner and Ghorpade [68] showed astrocytic TIMPs expression in inflammatory neurodegenerative diseases, which may have significant therapeutic relevance.

Colton et al. [69] showed the production of MMP-9 in activated microglial cells. Microglia activation reveals a biphasic process: an early phase of global activation is followed by a later phase in which microglia activation becomes increasingly focused in the lesions. During the early phase, expression of pro-inflammatory mediators like IL-1β TNF-α, and early growth response-1 (Erg-1) increases but is restricted to lesions [70]. Immunostaining of astrocytic and microglial cocultures showed the expression of MMP-3 by microglial cells but not by astrocytes. Further findings suggest that MMP-3 produced in microglia is important for the activation of MMP-9 produced in astrocytes during LPS stimulation [9].

Aminopeptidase N is also expressed in microglial cells. However, it is not a stable marker for microglia because its expression apparently is related to functional or morphological changes of microglia [71].

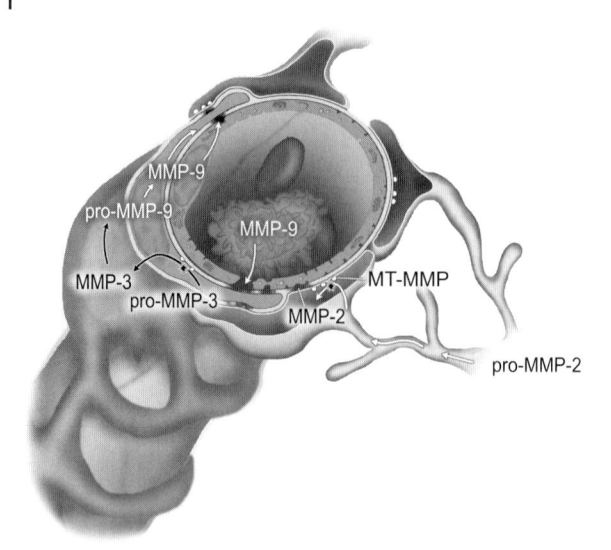

Fig. 14.2 Schematic view of the interaction of metalloproteinases (MPs) at the blood-brain barrier (BBB). Under normal conditions, there is a constitutive expression of pro-MMP-2 by astrocytic endfeet with the potential for activation by the membrane-type MMP (MT-MMP). Pericytes express aminopeptidase N and A. During cellular stress, some other MPs come into action. MMP-9 is secreted by lymphocytes and endothelial cells damaging the endothelial barrier and the basement membrane. Pericytes and microglial cells secrete pro-MMP-2, pro-MMP-3, and pro-MMP-9. Activation of pro-MMP-3 may be facilitated by the plasminogen/plasmin system in pericytic/microglial membranes. Possibly, activated MMP-3 could activate pro-MMP-9. This cumulative activation of MMPs could amplify BBB impairment and cause further brain dysfunction.

All in all, these data indicate that astrocytes, besides their physiological production of MMP-2, could be a major source of MMP-9 in the developing and inflamed brain. Thus, cytokines appear to be potent regulators of MMP production in astrocytes and microglia during inflammation. A model of the distribution of metalloproteinases in brain capillaries is given in Fig. 14.2.

14.5
Metalloproteinases and the Blood-Liquor Barrier

The epithelial cells of the choroid plexus form the structural basis of the barrier between the blood and cerebrospinal fluid. They are involved in transport processes from blood to cerebrospinal fluid (CSF) and vice versa. Expression of various metallopeptidases has been detected at the blood-CSF barrier [72]. Strazielle et al. [73] demonstrated that the choroid plexus is a source of pro- and active MMP-2 and MMP-9 in the brain and that pro-inflammatory cytokine treatment leads to an increase in choroidal MMP secretion at either the apical or basolateral membrane. Thus, during inflammation, the choroid plexuses could comprise a source of MMPs found in the cerebrospinal fluid, which facilitates leukocyte migration.

Concerning the distribution of further MPs, Schnabel et al. [74] showed a reduced activity of dipeptidylpeptidase IV, low activity of aminopeptidase A, and no activity of gamma-glutamyl transpeptidase in capillaries of the choroid plexuses. These authors described the presence of aminopeptidase M in leaky capillaries, but these findings were not substantiated by others [50].

14.6
Metalloproteinases and Brain Diseases

Blood-brain barrier disruption is common in many neurological diseases. Under pathological conditions the barrier becomes compromised, resulting in an intense cell trafficking from blood to brain, and the central nervous system becomes enriched with immune competent cells. Activated cytotoxic lymphocytes [75], macrophages, or certain types of metastatic cells [76] can even cross the intact BBB. In a first step, these cells obviously share the ability to recognize and bind to endothelial cells. Less clear in this context is the entrance mechanism by which these cells cross the intact BBB and how they manage to enter the brain parenchyma (see Chapters 4 and 18). A better understanding of these processes is needed to design therapeutic strategies with the aim to prevent monocyte infiltration across the BBB. Some investigators have claimed that the first transmigration step occurs transcellularly through endothelial cells [77], while subsequent excessive infiltration of monocytes by paracellular movement is accompanied by endothelial damage, resulting in the loss of tight junctions [75, 78]. In a further step, these cells can affect the function of neurovascular structures by attacking the basal lamina around the cerebral vessels, leading to hemorrhage and cerebral edema. Neuropathogenesis of HIV-1 infection demonstrates a facilitated transmigration of HIV-infected monocytes/macrophages by an increased production of MMP-9 activity. Furthermore, chemokines regulate the traffic of the infiltrated monocytes through the brain parenchyma [79].

Much has been learned recently about the role of MPs in neurological diseases, but our understanding of their multiple roles is still incomplete. Particularly, most MMPs are largely absent from the normal BBB and central nervous tissue. However, their upregulation has been reported in several neurological disorders, with an increase in capillary permeability and leukocyte migration into the central nervous system. Changes in the fine tuning of MMPs expression and TIMPs seem to affect extracellular matrix turnover and has been implicated in pathological and neuroinflammatory conditions of the central nervous system, including meningitis, encephalitis, and brain tumors (see Chapter 24), AIDS dementia and multiple sclerosis (MS). Other proteases are also increased at sites of secondary injury, e.g. plasminogen activators (PAs), and may act in concert to attack the extracellular matrix [80].

Synthetic inhibitors of MMPs have been developed for the treatment of cancer and other MMP-related diseases. These hydroxamate-based compounds have been shown to reduce injury in experimental allergic encephalomyelitis (EAE; a mouse disease model of human MS), experimental allergic neuritis (EAN), cerebral ischemia, intracerebral hemorrhage, and viral and bacterial infections of the central nervous system [9]. A better understanding of MP expression under various pathological conditions of the central nervous system will allow for the use of MMP inhibitors in the treatment of brain disorders.

14.6.1
Metalloproteinases and Cerebral Ischemia

MMPs are important mediators under stroke conditions, since they show significant upregulation after ischemia.

Chang et al. [81] showed that the matrix metalloproteinase MMP-2 participates in matrix degradation and disruption of the basal lamina of cerebral microvessels during focal cerebral ischemia. MMP-2 is induced in T cells bound to VCAM-1 on endothelial cells during inflammation when transmigration through the endothelial cell layer and basement membrane commences [82]. MMP-2 is secreted in a latent form (pro-MMP-2) and can be activated in a direct or indirect manner. Direct activators are the membrane-bound MT1-MMP and MT3-MMP, which are also constitutively expressed in brain, whereas plasminogen/plasmin system enzymes interact as indirect activators, such as urokinase-type plasminogen activator (uPA) and tissue-type plasminogen activator (t-PA). tPA also upregulates MMP-9 in vitro and in vivo. This response is mediated by the LDL receptor-related protein (LRP), which avidly binds tPA and possesses signaling properties [83]. Yepes et al. [84] demonstrated that, in the initial stages of cerebral ischemia, the opening of the BBB is mediated directly by tPA and that this activity is independent of MMP-9 but requires interaction with LRP. Furthermore, oxidative stress generated during stroke could mediate BBB disruption through local MMP-9 activation [85].

Asahi et al. [38], using knockout technology, showed that MMP-9 displays a central role in ischemic damage. MMP-9 knockout mice revealed smaller infarct volumes and exhibited protection against ischemic and traumatic brain injury. The protective effect of the MMP-9 gene knockout may be mediated by reduced proteolytic degradation of BBB components.

These studies indicate that MMP-9 may play a critical role in the broad pathophysiology of cerebral ischemia. However, MMPs might also provide beneficial effects after injury in the central nervous system (e.g. axonal regrowth, remyelinization, angiogenesis) [22].

14.6.2
Metalloproteinases and Brain Tumors

Many MMPs were first cloned from tumors or tumor cell lines, suggesting that an excess of MMP activity is common in many brain tumors. Enhanced MMP expression in cancer often correlates with poor survival prognosis. MMP secretion facilitates migration and invasion of tumor cells through the extracellular matrix for penetration of the surrounding tissue [86]. High expression of metalloproteinases has also been demonstrated in brain tumor angiogenesis [87]. However, endothelial cells in brain tumors dedifferentiated and lose BBB properties [88].

14.6.3
Metalloproteinases and Multiple Sclerosis

MMPs seem to be implicated in multiple sclerosis (MS), where one of their suggested roles is to facilitate the transmigration of circulating leukocytes into the brain tissue and to disrupt the integrity of the BBB.

Bar-Or et al. [89] investigated which of the 23 members of the human MMP family are critical to the MS disease process. They found a distinctive pattern of MMP expression in different cell populations, in which monocytes in particular show the strongest expression of different MMPs. This finding corresponds well with the rapid migration of monocytes across the BBB. Finally, the authors described higher levels of MMP-2 and MMP-14 in MS patients as compared to normal individuals. TIMP-2 levels were also elevated in monocytes from MS patients. The high expression of specific MMP members on monocytes can be used as a new target to design novel therapeutic strategies in MS.

Maeda and Sobel [90] reported an increased expression of MMP-1, MMP-2, MMP-3, and MMP-9 by macrophages in acute MS and necrotic lesions, whereas chronic MS lesions revealed fewer MMP-positive macrophages. These data indicate that acute multiple sclerosis MMP-mediated proteolysis may contribute to the breakdown of the BBB and leukocyte migration into the central nervous system. Overall, a common elevation of MMP-9 and MMP-12 is found in MS and in mouse and rat EAE, the animal model of MS [22]. During the early stages of EAE, MMP-2, MMP-3, MMP-7, and MMP-9 are present at elevated levels in rat brain.

In the course of acute EAE, Kunz et al. [91] demonstrated a dramatic downregulation of microvascular associated pAPN expression during the clinical peak of EAE. In addition, pAPN is transiently expressed by infiltrating macrophages and probably by a subpopulation of activated microglial cells in the white matter of lumbar spinal cord. Thus pAPN expression which under normal conditions is represented by pericytes is a sensitive marker for monitoring BBB damage.

14.6.4
MMPs and Migraine

It has been debated whether MMPs are involved in migraine aura, because activation of MMPs leads to leakage of the BBB and allows potassium, nitric oxide, adenosine, and other products released by a cortical spreading depression to reach and sensitize the dural perivascular trigeminal afferents [92].

Additionally, cortical spreading depression (CSD), a self-propagating wave of neuronal and glial depolarization, which is implicated in disorders of neurovascular regulation such as migraine, causes prolonged MMP-9 activity. In this case, MMP-9 activation initiates a cascade that disrupts the BBB [93].

14.7
Conclusion

The complexity of the role of MPs and MMPs at the BBB in health and disease is evident. MMPs are involved in tissue remodeling and repair during development, in tissue damage during inflammation with breakdown of the extracellular matrix around the cerebral blood vessels, and in recovery from injury. The roles of MPs include the degradation of neuropeptides.

The BBB components (endothelial cells, pericytes, astrocytes) are furnished with a specific set of MPs and MMPs, but the types and the stimuli that induce proteinase expression are different for the various cell types.

MMPs are rapidly upregulated after nearly all types of central nervous system dysfunctions; and inhibition of MMPs has been shown to prevent the progression of many neurological diseases. A goal for the future could be to design strategies in which the inhibition of MMP activities may be used to counteract their malignant effects occurring under disease conditions.

References

1 Rubin LL, Staddon JM **1999**, The cell biology of the blood-brain barrier, *Annu. Rev. Neurosci.* 22, 11–28.
2 Brownless J, Williams CH **1993**, Peptidases, peptides, and mammalian blood-brain barrier, *J. Neurochem.* 60, 793–803.
3 Bausback HH, Churchill L, Ward PE **1988**, Angiotensin metabolism by cerebral microvasculature aminopeptidase-A, *Biochem. Pharmacol.* 37, 155–160.
4 Churchill L, Bausback HH, Gerritsen ME, Ward PE **1987**, Metabolism of opioid peptides by cerebral microvasculature aminopeptidase M, *Biochim. Biophys. Acta* 923, 35–41.
5 Bausback HH, Ward PE **1988**, Kallidin and bradykinin metabolism by isolated cerebral microvessels, *Biochem. Pharmacol.* 37, 2973–2978.

6 Brecher P, Tercyak A, Chobanian AV **1981**, Properties of angiotensin-converting enzyme in intact cerebral microvessels, *Hypertension* 3, 198–204.

7 Brownson EA, Abbruscato TJ, Gillespie TJ, Hruby VJ, Davis TP **1994**, Effect of peptidases at the blood-brain barrier on the permeability of enkephalin, *J. Pharmacol. Exp. Ther.* 270, 675–680.

8 Lojda Z **1979**, Studies on dipeptidyl (amino)peptidase IV (glycylprolyl naphthylamidase) II. Blood vessels, *Histochemistry* 59, 153–166.

9 Rosenberg GA **2002**, Matrix metalloproteinases in neuroinflammation, *Glia* 39, 279–291.

10 Taylor A **1993**, Aminopeptidases: structure and function, *FASEB J.* 7, 290–298.

11 Mix KS, Mengshol JA, Benbow U, Vincenti MP, Sporn MB, Brinckerhoff CE **2001**, A synthetic triterpenoid selectively inhibits the induction of matrix metalloproteinases 1 and 13 by inflammatory cytokines, *Arthritis Rheum.* 44, 1096–1104.

12 Bernardo MM, Brown S, Li ZH, Fridman R, Mobashery S **2002**, Design, synthesis and characterization of potent, slow-binding inhibitors that are selective for gelatinases, *J. Biol. Chem.* 277, 11201–11207.

13 Egeblad M, Werb Z **2002**, New functions for the matrix metalloproteinases in cancer progression, *Nat. Rev. Cancer* 2, 161–174.

14 Nagase H, Woessner JF Jr **1999**, Matrix metalloproteinases, *J. Biol. Chem.* 274, 21491–21494.

15 Visse R, Nagase H **2003**, Matrix metalloproteinases and tissue inhibitors of metalloproteinases. Structure, function and biochemistry, *Circ. Res.* 92, 827–839.

16 Cuzner ML, Opdenakker G **1999**, Plasminogen activators and matrix metalloproteases, mediators of extracellular proteolysis in inflammatory demyelination of the central nervous system, *J. Neuroimmunol.* 94, 1–14.

17 Van Wart HE, Birkedal-Hansen H **1990**, The cyteine switch: a principle of regulation of metalloproteinase activity with potential applicability to the entire matrix metalloproteinase gene family, *Proc. Natl. Acad. Sci. USA* 87, 5578–5582.

18 Nagase H **1997**, Activation mechanism of matrix metalloproteinases, *Biol. Chem.* 378, 151–160.

19 Butler GS, Hutton M, Wattam BA, Williamson RA, Knauper V, Willenbrock F, Murphy G **1999**, The specificity of TIMP-2 for matrix metalloproteinases can be modified by single amino acid mutations, *J. Biol. Chem.* 274, 20391–20396.

20 Stratmann B, Farr M, Tschesche H **2001**, MMP-TIMP interaction depends on residue 2 in TIMP-4, *FEBS Lett.* 507, 285–287.

21 Woessner JF, Nagase H **2000**, Matrix Metalloproteinases and TIMPs, Oxford University Press, Oxford.

22 Wee Yong V, Power Ch, Forsyth P, Edwards DR **2001**, Metalloproteinases in biology and pathology of the nervous system, *Nat. Neurosci.* 2, 502–511.

23 Kinoshita T, Sato H, Okada A, Ohuchi E, Imai K, Okada Y, Seiki M **1998**, TIMP-2 promotes activation of progelatinase A by membrane-type 1 matrix metalloproteinase immobilized on agarose beads, *J. Biol. Chem.* 273, 16098–16103.

24 Morrison CJ, Butler GS, Bigg HF, Roberts CR, Soloway PD, Overall CM **2001**, Cellular activation of MMP-2 (gelatinase A) by MT2-MMP occurs via a TIMP-2-independent pathway, *J. Biol. Chem.* 276, 47402–47410.

25 Schlondorff J, Blobel CP **1999**, Metalloproteinase-disintegrins: molecular proteins capable of promoting cell-cell interactions and triggering signals by protein-ectodomain shedding, *J. Cell. Sci.* 112, 3603–3617.

26 Porter S, Clark IM, Kevorkian L, Edwards DR **2005**, The ADAMTS metalloproteinases, *Biochem. J.* 386, 15–27.

27 Tang BL **2001**, ADAMTs: a novel family of extracellular matrix proteases, *Int. J. Biochem. Cell. Biol.* 33, 33–44.

28 Yong VW, Power C, Forsyth P, Edwards DR **2001**, Metalloproteinases in biology and pathology of the nervous system, *Nat. Rev. Neurosci.* 2, 502–511.

29 Matyszak MK, Perry VH **1996**, Delayed-type hypersensitivity lesions in the central nervous system are prevented by inhibitors of matrix metalloproteinases, *J. Neuroimmunol.* 69, 141–149.

30 Lohmann C, Krischke M, Wegener J, Galla H-J **2004**, Tyrosine phosphatase inhibition induces loss of blood-brain barrier integrity by matrix metalloproteinase-dependent and -independent pathways, *Brain Res.* 995, 184–196.

31 Harkness KA, Adamson P, Sussman JD, Davies-Jones GA, Greenwood J, Woodroofe MN **2000**, Dexamethasone regulation of matrix metalloproteinase expression in CNS vascular endothelium, *Brain* 123, 698–709.

32 Franke H, Galla HJ, Beuckmann CT **1999**, An improved low-permeability in vitro model of the blood-brain barrier: transport studies on retinoids, sucrose, haloperidol, caffeine and mannitol, *Brain Res.* 818, 65–71.

33 Schneeberger EE, Lynch RD **1992**, Structures, function, and regulation of cellular tight junctions, *Am. J. Physiol.* 262, L647–L661.

34 Herron GS, Banda MJ, Clark EJ, Gavrilovic J, Werb Z **1986a**, Secretion of metalloproteinases by stimulated capillary endothelial cells. II. Expression of collagenase and stromelysin activities is regulated by endogenous inhibitors, *J. Biol. Chem.* 261, 2814–2818.

35 Herron GS, Werb Z, Dwyer K, Banda MJ **1986b**, Secretion of metalloproteinases by stimulated capillary endothelial cells. I: Production of procollagenase and prostromelysin exceeds expression of proteolytic activity, *J. Biol. Chem.* 261, 2810–2813.

36 Wachtel M, Frei K, Ehler E, Fontana A, Winterhalter K, Gloor SM **1999**, Occludin proteolysis and increased permeability in endothelial cells through tyrosine phosphatase inhibition, *J. Cell Sci.* 112, 4347–4356.

37 Wan H, Winton HL, Soeller C, Taylor GW, Gruenert DC, Thompson PJ, Cannell MB, Stewart GA, Garrod DR, Robinson C **2001**, The transmembrane protein occludin of epithelial tight junctions is a functional target for serine

peptidases from faecal pellets of *Dermatophagoides pteronyssinus*, *Clin. Exp. Allergy* 31, 279–294.

38 Asahi M, Wang X, Mori T, Sumii T, Jung JC, Moskowitz MA, Fini ME, Lo EH **2001**, Effects of metalloproteinase-9 gene knock-out on the proteolysis of blood-brain barrier and white matter components after cerebral ischemia, *J. Neurosci.* 21, 7724–7732.

39 Giebel SJ, Menicucci G, McGuire PG, Das A **2005**, Matrix metalloproteinases in early diabetic retinopathy and their role in alteration of the blood-brain barrier, *Lab. Invest.* 2, 14.

40 Behzadian MA, Wang XL, Windsor LJ, Ghaly N, Caldwell RB **2001**, TGF-beta increases retinal endothelial cell permeability by increasing MMP-9: possible role of glial cells in endothelial barrier function, *Invest. Ophthalmol. Vis. Sci.* 42, 853–859.

41 Miyamori H, Takino T, Kobayashi Y, Tokai H, Itoh Y, Seiki M, Sato H **2001**, Claudin promotes activation of pro-matrix metalloproteinase-2 mediated by membrane-type matrix metalloproteinases, *J. Biol. Chem.* 276, 28204–28211.

42 Herz J **2003**, LRP: a bright beacon at the blood-brain barrier, *J. Clin. Invest.* 112, 1483–1485.

43 Gottschall PE, Deb S **1996**, Regulation of matrix metalloproteinase expression in astrocytes, microglia and neurons, *Neuroimmunomodulation* 3, 69–75.

44 Hirschi KK, D'Amore PA **1996**, Pericytes in the microvasculature, *Cardiovasc. Res.* 32, 687–698.

45 Kunz J, Krause D, Kremer M, Dermietzel R **1994**, The 140-kDa protein of blood-brain barrier-associated pericytes is identical to aminopeptidase N, *J. Neurochem.* 62, 2375–2386.

46 Balabanov R, Dore-Duffy P **1998**, Role of the CNS microvascular pericyte in the blood-brain barrier, *J. Neurosci. Res.* 53, 637–644.

47 Risau W, Dingler A, Albrecht U, Dehouck M-P, Cecchelli R **1992**, Blood-brain barrier pericytes are the main source of γ-glutamyltranspeptidase activity in brain capillaries, *J. Neurochem.* 58, 667–672.

48 Healy DP, Wilk W **1993**, Localization of immunoreactive glutamyl aminopeptidase in rat brain. II. Distribution and correlation with angiotensin II, *Brain Res.* 606, 295–303.

49 Song L, Wilk E, Wilk S, Healy DP **1993**, Localization of immunoreactive glutamyl aminopeptidase in rat brain. I. Association with cerebral microvessels, *Brain Res.* 606, 286–294.

50 Krause D, Vatter B, Dermietzel R **1988**, Immunochemical and immunocytochemical characterization of a novel monoclonal antibody recognizing a 140 kDa protein in cerebral pericytes of the rat, *Cell Tissue Res.* 252, 543–555.

51 Mendelson FHO, Quirion R, Saavedra JM, Auilera G, Catt KL **1984**, Autoradiographic localization of angiotensin II receptor in rat brain, *Proc. Natl. Acad Sci. USA* 81, 1575–1579.

52 Solhonne B, Gros C, Pollard H, Schwartz J-C **1987**, Major localization of aminopeptidase M in rat microvessels, *Neuroscience* 22, 225–232.

53 Nehls V, Denzer K, Drenckhahn D **1992**, Pericyte involvement in capillary sprouting during angiogenesis in situ, *Cell Tissue Res.* 270, 469–474.

54 Ramsauer M, Krause D, Dermietzel R **2002**, Angiogenesis of the blood-brain barrier in vitro and the function of cerebral pericytes, *FASEB J.* 16, 1274–1276.

55 Nadal JA, Scicli GM, Carbini LA, Scicli AG **2002**, Angiotensin II stimulates migration of retinal microvascular pericytes: involvement of TGF-β and PDGF-BB, *Am. J. Physiol. Heart Circ. Physiol.* 282, H739–H748.

56 Chantrain CF, Shimada H, Jodele S, Groshen S, Ye W, Shalinsky DR, Werb Z, Coussens LM, DeClerck YA **2004**, Stromal matrix metalloproteinase-9 regulates the vascular architecture in neuroblastoma by promoting pericyte recruitment, *Cancer Res.* 64, 1675–1686.

57 Girolamo F, Virgintino D, Errede M, Capobianco C, Bernardini N, Bertossi M, Roncali L **2004**, Involvement of metalloprotease-2 in the development of human brain microvessels, *Histochem. Cell Biol.* 122, 261–270.

58 Janzer RC, Raff MC **1987**, Astrocytes induce blood brain barrier properties in endothelial cells, *Nature* 325, 253–257.

59 Maxwell K, Berliner JA, Cancilla PA **1987**, Induction of gamma glutamyl-transpeptidase in cultured cerebral endothelial cells by a product released in astrocytes, *Brain Res.* 410, 309–314.

60 Apodaca G, Rutka JT, Buohana K, Berens ME, Giblin JR, Rosenblum ML, McKerrow JH, Banda MJ **1990**, Expression of metalloproteinases and metalloproteinase inhibitors by fetal astrocytes and glioma cells, *Cancer Res.* 50, 2322–2329.

61 Gottschall PE, Yu X **1995**, Cytokines regulate gelatinase A and B (matrix metalloproteinase 2 and 9) activity in cultured rat astrocytes, *J. Neurochem.* 64, 1513–1520.

62 Vos CM, Sjulson L, Nath A, McArthur JC, Pardo CA, Rothstein J, Conant K **2000**, Cytotoxicity by matrix metalloproteinase-1 in organotypic spinal cord and associated neuronal cultures, *Exp. Neurol.* 163, 324–330.

63 Rosenberg GA, Cunningham LA, Wallace J, Alexander S, Estrada EY, Grossetete M, Razhagi A, Miller K, Gearing A **2001a**, Immunohistochemistry of matrix metalloproteinases in reperfusion injury to rat brain: activation of MMP-9 linked to stromelysin-1 and microglia in cell cultures, *Brain Res.* 893, 104–112.

64 Leveque T, Le Pavec G, Boutet A, Tardieu M, Dormont D, Gras G **2004**, Differential regulation of gelatinase A and B and Timp-1 and -2 by TNFalpha and HIV virions in astrocytes, *Microbes Infect.* 6, 157–163.

65 Witek-Zawada B, Koj A **2003**, Regulation of expression of stromelysin-1 by proinflammatory cytokines in mouse brain astrocytes, *J. Physiol. Pharmacol.* 54, 489–496.

66 Hsieh HL, Yen MH, Jou MJ, Yang CM **2004**, Intracellular signalings underlying bradykinin-induced matrix metalloproteinase-9 expression in rat brain astrocyte-1, *Cell Signal.* 16, 1163–1176.

67 Lee JE, Kim YJ, Kim JY, Lee WT, Yenari MA, Giffard RG **2004**, The 70 kDa heat shock protein suppresses matrix metalloproteinases in astrocytes, *Neuroreport* 15, 499–502.

68 Gardner J, Ghorpade A **2003**, Tissue inhibitor of metalloproteinase (TIMP)-1: the TIMPed balance of matrix metalloproteinases in the central nervous system, *J. Neurosci. Res.* 74, 801–806.

69 Colton CA, Keri JE, Chen WT, Monsky WL **1993**, Protease production by cultured microglia: substrate gel analysis and immobilized matrix degradation, *J. Neurosci. Res.* 35, 297–304.

70 Lynch NJ, Willis CL, Nolan CC, Roscher S, Fowler MJ, Weihe E, Ray DE, Schwaeble WJ **2004**, Microglial activation and increased synthesis of complement component C1q precedes blood-brain barrier dysfunction in rats, *Mol. Immunol.* 40, 709–716.

71 Lucius R, Sievers J, Mentlein R **1995**, Enkephalin metabolism by microglia aminopeptidase N (CD13), *J. Neurochem.* 64, 1841–1847.

72 Bourne A, Barnes K, Taylor BA, Turner AJ, Kenny AJ **1989**, Membrane peptidases in the pig choroid plexus and on other cell surface in contact with the cerebrospinal fluid, *Biochem. J.* 259, 69–80.

73 Strazielle N, Khuth ST, Murat A, Chalon A, Giraudon P, Belin MF, Ghersi-Egea JF **2003**, Pro-inflammatory cytokines modulate matrix metalloproteinase secretion and organic anion transport at the blood-cerebrospinal fluid barrier, *J. Neuropathol. Exp. Neurol.* 62, 1254–1264.

74 Schnabel R, Bernstein HG, Luppa H, Lojda Z, Barth A **1992**, Aminopeptidases in the centrumventricular organs of the mouse brain: a histochemical study, *Neuroscience* 47, 431–438.

75 Brown KA **2001**, Factors modifying the migration of lymphocytes across the blood-brain barrier, *Int. Immunopharmacol.* 1, 2043–2062.

76 Nicolson GL, Menter DG, Herrmann J, Cavanaugh P, Jia L, Hamada J, Yun Z, Nakajima M, Marchetti D **1994**, Tumor metastasis to brain: role of endothelial cells, neurotrophins, and paracrine growth factors, *Crit. Rev. Oncog.* 5, 451–471.

77 Faustmann PM, Dermietzel R **1985**, Extravasation of polymorphonuclear leukocytes from the cerebral microvasculature. Inflammatory response induced by alpha-bungarotoxin, *Cell Tissue Res.* 242, 399–407.

78 Sandig M, Negrou E, Rogers KA **1997**, Changes in distribution of LFA-1, catenins, and F-actin during transendothelial migration of monocytes in culture, *Trends Neurosci.* 14, 14–15.

79 Nottel HS **1999**, Interactions between macrophages and brain microvascular endothelial cells: role in pathogenesis of HIV-1 infection and blood-brain barrier function, *J. Neurovirol.* 5, 659–669.

80 Lukes A, Mun-Bryce S, Lukes M, Rosenberg GA **1999**, Extracellular matrix degradation by metalloproteinases and central nervous system diseases, *Mol. Neurobiol.* 19, 267–284.

81 Chang DI, Hosomi N, Lucero J, Heo JH, Abumiya T, Mazar AP, del Zoppo GJ **2003**, Activation systems for latent matrix metalloproteinase-2 are upregulated immediately after focal cerebral ischemia, *J. Cereb. Blood Flow Metab.* 23, 1408–1419.

82 Madri JA, Graesser D, Haas T **1996**, The roles of adhesion molecules and proteinases in lymphocyte transendothelial migration, *Biochem. Cell Biol.* 74, 749–757.

83 Wang X, Lee SR, Arai K, Lee SR, Tsuji K, Rebeck GW, Lo EH **2003**, Lipoprotein receptor-mediated induction of matrix-metalloproteinase by tissue plasminogen activator, *Nat. Med.* 9, 1313–1317.

84 Yepes M, Sandkvist M, Moore EG, Bugge TH, Strickland DK, Lawrence DA **2003**, Tissue-type plasminogen activator induces opening of the blood-brain barrier via LDL receptor-related protein, *J. Clin. Invest.* 112, 1533–1540.

85 Gasche Y, Copin JC, Sugawara T, Fujimura M, Chan PH **2001**, Matrix metalloproteinase inhibition prevents oxidative stress-associated blood-brain barrier disruption after transient focal cerebral ischemia, *J. Cereb. Blood Flow Metab.* 21, 1393–1400.

86 Coussens LM, Werb Z **1996**, Matrix metalloproteinase and the development of cancer, *Chem. Biol.* 3, 895–904.

87 Wang M, Wang T, Liu S, Yoshida D, Teramoto A **2003**, The expression of matrix metalloproteinase-2 and -9 in human gliomas of different pathological grades, *Brain Tumor Pathol.* 20, 65–72.

88 Wolburg H, Wolburg-Buchholz K, Kraus J, Rascher-Eggstein G, Liebner S, Hamm S, Duffner F, Grote EH, Risau W, Engelhardt B **2003**, Localization of claudin-3 in tight junctions of the blood-brain barrier is selectively lost during experimental autoimmune encephalomyelitis and human glioblastoma multiforme, *Acta Neuropathol.* 105, 586–592.

89 Bar-Or A, Nuttall RK, Duddy M, Alter A, Kim HJ, Ifergan I, Pennington CJ, Bourgoin P, Edwards DR, Yong VW **2003**, Analysis of all matrix metalloproteinase members in leukocytes emphasize monocytes as major inflammatory mediators in multiple sclerosis, *Brain* 126, 1–12.

90 Maeda A, Sobel RA **1996**, Matrix metalloproteinases in the normal human central nervous system, microglial nodules, and multiple sclerosis lesions, *J. Neuropathol. Exp. Neurol.* 55, 300–309.

91 Kunz J, Krause S, Gehrmann J, Dermietzel R **1995**, Changes in the expression pattern of blood-brain barrier-associated pericytic aminopeptidase N (pAP N) in the course of acute experimental autoimmune encephalomyelitis, *J. Neuroimmunol.* 59, 41–55.

92 Sanchez-del-Rio MA, Reuter UB **2004**, Migraine aura: new information and underlying mechanisms, *Curr. Opin. Neurol.* 17, 289–293.

93 Gursoy-Ozdemir Y, Qiu J, Matsuoka N, Bolay H, Bermpohl D, Jin H, Wang X, Rosenberg GA, Lo EH, Moskowitz MA **2004**, Cortical spreading depression activates and upregulates MMP-9, *J. Clin. Invest.* 113, 1447–1455.

Part IV
Culturing the Blood-Brain Barrier

15
Modeling the Blood-Brain Barrier

*Roméo Cecchelli, Caroline Coisne, Lucie Dehouck, Florence Miller,
Marie-Pierre Dehouck, Valérie Buée-Scherrer, and Bénédicte Dehouck*

15.1
Introduction

15.1.1
In Vitro BBB Model Interests

The first notions and the discovery of the blood-brain barrier (BBB) were done *in vivo* by Ehrlich [1] and Goldman [2]. Since their work, *in vivo* studies have attempted to evaluate the exchanges between blood and brain parenchyma. Since the middle of the 19th century, *in vitro* approaches were developed with, first, the use of isolated brain capillaries [3–5]. As these vessels are metabolically active, they allow numerous studies regarding enzymatic activity [6, 7]. However, many reasons, such as the presence of remnants astrocyte foot processes [8] and pericytes, as well as the rapid and serious depletion of ATP in these isolated vessels [9], encouraged researchers to establish brain capillary endothelial cell culture [10, 11]. In culture, brain capillary endothelial cells form an endothelium-like monolayer and retain many of the BBB properties; and thus they can be used as *in vitro* BBB models that allow going into further cellular and molecular mechanisms. Indeed, from that time, *in vitro* BBB models have played an important role in the development of our current understanding on the BBB. As simple and dynamic experimental systems, they can still be considered valuable tools for future investigations, such as cerebral endothelium disturbances observed in CNS infections and brain-targeting drug delivery.

This chapter focuses on some of the current methodologies for the establishment of *in vitro* BBB models, presenting the knowledge allowing "the culture of BBB", meaning the establishment of well characterized *in vitro* BBB models.

15.1.2
The BBB: Brain Capillary Endothelial Cells and Brain Parenchyma Cells

Since the end of the 19th century, brain capillary endothelial cells have been known to be responsible for the limited exchanges between blood and brain parenchyma. They form an endothelium which can be distinguished from the other vascular beds by the presence of continuous tight junctions and the absence of fenestration or channels [12]. Both of these characteristics reduce the nonspecific transport of molecules across the BBB. The blood-brain exchanges involve specific carrier-mediated transport systems that facilitate the uptake of nutrients. Thus, BBB characteristics are crucial for the maintenance of brain homeostasis.

Numerous studies reported that the brain endothelium phenotype is induced by surrounding brain cells, such as astrocytes, microglia, pericytes and neurons [13–19]. Although a few studies have shown that brain parenchyma cells are able to induce tight junction structures in aortic endothelial cells [20], the differentiation occurs only partially as compared to cerebral endothelial cells [21]. We have also shown the responsiveness of aortic endothelial cells to glial cell population influences [22]. However, it is worth mentioning that, in this study, glial cells had to be first cocultivated with brain capillary endothelial cells before being able to interact with aortic endothelial cells. As indicated in other studies, this suggests that specific bidirectional exchanges between brain parenchyma and brain endothelium are necessary for the establishment of BBB properties [15] and may explain why noncerebral endothelial cells are less responsive to brain parenchyma cells.

From all these studies, the idea emerges that the cerebral endothelial phenotype might contribute to a better responsiveness to brain parenchyma cells and explain why starting from brain microvessel endothelium is more successful for recreating the *in vivo* BBB situation *in vitro*.

15.2
Culturing Brain Capillary Endothelial Cells

15.2.1
Brain Capillary Endothelial Cell Isolation

15.2.1.1 Brain Capillary Endothelial Cell Isolation

Although everyone agrees that the BBB is localized at brain endothelium, a differentiation of endothelia derived from different parts of the brain should be done (Fig. 15.1). Indeed, a high level of heterogeneity in the properties of an endothelial cell population has been associated with the diverse locations that they occupy in the vascular tree from which they are isolated. Recently, DNA microarrays have indicated a significant variety in the gene expression pattern between endothelial cells from large vessels and microvessels [23]. Moreover,

15.2 Culturing Brain Capillary Endothelial Cells | 339

(a)
CORTEX
|
Mechanical Homogenization
and/or enzymatic digestion
↓
[MICROVESSELS
Arterioles/ Capillaries/ Venules]
|
Enzymatic digestion
↓
[Brain
Arterioles/ Capillaries/ Venules
Endothelial cells]

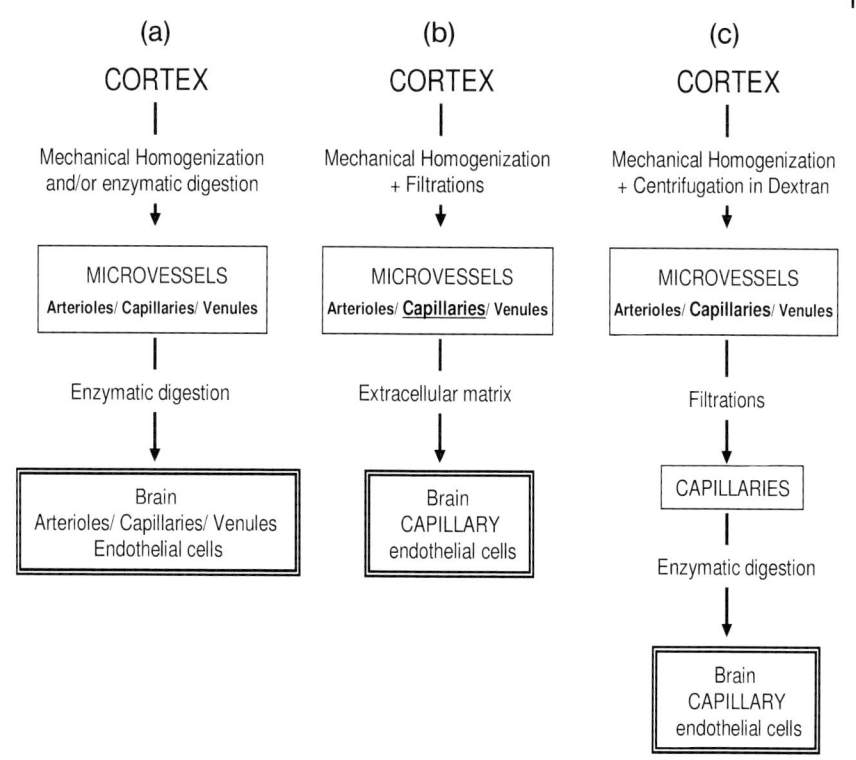

(b)
CORTEX
|
Mechanical Homogenization
+ Filtrations
↓
[MICROVESSELS
Arterioles/ **Capillaries**/ Venules]
|
Extracellular matrix
↓
[Brain
CAPILLARY
endothelial cells]

(c)
CORTEX
|
Mechanical Homogenization
+ Centrifugation in Dextran
↓
[MICROVESSELS
Arterioles/ **Capillaries**/ Venules]
|
Filtrations
↓
[CAPILLARIES]
|
Enzymatic digestion
↓
[Brain
CAPILLARY
endothelial cells]

Fig. 15.1 Isolation procedures of brain capillary endothelial cells.
(A) General procedure to isolate endothelial cells from cerebral microvessels.
(B) Isolation of bovine brain capillary endothelial cells (BBCECs), procedure from Meresse et al. [26].
(C) Isolation of mouse brain capillary endothelial cells (MBCECs), procedure from Coisne et al. [28].

Murugesan et al. [24] suggested that the induction of chemokines induced by lysophosphatidylcholine is different in microvascular endothelial cells compared to that of large vessel endothelial cells. Concerning brain vessels, Song and Pachter [25] showed a differential expression of BBB markers along isolated brain microvascular segments with obvious and progressive changes in the expression of these markers with the vessel size. Only vessels which can be considered as capillaries showed the BBB phenotype. Thus modeling the BBB requires a careful isolation and culture of primary endothelial cells coming from brain capillaries.

Several procedures and many modifications have been used to obtain endothelial cell primary cultures. The species and source of cerebral microvessels vary from one laboratory to another. Commonly, the first steps of such procedures use mechanical and/or enzymatic means to disperse brain tissue in order to collect brain vessels (Fig. 15.1A) [26–29]. However, enzymes should be used

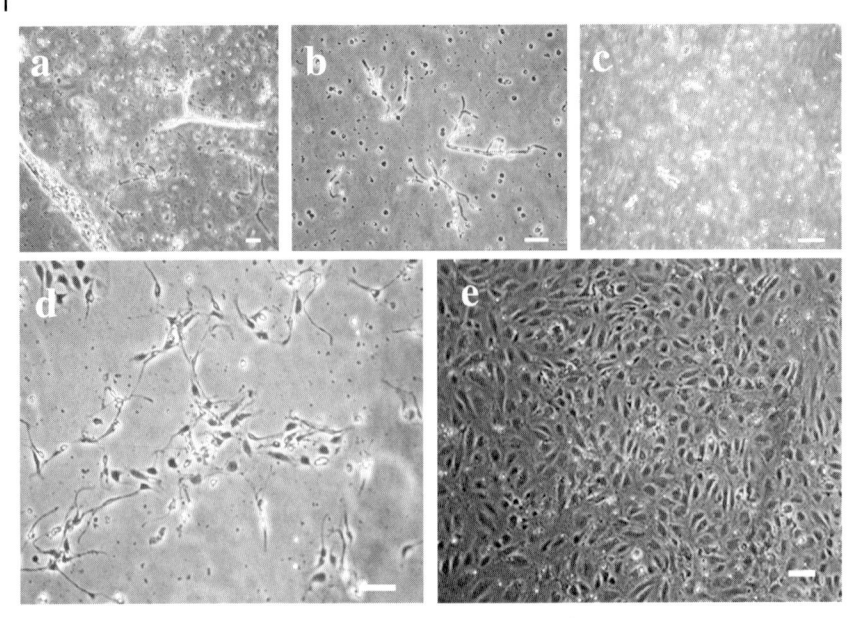

Fig. 15.3 Mouse brain capillary endothelial cell isolation procedure. Phase contrast micrographs of: (a) vascular components in mouse brain homogenates, (b) cerebral capillaries isolated from vascular components, (c) capillaries after collagenase/dispase digestion, (d, e) mouse brain capillary endothelial cells at 24 h (d) and 5 days (e) after seeding on matrigel-coated plastic dish. Bars = 50 µm. Micrographs were obtained from Coisne et al. [28].

Concerning mouse brain endothelial cells, viability and proliferation after trypsinization appear trickier and so far no real success in mouse endothelial cell subculture has been described. These cells show low rates of amplification, lack of development of real endothelium and loss of BBB phenotype. This fact explains why most attempts to model BBB model in rat and mouse use primary endothelial cells and why authors have used transfection to obtain immortalization [31]. Unfortunately, the cerebral phenotype can be modified in these conditions [32, 33].

15.2.1.4 Immortalization

Immortalized endothelial cell lines have been established by introducing genes, such as the SV40 large T-antigen gene, into primary cultured cells. The first obvious problem is the potential loss of contact inhibition after immortalization which, as explained below, can result in an incomplete cerebral phenotype. Many authors circumvent this problem by using conditional cell division being able to stop growth. Temperature-sensitive SV40 large T-antigen is inactivated by shifting the culture temperature from 33 to 37 °C, under which condition cell

Table 15.1 Examples of *in vitro* BBB models developed in commonly used species. Abbreviations used: *ACM* astrocyte conditioned medium, *γ-GT* γ-glutamyl transpeptidase, *mdr* multidrug resistance, *Pe studies* permeability studies, *SEM* scanning electron microscopy, *TEER* transendothelial electrical resistance, *TEM* transmission electron microscopy, *TJ*: tight junctions.

Species	Primary or subcultured endothelial cell	Isolation procedure	Endothelial cell	Coculture	BBB phenotype	Ref.
Cattle	Subcultured	Mechanical	From capillaries	With rat glial cells	ZO-1, occludin, P-gp, LDL and transferrin receptors, TEER, γ-GT, drug transport screening, Pe studies, *in vivo* correlation	70
Cattle	Subcultured	Mechanical and enzymatic	From capillaries	ACM	TEER, ZO-1	13
Pig	Subcultured	Enzymatic	From microvessels	No	TEER, Pe studies	29
Human	Primary	Enzymatic	From microvessels	No	TEER, P-gp, Pe studies	71
Macaque	Subcultured	Mechanical and enzymatic	From capillaries	With macaque astrocytes	GLUT-1, γ-GT, TJ in TEM, VCAM-1/E-selectin (±TNF-α)	69
Rat	Subcultured	Enzymatic	From microvessels	With rat astrocytes	TEER, γ-GT, OX-26, TJ in SEM	72
Mouse	Primary	Mechanical and enzymatic	From capillaries	With mouse glial cells	Occludin, claudins, JAM-A, Pe studies, TEER, Pg-P, CAMs (±LPS) MECA-32 negative, AHNAK	28
Mouse	Primary	Mechanical and enzymatic	From capillaries	No	Occludin, ZO-1	25
Mouse	Subcultured	Mechanical and enzymatic	From microvessels	No	GLUT-1, occludin, mdr, CAMs (±TNF-α)	39

growth is repressed or arrested [34]. Unfortunately, even so, a loss of BBB properties is observed. Is it the transfection method? Is it the disruption of expression of essential genes by the transgene? Or is it the fact that, at the beginning, these cells do not have any brain capillary endothelial cell phenotype? So far, no explanation has been given.

Recently, conditionally immortalized cells have been established by using transgenic mice [35] or rats [36] harboring the temperature-sensitive SV40 large T antigen. Using these transgenic animals, Hosoya et al. [37, 38] have established conditionally immortalized brain capillary endothelial cells (TM-BBB, TR-BBB). As reviewed by Terasaki et al. [31], these two cell lines showed more BBB features than traditionally transfected cells. However, TR-BBB and TM-BBB do not form rigid tight junctions and may not be used for permeability studies.

In conclusion, although immortalized cell lines possess obvious culture advantages, their use should be avoided for *in vitro* permeability studies. But, some of these cells do possess some *in vivo* functions, which, when determined, may be used for *in vitro* studies.

15.2.1.5 Purity

The problem of contamination with nonendothelial cells has been encountered by anyone isolating brain capillary endothelial cells. Indeed, depending on the method used, fibroblasts, astrocytes and mostly pericytes have been observed as contaminating cells. The first two cell types are usually discarded when endothelial cells are isolated from brain microvessels. However, the separation of endothelial cells from pericytes, which are inherent to the capillary basal lamina, is trickier.

The nonenzymatic method used for bovine brain capillary endothelial cell isolation in the laboratory allows the separation of endothelial cells and pericytes by microtrypsinization [14]. This separation step is even more important, as bovine endothelial cells will be subcultivated. Indeed, the presence of a few pericytes in the primary culture may lead to a real colonization of endothelial cells by pericytes after a few trypsinizations, as pericytes thrive under endothelial cell culture conditions.

Different methods have been described to get rid of pericytes. Song and Pachter [25] and Wu et al. [39] described an endothelial cell purification step using magnetic beads. Perrière et al. [40] used the fact that P-glycoprotein expression is much higher in brain capillary endothelial cells than in contaminating cells, so that the cell-toxic P-glycoprotein substrates enter the contaminating cell and cause their death. Other authors, working with primary endothelial cell subcultures, utilize a brief trypsinization step as the endothelial cells approach confluence. Thus, endothelial cells are readily detached, while pericytes or smooth muscle cells still adhere to the substratum [30, 41].

In our mouse procedure [28], collagenase/dispase digestion of isolated capillaries promoted the migration of endothelial cells. Hoechst staining and tight junction immunolabeling showed that mouse brain capillary endothelial cells

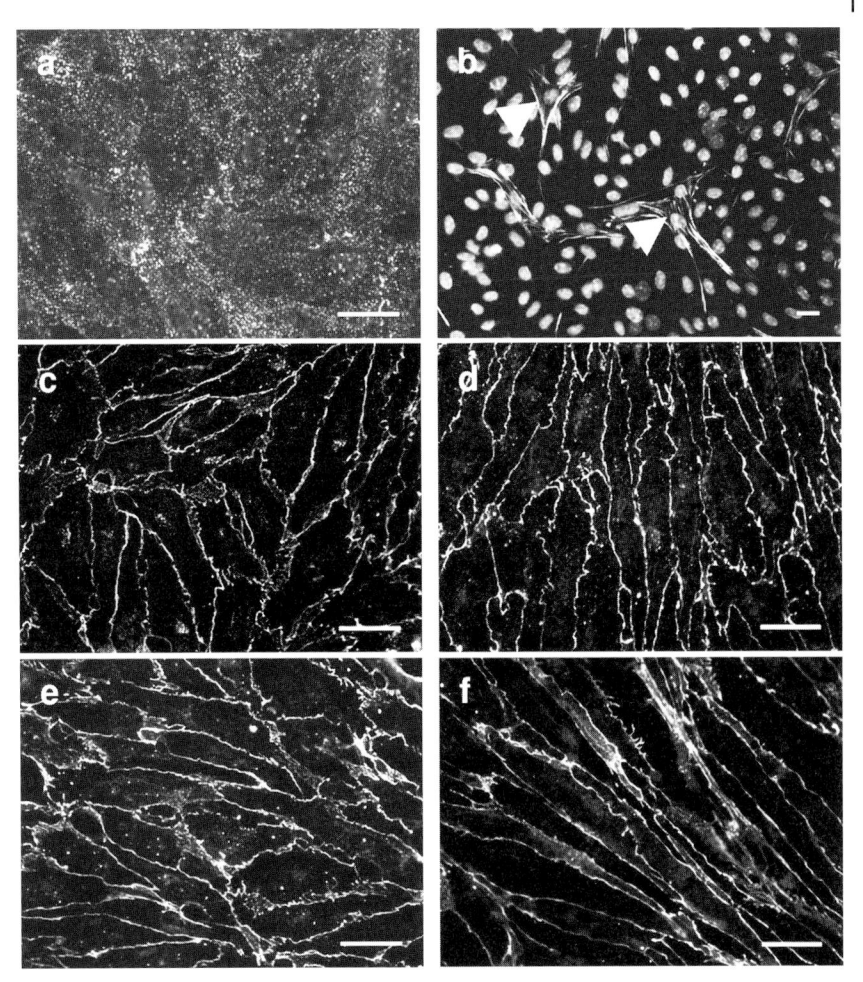

Fig. 15.4 Immunocytochemical characterization of confluent mouse brain capillary endothelial cell monolayer cocultivated with mouse glial cells. (a) PECAM-1: whole cell surface presents dense distribution of immunoreactive dots. (b) Nuclear Hoechst staining and alpha-actin labeling of pericytes (arrow). (c–f) Cell-border localization of tight junction-associated proteins: (c) claudin-5, (d) occludin, (e) claudin-3, (f) JAM-A. Bars = 25 μm. Micrographs were obtained from Coisne et al. [28].

formed a complete monolayer on matrigel-coated inserts (Fig. 15.4). The endothelial phenotype was confirmed by the expression of von Willebrand factor and PECAM-1 (Fig. 15.4a). However, the digestion of capillaries could generate some contaminating pericytes (Fig. 15.4b). FACS analysis showed that such cells were scarce compared to PECAM-1 positive cells (5% vs 95%). This low level of contamination was obtained without the use of any endothelial cell purification step and remained stable as experiments were made with primary endothelial cells. Furthermore, in spite of these contaminating cells, mouse brain capillary endothelial cells formed a well differentiated *in vitro* BBB (see below).

Over-proliferation of pericytes can disturb the investigation of BBB properties, such as studies of brain capillary endothelium permeability. However, efficient methods have been developed and are now available to remove this cell type from *in vitro* endothelial cell culture. Paradoxically, in parallel with the development of these methods, more and more studies have revealed pericyte influences on cerebral endothelium [16]. The development of endothelial cell coculture with pericytes and study of pericyte influences on BBB properties should be attentively followed up, since the role of pericytes may have been underestimated until now [42].

15.2.1.6 Species

As mentioned before, the success of *in vitro* BBB model establishment seems to be species-dependent. So far, no explanation has been given. However, two major hypotheses may be stated: (1) the longevity of each species may determine the capacity of the cells for *in vitro* cell division, (2) the culture conditions and reagents, e.g. the use of bovine serum, may currently be much better optimized for bovine than for murine species.

If the availability of tools and technical difficulties may influence the choice of the species, the decision has to be made according to what the model will be used for. If the establishment of a well differentiated human *in vitro* BBB model is one major priority, other species such as bovine, murine, macaque (which have so far allowed progress in BBB understanding) are still valuable tools for future investigations.

In our laboratory, the significance of developing an *in vitro* BBB model using murine tissue is supported by the fact that many *in vivo* models of chronic neuro-inflammation are developed in mouse and it also presents the possibility for creating a BBB model from gene-targeted (KO) animals, to complete *in vivo* studies.

15.2.2
Coculture

In this chapter entitled "Modeling the Blood-Brain Barrier", we must also mention the crucial role of brain capillary surrounding cells in the induction, and also in the maintenance and regulation of BBB properties. As mentioned above, numerous studies have shown the role of astrocytes, pericytes, microglia and recently, neurons [19, 43]. Recent data showed that a mixture of astrocytes and microglia was more efficient than purified astrocytes with regard to endothelium responsiveness, suggesting that intercommunication between each type of brain capillary cells may be important [15, 44].

Some studies also showed that endothelial cells are responsible for the astrocyte phenotype, emphasizing the reciprocal interrelations between coculture partners [45–48]. This fact should guide researchers in their choice of *in vitro* culture partners as it may determine the responsiveness of the system. In our laboratory, a coculture of bovine brain capillary endothelial cells and rat glial cells has been validated as an *in vitro* BBB model. Although useful, this heterologous model may lead to misinterpretations concerning coculture partner interrelations. For the study of inflammatory events, we have developed a syngenic *in vitro* model comprising of a coculture of mouse brain capillary endothelial cells with mouse glial cells.

When considering endothelial cells, attention should be paid to the choice and use of endothelial cell partners, such as glial cells. Immortalized cell lines may not be as efficient as primary cells in endothelial BBB property induction [49].

15.3
Characteristics Required for a Useful *In Vitro* BBB Model

15.3.1
Confluent Monolayer

As mentioned above, the BBB is first an endothelium in which cells have to be jointly packed and in which cell contact inhibition is essential for monolayer function. Regulation of receptor expression is usually dependent on cell-cell contact and apposition. For example, LDL receptor expression and function are regulated by confluence. Indeed, we described a switch of this receptor from an endothelial cell lipid supplier during cell growth to a receptor allowing the transport of LDL after the cerebral endothelium was found (Fig. 15.5) [50]. Confluence of endothelial cells is also necessary to evaluate the endothelial permeability.

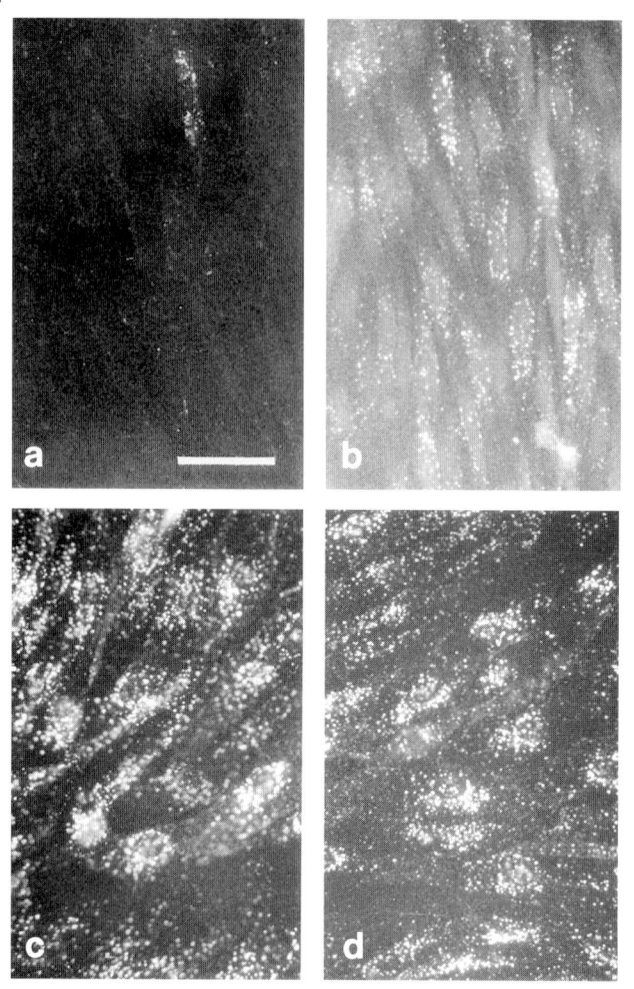

Fig. 15.5 Effect of filipin (a cholesterol-binding agent causing the disassembly of caveolae) on the DiI-LDL endocytosis by bovine brain capillary endothelial cells. Comparison of filipin effects on BBB differentiated endothelial cells (A, B) and endothelial cells in growing phase (C, D). A vs B shows reverse effect of filipin: (A) DiI-LDL endocytosis after filipin treatment, (B) DiI-LDL endocytosis after filipin removal. Bar=50 μm. C, D: DiI-LDL endocytosis was carried out with brain capillary endothelial cells without (C) or with (D) pretreatment with filipin. Micrographs were obtained from Dehouck et al. [50].

15.3.2
Tight Junctions and Paracellular Permeability

The presence of well differentiated tight junctions is one of the principal characteristics of the BBB, as these structures reduce paracellular transport, which can be considered as nonspecific transport through the cerebral endothelium. In this way, tight junctions participate actively in the selective barrier that contributes to the maintenance of brain homeostasis.

In another way, just as confluence is important for receptor expression and function, tight junctions are essential for the regulation of endothelial cell polarity, contributing to protein distribution at the cell surface and distinguishing the luminal and abluminal faces of the endothelium [51, 52]. *In vitro* cerebral endothelium polarity is particularly important to evaluate bidirectional and selective exchanges between brain and blood. For these reasons, the presence and efficiency of tight junctions must be evaluated before an *in vitro* endothelium is chosen as a BBB model.

Tight junctions are complex structures made of different proteins, such as occludin, claudin-5, claudin-3 and JAM-A (see Chapter 4) [53, 54]. To test the presence of tight junctions *in vitro*, immunostaining of these proteins can be assessed (Fig. 15.4). Caution must be taken, as the presence of several of these proteins on the cerebral endothelium is still in debate. This is the case for claudin-1 and claudin-3. *In vivo*, the claudins found in cerebral endothelial cells were first claudin-5 and claudin-1. Discrepancies concerning the presence of either claudin-1 or claudin-3 on cerebral endothelium were due to the use of a cross-reactive antibody [55].

The presence of occludin, claudin-5, claudin-3 and JAM-A on *in vitro* endothelium is a first step to study tight junctions; and the peripheral distribution of these proteins is one step further, because they were described as tight junctional transmembrane molecules believed to restrict permeability [54]. However, studies have shown that peripheral expression of these proteins is not sufficient and does not always correlate with the tightness of these structures [55, 56]. Indeed, as the role and regulation of tight junctions is not yet completely understood, their immunolocalization has to be completed with other studies, such as the evaluation of endothelium permeability for low molecular weight and hydrophilic molecules which do not interact actively with endothelial cells. The attenuated flux of these molecules (e.g. sucrose, inulin, Lucifer yellow, HRP) emphasizes the presence and efficiency of tight junctions on *in vitro* monolayers. Furthermore, the presence and efficiency of tight junctions reduce or control the exchange of charged molecules through the BBB. Transendothelial electrical resistance (TEER) directly corresponds to the "tightness" of *in vitro* endothelium. In our laboratory, this assessment is used during the establishment of the BBB barrier *in vitro* model. However, for recent experiments, permeability studies have replaced TEER as a measurement tool, as in our hands they are more sensitive, more reproducible and can easily be performed in the different laboratories working on this topic.

We should mention that coculture of endothelial cells with glial cells is essential for a good *in vitro* tight junction differentiation and that the *in vitro* presence of well differentiated tight junctions is important, not only to complete our knowledge of their structures and regulation but also to go further into BBB property investigation. Indeed, not only paracellular transport but also vesicular- and receptor-mediated transport depends on the expression of these tight junctions [14, 57].

15.3.3
Transcellular Transport, Receptor Mediated Transport

For reasons discussed in the former paragraph, *in vitro* transcellular transport studies should be performed on endothelia presenting well differentiated tight junctions.

Numerous receptors have been described as participating in nutrient transport through the BBB [58, 59]. The presence and function of these receptors can be checked to validate in vitro BBB models. *In vitro* expression of these proteins can either be used to study transcytosis of nutrients (Fig. 15.5) [50, 60], the transcytosis signalling pathway and/or to investigate cerebral drug delivery [61–63].

The next paragraphs list a few examples of receptor/protein expression depending on endothelium barrier function. As with tight junctions, most of the studies focus on astrocyte influence on receptor expression and vesicular transports. Recently, Berezowski et al. [16] have shown that pericytes also alter BBB features.

As mentioned before, P-glycoprotein (P-gp) is an important characteristic of BBB but is often lost in culture conditions, so the presence of this protein attests to the congruence of the model with the *in vivo* situation (Fig. 15.6). P-gp is a flippase involved in certain drug transport, thus its presence is a crucial issue when using *in vitro* models to study BBB drug transports. The functionality of the P-gp can be shown using the transport of vincristine, a P-gp substrate [64]. Also, multidrug resistance-associated proteins have been recently located in brain capillary endothelial cells and their expression is regulated by either astrocytes or pericytes [16]. Expression of these emerging proteins should be considered to study BBB drug transport (see Chapter 19).

Other proteins with unknown functions have been investigated in the BBB. The lack of MECA-32, a mouse antigen commonly described on nondifferentiated brain endothelial cells and noncerebral endothelial cells *in vivo*, suggests a good differentiation of the *in vitro* cerebral endothelium [65]. Recently, Gentil et al. [66] identify AHNAK as a protein marker of endothelial cells with barrier properties. Present *in vitro*, this protein relocates from the cytosol to the plasma membrane when endothelial cells acquire BBB properties (Fig. 15.6).

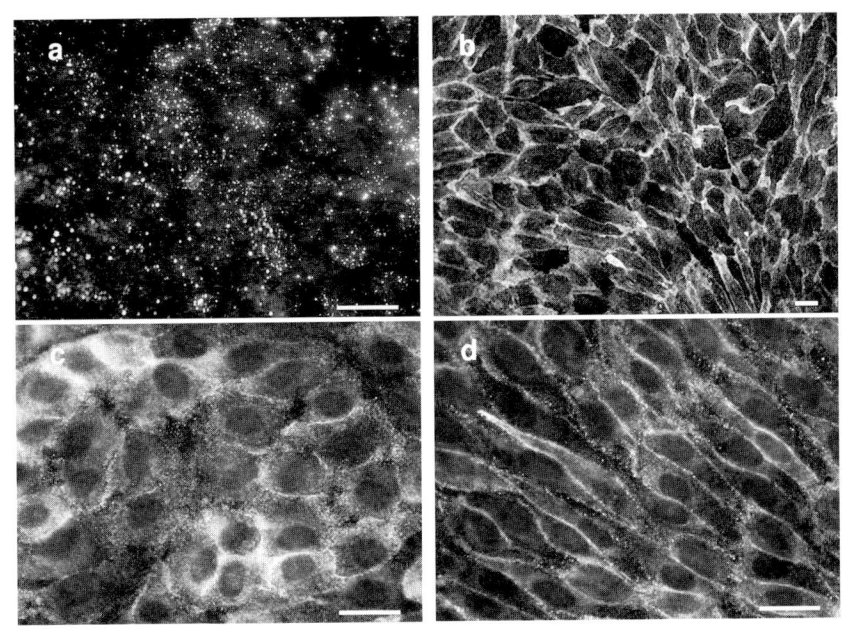

Fig. 15.6 (a, b): Characterisation of confluent mouse brain capillary endothelial cell monolayer cocultivated with glial cells: immunofluorescence of (a) P-gp (mAb: C219), (b) AHNAK, (c, d) AHNAK immunolabeling on confluent bovine brain capillary endothelial cell in "solo" culture (c) or cocultured with rat glial cells for 12 days (d) [66]. Bars = 25 μm.

15.3.4
Expression of Endothelial Adhesion Molecules/Vascular Inflammatory Markers

Cerebral endothelium disturbances observed in CNS infections emphasize the necessity for *in vitro* BBB models that mimic immune events. One approach to validate the relevance of a BBB model for studying conditions of inflammation is to examine the expression of adhesion molecules/vascular inflammatory markers on the *in vitro* endothelium. BBB responsiveness to inflammatory treatments can also be tested, focusing on endothelium permeability changes or the regulation of endothelial adhesion molecules. These *in vitro* studies should correspond to the literature regarding *in vivo* brain microvascular endothelium pathogenesis. Coisne et al. [28] have shown that LPS treatment of cocultures led to the upregulation of endothelial ICAM-1 and VCAM-1; and they referred to publications demonstrating the upregulation of ICAM-1 and VCAM-1 on cerebral endothelium in experimental autoimmune encephalomyelitis (EAE) [67], systemic lupus erythematosus (SLE) [68] and AIDS [69].

Furthermore, the establishment of an *in vitro* model of pathological BBB can be set up by either inducing a pathological environment [15] or isolating brain capillary endothelial cells from experimental disease-induced animal models or gene-targeted (KO) animals. These models will be valuable tools to complete the *in vivo* studies.

15.4
Conclusion

In vitro models are most useful when they closely mimic diverse *in vivo* situations. According to the literature, the establishment of a BBB model requires: (1) the provision of endothelial cells from cerebral capillaries, (2) the use of primary endothelial cells, (3) the coculture of these endothelial cells with glial cells. These three conditions are the most accurate ones, as they increase the success in obtaining a well differentiated endothelium which can be used as a BBB model. Because *in vitro* BBB have drawbacks and advantages, *in vitro* models are useful only if researchers are aware of the model limits. In any case, the use of *in vitro* models requires precautions; and checking the hypotheses *in vivo* still remains the most accurate proof. Under these conditions, modeling the BBB is possible and can be extremely valuable when correctly investigated with care.

References

1 P. Ehrlich **1885**, *Das Sauerstoff-Bedürfnis des Organismus. Eine farbenanalytische Studie*, Herschwald, Berlin.
2 E. E. Goldmann **1913**, *Beitrag zur Physiologie des Plexus Choroïdus und der Hirnhaute*, Herschwald, Berlin.
3 F. Joo, I. Karnushina **1973**, *Cytobios* 8, 41–48.
4 K. Brendel, E. Meezan, E. C. Carlson **1974**, *Science* 185, 953–955.
5 G. W. Goldstein, J. S. Wolinsky, J. Csejtey, I. Diamond **1975**, *J. Neurochem.* 25, 715–717.
6 H. M. Eisenberg, R. L. Suddith **1979**, *Science* 206, 1083–1085.
7 G. W. Goldstein, J. Csejtey, I. Diamond **1977**, *J. Neurochem.* 28, 725–728.
8 F. P. White, G. R. Dutton, M. D. Norenberg **1981**, *J. Neurochem.* 36, 328–332.
9 F. Lasbennes, J. Gayet **1984**, *Neurochem. Res.* 9, 1–10.
10 L. E. Debault, L. E. Kahn, S. P. Frommes, P. A. Cancilla **1979**, *In Vitro* 15, 473–487.
11 P. D. Bowman, A. L. Betz, D. Ar, J. S. Wolinsky, J. B. Penney, R. R. Shivers, G. W. Goldstein **1981**, *In Vitro* 17, 353–362.
12 T. S. Reese, M. J. Karnovsky **1967**, *J. Cell. Biol.* 34, 207–217.
13 L. L. Rubin, D. E. Hall, S. Porter, K. Barbu, C. Cannon, H. C. Horner, M. Janatpour, C. W. Liaw, K. Manning, J. Morales **1991**, *J. Cell. Biol.* 115, 1725–1735.
14 M. P. Dehouck, S. Meresse, P. Delorme, J. C. Fruchart, R. Cecchelli **1990**, *J. Neurochem.* 54, 1798–1801.

15 L. Descamps, C. Coisne, B. Dehouck, R. Cecchelli, G. Torpier **2003**, *Glia* 42, 46–58.
16 V. Berezowski, C. Landry, M. P. Dehouck, R. Cecchelli, L. Fenart **2004**, *Brain Res.* 1018, 1–9.
17 K. Hayashi, S. Nakao, R. Nakaoke, S. Nakagawa, N. Kitagawa, M. Niwa **2004**, *Regul. Peptides* 123, 77–83.
18 S. Hori, S. Ohtsuki, K. Hosoya, E. Nakashima, T. Terasaki **2004**, *J. Neurochem.* 89, 503–513.
19 G. Savettieri, I. Di Liegro, C. Catania, L. Licata, G. L. Pitarresi, S. D'Agostino, G. Schiera, V. de Caro, G. Giandalia, L. I. Giannola, A. Cestelli **2000**, *Neuroreport* 11, 1081–1084.
20 K. A. Stanness, E. Guatteo, D. Janigro **1996**, *Neurotoxicology* 17, 481–496.
21 N. Bernoud, L. Fenart, P. Moliere, M. P. Dehouck, M. Lagarde, R. Cecchelli, J. Lecerf **1999**, *J. Neurochem.* 72, 338–345.
22 B. Dehouck, M. P. Dehouck, J. C. Fruchart, R. Cecchelli **1994**, *J. Cell. Biol.* 126, 465–473.
23 J. T. Chi, H. Y. Chang, G. Haraldsen, F. L. Jahnsen, O. G. Troyanskaya, D. S. Chang, Z. Wang, S. G. Rockson, M. Van de Rijn, D. Botstein, P. O. Brown **2003**, *Proc. Natl Acad. Sci. USA* 100, 10623–10628.
24 G. Murugesan, M. R. Sandhya Rani, C. E. Gerber, C. Mukhopadhyay, R. M. Ransohoff, G. M. Chisolm, K. Kottke Marchant **2003**, *J. Mol. Cell Cardiol.* 35, 1375–1384.
25 L. Song, J. S. Pachter **2003**, *In Vitro Cell. Dev. Biol. Anim.* 39, 313–320.
26 S. Meresse, M. P. Dehouck, P. Delorme, M. Bensaid, J. P. Tauber, C. Delbart, J. C. Fruchart, R. Cecchelli **1989**, *J. Neurochem.* 53, 1363–1371.
27 U. Tontsch, H. C. Bauer **1989**, *Microvasc. Res.* 37, 148–161.
28 C. Coisne, L. Dehouck, C. Faveeuw, Y. Delplace, F. Miller, C. Landry, C. Morissette, L. Fenart, R. Cecchelli, P. Tremblay, B. Dehouck **2005**, *Lab. Invest.* (in press).
29 H. Franke, H. Galla, C. T. Beuckmann **2000**, *Brain Res. Brain Res. Protocol.* 5, 248–256.
30 A. G. MacLean, M. S. Orandle, X. Alvarez, K. C. Williams, A. A. Lackner **2001**, *J. Neuroimmunol.* 118, 223–232.
31 T. Terasaki, S. Ohtsuki, S. Hori, H. Takanaga, E. Nakashima, K. Hosoya **2003**, *Drug Discov. Today* 8, 944–954.
32 R. K. Rohnelt, G. Hoch, Y. Reiss, B. Engelhardt **1997**, *Int. Immunol.* 9, 435–450.
33 Y. Omidi, L. Campbell, J. Barar, D. Connell, S. Akhtar, M. Gumbleton **2003**, *Brain Res.* 990, 95–112.
34 D. Lechardeur, B. Schwartz, D. Paulin, D. Scherman **1995**, *Exp. Cell. Res.* 220, 161–170.
35 M. Obinata **1997**, *Genes Cells* 2, 235–244.
36 R. Takahashi, M. Hirabayashi, N. Yanai, M. Obinata, M. Ueda **1999**, *Exp. Anim.* 48, 255–261.

37 K. I. Hosoya, T. Takashima, K. Tetsuka, T. Nagura, S. Ohtsuki, H. Takanaga, M. Ueda, N. Yanai, M. Obinata, T. Terasaki **2000**, *J. Drug Target.* 8, 357–370.

38 K. Hosoya, K. Tetsuka, K. Nagase, M. Tomi, S. Saeki, S. Ohtsuki, H. Takanaga, N. Yanai, M. Obinata, A. Kikuchi, T. Okano, T. Terasaki **2000**, *AAPS PharmSci.* 2, E27.

39 Z. Wu, F. M. Hofman, B. V. Zlokovic **2003**, *J. Neurosci. Methods* 130, 53–63.

40 N. Perriere, P. H. Demeuse, E. Garcia, A. Regina, M. Debray, J. P. Andreux, P. Couvreur, J. M. Scherrmann, J. Temsamani, P. O. Couraud, M. A. Deli, F. Roux, *Journal of Neurochemistry* **2005** (in press).

41 P. J. Gaillard, L. H. Voorwinden, J. L. Nielsen, A. Ivanov, R. Atsumi, H. Engman, C. Ringbom, A. G. de Boer, D. D. Breimer **2001**, *Eur. J. Pharm. Sci.* 12, 215–222.

42 M. Ramsauer, D. Krause, R. Dermietzel **2002**, *FASEB J.* 16, 1274–1276.

43 G. Schiera, E. Bono, M. P. Raffa, A. Gallo, G. L. Pitarresi, I. Di Liegro, G. Savettieri **2003**, *J. Cell. Mol. Med.* 7, 165–170.

44 V. Balasingam, K. Dickson, A. Brade, V. W. Yong **1996**, *Glia* 18, 11–26.

45 J. H. Tao Cheng, M. W. Brightman **1988**, *Int. J. Dev. Neurosci.* 6, 25–37.

46 C. Estrada, J. V. Bready, J. A. Berliner, W. M. Pardridge, P. A. Cancilla **1990**, *J. Neuropathol. Exp. Neurol.* 49, 539–549.

47 R. G. Ladenheim, I. Lacroix, N. Foignant Chaverot, A. D. Strosberg, P. O. Couraud **1993**, *J. Neurochem.* 60, 260–266.

48 E. J. Yoder **2002**, *Glia* 38, 137–145.

49 M. Boveri, V. Berezowski, A. Price, S. Slupek, A. M. Lenfant, C. Benaud, T. Hartung, R. Cecchelli, P. Prieto, M. P. Dehouck **2005**, *Glia* (in press).

50 B. Dehouck, L. Fenart, M. P. Dehouck, A. Pierce, G. Torpier, R. Cecchelli **1997**, *J. Cell. Biol.* 138, 877–889.

51 C. L. Farrell, W. M. Pardridge **1991**, *Proc. Natl Acad. Sci. USA* 88, 5779–5783.

52 E. Beaulieu, M. Demeule, L. Ghitescu, R. Beliveau **1997**, *Biochem. J.* 326, 539–544.

53 J. D. Huber, R. D. Egleton, T. P. Davis **2001**, *Trends Neurosci.* 24, 719–725.

54 H. Wolburg, A. Lippoldt **2002**, *Vascul. Pharmacol.* 38, 323–337.

55 S. Hamm, B. Dehouck, J. Kraus, K. Wolburg Buchholz, H. Wolburg, W. Risau, R. Cecchelli, B. Engelhardt, M. P. Dehouck **2004**, *Cell Tissue Res.* 315, 157–166.

56 P. Gao, R. R. Shivers **2004**, *J. Submicrosc. Cytol. Pathol.* 36, 7–15.

57 T. J. Raub, S. L. Kuentzel, G. A. Sawada **1992**, *Exp. Cell. Res.* 199, 330–340.

58 Q. R. Smith **2000**, *J. Nutr.* 130, 1016S–1022S.

59 W. M. Pardridge, R. J. Boado, C. R. Farrell **1990**, *J. Biol. Chem.* 265, 18035–18040.

60 L. Descamps, M. P. Dehouck, G. Torpier, R. Cecchelli **1996**, *Am. J. Physiol.* 270, H1149–H1158.

61 T. Sakaeda, T. J. Siahaan, K. L. Audus, V. J. Stella **2000**, *J. Drug Target.* 8, 195–204.

62 B. W. Song, H. V. Vinters, D. Wu, W. M. Pardridge **2002**, *J. Pharmacol. Exp. Ther.* 301, 605–610.

63 C. C. Visser, S. Stevanovic, L. Heleen Voorwinden, P. J. Gaillard, D. J. Crommelin, M. Danhof, A. G. de Boer **2004**, *J. Drug Target.* 12, 145–150.

64 L. Fenart, V. Buee Scherrer, L. Descamps, C. Duhem, M. G. Poullain, R. Cecchelli, M. P. Dehouck **1998**, *Pharm. Res.* 15, 993–1000.

65 R. Hallmann, D. N. Mayer, E. L. Berg, R. Broermann, E. C. Butcher **1995**, *Dev. Dyn.* 202, 325–332.

66 B. J. Gentil, C. Benaud, C. Delphin, C. Remy, V. Berezowski, R. Cecchelli, O. Feraud, D. Vittet, J. Baudier **2004**, *J. Cell. Physiol.* 203, 362–371.

67 B. J. Steffen, E. C. Butcher, B. Engelhardt **1994**, *Am. J. Pathol.* 145, 189–201.

68 A. Zameer, S. A. Hoffman **2003**, *J. Neuroimmunol.* 142, 67–74.

69 A. G. MacLean, M. S. Orandle, J. MacKey, K. C. Williams, X. Alvarez, A. A. Lackner **2002**, *J. Neuroimmunol.* 131, 98–103.

70 R. Cecchelli, B. Dehouck, L. Descamps, L. Fenart, V. Buee Scherrer, C. Duhem, S. Lundquist, M. Rentfel, G. Torpier, M. P. Dehouck **1999**, *Adv. Drug Deliv. Rev.* 36, 165–178.

71 D. Biegel, D. D. Spencer, J. S. Pachter **1995**, *Brain Res.* 692, 183–189.

72 P. Demeuse, A. Kerkhofs, C. Struys Ponsar, B. Knoops, C. Remacle, P. Van Den Bosch De Aguilar **2002**, *J. Neurosci. Methods* 121, 21–31.

16
Induction of Blood-Brain Barrier Properties in Cultured Endothelial Cells

Alla Zozulya, Christian Weidenfeller, and Hans-Joachim Galla

16.1
Introduction

Cerebral endothelial cells build up the inner microvascular wall, separating the blood from the brain parenchyma. These cells form a functional barrier limiting free exchange between the blood and the brain tissue proper. This structure is commonly known as the blood-brain barrier (BBB). The BBB provides homeostasis to the brain interstitium, and is essential for optimal neuronal functioning through the network of tight cell-cell contacts (tight junctions) between capillary endothelial cells. Therefore, the morphological substrate of the BBB is attributed to the specialized endothelial cells that strictly control the transcellular pathway of toxic and xenobiotic substances from the blood to the interstitial fluid of the brain [1]. The interendothelial tight junctions are specialized connections (see Chapter 4) that prevent the intercellular leakage of blood-borne substances into the brain parenchyma [2]. Tight junctions seal the interendothelial cleft; and as a consequence no paracellular shunt of drugs or solutes diffusion from the blood site into the CNS emerges. Due to these structural and morphological features, drug delivery to the brain is challenged (see Part V). To allow a better design of new pharmaceuticals and noninvasive strategies to delivery drugs to the brain tissue, a better understanding of the mechanisms which regulate the formation of the BBB is important.

Although the endothelium is the principle barrier and communicating interface, the local microenvironment modulated by associated cells, such as astrocytes and pericytes, contributes to BBB function. Brain microvascular endothelial cells depend on constant support from both astrocytes and pericytes in order to maintain their BBB ability. The associated cells are in close contact with endothelial cells separated by basement membranes (BM) [3], which cover the abluminal surface of the endothelium. The BM is formed by extracellular matrix (ECM) proteins which are involved in the regulation of specific endothelial functions. The molecular components of ECM derive from the surrounding glia, pericytes, macrophages and endothelial cells, and seem to play an important

Blood-Brain Interfaces: From Ontogeny to Artificial Barriers.
Edited by R. Dermietzel, D.C. Spray, M. Nedergaard
Copyright © 2006 WILEY-VCH Verlag GmbH & Co. KGaA, Weinheim
ISBN: 3-527-31088-6

role in barrier formation. However, the molecular mechanism of ECM regulation at the BBB is not yet well understood.

Cell cultures are an excellent tool to investigate molecular mechanism that underlie the formation of the BBB. The successful production of a BBB in vitro is strongly dependent on an adequate mimicking of the in vivo situation.

In our laboratory, we established a primary culture of pig microvascular endothelial cells and very recently a second in vitro model based on a serum-free mouse endothelial culture system [4]. The pig system, with its high transendothelial electrical resistance (TEER) values up to $2000\,\Omega\,cm^{-2}$ and low permeability of about 2×10^{-7} cm s for sucrose, allows us to investigate transport processes to quantify the access of pharmacological substrates across the BBB into the brain. The newly developed mouse BBB model facilitates the application of molecular biological strategies currently available due to access to the mouse genome data. This allows in silico studies to design molecular tools like small interfering RNA, and antisense oligonucleotides, as well as knockout mice to elucidate the role of specific genes and proteins in BBB function.

The application of the glucocorticoid hormone hydrocortisone (HC) in chemically defined serum-free medium was found to improve the BBB properties of both pig and mouse microvascular endothelial cell cultures [4, 5], leading to high values in TEER and low sucrose permeability. Obviously, the junction tightness is drastically improved under the influence of HC. This is important to note since such high barrier properties as described in our system have not been achieved by astrocyte/endothelial cell coculture systems so far. Moreover, we found that pig cerebral endothelial cells exhibit a reduced activity of matrix metalloproteinases (MMPs) when substituted with physiological concentrations of HC [6]. These proteinases (see Chapter 14) were found to be secreted by diverse cell types and it was also shown that MMPs degrade the major macromolecules of the brain ECM [7–9]. Serum was found to contain MMP-2 and MMP-9 molecules. Due to the high MMP activity in serum, the disruption of cell-cell contacts and proteolysis of the tight junction protein occludin was to be expected within in vitro systems [10]. Since MMPs balance the proteolysis, synthesis and assembly of the extracellular matrix, it became obvious that these compounds are involved in the differentiation of the BBB phenotype. New evidence will be described in this chapter indicating that the formation of cell-cell and cell-substrate contacts are reinforced by the endogenously produced extracellular matrix proteins derived from glial cells such as astrocytes and from pericytes.

In essence, here we give a report, in conjunction with Chapter 15, on the establishment of in vitro BBB systems with the focus on pig and mouse in vitro models as well as on the involvement of ECM and MMPs to build up barrier properties. The described models can be used to study neurological disorders, inflammatory reactions of the CNS and the recruitment and transmigration of immune cells in diseases related to multiple sclerosis and human immunodeficiency virus (HIV)-associated syndromes.

16.2
In Vitro BBB Models

The unique features of cerebral microvascular endothelial cells, as compared to peripheral blood vessels, are the presence of tight intercellular junctions, the paucity of pinocytotic activity [11, 12], high number of mitochondria [13], the absence of fenestrations [14] and the expression of specific marker enzymes like transferrin receptors [15] and γ-glutamyl transpeptidase [16]. Among the variety of specialized properties, the most important feature is an efficient endothelial barrier with a low paracellular permeability, which can be easily probed by monitoring the passage of small solutes (for instance, sucrose or inulin) across the cellular monolayer. However, tracer methods have often limited time resolution. Therefore, tight junction complexity and barrier function are most sensitively and readily quantified by measuring the TEER [17]. Further progress in this field was made possible by the development of a new technique which allows the monitoring of cell-cell and cell-substrate contacts in vitro. This method has been coined electric cell-substrate impedance sensing (ECIS) [18].

In 1978, Panula and coworkers demonstrated that isolated rat brain microvascular endothelial cells can be cultured under in vitro conditions [19]. Today, the preferential sources of endothelial cells [20] derive from rat [21, 22], dog [23], cow [24–28], pig [29, 30] and primates [31]. Properly cultured cells exhibit typical features of microvascular endothelial cells, which include attenuated pinocytosis, lack of fenestrations and tight junctions [24, 32, 33], in addition to a typical protein expression pattern. However, higher reproducibility, more homogeneity, ease of propagation and a short population-doubling time were the reasons to develop cell lines derived from rat [34–39], cow [40, 41], pig [42] and human brain microvessels [43]. Nevertheless, due to the lack of barrier properties, none of these immortalized cell lines express all the features necessary for reliably modelling a BBB phenotype. Attempts to induce BBB differentiation in endothelial cell lines [36] have been performed, but most of them failed. Some cerebral endothelial cell lines have been shown to restrict the passage of macromolecules and have been used in studies of paracellular permeability.

To overcome this problem, considerable efforts have been made to reinduce BBB properties in endothelial cells by coculturing them with astrocytes or glioma cells [28, 44–47], or by applying brain-derived differentiation factors. These studies indicate that the establishment and maintenance of the unique cerebral endothelial phenotype results from the neural milieu, in particular the specific interactions provided by astrocytes which surround the brain microvessels [48–50]. Whether physical cell-cell contacts are necessary for the induction of the BBB properties is not clear yet. Also, the use of conditioned media obtained from cultures of rat brain astrocytes or C6 glioma cells reveal some BBB-supporting capacity [28, 46, 51–56]. However, neither cocultures with astrocytes, nor addition of conditioned media induce the entire complement of BBB features found in vivo. Therefore, a better knowledge and appropriate technologies are essential to define new in vitro models for a better understanding of the molecular and biochemical features of the normal BBB.

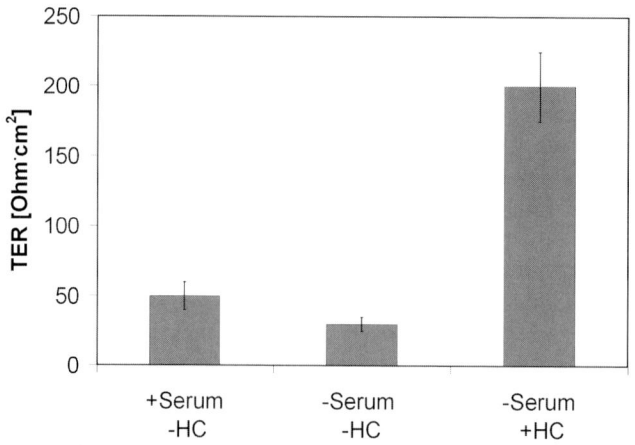

Fig. 16.2 The combination of serum withdrawal with the addition of HC improves the transendothelial electrical resistance (TEER) of mouse brain capillary endothelial cells (MBCEC). Cells cultured with HC show a 5-fold higher TER than cells cultured in serum-containing medium, while the TEER of cells cultured without both supplements reveal the lowest values of resistance, compared to supplemented conditions. Standard deviation is expressed for $n=5$.

properties. This lack of tightness may be the consequence of an incomplete tight junction formation in culture caused by a dedifferentiation in response to the in vitro situation. However, the measured low values of transendothelial resistance in MBCE were in the range of other reported cell culture systems [5, 54, 70]. Additionally, differentiation of MBCEC in vitro has been described to be different from the in vivo situation [70, 71].

16.4
The Involvement of Serum Effects

Cell cultures require a number of different factors that are normally supplied by the added serum. It is well known that fibronectin, growth factors, hormones and extracellular matrix proteins are common components of serum-containing media which can influence cell proliferation and differentiation. It is thus rendered difficult to unveil cell biological aspects of differentiation in the presence of serum. As mentioned before, the withdrawal of serum significantly reinforces the BBB properties of pig cerebral endothelial cells (Fig. 16.1). The inhibition of serum factors favoring the formation of tight cell-cell contacts was recently described by us [60]. This finding is supported by other groups which corroborate the ability of serum to alter both the permeability and electrical resistance of epithelial and endothelial monolayers. As reported by Chang et al. [72], serum induces a breakdown in the junction tightness of cultured retinal-pigmented epithelial cells. Alexander et al. [73] have further shown that lysophosphatidic acid (LPA) decreases the permeability for cyanocobalamine in bovine pulmonary aortic endothelial cells (BPAEC). English et al. [74] looked for an effect of LPA on TEER and observed an increased value of electrical resistances in BPAEC treated with LPA [74]. In contrast to this, Schulze et al. [75] noticed that LPA under physiological conditions decreases TEER, and increases the permeability in pig cultured endothelial cells [75]. Vascular endothelial growth factor (VEGF), a secreted highly endothelial-specific mitogen [76] and a potent angiogenic factor [77] (see Chapter 2), is also known to influence the permeability and electrical resistances of vascular endothelial cells [78–80]. Additionally, previous experiments have shown that VEGF causes an increase in the albumin permeability of cultured cow retinal microvascular endothelial cells [81], increases retinal vascular permeability of fluorescein in vivo [82], stimulates an increase in water permeability in cow aortic endothelial cell cultures and induces an increase in paracellular permeability of human umbilical vein endothelial cells [83]. Further data obtained by us revealed that both VEGF and LPA decrease TEER in PBCEC in vitro [60]. We identified several compounds of serum which are able to influence TEER in vitro [60]. Among them was a 67-kDa fraction which could be detected by MALDI and which was proven to significantly decrease the PBCEC TEER in vitro. We also observed that heat-inactivated serum induces a decrease in TEER similar to nonheat-inactivated serum, although to a minor extent. At that time, the 67-kDa factor found by Nitz et al. [60] in serum was not completely analyzed. However, this molecule was further identified as matrix metalloproteinase type 2 [6].

When discussing the effect of serum on cultured brain microvascular endothelial cells, it should be taken into account that sera from different species evoke different effects. Fetal calf serum and ox serum cause a stronger effect on PBCEC as compared to pig serum. Horse and human sera have almost no effect on PBCEC. These data indicate that species differences do exist, a fact that must be considered when PBCECs are exploited.

16.5
Hydrocortisone Improves the Culture Substrate by Suppressing the Expression of Matrix Metalloproteinases In Vitro

16.5.1
ECIS Analysis of Improved Endothelial Cell-Cell and Cell-Substrate Contacts in HC-Supplemented Medium

As mentioned above, the electrical resistance provides a measure of junction tightness and endothelial monolayer integrity. Generally, establishment of TEER is considered without taking into account cell-cell and cell-substrate interactions. This fact may lead to false conclusions with respect to the mechanism of barrier formation, since some mediators seem to influence TEER by acting on the cell-matrix interaction and/or the cell-cell contacts. The application of a newly devel-

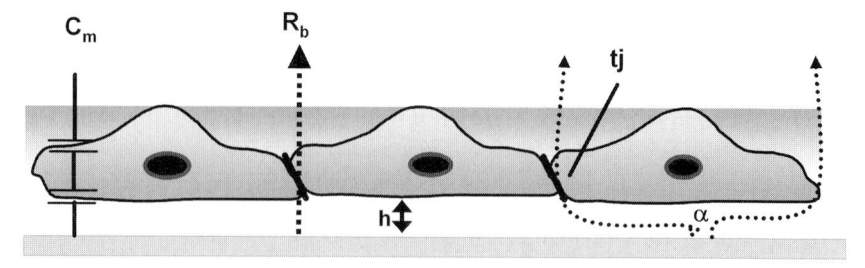

Fig. 16.3 Schematic representation of the model of Giaever and Keese [89] to calculate cell topology, including the formation of cell-cell (R_b) and cell-substrate (α) interactions, as well as the capacitance of the cell membrane (C_m). These parameters allow a description of the barrier and adhesion properties of cells.

oped ECIS technique allows us not only to measure the dynamics of cell adhesion, spreading and tight junction formation on electrode surfaces, but also to discriminate between different morphological parameters of the cell monolayer [18, 84–87]. The parameters for cell-substrate contact (α) and cell-cell contact (R_b; Fig. 16.3) were defined as essential values which determine TEER. A change in R_b detected by ECIS indicates an alteration in paracellular resistance, while increasing values for α indicate a closer cell-substrate contact. We applied this technique to our BBB in vitro model to assess the effects of HC and serum

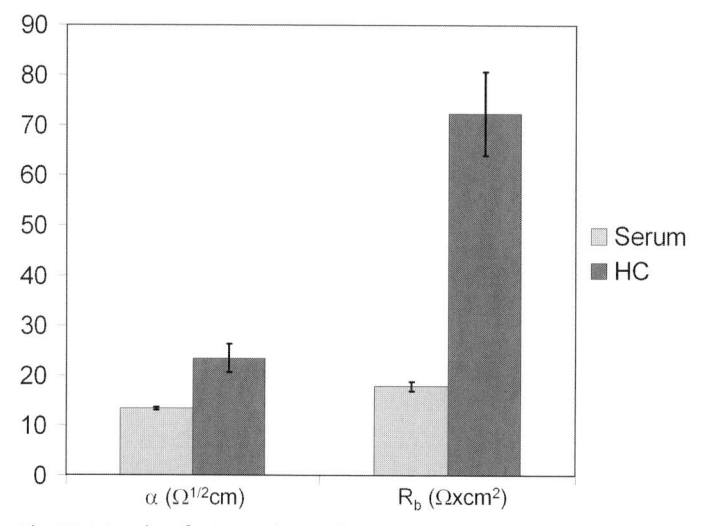

Fig. 16.4 Results of ECIS readings. The parameters α and R_b are presented as weighted average values for $n=5$. HC addition increases parameter α compared to serum-treated cells, indicating that the formation of cell-substrate interaction is influenced by the glucocorticoid hormone. Likewise, R_b values point to a strengthening of cell-cell contacts induced by HC.

on the formation of cell-cell and cell-substrate contacts. Endothelial cells closely attached to the substrate inhibit the flow of ions underneath. An increase in extracellular matrix proteins reduces the electrical current beneath the cells. In cell cultures with low electrical resistances, like mouse cerebral endothelial cells, an increase in a becomes more important and contributes more to overall TEER. These data clearly indicate that both a and R_b have to be taken into account when effects on TEER are considered.

Figure 16.4 shows that higher values for both a and R_b can be detected in HC-containing medium. This observation shows that the paracellular electrical permeability of MBCEC is reduced mainly by R_b, which provides evidence for improved cell-cell contacts. The increase in a after HC treatment may be multifactorial and we will discuss this in the following.

16.5.2
Low Degradation of ECM in HC-Supplemented Medium Leads to Improved Cell-Substrate Contacts of Cerebral Endothelial Cells

It is well known that glucocorticoids increase the barrier properties of vascular cells [88–90]. It has also been shown that HC increases barrier properties in terms of TEER and permeability for ^{14}C-mannitol in epithelial cells [91]. A similar effect was found in a breast cancer tumor cell line [92]. Glucocorticoids are able to influence the expression of MMPs. These enzymes play an important role in the remodeling of the extracellular matrix proteins [7–9] by proteolytic degradation (see Chapter 14). There is evidence that the expression of MMPs can be down-regulated by medium supplemented with glucocorticoids [93].

Our recent studies provide evidence that PBCECs are involved in the secretion of MMPs at the BBB. HC was shown to be able to suppress endothelial MMP secretion, resulting in increased barrier properties [6]. This finding strongly suggests a role for MMPs in the regulation of BBB permeability. In accordance with this result, we presume that a from ECIS readings can be increased by optimizing the integrity of ECM under HC treatment. Higher endothelial MMP secretion in HC-free cultures would lead to a consecutive degradation of ECM proteins, resulting in fewer cell-substrate contacts and consequently a lower a value. Immunocytochemical analysis showed that MMP molecules remain firmly attached to the extracellular matrix substrate when cells are depleted by lysis. We also observed a significant decrease in the amount of MMPs after incubation of endothelial cells in HC-supplemented medium. Furthermore, we discovered that increased endothelial MMP activity at the BBB leads to MMP-dependent cleavage of the tight junction protein-occluding and a profound disruption of cell-cell contacts [10]. This could explain the lower R_b value derived from ECIS readings. The described effects can be taken as evidence that HC improves both the cell-cell (tight junctions) and the cell-substrate interaction by partly inhibiting MMP activity at the BBB.

To prove the ability of HC to modify cell-substrate contacts by suppressing the expression of MMPs, we investigated the basolateral MMP secretion by endothelial cells, using immunostaining. We found that MMPs remain bound to the extracellular matrix of endothelial cells when secreted. Thus, the endogenous ECM derived from first generation of endothelial cells grown in HC-free medium are likely to be different from ECM derived from cells cultured in HC-supplemented medium. We then studied the influence of modified endogenously produced ECM by secreted MMPs on the formation of cell-cell and cell-substrate contacts, using ECIS readings. The ECIS data were recorded as a function of time (unit: hours) and are depicted in Fig. 16.5.

The capacitance recorded at 40 kHz was used as a measure for endothelial cell-substrate contacts. The cells attach to the substrate with an almost spherical morphology (Fig. 16.5 A, C). Therefore, the area of contact between the basement membrane and the cell surface is remarkably small. It increases continuously while the cells spread out. Due to the increase in cell attachment and spreading, the capacitance continuously drops down. Thus, the dynamics of the reduction in capacitance could be read as the process of increasing adhesion, which leads to further attachment and spreading of endothelial cells on the electrode surface. The resistance of the system at 400 Hz was taken as a measure for endothelial cell-cell contacts. Cultured cells with highly developed tight junctions exhibit elevated electrical resistance values (Fig. 16.5 B, D). Details of this experiment have been described elsewhere [18].

To investigate the influence of HC on ECM expression, BCEC were first cultured in serum-free medium with or without HC. The time-course of capacitance and resistance of PBCEC measured by ECIS under these conditions is shown in Fig. 16.5 A, B. The cells became completely confluent in less than 10 h after plating and they adhered in the same manner under both culture conditions, with and without HC. As mentioned before, HC improves the barrier properties of brain endothelial cells and thus increases the transendothelial electrical resistance (Fig. 16.5 B). After 30 h of ECIS measurements, the cells were lysed, leaving the ECM attached to the electrode surfaces. By doing this, two different endogenously derived types of ECM were coated on the electrode surfaces: the plus-HC ECM and the minus-HC ECM. A second generation of primary cultured cerebral endothelial cells in serum-free medium was then seeded onto the endogenously derived ECM of both plus-HC and minus-HC conditions. ECIS readings of the capacitance and resistance over time for the second PBCEC generations in serum-free medium without HC supplementation are shown in Fig. 16.5 C, D, respectively. In the case of the endogenously derived ECM from endothelial cells cultured without HC supplementation, there is a delay in the time-course of cell attachment and spreading compared to PBCEC seeded on the ECM derived from the cells cultured before exposure to HC (Fig. 16.5 C). A plausible explanation is that MMP molecules secreted by the first generation of endothelial cells remained present on the ECM of cells cultured in HC-free medium. These MMPs apparently modified the matrix proteins in a way that the second generation of BCEC could not attach and spread

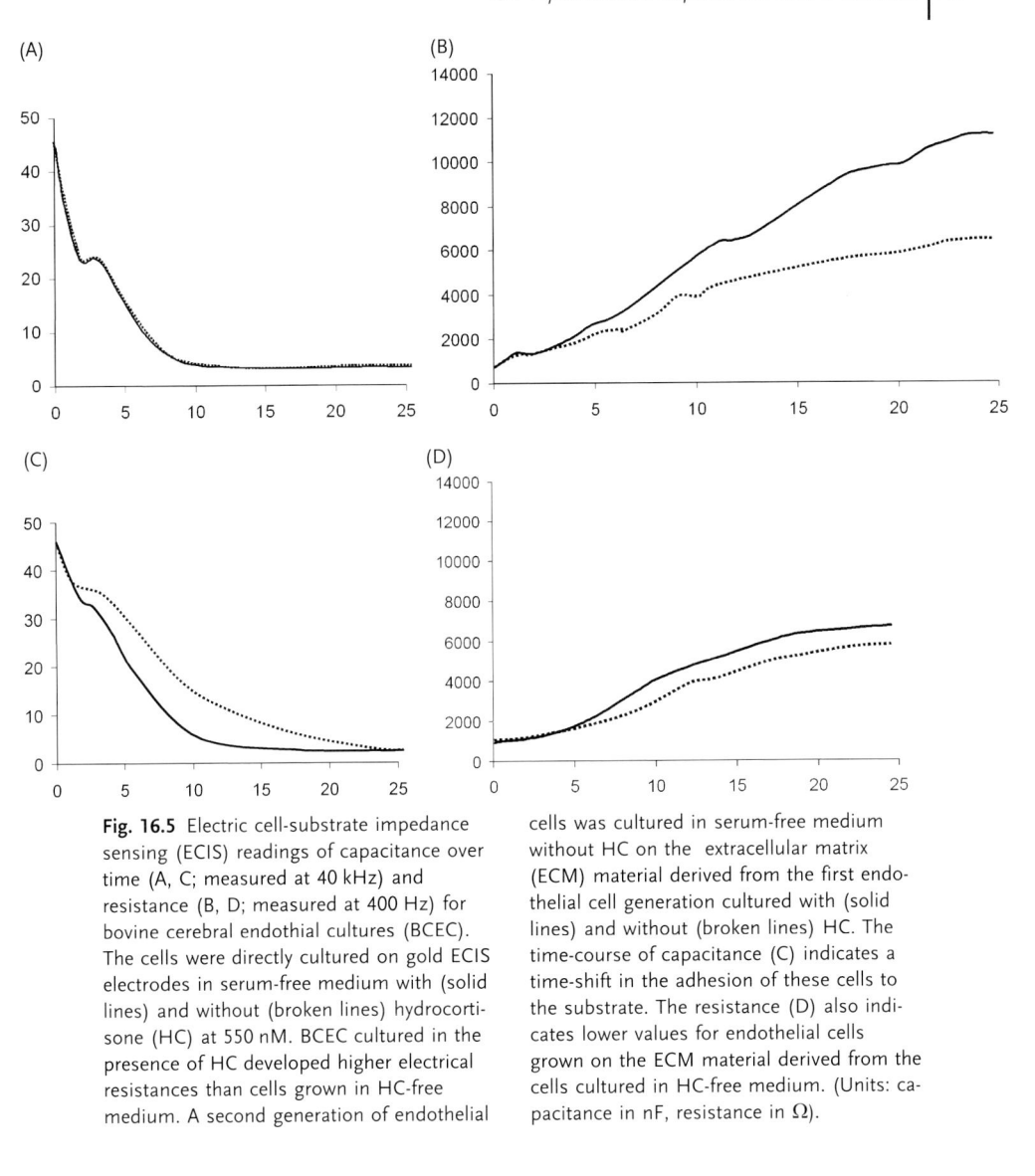

Fig. 16.5 Electric cell-substrate impedance sensing (ECIS) readings of capacitance over time (A, C; measured at 40 kHz) and resistance (B, D; measured at 400 Hz) for bovine cerebral endothial cultures (BCEC). The cells were directly cultured on gold ECIS electrodes in serum-free medium with (solid lines) and without (broken lines) hydrocortisone (HC) at 550 nM. BCEC cultured in the presence of HC developed higher electrical resistances than cells grown in HC-free medium. A second generation of endothelial cells was cultured in serum-free medium without HC on the extracellular matrix (ECM) material derived from the first endothelial cell generation cultured with (solid lines) and without (broken lines) HC. The time-course of capacitance (C) indicates a time-shift in the adhesion of these cells to the substrate. The resistance (D) also indicates lower values for endothelial cells grown on the ECM material derived from the cells cultured in HC-free medium. (Units: capacitance in nF, resistance in Ω).

as fast on the ECM generated by endothelial cells cultured in HC-supplemented medium. The time-shift detected for the endothelial cells in the form of a delayed attachment phase on minus-HC ECM coating could be explained by the increased time needed by the second generation of cells to produce their own extracellular matrix components. Thus, HC was again shown to suppress the degradation of a proper extracellular matrix, probably by an inhibited secretion of MMPs. This is also indicated by the electrical resistance (Fig. 16.5 D). The second generation of endothelial cells grown on matrix material derived from the first generation of cells grown in HC-supplemented medium developed a higher resistance in a shorter time (Fig. 16.5 D).

16.6
The Role of Endogenously Derived ECM for the BBB Properties of Cerebral Endothelial Cells In Vitro

As described, astrocytic endfeet are in close proximity with capillary endothelial cells (see Chapter 9) and are separated from the vessels only by pericytes and the BM. The close proximity of astrocytic endfeet to brain capillaries attracts continuous interest and has led to the idea that astrocytes convey a modulator function to BBB properties [63, 94]. Pericytes comprise a further class of cells found in close association with endothelial cells (see Chapter 5). They may be involved in the regulation of vascular permeability and transport processes across the BBB [1]. Further studies have indicated that signals provided by the interactions between capillary endothelial cells and the surrounding neuroglia, including the ECM, induce the BBB phenotype [95]. By measuring the electrical properties of cerebral endothelial cells cultured on filters covered with different purified extracellular matrix proteins, it was suggested that the BM placed between the endothelium and surrounding astrocytes may be involved in the differentiation of the BBB phenotype [96]. The molecular mechanisms through which both ECM and glial cells affect the BBB are still poorly understood. To get a better insight into this interaction we aimed to investigate the role of ECM proteins endogenously derived from astrocytes and pericytes.

For these purposes, barrier formation by BCEC cultured on astrocyte- and pericyte-derived ECM was studied by ECIS and compared to resistance values obtained on ECMs produced by cerebral and aorta endothelial cells. Aorta endothelial cells lack barrier properties and are leaky. The extracellular matrix derived from these cells was used as a control substrate, representing a type of ECM differing from that derived from brain endothelial cells. The rationale for this experimental design derived from observations which showed significant differences in the barrier development of BCEC on endogenously produced ECM as compared to purified individual ECM molecules [97]. The resulting resistance values obtained in these experiments are summarized in Fig. 16.6. Note that the resistances given here are measured at 400 Hz and should be not confused with normally reported ($\Omega\,cm^{-2}$) resistance values. The functional significance of purified ECM molecules for the development of electrical resistances increases from laminin ($5\pm1\times10^3\,\Omega$) to fibronectin ($19\pm2.5\times10^3\,\Omega$), with fibronectin being the best exogenous substrate for the promotion of cell-cell interaction. Although the PBCEC cultured on fibronectin acquired high resistance, endothelial cells grown on cellular basement membranes derived from precultured tight BCEC developed an even better barrier. When PBCEC were cultured on ECM derived from pericytes, PBCECs developed an electrical resistance of $25\pm2\times10^3\,\Omega$. On an astrocytic extracellular matrix, BCEC developed more than

16.6 The Role of Endogenously Derived ECM for the BBB Properties of Cerebral Endothelial Cells

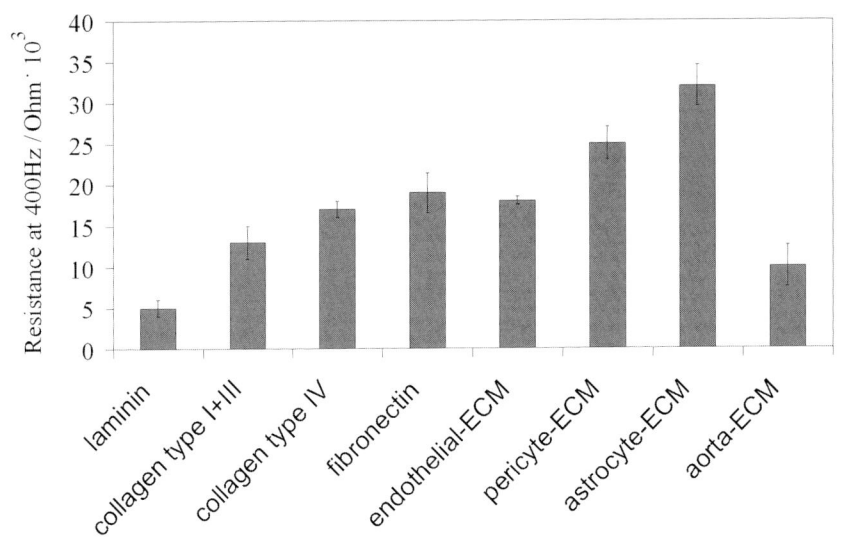

Fig. 16.6 The influence of purified individual ECM molecules (laminin, collagen type I+III, collagen type IV, fibronectin) and endogenously derived ECM from brain cells (endothelial, pericytes, astrocytes) compared to nonbrain ECM (aorta ECM) on the transendothelial electrical resistance of BCEC monolayers. The cells were cultured in serum-free medium supplied with HC. The values for electrical resistance measured at 400 Hz are depicted after 55 h of culturing the BCEC. Data given are means ±SD for one out of two ECIS assays, all of which were conducted in duplicate.

$33 \pm 2.5 \times 10^3 \, \Omega$. Interestingly, the endogenously produced ECM derived from aorta endothelial cells only slightly improved the tightness of PBCEC (cell-cell contacts). Although the lowest electrical resistances for endogenously derived ECM was found in aortic ECM ($10 \pm 2.5 \times 10^3 \, \Omega$), these values were still significantly higher than that assessed for the cells cultured on laminin coating ($5 \pm 1 \times 10^3 \, \Omega$). A feasible explanation of this finding is that the complexity of endogenous ECM can not be mimicked by a single-component coating of ECM substrates. From all endogenously derived substrates, the extracellular matrix derived from astrocytes was the best ECM for PBCEC in terms of its attachment and capacity to form tight junctions.

With respect to the methodological approach, ECIS is a useful tool in measuring the potentials of endogenously derived ECMs on brain endothelial cells. The production of some common integrins and ECM molecules in astrocyte-endothelial cell cocultures has been already described by Wagner et al. [98]. They found that astrocytes secrete laminin-5, which is bound to the basement membrane of cerebral microvessels by its integrin receptor $\alpha 6\beta 1$. A physical contact possibly involving the binding of endothelial $\alpha 6\beta 1$ integrin receptors and laminin-5 was found to be responsible for the induction of astrocytic laminin [98]. Based on our ECIS experiments, it is tempting to speculate that astrocytic laminin-5, among other ECM molecules, may reside on the electrode surface after the lysis of astrocytes during ECM denudation. Since PBCEC were found to secrete an elevated level of $α6β1$ receptor as found by flow cytometry (own unpublished observations), the formation of high resistance in PBCEC monolayers on astrocyte-derived ECM can be explained by the secretion of integrin receptors on endothelial and/or astrocytic cell surfaces and their interactions with matrix proteins. Additionally, we found that astrocytes express inhibitors for MMPs and exhibit low expression levels of MMP-2 and MMP-9 enzymes. Schiera et al. [99] previously reported that cocultures of brain endothelial cells with astrocytes and neurons leads to the stabilization of tight junctions and the formation of a functional BBB. We suggest that MMPs released by astrocytes as well as by pericytes may activate growth factors bound to the ECM, which then exert their influence through specific receptor complements on the differentiation and growth of BCECs. Obviously, a complex interplay of a variety of components including suitable adhesion molecules and their complementary receptors, as well as matrix proteinases in conjunction with their specific inhibitors, are orchestrated in a way that leads to a BCEC phenotype which mimics a suitable BBB culture model.

16.7
Conclusions

Two serum-free BBB models derived from pig and mouse brains are described for the investigation of the BBB function. These cells depend on serum in the growth medium to proliferate. Serum, however, inhibits the formation of cell-cell contacts, as demonstrated by the decrease in the transendothelial electrical resistances in monolayers. Withdrawal of serum from monolayers significantly improves the barrier properties of pig capillary endothelial cells, but not of cells obtained from mouse brains. Both pig and mouse brain endothelial cultures express higher TEER values when incubated in serum-free medium and in the presence of hydrocortisone [4, 30]. Besides serum factors, the BM conveys some important function to the BBB, since its degradation by MMPs leads to a disruption of the BBB integrity. Improved BBB properties observed under the influence of HC hints at the ability of this hormone to suppress the secretion of MMPs. The level of secretion of these enzymes was found to correlate with the integrity of the BBB in the form of its electrical resistance [10]. Extracellular matrix proteins secreted by the endothelial cells in HC-supplemented medium seem to be less degraded, thereby improving the cell-substrate contacts of cerebral endothelial cells. The importance of the ECM synthesized by glial cells is also discussed. The extracellular matrixes derived from both astrocytes and pericytes were found to be most efficient with respect to the development of an endothelial barrier. The ECM components seem to contribute considerably to the development of the BBB properties of endothelial cells in vitro [97].

References

1 G. Allt, J. G. Lawrenson **2000**, *Brain Res. Bull.* 52, 1.
2 G. A. Grant, N. J. Abbott, D. Janigro **1998**, *News Physiol. Sci.* 13, 287.
3 G. W. Goldstein, A. L. Betz **1986**, *Sci. Am.* 255, 74.
4 C. Weidenfeller, S. Schrot, A. Weidenfeller, H. J. Galla **2005**, *Brain Res.* 1053, 162–174.
5 R. D. Hurst, J. B. Clark **1998**, *Neurochem. Res.* 23, 149.
6 T. Lohmann **2003**, Westfälische Wilhelms-Universität, Münster.
7 E. Pirila, A. Sharabi, T. Salo, V. Quaranta, H. Tu, R. Heljasvaara, N. Koshikawa, T. Sorsa, P. Maisi **2003**, *Biochem. Biophys. Res. Commun.* 303, 1012.
8 K. Nishino, K. Yamanouchi, K. Naito, H. Tojo **2002**, *Dev. Growth Differ.* 44, 35.
9 M. D. Sternlicht, Z. Werb **2001**, *Annu. Rev. Cell Dev. Biol.* 17, 463.
10 C. Lohmann, M. Krischke, J. Wegener, H. J. Galla **2004**, *Brain Res.* 995, 184.
11 F. Joo **1971**, *Br. J. Exp. Pathol.* 52, 646.
12 M. W. Brightman, T. S. Reese **1969**, *J. Cell Biol.* 40, 648.
13 W. H. Oldendorf, M. E. Cornford, W. J. Brown **1977**, *Ann. Neurol.* 1, 409.
14 P. A. Stewart, K. Hayakawa, C. L. Farrell **1994**, *Microsc. Res. Tech.* 27, 516.
15 W. A. Jefferies, M. R. Brandon, S. V. Hunt, A. F. Williams, K. C. Gatter, D. Y. Mason **1984**, *Nature* 312, 162.
16 M. Orlowski, G. Sessa, J. P. Green **1974**, *Science* 184, 66.
17 E. E. Schneeberger, R. D. Lynch **1992**, *Am. J. Physiol.* 262, L647.
18 J. Wegener, C. R. Keese, I. Giaever **2000**, *Exp. Cell Res.* 259, 158.
19 P. Panula, F. Joo, L. Rechardt **1978**, *Experientia* 34, 95.
20 K. L. Audus, L. Ng, W. Wang, R. T. Borchardt **1996**, *Pharm. Biotechnol.* 8, 239.
21 N. J. Abbott, C. C. Hughes, P. A. Revest, J. Greenwood **1992**, *J. Cell Sci.* 103, 23.
22 M. A. Barrand, K. J. Robertson, S. F. von Weikersthal **1995**, *FEBS Lett.* 374, 179.
23 R. C. Speth, S. I. Harik **1985**, *Proc. Natl Acad. Sci. USA* 82, 6340.
24 K. L. Audus, R. T. Borchardt **1987**, *Ann. NY Acad. Sci.* 507, 9.
25 R. Cecchelli, B. Dehouck, L. Descamps, L. Fenart, V. V. Buee-Scherrer, C. Duhem, S. Lundquist, M. Rentfel, G. Torpier, M. P. Dehouck **1999**, *Adv. Drug Deliv. Rev.* 36, 165.
26 M. P. Dehouck, S. Meresse, P. Delorme, J. C. Fruchart, R. Cecchelli **1990**, *J. Neurochem.* 54, 1798.
27 S. Meresse, M. P. Dehouck, P. Delorme, M. Bensaid, J. P. Tauber, C. Delbart, J. C. Fruchart, R. Cecchelli **1989**, *J. Neurochem.* 53, 1363.
28 L. L. Rubin, D. E. Hall, S. Porter, K. Barbu, C. Cannon, H. C. Horner, M. Janatpour, C. W. Liaw, K. Manning, J. Morales, et al. **1991**, *J. Cell Biol.* 115, 1725.
29 U. Mischek, J. Meyer, H.-J. Galla **1989**, *Cell Tissue Res.* 256, 221.
30 D. Hoheisel, T. Nitz, H. Franke, J. Wegener, A. Hakvoort, T. Tilling, H. J. Galla **1998**, *Biochem. Biophys. Res. Commun.* 244, 312.

31 F. Shi, K. L. Audus, *Neurochem. Res.* **1994**, *19*, 427.
32 K. L. Audus, R. T. Borchardt **1986**, *J. Neurochem.* 47, 484.
33 P. D. Bowman, S. R. Ennis, K. E. Rarey, A. L. Betz, G. W. Goldstein **1983**, *Ann. Neurol.* 14, 396.
34 O. Durieu-Trautmann, N. Foignant-Chaverot, J. Perdomo, P. Gounon, A. D. Strosberg, P. O. Couraud **1991**, *In Vitro Cell Dev. Biol.* 27A, 771.
35 J. Greenwood, G. Pryce, L. Devine, D. K. Male, W. L. dos Santos, V. L. Calder, P. Adamson **1996**, *J. Neuroimmunol.* 71, 51.
36 D. Lechardeur, B. Schwartz, D. Paulin, D. Scherman **1995**, *Exp. Cell Res.* 220, 161.
37 F. Roux, O. Durieu-Trautmann, N. Chaverot, M. Claire, P. Mailly, J. M. Bourre, A. D. Strosberg, P. O. Couraud **1994**, *J. Cell Physiol.* 159, 101.
38 K. H. Tan, M. S. Dobbie, R. A. Felix, M. A. Barrand, R. D. Hurst **2001**, *Neuroreport* 12, 1329.
39 O. D. Trautmann, S. Bourdoulous, F. Roux, J. M. Bourre, A. D. Strosberg, P.-O. Couraud **1993**, *Immortalized Rat Brain Microvessel Endothelial Cells: II. Pharmacological Characterization*, Plenum Press, New York.
40 K. Sobue, N. Yamamoto, K. Yoneda, M. E. Hodgson, K. Yamashiro, N. Tsuruoka, T. Tsuda, H. Katsuya, Y. Miura, K. Asai, T. Kato **1999**, *Neurosci. Res.* 35, 155.
41 M. F. Stins, N. V. Prasadarao, J. Zhou, M. Arditi, K. S. Kim **1997**, *In Vitro Cell Dev. Biol. Anim.* 33, 243.
42 M. Teifel, P. Friedl **1996**, *Exp. Cell Res.* 228, 50.
43 A. Muruganandam, L. M. Herx, R. Monette, J. P. Durkin, D. B. Stanimirovic **1997**, *FASEB J.* 11, 1187.
44 P. J. Gaillard, I. C. van der Sandt, L. H. Voorwinden, D. Vu, J. L. Nielsen, A. G. de Boer, D. D. Breimer **2000**, *Pharm. Res.* 17, 1198.
45 P. J. Gaillard, L. H. Voorwinden, J. L. Nielsen, A. Ivanov, R. Atsumi, H. Engman, C. Ringbom, A. G. de Boer, D. D. Breimer **2001**, *Eur. J. Pharm. Sci.* 12, 215.
46 R. D. Hurst, I. B. Fritz **1996**, *J. Cell Physiol.* 167, 81.
47 F. E. Arthur, R. R. Shivers, P. D. Bowman **1987**, *Brain Res.* 433, 155.
48 R. Dermietzel, D. Krause **1991**, *Int. Rev. Cytol.* 127, 57.
49 R. C. Janzer, M. C. Raff **1987**, *Nature* 325, 253.
50 P. A. Stewart, M. J. Wiley **1981**, *Dev. Biol.* 84, 183.
51 R. Cancilla, J. Bready, J. Berliner **1993**, in *Astrocytes: Pharmacology and Function*, ed. G. Murphy, Academic, San Diego, p. 383.
52 B. Dehouck, M. P. Dehouck, J. C. Fruchart, R. Cecchelli **1994**, *J. Cell Biol.* 126, 465.
53 Y. Igarashi, H. Utsumi, H. Chiba, Y. Yamada-Sasamori, H. Tobioka, Y. Kamimura, K. Furuuchi, Y. Kokai, T. Nakagawa, M. Mori, N. Sawada **1999**, *Biochem. Biophys. Res. Commun.* 261, 108.
54 P. V. Ramsohoye, I. B. Fritz **1998**, *Neurochem. Res.* 23, 1545.
55 T. J. Raub **1996**, *Am. J. Physiol.* 271, C495.
56 R. J. Rist, I. A. Romero, M. W. Chan, P. O. Couraud, F. Roux, N. J. Abbott **1997**, *Brain Res.* 768, 10.

57 C. Crone, S. P. Olesen **1982**, *Brain Res.* 241, 49.
58 V. A. Levin **1980**, *J. Med. Chem.* 23, 682.
59 T. Eisenblatter, H. J. Galla **2002**, *Biochem. Biophys. Res. Commun.* 293, 1273.
60 T. Nitz, T. Eisenblatter, K. Psathaki, H. J. Galla **2003**, *Brain Res.* 981, 30.
61 T. Eisenblatter, S. Huwel, H. J. Galla **2003**, *Brain Res.* 971, 221.
62 H. Franke, H. Galla, C. T. Beuckmann **2000**, *Brain Res. Brain Res. Protocol* 5, 248.
63 H. Wolburg, J. Neuhaus, U. Kniesel, B. Krauss, E. M. Schmid, M. Ocalan, C. Farrell, W. Risau **1994**, *J. Cell Sci.* 107, 1347.
64 H. Franke, H. J. Galla, C. T. Beuckmann **1999**, *Brain Res.* 818, 65.
65 M. A. Deli, C. S. Abraham, M. Niwa, A. Falus **2003**, *Inflamm. Res.* 52 [Suppl. 1], S39.
66 D. D. Wagner, V. J. Marder **1984**, *J. Cell Biol.* 99, 2123.
67 I. L. Karnushina, J. M. Palacios, G. Barbin, E. Dux, F. Joo, J. C. Schwartz **1980**, *J. Neurochem.* 34, 1201.
68 M. A. Gimbrone, Jr., G. R. Majeau, W. J. Atkinson, W. Sadler, S. A. Cruise **1979**, *Life Sci.* 25, 1075.
69 A. M. Butt, H. C. Jones, N. J. Abbott **1990**, *J. Physiol.* 429, 47.
70 P. M. Reardon, K. L. Audus **1993**, *Pharm. Sci.* 3, 63.
71 F. Joo **1992**, *J. Neurochem.* 58, 1.
72 C. Chang, X. Wang, R. B. Caldwell **1997**, *J. Neurochem.* 69, 859.
73 J. S. Alexander, W. F. Patton, B. W. Christman, L. L. Cuiper, F. R. Haselton **1998**, *Am. J. Physiol.* 274, H115.
74 D. English, A. T. Kovala, Z. Welch, K. A. Harvey, R. A. Siddiqui, D. N. Brindley, J. G. Garcia **1999**, *J. Hematother. Stem Cell Res.* 8, 627.
75 C. Schulze, C. Smales, L. L. Rubin, J. M. Staddon **1997**, *J. Neurochem.* 68, 991.
76 D. W. Leung, G. Cachianes, W. J. Kuang, D. V. Goeddel, N. Ferrara **1989**, *Science* 246, 1306.
77 W. Risau **1990**, *Prog. Growth Factor Res.* 2, 71.
78 S. Esser, M. G. Lampugnani, M. Corada, E. Dejana, W. Risau **1998**, *J. Cell Sci.* 111, 1853.
79 P. J. Keck, S. D. Hauser, G. Krivi, K. Sanzo, T. Warren, J. Feder, D. T. Connolly **1989**, *Science* 246, 1309.
80 D. R. Senger, S. J. Galli, A. M. Dvorak, C. A. Perruzzi, V. S. Harvey, H. F. Dvorak **1983**, *Science* 219, 983.
81 T. Marumo, T. Noll, V. B. Schini-Kerth, E. A. Harley, J. Duhault, H. M. Piper, R. Busse **1999**, *J. Vasc. Res.* 36, 510.
82 L. P. Aiello, S. E. Bursell, A. Clermont, E. Duh, H. Ishii, C. Takagi, F. Mori, T. A. Ciulla, K. Ways, M. Jirousek, L. E. Smith, G. L. King **1997**, *Diabetes* 46, 1473.
83 J. A. Yaccino, Y. S. Chang, T. M. Hollis, T. W. Gardner, J. M. Tarbell **1997**, *Curr. Eye Res.* 16, 761.
84 I. Giaever, C. R. Keese **1991**, *Proc. Natl Acad. Sci. USA* 88, 7896.
85 I. Giaever, C. R. Keese **1993**, *Nature* 366, 591.

86 C. M. Lo, C. R. Keese, I. Giaever **1995**, *Biophys. J.* 69, 2800.
87 P. Mitra, C. R. Keese, I. Giaever **1991**, *Biotechniques* 11, 504.
88 F. D. Ingraham, D. D. Matson, R. L. McLaurin **1952**, *N. Engl. J. Med.* 15, 568.
89 J. O. Jarden, V. Dhawan, J. R. Moeller, S. C. Strother, D. A. Rottenberg **1989**, *Ann. Neurol.* 25, 239.
90 K. Yamada, Y. Ushio, T. Hayakawa, N. Arita, N. Yamada, H. Mogami **1983**, *J. Neurosurg.* 59, 612.
91 K. S. Zettl, M. D. Sjaastad, P. M. Riskin, G. Parry, T. E. Machen, G. L. Firestone **1992**, *Proc. Natl Acad. Sci. USA* 89, 9069.
92 P. Buse, P. L. Woo, D. B. Alexander, H. H. Cha, A. Reza, N. D. Sirota, G. L. Firestone **1995**, *J. Biol. Chem.* 270, 6505.
93 C. Jonat, H. J. Rahmsdorf, K. K. Park, A. C. Cato, S. Gebel, H. Ponta, P. Herrlich **1990**, *Cell* 62, 1189.
94 I. Isobe, T. Watanabe, T. Yotsuyanagi, N. Hazemoto, K. Yamagata, T. Ueki, K. Nakanishi, K. Asai, T. Kato **1996**, *Neurochem. Int.* 28, 523.
95 S. Fischer, M. Wobben, J. Kleinstuck, D. Renz, W. Schaper **2000**, *Am. J. Physiol. Cell Physiol.* 279, C935.
96 T. Tilling, D. Korte, D. Hoheisel, H. J. Galla **1998**, *J. Neurochem.* 71, 1151.
97 A. Zozulya **2004**, Westfälische Wilhelms-Universität, Münster.
98 S. Wagner, H. Gardner **2000**, *Neurosci. Lett.* 284, 105.
99 G. Schiera, E. Bono, M. P. Raffa, A. Gallo, G. L. Pitarresi, I. Di Liegro, G. Savettieri **2003**, *J. Cell Mol. Med.* 7, 165.

17
Artificial Blood-Brain Barriers

Luca Cucullo, Emily Oby, Kerri Hallene, Barbara Aumayr, Ed Rapp, and Damir Janigro

17.1
Introduction

The current view of the BBB has shifted from a purely anatomic concept to a more physiological and dynamic vision. This change was brought about by evidence that the BBB actively regulates the transport of substances from blood to brain and provides active shielding to many potentially noxious xenobiotic substances. Initial studies of the BBB were performed in vivo, mostly to determine the permeability of compounds across the brain endothelium. This approach offered valuable information about the behavior of different classes of compounds and helped identify specific BBB transport systems (e.g., injection single-pass method to measure cerebrovascular transport and permeability introduced by Oldendorf in 1970). However, further characterization of the BBB, especially at the cellular and molecular level, was delayed due to the complex environmental conditions present in vivo. These complexities stimulated the development of in vitro experimental approaches, such as new cell culture techniques and improved technologies to monitor BBB function (see Chapters 15 and 16). Thanks to these advancements, the BBB has been characterized in detail both physiologically and morphologically [1–6].

The BBB is found in the brain of all vertebrates; and in humans it is formed in the first trimester of fetal life [7]. Morphologically, the BBB is formed by specialized ECs lining the intraluminal portion of brain microvessels together with the closely associated astrocytic endfeet processes sharing the basal lamina and enveloping more than 98% of the BBB endothelium [8]. Astrocyte interaction with the cerebral endothelium determines BBB function, regulates protein expression, modulates endothelium differentiation and appears to be critical for the induction and maintenance of the tight junctions and BBB properties [9–14]. The microvascular endothelium at the BBB level is characterized by the presence of tight junctions (zonulae occludentes), a lack of fenestrations and minimal pinocytotic vesicles. *BBB EC characteristics* include:

- few pinocytic vesicles,
- tight junctions with small hydrophilic pores (15–20 Å),
- fractional area: 0.01%,
- high negative charge density at the surface (sulfated glycoproteins),
- low permeability to water soluble molecules (electrolytes),
- 15% (by weight) of BBB cell membranes is lipid (other cells: 50%),
- P-glycoprotein efflux system.

These distinct morphological properties of the BBB account for the "restraining" nature of brain capillary ECs. In particular, tight junctions between the cerebral ECs form a diffusion barrier (Fig. 17.1) which selectively excludes most blood-borne substances and xenobiotics from entering the brain, protecting it from systemic influences. However, the BBB does not act as an absolute barrier, since non-polar molecules (lipid-soluble substances such as alcohol, narcotics, anti-convulsants) pass with ease. The transit across the BBB involves translocation through the capillary endothelium by carrier-mediated transport systems. The BBB also provides specialized transport systems for nutrients and other biologically important substances (i.e., D-glucose, lactate, phenylamine, choline, adenosine, arginine, adenine) from the peripheral circulation to neurons in the parenchyma. The transport systems at the BBB include:

1. Uptake at the BBB
- hexoses (e.g., glucose),
- amino acids,
- peptides,
- acetylated low density lipoprotein (scavenger receptor),
- transferrin (transferrin receptor),
- organic cations,
- monocarboxylic acids,
- nucleosides.

2. Efflux at the BBB
- neurotransmitter metabolites,
- uremic toxins,
- organic anions,
- immunoglobulin γ (immunoglobulin fc receptor),
- P-glycoprotein,
- multidrug resistance proteins.

Because of its selectivity, the BBB plays a crucial role in the determination of neurotoxicity and its prevention by specific transport mechanisms [15, 16].

Unfortunately, some of these efflux transport systems that developed to protect the brain from potentially dangerous substances may also contribute to the phenomenon known as multiple drug resistance (MDR) during the treatment of several CNS disorders, such as drug refractory epilepsy or intractable brain tumors [17–19]. The necessity to develop alternative pharmaceutical strategies that bypass the shielding of brain parenchyma and to study new therapeutic approaches for many neurological diseases demanded the rapid development of cell culture-based in vitro models that are able to reproduce the physiological, anatomical and functional characteristics of the BBB. These cell culture systems have the potential to be valuable tools in unraveling the complex molecular interactions underlying and regulating the permeability of the cerebral endothelium in normal and pathological conditions, as well as allowing for transendothelial permeability screening and a better prediction of drug penetration across the BBB. This explains why the aim of in vitro models is to functionally express as many unique characteristics of the BBB cerebral endothelium in vivo as possible.

The understanding of the BBB physiology is critical for complex issues such as pathogenesis of neurological diseases involving BBB dysfunction (e.g., brain tumors, ischemia, hypoxia, brain edema, multiple sclerosis, meningitis) and the development of pharmaceuticals that can cross the BBB. In fact, CNS drug design (e.g., antineoplastics, antivirals, antiepileptics) cannot rely entirely upon physico-chemical properties to cross the BBB. For example, lipophilicity alone is a poor predictor for drug penetration into the CNS since it relies on a passive, diffusional type of uptake. Many lipophylic drugs are potential substrates for efflux carriers of the BBB (particularly P-glycoprotein) that can drastically reduce their penetration into the brain [20].

In the following paragraphs, we provide an overview of the characteristics of cell culture-based models of the BBB available today and insights into ongoing development of future devices.

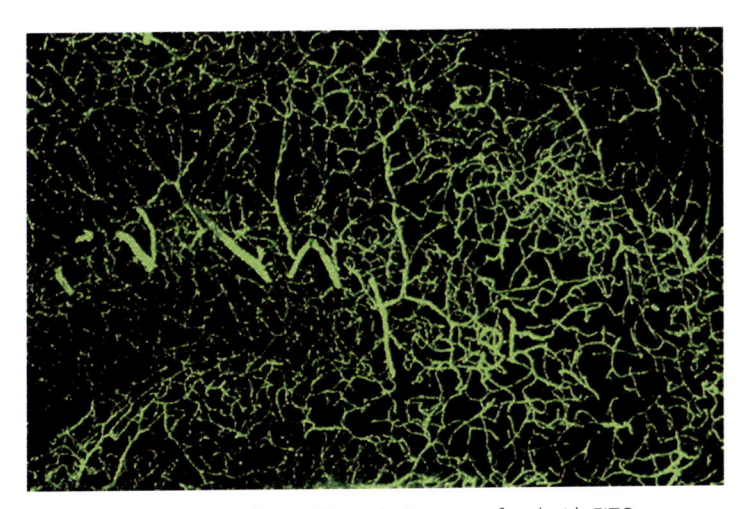

Fig. 17.1 Brain microvasculature. Mouse brain was perfused with FITC-albumin. Note the lack of fluid extravasations due to the BBB.

17.2
Requirements for a Good BBB Model

There are significant quantitative and qualitative differences between the various BBB cell culture systems, but there are specific requirements common to all that need to be addressed in order for the model to be as functional and accurate as possible. The ingredients of a good *in vitro BBB model* include:

- expression of tight junctions between ECs;
- low permeation to sucrose or electric current flow;
- selective and asymmetric permeability to physiologically relevant ions, such as Na^+ and K^+;
- selective permeability to molecules, based on their oil/water partition coefficient and molecular weight;
- expression of drug-metabolizing enzymes;
- functional expression of mechanisms of active extrusion such as multidrug resistance proteins;
- cell-cell interaction and relative exposure to as yet unidentified "permissive" or "promoting" factors (presumably secreted by astrocytes);
- exposure of the apical membrane to shear stress to promote growth inhibition and differentiation of ECs;
- similarity in responsiveness to permeation modulators;
- ease of culture and low cost.

The expression of tight junctions between ECs (determining a restrictive paracellular permeability) and low permeation to polar molecules that do not have a specific carrier-mediated transport system (e.g., sucrose) are mandatory conditions. Moreover, a functional BBB model must guarantee a selective permeability to molecules, based on their oil/water partition coefficient and molecular weight, and selective and asymmetric permeability to physiologically relevant ions, such as Na^+ and K^+. Furthermore, in order to obtain a functional BBB, the model should allow for the development and maintenance of mechanisms of active extrusion such as multidrug resistance proteins (e.g., Mdr1, Pgp) naturally present in vivo and for the expression of "drug" metabolizing enzymes. Cell-cell interactions leading to physiologically more natural structural design and relative exposure to "permissive" or "promoting" factors, mostly released by surrounding glia, are also required. Also, in order to be an effective research tool, the in vitro BBB should be easy to culture, able to assure data reproducibility and be as inexpensive as possible. Finally, a new wave of research has demonstrated that circulating blood cells are critical elements for the establishment of a functional in vitro BBB environment, capable of reproducing CNS-impairment-related phenomenons (e.g., inflammation, stroke, brain trauma, etc.) that occur in vivo.

17.3
Immobilized Artificial Membranes

Immobilized artificial membrane (IAM) chromatography is a technique used for the analysis and purification of many biological molecules. IAM has been proposed as a model for predicting drug permeability across the blood-brain barrier [21]. Originally developed by Charles Pidgeon at Purdue University, the IAM stationary phase consists of a monolayer of phospholipid covalently immobilized on an inert silica support. The resulting IAM surface is a chemically stable chromatographic material which simulates the exterior of a biological cell membrane. IAM chromatography can be used for the analytical and preparative separation of membrane-associated proteins and for the non-covalent immobilization of membrane associated proteins. IAM chromatography has gained acceptance for the chromatographic estimation of the membrane permeability of small-molecule drugs [22, 23]. In particular, IAM chromatography can be used to estimate drug permeability and to measure phospholipophilicity. However, this system has several limitations: (a) the IAM does not mimic the dynamics of fluid membranes, in particular lateral diffusion, (b) it is impossible to simulate diffusion across a membrane bilayer, (c) the IAM can model only drug permeation through cells where equilibrium across the membrane is the rate-limiting step, and (d) the IAM cannot mimic active extrusion.

17.4
Cell Culture-Based In Vitro BBB Models

In the early 1990s, in vitro models of the BBB started to emerge which had the potential to be valuable tools in unraveling the complex molecular interactions underlying and regulating the permeability of the cerebral endothelium in normal and pathological conditions and allowing for transendothelial permeability screening. These in vitro methods are based on isolated brain capillaries or the isolation and subsequent culture of brain capillary ECs in static, and more recently, in flow-based (dynamic) in vitro apparati. Other approaches implicate the use of immortalized cell lines (e.g., RBE4, MDCK) [24, 25]. All these models attempt to mimic the complexity of the mammalian BBB, but each of them is characterized by a different selective permeability to different compounds and may manifest a different range of trans-endothelial electrical resistance (TEER) values. TEER is a measure of the electric (ionic) conductance of the monolayer and is a useful measure of the "tightness" of the monolayer. Before entering into a detailed description of the cell culture-based models of the BBB, we want to provide a general overview of the cell types currently used to establish an in vitro BBB.

17.4.1
Cell Lines

Several cell lines have been used as biological substrates for BBB modeling. Usually, one favors cell lines that are easier to expand instead of relying on more time-consuming primary culture of brain ECs.

17.4.1.1 Immortalized Rat Brain Endothelial Cells

The immortalized rat brain endothelial (RBE4) cell line is probably the most extensively characterized cell line used for BBB modeling. Unfortunately, no data are available on how RBE4 behave when cultured under dynamic conditions, since all data available were obtained under static conditions. These ECs have been transfected with a plasmid containing the E1A adenovirus gene. The RBE4 has been shown to functionally express a number of BBB transporters and mechanisms of active extrusion (e.g., Pgp, Mrp1) [26–32]. An immortalized cell line has the advantage that cell cultures can be expanded at will, which drastically reduces cost and labor. However, some disadvantages of using an immortalized cell line include the fact that the cells have been genetically manipulated and generally form incomplete tight junctions, thus leading to the establishment of a less functional and less selective BBB. This drawback is evident when looking at permeability data for a known blood volume marker molecule, such as sucrose. These data indicate that cell lines form very "leaky" monolayers and lack the necessary paracellular barrier properties to be considered a BBB permeability screen [33]. One possible explanation may be that ease of expansion is obtained by molecular strategies that alter cell cycle checkpoints. Desai et al. [34] have shown that reduced proliferation is a determinant of endothelium differentiation perhaps allowing for astrocytic influences to "tighten" the BBB.

17.4.1.2 Other Cells From Non-Cerebral Sources

Mardin-Darby canine kidney (MDCK) cell lines are widely used in the establishment of an in vitro BBB. BBB models based on MDCK cell monolayers show low permeability to sucrose, and because of the ease with which these cells can be grown; they are used for the screening of passively transported CNS compounds. The MDCK cell line is comprised of different clones with different properties. The MDCK-I clone achieves the lowest permeability to sucrose but there are also MDCK cells available that are transfected with the human MDR-1 gene (MDCKmdr-1 cell line) which lead to a polarized overexpression of P-glycoprotein [35]. However, these non-cerebral cells differ from brain ECs (BCEC) with respect to morphology, transport properties, metabolism and growth (e.g., lack of contact-induced inhibition of the cell proliferation mechanism that is present in brain ECs). Morphological differences compared to cerebral ECs are likely to be reflected in drug-cell membrane interactions resulting in not only different transport-mediated permeation but also transfer mediated by passive diffusion [36, 37]. The use of intestine derived cell line Caco-2 in the establishment of an in vitro BBB has also been taken into consideration [38]. However, differences in drug permeability between the intestine and brain-derived cells are significant, thus limiting the use of Caco-2 cells [39].

17.4.1.3 Brain Capillary Endothelial Cell Cultures

Highly purified populations of cultured microvascular cells for the study of the developmental and pathophysiological processes of the BBB became available in the early 1980s. Brain microvessel ECs can be isolated by mechanical dispersion (homogenization, filtration, sieving, centrifugation), enzymatic procedures using collagenase or a combination of both mechanical dispersion and enzymatic digestion from brain microvessels in culture [40–42]. With these techniques, a viable and homogeneous population of brain capillary ECs can be obtained for the establishment of a BBB culture system. Primary or low-passage brain capillary cells generally provide the closest phenotypic resemblance to the in vivo BBB phenotype such as *bovine brain* ECs. The immediate availability of bovine brain tissue in combination with the relative ease with which pericyte free clones can be obtained and subcultured have led to the extensive use of bovine brain as a source for ECs. Bovine brain ECs are commonly used to establish BBB models as monoculture or co-cultured with astrocytes. Another convenient source of brain ECs are derived from *pig brain* (PBEC) and yield a sufficient number of brain ECs to allow extensive permeability testing. *Rat brain* ECs have also been utilized but have the main disadvantage of low yield of capillary fragments (due to the relatively small brain) and a tendency for a negative impact of pericytes and other contaminants on the permeability of the monolayer, as compared to cells from other tissue sources. Furthermore, while a number of drug resistance molecules are expressed in rodent brain, the human brain appears to be much more complicated as evidenced by both clinical (e.g., multiple drug resistance in epileptics) [43] and in vitro studies (expression of various drug resistance molecules in different tissues, expression of brain tumor markers in known brain tumors, epileptic brain, etc.) [44].

Given the complexity and specificity of the interactions at the BBB level, the use of primary cultures of *human brain* ECs (in coculture with human astrocytes) rather than cell lines can provide important insights (e.g., improved accuracy of drug permeability values due to maintained substrate/transporter specificity) in the study of drug refractory brain disease and in the prescreening and optimization of new drug formulations. However, the availability and ease of obtaining human specimens in order to isolate cells can be a challenging obstacle; and the large number of cells to be isolated is crucial to run a successful experiment. Thus, for industrial purposes, human cells are still an unrealistic alternative. Although these pitfalls seem major, the benefits of using primary human cultures far outweigh the costs. Human specimens provide research opportunities for a variety of etiologies that otherwise may be impossible to recapitulate in cell lines or even primary culture from rodent, pig or cattle CNS. The establishment of such humanized in vitro models can provide accurate representations of many neurological diseases involving BBB dysfunction in drug refractory patients.

17.4.2
Monoculture-Based In Vitro BBB Models

Cell culture systems have been developed to reproduce the key properties of the intact BBB and to allow for testing of the mechanisms of transendothelial drug permeation. The most common and easy apparatus to culture EC is the Transwell system. This apparatus was developed based on a simplified view of the BBB. The Transwell system consists of a porous membrane support (polycarbonate or polyester) submerged in culture medium. This model is characterized by a side-by-side vertical diffusion system, where cultures of ECs derived from various sources are grown on semi-permeable membranes to allow for nutrient exchange (Fig. 17.2 A, B). Attractive features of this model (Table 17.1) are its simplicity (easy to culture with minimal skill requirements) and the ability to perform multiple experiments at the same time (one type of drug per well, or different concentrations of the same drug in each well). Furthermore, this model is ideal for Michaelis-Menten kinetics of transport due to fixed volumes in each compartment and is very cost-efficient [45]. However, BBB properties in vivo (e.g., expression of tight junctions) are bestowed on ECs by the surrounding physiological environment (e.g., shear stress, astrocytes, pericytes, blood cells) [46–48]. In the absence of these physiological stimuli, the ECs lose their "BBB properties", limiting this model to results with little predictive value (Fig. 17.2 C).

For example, it has been shown that ECs grown in monoculture under static conditions may lack the expression of specific transporters and tight junctions, leading to abnormal permeability across the EC layer [49]. Thus, the sucrose permeability ($P_{sucrose}$) of these monolayers ranged from 10^{-4} cm s^{-1} to 10^{-5} cm s^{-1}, compared with 10^{-6} cm s^{-1} to 10^{-8} cm s^{-1} in vivo [5]. As a consequence of an elevated $P_{sucrose}$, BBB permeability values of other compounds can also be overestimated (Fig. 17.2 C, D). The TEER in vivo is estimated to be ca. 1500 Ω cm^{-2}, whereas the measurements in endothelial monolayers cultured in vitro vary over 20–200 Ω cm^{-2}. However, it has been demonstrated that the addition of hydrocortisone or dexamethasone to the culture media seems to provide significant BBB "tightening" effects even in the absence of glia, thus improving barrier function [50, 51].

Endothelial cells at the BBB level in vivo present a functional specialization of the luminal and abluminal membranes [52]. This polarity was initially demonstrated by Betz and Goldstein in 1978 and is reflected by the preferential expression and distribution of membrane transporters and enzymes that act to protect the CNS and to promote substrate delivery to the brain parenchyma maintaining brain homeostasis [53–55]. Another factor to be considered is the fact that in vivo only the luminal side of ECs is exposed to serum proteins, while the ab-

Table 17.1 Advantages and disadvantages of current BBB models.

	Mono-dimensional	Bi-dimensional (coculture)	Tri-dimensional flow-based (DIV-BBB)
Pros	Ease of use	Coculture allows induction of BBB properties	Artificial capillaries connected by gas-permeable tubing allows source of growth medium, exchange of $O_2 + CO_2$, exposure to flow
	Ideal for Michaelis-Menten kinetics of transport due to fixed volumes in each compartment	Molecule flux across is lower than mono-dimensional	Induction of BBB properties
	Can do several experiments at same time	Cost-efficient	High TEER
	Cells are easier to isolate for further study		Low permeability to sucrose
	Minimum skills		Stereoselective transport
	Cost-efficient		Can mimic I.V./I.A. vs P.O. delivery
			Long-term studies possible
			Presence of drug extrusion mechanisms
Cons	Lacks physiological flow	Lacks physiological flow	Linear kinetic studies more difficult, due to flow
	High permeability to sucrose	High permeability to sucrose	Cost
	Very low TEER	Low TEER (to a lesser extent)	Skills
	Exposure to serum on both sides (luminal + abluminal)		Number of cells to be loaded/used

17 Artificial Blood-Brain Barriers

A

B
Cover
Bottom Chamber (ECS)
Tranwell
Upper Chamber (Lumen)
Pipet tip access port
Endothelial cells
Lumen
ECS
1 mm
Microporous membrane
Cluster plate

C

Compounds	Permeability (In Vivo)	Permeability (Transwell)
D-Glucose	$1.8E^{-5}$	$7.0E^{-4}$
Dilantin	$1.3E^{-5}$	$1.3E^{-5}$
L-Asp	$1.1E^{-5}$	$5.0E^{-4}$
Morphine	$9.7E^{-7}$	$1.0E^{-4}$
Mannitol	$1.9E^{-7}$	$2.0E^{-4}$
D-Asp	$1.4E^{-7}$	$8.0E^{-4}$
Sucrose	$2.0E^{-8}$	$2.0E^{-5}$

D
Permeability (Transwell) vs Permeability (In Vivo)
R = 0.1937
Slope = 0.1220

E
Blood-brain barrier- *in vivo*

Apical (luminal) side

pH: Normally 7.4
Serum protein content: 6 - 8%

Capillary blood flow

Endothelial cells
Extracellular matrix
(Filter support)
Astrocytic endfeet

pH: Normally 7.3
ISF/CSF protein content: <0.5%

Basolater (abluminal) side

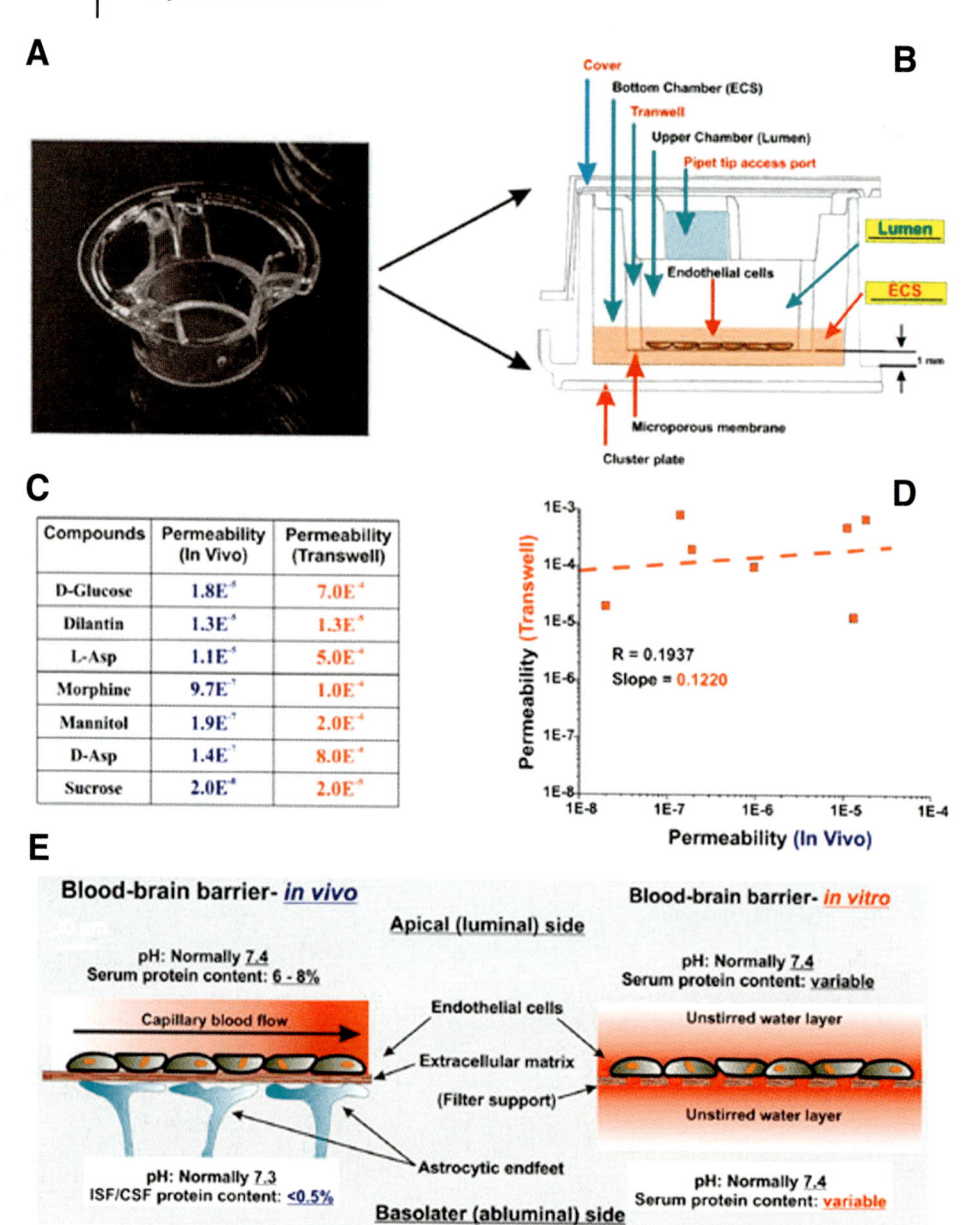

Blood-brain barrier- *in vitro*

pH: Normally 7.4
Serum protein content: variable

Unstirred water layer

Unstirred water layer

pH: Normally 7.4
Serum protein content: variable

Fig. 17.2 In vivo versus in vitro mono-culture-based BBB: the Transwell apparatus. (A) Shows a typical Transwell apparatus, while (B) is a schematic representation of a Transwell diffusion chamber. The microporous membrane allows for the free passage of nutrients and diffusible factors between the luminal and abluminal compartments. (C) Lists the permeability values of clinically relevant compounds both in vivo and in vitro (Transwell). Note how the permeability values obtained in a Transwell apparatus (D) are often misleading (significantly higher). (E) Is a schematic representation of in vivo BBB versus a monoculture-based in vitro BBB. Note that in vivo, only the luminal side of ECs is exposed to serum proteins, while the abluminal one is exposed to perivascular influence (e.g., astrocytes).

luminal part is either exposed to glial influence, basal lamina or cerebrospinal fluid (Fig. 17.2 E). ECs grown as a mono-dimensional layer upon a porous membrane are exposed to serum, on both the luminal (intravascular) and the abluminal (parenchymal) side. This non-physiological condition may accelerate the dedifferentiating process that the ECs experience and further enhance the loss of the BBB characteristics with serial cell passage. Furthermore, an increased cell cycle rate due to lack of antimitotic influences by laminin and flow will cause ECs to pile up in a multilayer fashion [56, 57]. EC grown in static monoculture-based models generally have a short lifespan compared to flow-based culture models. In addition, the occurrence of irregular patterns of cell adhesion or "edge effect" in Transwells can seriously hamper the measurement of BBB permeability.

17.4.3
Coculture-Based In Vitro BBB Models

Although primary cultures of brain endothelium alone may form tight intercellular junctions, coculture with astrocytes results in the increased formation and complexity of endothelial tight junctions and expression of specific BBB markers [58], including the physiologic glucose transporter isotype (GLUT-1), P-glycoprotein, GGTP and OX-26 [5]. The presence of glia and the establishment of glial-endothelial interactions has also been shown to increase the expression of brain endothelial marker enzymes (e.g., γ-GTP, alkaline phosphatase, acetylcholinesterase, Na^+/K^+-ATPase), MDR proteins (e.g., P-glycoprotein) and tight junctions and to help create a phenotype more closely resembling that found in vivo. A recent study performed in rat brain capillaries has provided evidence that functional expression of a typical ATP-binding cassette, G2 (rat homologues rABCG2) on the luminal side of brain capillaries is up-regulated by astrocyte-derived soluble factor(s) concomitantly with the induction of its specific mRNA [59]. Coculture with astrocytes also increase barrier tightness, as shown by higher TEER values compared to monoculture models. This increase is even higher when dexamethasone or L-α-glycero-phospho-D-mio-inositol (GPI) is added to the culture medium [60]. By the use of the Transwell apparatus previously described, the coculture can be established to enable cell-cell contact via astrocytic endfeet (astrocytes and ECs are seeded on opposite sides of the porous support) or without any contact by seeding the astrocytes at the bottom of the well and the ECs on the porous support. The coculture model is useful for studying the functionality of the BBB as well as transport-related processes and interactions between ECs and glia [61]. The major advantage of the bi-dimensional model compared to the mono-dimensional one is the establishment of conditions more closely resembling the brain microanatomy (Table 17.1). Finally, recent studies have shown that the use of conditionally immortalized brain capillary EC lines (derived from transgenic animals harboring temperature-sensitive SV40 large T-antigen gene) in coculture with pericytes and/or astrocyte cell lines, can provide a BBB model capable of retaining the in vivo transport rate of several compounds and various forms of gene expression [62]. However, coculture models of BBB still lack the presence of shear stress (intraluminal blood flow), which has been demonstrated to further differentiate the ECs and to play a central role in the cerebrovascular system by promoting the differentiation and maintenance of the BBB phenotype [5, 6, 10, 34, 56, 63–69].

17.5
Shear Stress and Cell Differentiation

Endothelial cells in vivo are continuously exposed to shear stress, a tangential force generated by the flow of blood across their apical surfaces. Shear stress affects EC structure and function, such as cell orientation with flow direction, distribution of cell fibers, induction/suppression of genes [70–72], production of vasoactive substances and improved cell adhesion [73–75] and mitotic arrest in ECs [34, 57, 76] and is also able to induce metabolic changes [34, 64]. Turbulent, but not laminar shear stress stimulates EC turnover [77]. An acute increase in shear stress induces increased release of prostacyclin and nitric oxide [78, 79] and a decrease in endothelin-1 mRNA [80]. Microvascular ECs cultured under flow are larger in volume and show an abundance of microfilaments, endocytic vesicles and clathrin-coated pits [74]. In addition, steady laminar shear stress inhibits DNA synthesis of ECs. The inhibition of the cell transition from the G0/G1 to the S phase of the cell cycle is due to up-regulation of the cdk inhibitor p21 [81]. Accordingly, in vivo the DNA synthesis of EC preferentially occurs at branch orifices, with low flow rates, for example where atherosclerotic processes are initiated. High laminar shear stress promotes glycosaminoglycan synthesis, tight junction formation and the expression of junction-related proteins [82, 83]. Thus, it is clear that biomechanical forces generated by blood flow play a role in the induction of many BBB properties. Interestingly, EC are not the only cells responsive to shear stress. All cells in the circulation respond to physiological shear stress, including erythrocytes, platelets and leukocytes [84–86]. Given the considerations summarized above, it is not surprising that attempts were made to culture ECs under the influence of shearing forces. An EC monolayer (even from non-brain origin) can be exposed to a quasi-laminar or pulsatile shear stress by the use of a purpose-built cone and plate viscometer. These systems can be modified to accept culture plates of different sizes. Cells are seeded on the bottom of the plate and the shear stress is generated by a rotating cone which transmits the shear force to the cells through the culture media (Fig. 17.3). The angular velocity and the cone angle determine the level of shear stress generated. Pulsing and reversing the angular velocity of the cone can be used to achieve backflow or pulsatile shear stress.

Cone and Plate Shear Apparatus

$$T\alpha = \frac{\mu\omega}{\alpha}$$

$$\tilde{R} = \frac{r^2\omega\alpha^2}{12\upsilon}$$

Oscillatory Shear

Unidirectional Laminar Shear

Endothelial Cells   Culture Plate

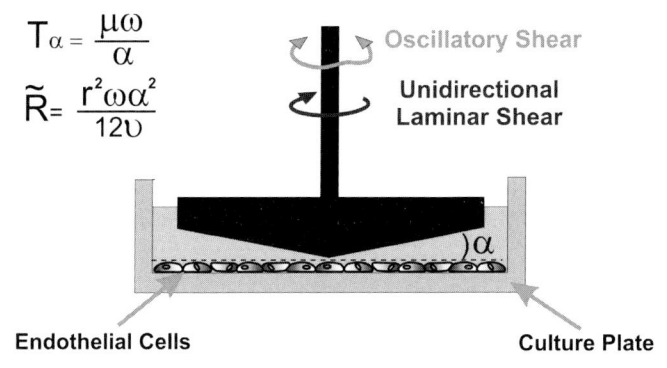

Fig. 17.3 Schematic representation of a cone and plate viscometer. An EC monolayer can be exposed to a quasi-laminar or pulsatile shear stress by the use of a purpose-built cone and plate viscometer. Note that μ is viscosity, ν is kinematic viscosity, ω is angular velocity and α is cone angle. The angular velocity and cone angle determine the level of shear stress generated.

17.6
Flow-Based In Vitro BBB Systems

17.6.1
Dynamic In Vitro BBB: Standard Model

The dynamic in vitro model of BBB (DIV-BBB) originates from a modification of a cell culture system used for hybridoma cell expansion. This in vitro "cell differentiation factory" provides quasi-physiological experimental conditions for culturing ECs and astrocytes in a capillary-like structure and is able to functionally and anatomically mimic the brain microvasculature. The DIV-BBB is characterized by a pronectin-coated microporous polypropylene hollow fiber structure that enables coculturing of EC (intraluminally) with glia (abluminally). The entire system is connected to a media reservoir via gas permeable silicon tubes that allow for the exchange of O_2 and CO_2. A servo-controlled variable-speed pulsatile pump generates flow from the media source through the capillary bundle and back. One-way valves positioned on either side of the "pump" ensure unidirectional flow (Fig. 17.4A). The pumping mechanism is capable of generating flow levels of 1–50 ml min^{-1} with associated shear stress levels of 1–200 µN cm^{-2}. Shear stress levels are estimated to be in the range of 1–5 µN cm^{-2} at the capillary level in vivo but the pump allows changes in shear stress levels at any time point without changing oxygen tension and/or glucose levels. The DIV-BBB also enables realtime continuous monitoring of BBB function by measurement of TEER across the barrier via electrodes inserted in the luminal and abluminal (*ecs*) compartments. Exposure to

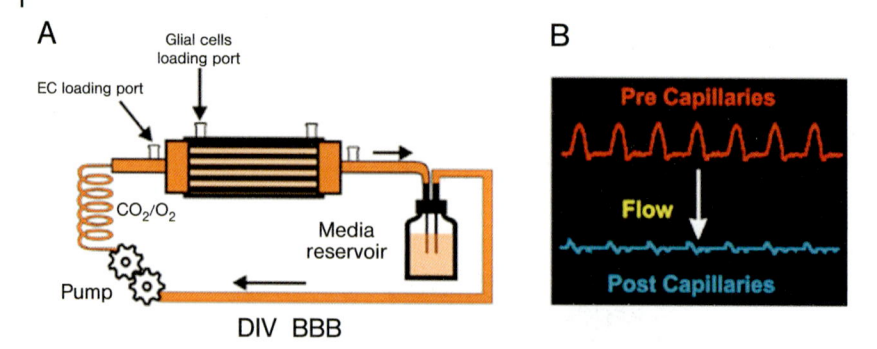

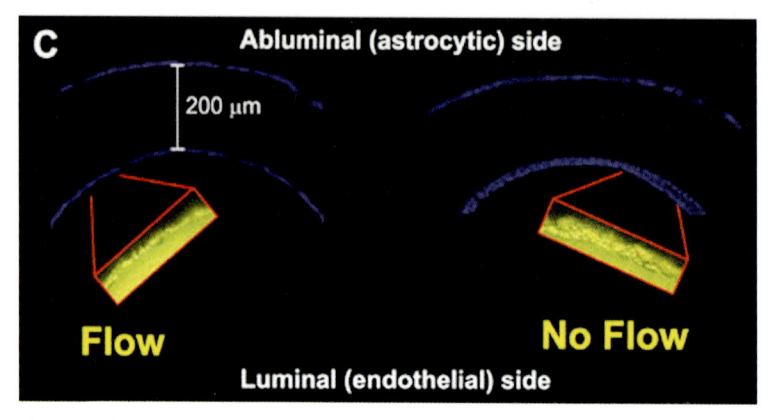

Fig. 17.4 Flow-based in vitro BBB (DIV-BBB).
(A) Shows a schematic representation of a typical DIV-BBB. Endothelial cells are seeded intraluminally in fibronectin-coated hollow fibers, while astrocytes are cultured extraluminally.
(B) Shows waveforms of pressure changes in pre- and post-DIV-BBB capillary segments mimicking the physiological changes in vivo.
(C) Shows a cross-sectional view of artificial capillaries employed in our model. Note that ECs grow in a multilayer fashion in the absence of luminal flow (right side). When flow is present (left side), ECs grow in a typical monolayer comparable to what observed in brain microvasculature.

controlled pumping rates makes the model suited for the study of EC responses to levels of shear stress corresponding to large arteries (ca. 70 μN cm^{-2}).

Note, however, that as implemented today, the system does not allow studying the effects of turbulent or otherwise altered flow.

The pulsatile flow generated consists of a complex waveform with a substantial drop of pressure occurring at the end of the capillaries, giving these models the ability to reproduce the hemodynamic conditions observed in vivo (Fig. 17.4B). Under these conditions, ECs develop a morphology that closely resembles the endothelial phenotype in situ [73], demonstrating that ECs grown with flow develop greater differentiation than after static culture (Fig. 17.4C). This improved cellular differentiation is also reflected by much higher TEER

A

Compounds	Permeability (In Vivo)	Permeability (DIV-BBB)
D-Glucose	$1.8E^{-5}$	$1.8E^{-5}$
Dilantin	$1.3E^{-5}$	$1.5E^{-5}$
L-Asp	$1.1E^{-5}$	$1.9E^{-5}$
Morphine	$9.7E^{-7}$	$5.3E^{-7}$
Mannitol	$1.9E^{-7}$	$9.7E^{-7}$
D-Asp	$1.4E^{-7}$	$4.8E^{-7}$
Sucrose	$2.0E^{-8}$	$2.9E^{-7}$

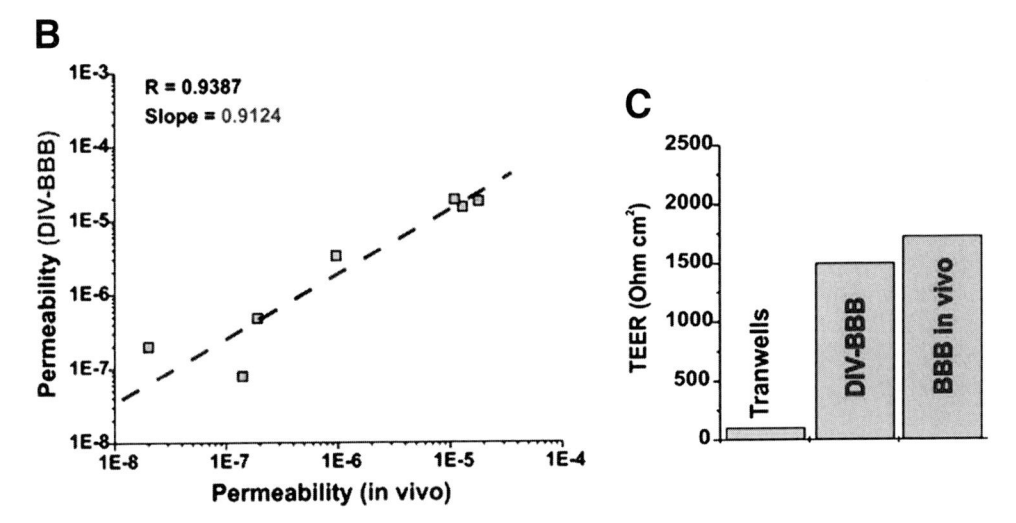

Fig. 17.5 BBB permeability in vivo vs in vitro.
(A) Shows the permeability of clinically relevant compounds in vivo and in a DIV-BBB. Note how the permeability values obtained in a DIV-BBB (B) are comparable to permeability in vivo. (C) Shows typical TEER values measured in vivo, in the DIV-BBB and in Transwells. Note how the DIV-BBB model closely compares to the in vivo situation.

($>1200\ \Omega\ cm^{-2}$) and drug permeability values that reproduce the in vivo scenario (Fig. 17.5). Typical BBB properties in ECs grown in hollow fibers under dynamic condition include low permeability to intraluminal potassium, negligible extravasation of proteins and the expression of a glucose transporter. In addition, culturing ECs with glia affects the overall morphology of the cells and induces the expression of BBB-specific ion channels [56, 64, 87, 88]. However, this model has limited applicability due to its design. The compartmentalization of

ECs and glia in the hollow fiber system in the presence of pulsatile flow makes the study of linear kinetics more complex. The cylindrical shape of the apparatus does not allow for visualization of either compartment to assess morphologic and/or phenotypic changes of the cells of interest. Additionally, the cell inoculation volume and media requirements are enormous, considering the fact that in general this model is not reusable. Finally, the volume and physical access to the ECs allow only for introduction of cell suspensions and not tissue slice preparations. In spite of these limitations, several reports have described the usefulness of the DIV-BBB for molecular [89], pharmaceutical [65, 90–93], morphological and functional studies [10, 34, 56, 64, 66, 67, 94, 95].

17.6.2
Dynamic In Vitro BBB: New Model

The new DIV-BBB (NDIV-BBB), recently developed, maintains all the dynamic characteristics that are attractive in the standard DIV-BBB but also includes novel features unavailable under the previous configuration (see Fig. 17.6 panels A and B for a side-by-side comparison and a detailed technical description). The apparatus consists of a rectangular polycarbonate hollow chamber enclosed by a permanent glass bottom and a removable clear acrylic top. The chamber contains an adjustable number of artificial polypropylene capillaries. These capillaries have an inner diameter of 600 µm, a wall thickness of 200 µm and 0.64-µm trans-capillary pores that allow free diffusion of solutes from the abluminal to the luminal compartment and vice versa. Several improvements in the design of the dynamic BBB model allow for longitudinal studies of the effects of flow and coculture in a controlled and fully recyclable environment that simultaneously permits visual inspection of the abluminal compartment and manipulation of individual capillaries. These modifications have expanded capabilities for research involving the central nervous system and neurological diseases. This model in particular has several advantages compared to the previous one:

1. The ability to expose and remove single or multiple capillaries during the course of an experiment for cell isolation (luminal and/or abluminal), morphological and/or phenotypic studies [56].
2. The possibility of visualizing cell-cell interactions with inverted light microscopy as well as techniques involving fluorescence imaging.
3. The number of cells required to establish a functional BBB is one-fourth of the standard model.
4. It is not limited to EC and astrocyte coculture, since the model allows for inclusion of other relevant cell types (e.g., neurons) that can be seeded on the bottom of the abluminal chamber.
5. The model can be customized to accommodate different numbers of capillaries.

17.6 Flow-Based In Vitro BBB Systems

A DIV BBB

B New DIV BBB

Technical characteristics

- Dimensions: 135 mm x 9 mm x 9 mm
- Capillary number – 50
- Capillary length: 13 cm
- Pore size: 0.5 µm
- Capillary surface area: 128.5 cm^2
- Lumen surface area: 67.35 cm^2
- ECS volume: 1.4 mL
- Cell inoculation (ECS): 8 - 10 x 10^6 cells
- Cell inoculation (Lumen): 6 x 10^6 cells
- ECS sampling port: 2
- Access to ECS: Limited
- Possibility to isolate capillaries: NO
- Visualization of cells in the ECS by
- inverted light microscopy: NO
- Capillary type : fixed
- Not recyclable

- Dimensions: 62 mm x 34 mm x 5 mm
- Capillary number: variable (1 to 12)
- Capillary length: 8.6 cm
- Pore size: 0.64 µm
- Capillary surface area: 32.4 cm^2
- Lumen surface area – 19.45 cm^2
- ECS (functional) volume: variable
- Cell inoculation (ECS): 2 x 10^6 cells
- Cell inoculation (Lumen): 1.5 x 10^6 cells
- ECS sampling port: 4
- Access to ECS: Unlimited
- Possibility to isolate capillaries: YES
- Visualization of cells in the ECS by
- inverted light microscopy: YES
- Capillary type : Customizable
- Recyclable

Standard DIV-BBB

New DIV-BBB
Multi Capillary   Mono Capillary

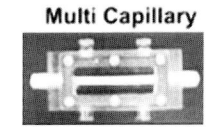

Fig. 17.6 Cross-comparison between DIV-BBB model technical specifications.

However, this model is not designed for industrial use where numerous tests have to be quickly performed on a large scale (e.g., pharmaceutical companies). Furthermore, the system setup requires greater technical skills and more time than any other model previously described, limiting the use of this apparatus.

companies with cost-effective flow-based systems implemented with monitoring and analysis features, suited for both large-scale and laboratory use. The cell types currently used have also been discussed, setting an inevitable choice between fast data acquisition versus accuracy and reproducibility. We also want to direct attention toward the need for human cell-based models, since several pathological conditions cannot be reproduced in cell line-derived or rodent-based BBB models. Furthermore, with the rather novel prospect of stem cell research and expanding our understanding of the mechanisms behind cell-cell interactions and cross-communication at the BBB level, the future moves into a direction where differentiating factors will be successfully used to induce BBB properties in omnipotent stem cells.

Acknowledgments

This work was supported by NIH-2RO1 HL51614, NIH-RO1 NS 43284, NIH-RO1 NS38195 to Damir Janigro and by The Alternatives Research & Development Foundation grant award to Luca Cucullo. We would also like to thank those who have in the past contributed to the development of the models described here: Liljiana Krizanac-Bengez, PhD, Kathe Stannes, BS, Eleonora Fornaciari, PhD, Patrizia Mascagni, PhD, Francesco Macchia, PhD MBA, and Francesca Salvetti, PhD.

References

1 Bradbury, M.W. **1993**, The blood-brain barrier, Exp. Physiol. 78, 453–472.
2 Dermietzel, R., Krause, D. **1991**, Molecular anatomy of the blood-brain barrier as defined by immunocytochemistry, Int. Rev. Cytol. 127, 57–109.
3 Rubin, L.L., Staddon, J.M. **1999**, The cell biology of the blood-brain barrier, Annu. Rev. Neurosci. 22, 11–28.
4 Engelhardt, B. **2003**, Development of the blood-brain barrier, Cell Tissue Res. 314, 119–129.
5 Grant, G.A., Abbott, N.J., Janigro, D. **1998**, Understanding the physiology of the blood-brain barrier: in vitro models, *News Physiol. Sci.* 13, 287–293.
6 Grant, G.A., Janigro, D. **2004**, The blood-brain barrier, in *Youmans Neurological Surgery*, vol. 1, ed. Winn, H.R., Saunders, Philadelphia, p. 153–174.
7 Virgintino, D., Errede, M., Robertson, D., Capobianco, C., Girolamo, F., Vimercati, A., Bertossi, M., Roncali, L. **2004**, Immunolocalization of tight junction proteins in the adult and developing human brain, *Histochem. Cell Biol.* 122, 51–59.
8 Emmi, A., Wenzel, H.J., Schwartzkroin, P.A., Taglialatela, M., Castaldo, P., Bianchi, L., Nerbonne, J., Robertson, G.A., Janigro, D. **2000**, Do glia have heart? Expression and functional role for ether-a-go-go currents in hippocampal astrocytes, *J. Neurosci.* 20, 3915–3925.

9 Liberto, C. M., Albrecht, P. J., Herx, L. M., Yong, V. W., Levison, S. W. **2004**, Pro-regenerative properties of cytokine-activated astrocytes, *J. Neurochem.* 89, 1092–1100.

10 Stanness, K. A., Westrum, L. E., Fornaciari, E., Mascagni, P., Nelson, J. A., Stenglein, S. G., Myers, T., Janigro, D. **1997**, Morphological and functional characterization of an in vitro blood-brain barrier model, *Brain Res.* 771, 329–342.

11 Mizuguchi, H., Utoguchi, N., Mayumi, T. **1997**, Preparation of glial extracellular matrix: a novel method to analyze glial-endothelial cell interaction, *Brain Res. Brain Res. Protocol* 1, 339–343.

12 Hamm, S., Dehouck, B., Kraus, J., Wolburg-Buchholz, K., Wolburg, H., Risau, W., Cecchelli, R., Engelhardt, B., Dehouck, M. P. **2004**, Astrocyte mediated modulation of blood-brain barrier permeability does not correlate with a loss of tight junction proteins from the cellular contacts, *Cell Tissue Res.* 315, 157–166.

13 Hori, S., Ohtsuki, S., Tachikawa, M., Kimura, N., Kondo, T., Watanabe, M., Nakashima, E., Terasaki, T. **2004**, Functional expression of rat ABCG2 on the luminal side of brain capillaries and its enhancement by astrocyte-derived soluble factor(s), *J. Neurochem.* 90, 526–536.

14 Kramer, S. D., Schutz, Y. B., Wunderli-Allenspach, H., Abbott, N. J., Begley, D. J. **2002**, Lipids in blood-brain barrier models in vitro II: influence of glial cells on lipid classes and lipid fatty acids, *In Vitro Cell Dev. Biol. Anim.* 38, 566–571.

15 De Boer, A. B., De Lange, E. L., Breimer, D. D. **1998**, Transporters and the blood-brain barrier (BBB), *Int. J. Clin. Pharmacol. Ther.* 36, 14–15.

16 Hagenbuch, B., Gao, B., Meier, P. J. **2002**, Transport of xenobiotics across the blood-brain barrier, *News Physiol. Sci.* 17, 231–234.

17 Abbott, N. J., Khan, E. U., Rollinson, C. M., Reichel, A., Janigro, D., Dombrowski, S. M., Dobbie, M. S., Begley, D. J. **2002**, Drug resistance in epilepsy: the role of the blood-brain barrier, *Novartis Found. Symp.* 243, 38–47.

18 Cox, D. S., Scott, K. R., Gao, H., Raje, S., Eddington, N. D. **2001**, Influence of multidrug resistance (MDR) proteins at the blood-brain barrier on the transport and brain distribution of enaminone anticonvulsants, *J. Pharm. Sci.* 90, 1540–1552.

19 Regina, A., Demeule, M., Laplante, A., Jodoin, J., Dagenais, C., Berthelet, F., Moghrabi, A., Beliveau, R. **2001**, Multidrug resistance in brain tumors: roles of the blood-brain barrier, *Cancer Metastasis Rev.* 20, 13–25.

20 Mahar Doan, K. M., Humphreys, J. E., Webster, L. O., Wring, S. A., Shampine, L. J., Serabjit-Singh, C. J., Adkison, K. K., Polli, J. W. **2002**, Passive permeability and P-glycoprotein-mediated efflux differentiate central nervous system (CNS) and non-CNS marketed drugs, *J. Pharmacol. Exp. Ther.* 303, 1029–1037.

21 Reichel, A., Begley, D. J. **1998**, Potential of immobilized artificial membranes for predicting drug penetration across the blood-brain barrier, *Pharm. Res.* 15, 1270–1274.

22 Ong, S., Liu, H., Pidgeon, C. **1996**, Immobilized-artificial-membrane chromatography: measurements of membrane partition coefficient and predicting drug membrane permeability, *J. Chromatogr. A* 728, 113–128.

23 Pidgeon, C., Ong, S., Liu, H., Qiu, X., Pidgeon, M., Dantzig, A. H., Munroe, J., Hornback, W. J., Kasher, J. S., Glunz, L. **1995**, IAM chromatography: an in vitro screen for predicting drug membrane permeability, *J. Med. Chem.* 38, 590–594.

24 Wang, Q., Rager, J. D., Weinstein, K., Kardos, P. S., Dobson, G. L., Li, J., Hidalgo, I. J. **2005**, Evaluation of the MDR-MDCK cell line as a permeability screen for the blood-brain barrier, *Int. J. Pharm.* 288, 349–359.

25 Reichel, A., Abbott, N. J., Begley, D. J. **2002**, Evaluation of the RBE4 cell line to explore carrier-mediated drug delivery to the CNS via the L-system amino acid transporter at the blood-brain barrier, *J. Drug Target.* 10, 277–283.

26 Begley, D. J., Lechardeur, D., Chen, Z. D., Rollinson, C., Bardoul, M., Roux, F., Scherman, D., Abbott, N. J. **1996**, Functional expression of P-glycoprotein in an immortalised cell line of rat brain endothelial cells, RBE4, *J. Neurochem.* 67, 988–995.

27 Chishty, M., Begley, D. J., Abbott, N. J., Reichel, A. **2003**, Functional characterisation of nucleoside transport in rat brain endothelial cells, *Neuroreport* 14, 1087–1090.

28 Bendayan, R., Lee, G., Bendayan, M. **2002**, Functional expression and localization of P-glycoprotein at the blood brain barrier, *Microsc. Res. Tech.* 57, 365–380.

29 Cestelli, A., Catania, C., D'Agostino, S., Di, L., Licata, L., Schiera, G., Pitarresi, G. L., Savettieri, G., De, C., Giandalia, G., Giannola, L. I. **2001**, Functional feature of a novel model of blood brain barrier: studies on permeation of test compounds, *J. Control Release* 76, 139–147.

30 Regina, A., Koman, A., Piciotti, M., El Hafny, B., Center, M. S., Bergmann, R., Couraud, P. O., Roux, F. **1998**, Mrp1 multidrug resistance-associated protein and P-glycoprotein expression in rat brain microvessel endothelial cells, *J. Neurochem.* 71, 705–715.

31 Chat, M., Bayol-Denizot, C., Suleman, G., Roux, F., Minn, A. **1998**, Drug metabolizing enzyme activities and superoxide formation in primary and immortalized rat brain endothelial cells, *Life Sci.* 62, 151–163.

32 El Hafny, B., Cano, N., Piciotti, M., Regina, A., Scherrmann, J. M., Roux, F. **1997**, Role of P-glycoprotein in colchicine and vinblastine cellular kinetics in an immortalized rat brain microvessel endothelial cell line, *Biochem. Pharmacol.* 53, 1735–1742.

33 Rist, R. J., Romero, I. A., Chan, M. W., Couraud, P. O., Roux, F., Abbott, N. J. **1997**, F-actin cytoskeleton and sucrose permeability of immortalised rat brain microvascular endothelial cell monolayers: effects of cyclic AMP and astrocytic factors, *Brain Res.* 768, 10–18.

34 Desai, S. Y., Marroni, M., Cucullo, L., Krizanac-Bengez, L., Mayberg, M. R., Hossain, M. T., Grant, G. G., Janigro, D. **2002**, Mechanisms of endothelial survival under shear stress, *Endothelium* 9, 89–102.

35 Smith, B.J., Doran, A.C., McLean, S., Tingley, F.D. III, O'Neill, B.T., Kajiji, S.M. **2001**, P-glycoprotein efflux at the blood-brain barrier mediates differences in brain disposition and pharmacodynamics between two structurally related neurokinin-1 receptor antagonists, *J. Pharmacol. Exp. Ther.* 298, 1252–1259.

36 Garberg, P., Ball, M., Borg, N., Cecchelli, R., Fenart, L., Hurst, R.D., Lindmark, T., Mabondzo, A., Nilsson, J.E., Raub, T.J., Stanimirovic, D., Terasaki, T., Oberg, J.O., Osterberg, T. **2005**, In vitro models for the blood-brain barrier, *Toxicol. In Vitro* 19, 299–334.

37 Wang, Q., Rager, J.D., Weinstein, K., Kardos, P.S., Dobson, G.L., Li, J., Hidalgo, I.J. **2005**, Evaluation of the MDR-MDCK cell line as a permeability screen for the blood-brain barrier, *Int. J. Pharm.* 288, 349–359.

38 Zhu, Z.B., Makhija, S.K., Lu, B., Wang, M., Rivera, A.A., Preuss, M., Zhou, F., Siegal, G.P., Alvarez, R.D., Curiel, D.T. **2004**, Transport across a polarized monolayer of Caco-2 cells by transferrin receptor-mediated adenovirus transcytosis, *Virology* 325, 116–128.

39 Lohmann, C., Huwel, S., Galla, H.J. **2002**, Predicting blood-brain barrier permeability of drugs: evaluation of different in vitro assays, *J. Drug Target.* 10, 263–276.

40 Dallaire, L., Tremblay, L., Beliveau, R. **1991**, Purification and characterization of metabolically active capillaries of the blood-brain barrier, *Biochem. J.* 276, 745–752.

41 Biegel, D., Spencer, D.D., Pachter, J.S. **1995**, Isolation and culture of human brain microvessel endothelial cells for the study of blood-brain barrier properties in vitro, *Brain Res.* 692, 183–189.

42 Franke, H., Galla, H., Beuckmann, C.T. **2000**, Primary cultures of brain microvessel endothelial cells: a valid and flexible model to study drug transport through the blood-brain barrier in vitro, *Brain Res. Brain Res. Protocol* 5, 248–256.

43 Dombrowski, S., Desai, S., Marroni, M., Cucullo, L., Bingaman, W., Mayberg, M.R., Bengez, L., Janigro, D. **2001**, Overexpression of multiple drug resistance genes in endothelial cells from patients with refractory epilepsy, *Epilepsia* 42, 1504–1507.

44 Marroni, M., Marchi, N., Cucullo, L., Abbott, N.J., Signorelli, K., Janigro, D. **2003**, Vascular and parenchymal mechanisms in multiple drug resistance: a lesson from human epilepsy, *Curr. Drug Target.* 4, 297–304.

45 Berezowski, V., Landry, C., Lundquist, S., Dehouck, L., Cecchelli, R., Dehouck, M.P., Fenart, L. **2004**, Transport screening of drug cocktails through an in vitro blood-brain barrier: is it a good strategy for increasing the throughput of the discovery pipeline? *Pharm. Res.* 21, 756–760.

46 Hori, S., Ohtsuki, S., Hosoya, K., Nakashima, E., Terasaki, T. **2004**, A pericyte-derived angiopoietin-1 multimeric complex induces occludin gene expression in brain capillary endothelial cells through Tie-2 activation in vitro, *J. Neurochem.* 89, 503–513.

47 Hamm, S., Dehouck, B., Kraus, J., Wolburg-Buchholz, K., Wolburg, H., Risau, W., Cecchelli, R., Engelhardt, B., Dehouck, M. P. **2004**, Astrocyte mediated modulation of blood-brain barrier permeability does not correlate with a loss of tight junction proteins from the cellular contacts, *Cell Tissue Res.* 315, 157–166.

48 Toimela, T., Maenpaa, H., Mannerstrom, M., Tahti, H. **2004**, Development of an in vitro blood-brain barrier model: cytotoxicity of mercury and aluminum, *Toxicol. Appl. Pharmacol.* 195, 73–82.

49 Berezowski, V., Landry, C., Dehouck, M. P., Cecchelli, R., Fenart, L. **2004**, Contribution of glial cells and pericytes to the mRNA profiles of P-glycoprotein and multidrug resistance-associated proteins in an in vitro model of the blood-brain barrier, *Brain Res.* 1018, 1–9.

50 Grabb, P. A., Gilbert, M. R. **1995**, Neoplastic and pharmacological influence on the permeability of an in vitro blood-brain barrier, *J. Neurosurg.* 82, 1053–1058.

51 Hoheisel, D., Nitz, T., Franke, H., Wegener, J., Hakvoort, A., Tilling, T., Galla, H. J. **1998**, Hydrocortisone reinforces the blood-brain barrier properties in a serum free cell culture system, *Biochem. Biophys. Res. Commun.* 244, 312–316.

52 Betz, A. L., Goldstein, G. W. **1978**, Polarity of the blood-brain barrier: neutral amino acid transport into isolated brain capillaries, *Science* 202, 225–227.

53 Betz, A. L., Firth, J. A., Goldstein, G. W. **1980**, Polarity of the blood-brain barrier: distribution of enzymes between the luminal and antiluminal membranes of brain capillary endothelial cells, *Brain Res.* 192, 17–28.

54 Farrell, C. L., Pardridge, W. M. **1991**, Blood-brain barrier glucose transporter is asymmetrically distributed on brain capillary endothelial lumenal and ablumenal membranes: an electron microscopic immunogold study, *Proc. Natl Acad. Sci. USA* 88, 5779–5783.

55 Pardridge, W. M., Golden, P. L., Kang, Y. S., Bickel, U. **1997**, Brain microvascular and astrocyte localization of P-glycoprotein, *J. Neurochem.* 68, 1278–1285.

56 Cucullo, L., McAllister, M. S., Kight, K., Krizanac-Bengez, L., Marroni, M., Mayberg, M. R., Stanness, K. A., Janigro, D. **2002**, A new dynamic in vitro model for the multidimensional study of astrocyte-endothelial cell interactions at the blood-brain barrier, *Brain Res.* 951, 243–254.

57 Ziegler, T., Nerem, R. M. **1994**, Effect of flow on the process of endothelial cell division, *Arterioscler. Thromb.* 14, 636–643.

58 Goldstein, G. W. **1988**, Endothelial cell-astrocyte interactions. A cellular model of the blood-brain barrier, *Ann. NY Acad. Sci.* 529, 31–39.

59 Hori, S., Ohtsuki, S., Tachikawa, M., Kimura, N., Kondo, T., Watanabe, M., Nakashima, E., Terasaki, T. **2004**, Functional expression of rat ABCG2 on the luminal side of brain capillaries and its enhancement by astrocyte-derived soluble factor(s), *J. Neurochem.* 90, 526–536.

60 Cucullo, L., Hallene, K., Dini, G., Dal Toso, R., Janigro, D. **2004**, Glycerophosphoinositol and dexamethasone improve transendothelial electrical resistance in an in vitro study of the blood-brain barrier, *Brain Res.* 997, 147–151.

61 Cucullo, L., Marchi, N., Marroni, M., Fazio, V., Namura, S., Janigro, D. **2003**, Blood-brain barrier damage induces release of α2-macroglobulin, *Mol. Cell Proteomics* 2, 234–241.

62 Terasaki, T., Ohtsuki, S., Hori, S., Takanaga, H., Nakashima, E., Hosoya, K. **2003**, New approaches to in vitro models of blood-brain barrier drug transport, *Drug Discov. Today* 8, 944–954.

63 Salvetti, F., Cecchetti, P., Janigro, D., Lucacchini, A., Benzi, L., Martini, C. **2002**, Insulin permeability across an in vitro dynamic model of endothelium, *Pharm. Res.* 19, 445–450.

64 McAllister, M. S., Krizanac-Bengez, L., Macchia, F., Naftalin, R. J., Pedley, K. C., Mayberg, M. R., Marroni, M., Leaman, S., Stanness, K. A., Janigro, D. **2001**, Mechanisms of glucose transport at the blood-brain barrier: an in vitro study, *Brain Res.* 904, 20–30.

65 Janigro, D., Leaman, S. M., Stanness, K. A. **1999**, Dynamic modeling of the blood-brain barrier: a novel tool for studies of drug delivery to the brain, *News Physiol. Sci.* 12, 7–12.

66 Stanness, K. A., Neumaier, J. F., Sexton, T. J., Grant, G. A., Emmi, A., Maris, D. O., Janigro, D. **1999**, A new model of the blood-brain barrier: co-culture of neuronal, endothelial and glial cells under dynamic conditions, *Neuroreport* 10, 3725–3731.

67 Krizanac-Bengez, L., Kapural, M., Parkinson, F., Cucullo, L., Hossain, M., Mayberg, M. R., Janigro, D. **2003**, Effects of transient loss of shear stress on blood-brain barrier endothelium: role of nitric oxide and IL-6, *Brain Res.* 977, 239–246.

68 Ballermann, B. J., Dardik, A., Eng, E., Liu, A. **1998**, Shear stress and the endothelium, *Kidney Int. Suppl.* 67, S100–S108.

69 Haidekker, M. A., L'Heureux, N., Frangos, J. A. **2000**, Fluid shear stress increases membrane fluidity in endothelial cells: a study with DCVJ fluorescence, *Am. J. Physiol. Heart Circ. Physiol.* 278, H1401–H1406.

70 Wasserman, S. M., Topper, J. N. **2004**, Adaptation of the endothelium to fluid flow: in vitro analyses of gene expression and in vivo implications, *Vasc. Med.* 9, 35–45.

71 Krizanac-Bengez, L., Mayberg, M. R., Janigro, D. **2004**, The cerebral vasculature as a therapeutic target for neurological disorders and the role of shear stress in vascular homeostasis and pathophysiology, *Neurol. Res.* 26, 846–853.

72 Brooks, A. R., Lelkes, P. I., Rubanyi, G. M. **2004**, Gene expression profiling of vascular endothelial cells exposed to fluid mechanical forces: relevance for focal susceptibility to atherosclerosis, *Endothelium* 11, 45–57.

73 Ott, M. J., Ballermann, B. J. **1995**, Shear stress-conditioned, endothelial cell-seeded vascular grafts: improved cell adherence in response to in vitro shear stress, *Surgery* 117, 334–339.

74 Ballermann, B. J., Ott, M. J. **1995**, Adhesion and differentiation of endothelial cells by exposure to chronic shear stress: a vascular graft model, *Blood Purif.* 13, 125–134.

75 Ando, J., Kamiya, A. **1996**, Flow-dependent regulation of gene expression in vascular endothelial cells, *Jpn Heart J.* 37, 19–32.

76 Lin, K., Hsu, P. P., Chen, B. P., Yuan, S., Usami, S., Shyy, J. Y., Li, Y. S., Chien, S. **2000**, Molecular mechanism of endothelial growth arrest by laminar shear stress, *Proc. Natl. Acad. Sci. USA* 97, 9385–9389.

77 Davies, P. F., Remuzzi, A., Gordon, E. J., Dewey, C. F. Jr., Gimbrone, M. A. Jr. **1986**, Turbulent fluid shear stress induces vascular endothelial cell turnover in vitro, *Proc. Natl. Acad. Sci. USA* 83, 2114–2117.

78 Grabowski, E. F., Jaffe, E. A., Weksler, B. B. **1985**, Prostacyclin production by cultured endothelial cell monolayers exposed to step increases in shear stress, *J. Lab. Clin. Med.* 105, 36–43.

79 Buga, G. M., Gold, M. E., Fukuto, J. M., Ignarro, L. J. **1991**, Shear stress-induced release of nitric oxide from endothelial cells grown on beads, *Hypertension* 17, 187–193.

80 Sharefkin, J. B., Diamond, S. L., Eskin, S. G., McIntire, L. V., Dieffenbach, C. W. **1991**, Fluid flow decreases preproendothelin mRNA levels and suppresses endothelin-1 peptide release in cultured human endothelial cells, *J. Vasc. Surg.* 14, 1–9.

81 Akimoto, S., Mitsumata, M., Sasaguri, T., Yoshida, Y. **2000**, Laminar shear stress inhibits vascular endothelial cell proliferation by inducing cyclin-dependent kinase inhibitor p21(Sdi1/Cip1/Waf1), *Circ. Res.* 86, 185–190.

82 Arisaka, T., Mitsumata, M., Kawasumi, M., Tohjima, T., Hirose, S., Yoshida, Y. **1995**, Effects of shear stress on glycosaminoglycan synthesis in vascular endothelial cells, *Ann. NY Acad. Sci.* 748, 543–554.

83 Yoshida, Y., Okano, M., Wang, S., Kobayashi, M., Kawasumi, M., Hagiwara, H., Mitsumata, M. **1995**, Hemodynamic-force-induced difference of interendothelial junctional complexes, *Ann. NY Acad. Sci.* 748, 104–120.

84 Johnson, R. M. **1994**, Membrane stress increases cation permeability in red cells, *Biophys. J.* 67, 1876–1881.

85 Kroll, M. H., Hellums, J. D., McIntire, L. V., Schafer, A. I., Moake, J. L. **1996**, Platelets and shear stress, *Blood* 88, 1525–1541.

86 Moazzam, F., DeLano, F. A., Zweifach, B. W., Schmid-Schonbein, G. W. **1997**, The leukocyte response to fluid stress, *Proc. Natl Acad. Sci. USA* 94, 5338–5343.

87 Janigro, D., Nguyen, T. S., Gordon, E. L., Winn, H. R. **1996**, Physiological properties of ATP-activated cation channels in rat brain microvascular endothelial cells, *Am. J. Physiol.* 270, H1423–H1434.

88 Stanness, K. A., Guatteo, E., Janigro, D. **1996**, A dynamic model of the blood-brain barrier "in vitro", *Neurotoxicology* 17, 481–496.

89 Marroni, M., Marchi, N., Cucullo, L., Abbott, N. J., Signorelli, K., Janigro, D. **2003**, Vascular and parenchymal mechanisms in multiple drug resistance: a lesson from human epilepsy, *Curr. Drug Target.* 4, 297–304.

90 Parkinson, F. E., Friesen, J., Krizanac-Bengez, L., Janigro, D. **2003**, Use of a three-dimensional in vitro model of the rat blood-brain barrier to assay nucleoside efflux from brain, *Brain Res.* 980, 233–241.

91 Salvetti, F., Cecchetti, P., Janigro, D., Lucacchini, A., Benzi, L., Martini, C. **2002**, Insulin permeability across an in vitro dynamic model of endothelium, *Pharm. Res.* 19, 445–450.

92 Sinclair, C. J., Krizanac-Bengez, L., Stanness, K. A., Janigro, D., Parkinson, F. E. **2001**, Adenosine permeation of a dynamic in vitro blood-brain barrier inhibited by dipyridamole, *Brain Res.* 898, 122–125.

93 Strelow, L., Janigro, D., Nelson, J. A. **2002**, Persistent SIV infection of a blood-brain barrier model, *J. Neurovirol.* 8, 270–280.

94 Pekny, M., Stanness, K. A., Eliasson, C., Betsholtz, C., Janigro, D. **1998**, Impaired induction of blood-brain barrier properties in aortic endothelial cells by astrocytes from GFAP-deficient mice, *Glia* 22, 390–400.

95 Stanness, K. A., Guatteo, E., Janigro, D. **1996**, A dynamic model of the blood-brain barrier "in vitro", *Neurotoxicology* 17, 481–496.

Table 18.1 Compounds used for in silico models of BBB permeation as given in [10].

Compound	LogBB	Compound	LogBB
Methane	0.04	5 [a]	−1.06
Pentane	0.76	Clonidine (6) [a]	0.11
Hexane	0.8	Mepyramine (7) [a]	0.49
2-Methylpentane	0.97	Imipramine (8) [a]	1.06
3-Methylpentane	1.01	Rantidine (9) [a]	−1.23
2,2-Dimethylbutane	1.04	Tiotidine (10) [a]	−0.82
Heptane	0.81	11 [a]	−1.17
3-Methylhexane	0.9	12 [a]	−2.15
Cyclopropane	0.0	13 [a]	−0.67
Cyclohexane	0.92	14 [a]	−0.66
Methylcyclopentane	0.93	15 [a]	−0.12
Dichloromethane	−0.11	16 [a]	−0.18
Trichloromethane	0.29	17 [a]	−1.15
1,1,1-Trichloroethane	0.4	18 [a]	−1.57
Trichloroethylene	0.34	19 [a]	−1.54
1,1,1-Trifluoro-2-chloroethane	0.08	20 [a]	−1.12
Halothane	0.35	21 [a]	−0.73
Teflurane	0.27	22 [a]	−0.27
Diethyl ether	0.0	23 [a]	−0.28
Divinyl ether	0.11	24 [a]	−0.46
Methoxyflurane	0.25	25 [a]	−0.24
Isoflurane	0.42	26 [a]	−0.02
Enflurane	0.24	27 [a]	0.69
Fluorohexene	0.13	28 [a]	0.44
Propanone	−0.15	29 [a]	0.14
Butanone	−0.08	30 [a]	0.22
Ethanol	−0.16	31 [a]	0
Propan-1-ol	−0.16	36 [a]	0.89
Propan-2-ol	−0.15	Y-G14	−0.3
2-Methylpropan-1-ol	−0.17	Y-G15	−0.06
Benzene	0.37	Y-G16	−0.42
Toluene	0.37	Y-G19	−1.3
Ethylbenzene	0.2	Y-G20	−1.4
p-Xylene	0.31	SKF 89124	−0.43
m-Xylene	0.29	SKF 101468	0.25
o-Xylene	0.37	Acetylsalicylic acid	−0.5
Cimetidin (1) [a]	−1.42	Valproic acid	−0.22
2 [a]	−0.04	Theophylline	−0.29
3 [a]	−2.0	Caffeine	−0.05
4 [a]	−1.3	Antipyrine	−0.1
Salicylic acid	−1.1	Diazepam	0.52
Acetaminophen	−0.31	Phenytoin	−0.04
Ibuprofen	−0.18	Hexobarbital	0.1

Table 18.1 (continued)

Compound	LogBB	Compound	LogBB
Codeine	0.55	Aminobarbital	0.04
Pentobarbital	0.12	Phenylbutazone	−0.52
Alprazolam	0.04	Aminopyrine	0
Indomethacin	−1.26	Desmethyldesipramine	1.06
Oxazepam	0.61	Bretazenil	−0.09
Hydroxyzine	0.39	Flumazenil	−0.29
Desipramine	1.2	RO19-4603	−0.25
Midazolam	0.36	Paraxanthine	0.06
Promazine	1.23	Quinidine	−0.46
Chlorpromazine	1.06	Salicyluric acid	−0.44
Trifluoperazin	1.44	Fluphenazine	1.51
Thioridazine	0.24	Haloperidol	1.34
32 [a]	−0.34	Mesoridazine	−0.36
33 [a]	−0.3	Sulforidazine	0.18
34 [a]	−1.34	Bromperidol	1.38
35 [a]	−1.82	Morthioridazine	0.75
Mianserin	0.99	Nor-1-chlorpromazine	1.37
Org4428	0.82	Nor-2-chlorpromazine	0.97
Org5222	1.03	Desmonomethylpromazine	0.59
Org16962	1.64	Desmethyldiazepam	0.5
Org13011	0.16	1-Hydroxymidazolam	−0.07
Org32104	0.52	4-Hydroxymidazolam	−0.3
Org30526	0.39	Triazolam	0.74
Mirtazapine	0.53	Clobazam	0.35
Tibolone	0.4	Flunitrazepam	0.06
Org34167	0.0	Desmethylcobazam	0.36
Risperidone	−0.02	Thiopental	−0.14
Risperidone-9-OH	−0.67	Methohexital	−0.06
Theobromine	−0.28	Didanosine	−1.3
Morphine	−0.16	Indinavir	−0.74
Propanolol	0.64	Nevirapine	0
Atenolol	−1.42	Zidovudine	−0.72

a) Chemical structures can be obtained from the authors on request.

Within the past decade, data from different sources have been collected and widely used for computational studies. However, when comparing the different sets used, it becomes evident that in total about 150 different compounds are present including also nondrug-like chemicals (Table 18.1) [10].

The most commonly used in vitro model is the Transwell system based on bovine brain microvascular endothelial cells (BBMEC). The system consists of a porous membrane support submerged in culture medium and is, therefore, characterized by a horizontal or vertical side-by-side diffusion system (Fig. 18.1). Co-culturing of these endothelial cells with primary astrocytes or the use of astrocyte-conditioned medium induces characteristics which are normally asso-

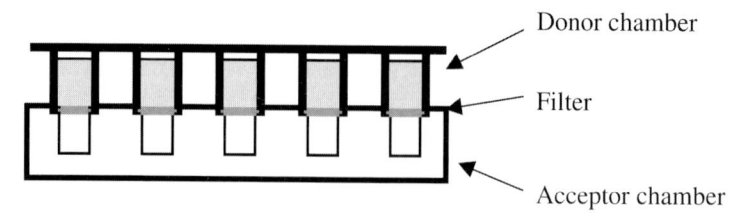

Fig. 18.1 Schematic representation of the Transwell system.

ciated with the BBB, such as the presence of tight junctions and polarized expression of transport proteins. This leads to good correlations between in vitro and in vivo brain permeation data.

However, there are several drawbacks which might be the reason for the lack of in silico studies utilizing permeability values based on data obtained from the Transwell system. First, most of the studies published are focused on technological and molecular biological issues. This implies that very often only small sets of structurally highly diverse molecules are investigated. The main question very often is to determine the tightness of the cell layer and to correlate the permeability of a given reference set of compounds to biological marker proteins. It normally does not address the pharmaceutical dimension of the problem, i.e. is there a statistically significant difference between structurally related compounds synthesized during a lead optimization program. Second, data published suffer from high inter-experimental variation which renders the permeability values rather noisy. This problem has been recently addressed by our group in order to generate data suitable for computational studies. In order to decrease inter-experimental variability, we simultaneously tested small groups of up to five compounds, including one reference compound (diazepam). Permeability values obtained were set in relation to the internal standard. This remarkably improved reproducibility and allowed the use of standardized logPE values as a dependent variable in regression analysis. The model obtained for a combined set of benzodiazepines and nonsteroidal antiinflammatory drugs showed a clear bilinear dependency of logPE values from calculated logP values (Fig. 18.2) [11]. This might become a standard method in our laboratory to generate BBB permeation data suitable for QSAR studies on homologous series of compounds.

Several authors proposed a third method for generating data sets, especially for the use of large databases of known drugs. The underlying assumption of this approach is that compounds active on the CNS (CNS$^+$) penetrate the BBB, whereas CNS$^-$ do not. While the first statement evidently is true, the second one is less reliable. Lack of CNS activity does not necessarily mean that a compound is not able to penetrate the BBB. It might also be the case that a CNS$^-$ compound is simply inactive against the respective molecular targets in the brain. This classification is thus suitable for developing predictive systems for CNS activity (yes/no), but cannot serve as basis for in silico BBB models.

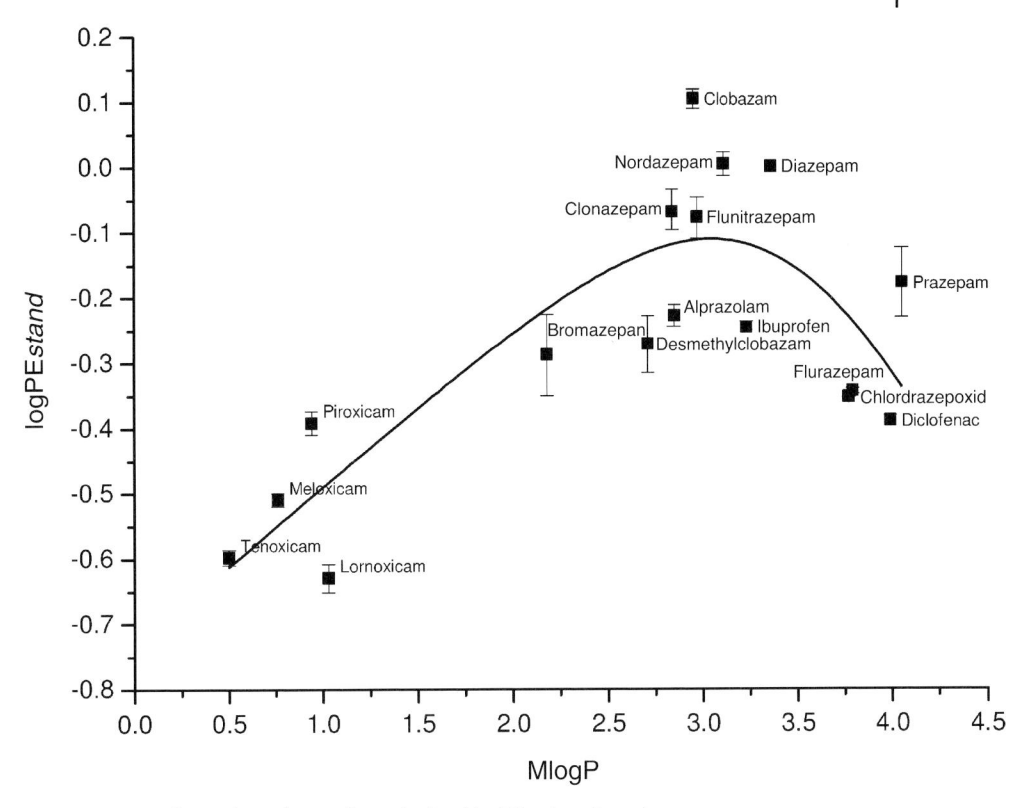

Fig. 18.2 Bilinear dependency of standardized logPE values from the calculated Moriguchi logP for a set of benzodiazepines and NSAIDs [11].

To summarize, the major problem with regard to available data sets is the low number of compounds with accurately measured logBB values. Considering the fact that, via ChemNavigator, a screening library of almost 20 000 000 commercially available, drug-like compounds is offered, models based on at maximum 150 molecules are clearly not able to serve the whole chemical space. Additionally, almost no attention has been drawn to carefully monitor the main permeation driving/limiting force, i.e. passive diffusion, active influx or active efflux. Thus, the data available mostly represent the combined, overall picture and are thus only of limited value for in silico models which try to consider both passive and active transport phenomena.

18.4
Computational Models

As outlined above, computational models have to consider two different categories of properties of the compounds under consideration. First, there are properties related to passive diffusion-controlled permeability, which are mainly influenced by the inherent physicochemical attributes of the compounds, such as logP, solubility and surface area (to mention just a few). For this, molecular descriptor-based methods have been used to generate predictive models. Second, we have to consider ligand-receptor, or in case of multispecific efflux transporters, ligand-antitarget interactions. These are normally targeted using pharmacophore-based approaches or classification systems. Obviously, a predictive in silico model suitable for both the lead identification and the lead optimization processes should include both categories. Although a broad, generic approach, which allows implementation of a single software tool for reliable prediction of BBB permeation taking into account passive and active phenomena, would be the optimum, until now this seems beyond reality.

18.5
Passive Diffusion

18.5.1
Regression Models

Although available data are rather limited, a number of in silico BBB models have been published in the literature. Most of them are based on classic regression equations. In the early 1980s, Levin published a correlation between the brain capillary permeability coefficient and $\log(P[MW]^{-1/2})$ for molecules with MW < 400 (Eq. (18.2)).

$$\log P_c = 0.4115 \log(P[MW]^{-1/2}) - 4.605 \tag{18.2}$$

$r = 0.91, n = 22$

Despite being one of the first QSAR studies on BBB permeation data, the most interesting items are the four outlyers adriamycine, epipodophyllotoxin, vincristine and bleomycine (Fig. 18.3). All three compounds showed lower permeability coefficients than expected from their $\log(P[MW]^{-1/2})$ values. The author tried to rationalize this behavior by proposing that the compounds due to their high lipophilicity may penetrate and distribute into – but not through – the brain capillary endothelia. Nowadays, we know that all these natural product toxins are excellent substrates for P-gp.

Later on, more advanced regression models were published using additional descriptors such as ΔlogP, lipoaffinity, polar and accessible surface area, polariz-

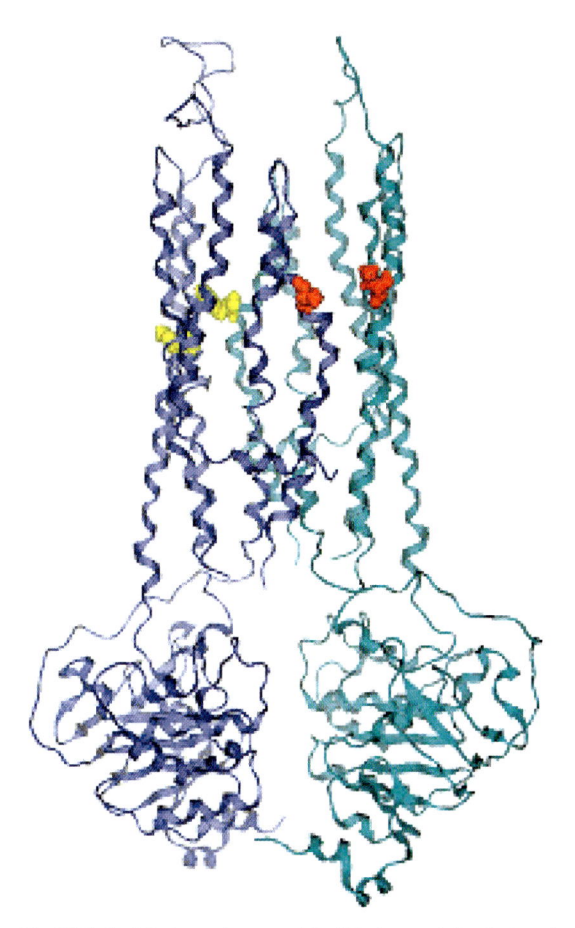

Fig. 18.3 Protein homology model of P-glycoprotein. Red and yellow: regions of highest labeling with propafenone-type photoaffinity ligands [53].

ability, free energy of solvation and H-bonding properties. Although the octanol/water partition coefficient plays a predominant role in many regression models, in some cases it shows only a weak correlation with logBB. Ter Laak et al. found that the brain permeability of a series of structurally diverse histamine H1 receptor antagonists was better explained by logD rather than by logP, which also takes account of the dissociation properties of the respective compounds. Kaliszan included the molecular weight as additional descriptor for molecular bulkiness (Eq. 18.3).

$$\log BB = 0.272 \Delta \log P - 0.00112 M_m - 0.088 \tag{18.3}$$

$n = 33, r^2 = 0.90, s = 0.126, F = 131.1$

Taking into account also descriptors related with molecular surface properties, both Kelder (Eq. 18.4) and Clark (Eq. 18.5) published equations using the polar surface area (PSA) as the only descriptor.

$$\log BB = 1.33 - 0.032\, PSA \tag{18.4}$$

$$n = 45, r^2 = 0.84, F = 229$$

$$\log BB = 0.55 - 0.016\, PSA \tag{18.5}$$

$$n = 55, r^2 = 0.71, s = 0.41, F = 128$$

However, the drawback of using PSA alone is the inability to distinguish the differences of hydrocarbon compounds. Thus, Clark combined both hydrophobicity and PSA, which improved the model (Eq. 18.6).

$$\log BB = 0.152\, ClogP - 0.148\, PSA + 0.139 \tag{18.6}$$

$$n = 55, r^2 = 0.79, s = 0.35, F = 95.8$$

More complex models include the dipolarity/polarizability parameters, calculation of free energy values and the solvent-accessible surface area (SASA) calculated via Monte Carlo simulations of compounds in water. This approach is based on the assumption that the solvation of compounds in water and a lipid phase might be accompanied by conformational changes, which will be more pronounced for large flexible molecules. In an ideal setting, one might perform a simulation of the diffusion of molecules through a lipid bilayer, applying molecular dynamics simulations. Unfortunately, the timescales of diffusion of small molecules are in the range of several microseconds, whereas MD simulations normally only span a few nanoseconds. Therefore, in an extension of the work of Jorgensen, Kaznessis et al. used Monte Carlo simulations of a set of 85 compounds in water to calculate diffusion-related properties, such as the SASA, the hydrophobic, hydrophilic and aromatic component of SASA, the dipole moment and both the Coulomb and Lenard-Jones energies between solutes and solvent. Although several highly predictive models were obtained ($r=0.97$), it has to be noted that computation times were rather high (1.6 CPU hours per compound on a 600 MHz Pentium PC).

Hutter used information derived from quantum chemical calculations for a model based on a training set of 90 compounds [12]. In total, 41 descriptors were calculated using the semiempirical VAMP program package. In addition to several atom counts, the number of ionizable groups, number of aromatic rings and number of H-bond donors and acceptors, a set of descriptors which account for the molecular shape and expansion of the respective molecules were also used (PCGA, PCGB, PCGC). These descriptors are derived from a principal component analysis (PCA) of the molecular geometry and are computed as the square root of the principal moments of the cartesian coordinates. The first principal component (PC) corresponds to the largest extension of the molecule, the second PC to the largest extension perpendicular to the first PC and the third PC to the largest extension perpendicular to the first and second PC, respectively. The magnitude of these dimensions is assumed to be related to the likelihood of penetrating the membrane. The final equation obtained had a r^2 value of 0.865 and contained several count-based descriptors (H-bond donors, halogens, NO_2, sulfur atoms, six-membered aromatic rings, rotable bonds) and shape descriptors PCGA and PCGC, as well as quantum chemically derived ones. These include values based on the molecular electrostatic potential and the covalent hydrogen bond basicity. Outlyers were shown to underlie the active transport processes. Computation time for the whole data set was about 3 h on a single CPU in a Compaq Alpha Server ES40 (667 MHz), which corresponds to an average of one molecule per minute.

Iyer et al. followed an even more demanding computational approach in predicting BBB partitioning using membrane interaction QSAR analysis (MI-QSAR) [13]. In MI-QSAR, structure-based design methodology is combined with classic ligand-based, descriptor-driven QSAR analysis to model chemically and structurally diverse compounds interacting with cellular membranes [14, 15]. In the present study, a training set of 56 compounds was used and molecular dynamics simulations were performed to determine the explicit interaction of each test compound with a model DMPC monolayer membrane model. This gave a series of solute-membrane interaction descriptors. Additionally, a series of dissolution and solvation solute descriptors (aqueous solvation free energy, 1-octanol solvation free energy, logP, cohesive packing energy of solute molecules, hypothetical crystal-melt transition temperature, hypothetical glass transition temperature of the solute) and a set of 18 general intramolecular solute descriptors, such as HOMO, LUMO, dipole moment, molecular volume, molecular surface area, MW and PSA, were also calculated and combined with the membrane-solute descriptors using multiple linear regression (MLR) analysis and genetic function approximation (GFA). Validation of the models was performed both using Y-scrambling and an external test set. Judged on predictive power q^2, the best model obtained included five descriptors (Eq. 18.7).

$$\log BB = 0.1591 \, \text{ClogP} - 0.0231 \, \text{PSA} - 0.0071 E_{MS}(\text{chg-hbd}) \\ + 0.0346 E_{SS}(\text{tor}) + 0.0075 \Delta E_{TT}(1-4) + 0.0156 \quad (18.7)$$

$n = 56, r^2 = 0.845, q^2 = 0.795$

E_{MS}(chg-hbd) represents the total intermolecular electrostatic and hydrogen bonding interaction energy between the solute and the DMPC monolayer. E_{SS}(tor) is the torsion energy of the solute for the solute located at the position corresponding to the lowest solute-membrane interaction energy state of the model system. ΔE_{TT}(1–4) is the change in the 1,4-nonbonded interaction energy of the system due to the uptake of the solute from free space to the position corresponding to the lowest solute-membrane interaction energy state of the model system. According to the logBB MI-QSAR models obtained, the authors proposed the following physicochemical factors influencing BBB permeation.

- The relative polarity of the solute, as represented by PSA and logP. In general, less polar, more lipophilic compounds partition more readily into the membrane.
- The strength of interaction (binding) of the solute with the membrane. The greater the binding, the higher the BBB partitioning.
- The conformational flexibility of the solute in the membrane and the conformational flexibility of the membrane-solute complex. Increasing flexibility corresponds to increasing logBB values.

Unfortunately no information on CPU time needed is given, so this very interesting approach cannot be judged on the basis of computational costs.

Lobell et al. followed two different lines, one suitable for a highly accurate prediction in a low throughput mode and another optimized for ultra-high-throughput processing [16]. For the first one, they calculated 34 descriptors, such as logP, molar refractivity (MR), desolvation energies in water and octanol, the total polar surface area, MW, number of rotable bonds, H-bond donors and acceptors, kappa indices, several indicator variables for charge and ten so-called shadow descriptors. These are a set of geometric descriptors that help to characterize the shape of molecules. Shadow descriptors are calculated by projecting the molecular surface on three mutually perpendicular planes, thus signifying the ratio of the largest to the smallest dimension. The coefficient of these descriptors in the final equation indicated that spherical shapes have a small advantage compared with rod-like shapes, with regard to transport across the BBB. To capture the charge influences, predicted pK_a values were used to classify the compounds depending on the pH range when these groups obtain a predominantly positive or negative charge. This is based on Seelig et al., who have already shown that compounds that contain an acidic group with $pK_a < 4$ or a basic group with $pK_a > 10$ do not cross the BBB by passive diffusion [17]. Thus, Lobell et al. defined six classes and compounds were either assigned 0 or 1 for each class. However, this was done manually which renders the method rather slow. For high-throughput in silico calculations, the C2-ADME logBB module implemented in the Cerius2 software package was used. Comparing the results with other established methods, such as those published by Lombardo, Feher, Clark, Kelder and the authors themselves, proved the accuracy of the models.

Hou and Xu applied a genetic algorithm (GA) to a large set of descriptors and published a model for a set of 96 structurally diverse compounds [18]. Taking into account the difficulty in selecting the appropriate descriptors for a problem as complex as permeation through the BBB, GAs might be the method of choice. GAs are expected to find a group of reliable QSAR or QSPR models from a large number of samples very efficiently. This method was first proposed by Rogers and Hopfinger [19] and Kubinyi [20]. Using 27 descriptors, 100 models with three descriptors and 100 models with four descriptors were obtained. The best model is shown in Eq. (18.8).

$$\log BB = 0.32\,\Delta\log P - 0.11\,\text{RotBonds} + 0.0024\,\text{Jurs-PNSA-2}$$
$$+ 0.35\,\text{RadofGyration} - 0.62 \qquad (18.8)$$

$n = 59,\ r = 0.87,\ q^2 = 0.84,\ sd = 0.41,\ F = 42.14$

In an extension of this work and taking into account the importance of logP for BBB models, the authors introduced SLOGP, a new parameter for lipophilicity. SLOGP is based on simple atomic addition and thus quickly and efficiently calculates lipohilicity values. The values obtained by SLOGP correspond well to those from other algorithms, such as CLOGP, ALOGP, ALOGP98 and HINT. The generation of several models via stepwise inclusion of additional descriptors, such as the highly charged polar surface area (HCPSA) and a parameter for molecular bulkiness (MW-360) yielded the highly significant Eq. (18.9):

$$\log BB = 0.197\,\text{SLOGP} - 0.0135\,\text{HCPSA} - 0.0140\,(\text{MW}-360) + 0.00845 \quad (18.9)$$

$n = 78,\ r = 0.876,\ s = 0.364,\ F = 81.5$

HCPSA reflects the polar surface area given by polar atoms with absolute partial charges larger than 0.1 as calculated according to Gasteiger. Use of (MW-a) introduces a special spline model for bulkiness. Thus, if the molecular weight (MW) is lower than a given value a, the term (MW-a) is set to 0. Otherwise the term is used as calculated. This regression with splines allows the incorporation of features that do not have a linear effect over their entire range. To determine the best value of a, a systematic search was used changing this value from 100 to 400 using a step of 10. The best equation was obtained for $a = 360$. The use of these three descriptors gives a meaningful physical picture of the molecular mechanisms involved in BBB permeation. First, hydrophobic molecules can permeate the blood-brain barrier more easily than hydrophilic ones. Second, the larger the polar surface area is, the more negatively this accounts for BBB permeation. However, this contribution is limited to those atoms with high charge densities (HCPSA). Third, large molecules also will show limited BBB permeation, but this bulk effect may become effective only when the MW is higher than 360. To improve efficiency and speed of calculation, the authors also reparameterized the SLOGP algorithm and developed a new set of parameters to calculate a topological highly charged polar surface area (HCTPSA). This allows the prediction of logBB values solely on basis of the topological structure of the molecules and thus renders the method a high-throughput logBB prediction method. Corresponding software packages are available on request from the authors.

A completely different, but also atom-based approach was published by Sun. In a search for a generic molecular descriptor system which allows the prediction of most of the relevant properties of a molecule (logP, solubility, logBB, intestinal absorption), the author developed an atom classification system. First, a primary classification tree based on experience and chemical intuition was implemented using the Daylight Toolkits. Then, the logP data set from Starlist was used to determine more accurately where to split and where to stop splitting the tree. The Starlist contains 10 974 compounds which span a logP range of 16 orders of magnitude and include chemical entities from simple alcohols to complex peptides. The atom types identified from the original classification tree served as molecular descriptors to generate the predictive model for logP calculation. Thorough analysis of the errors and standard deviations of the respective coefficients yielded clues for further modification of the classification tree. Repeated optimization cycles finally led to a final model for the calculation of logP values with a r^2 of 0.912. In total, 218 atom types were identified, comprising 88 types of C, seven types of H, 55 types of N, 31 types of O, eight types of halides, 24 types of S, six types of P and 26 correction factors (no aromatic rings, intramolecular hydrogen bond, etc.). For the BBB permeation model, the data set of Abraham (57 compounds) was used and a total of 94 different atom types were identified for this data set. Additionally, half of these atom types occurred only once or twice in the data set. This imbalance in the number of descriptors to number of cases made it difficult to properly calculate the contribution factors for each atom type. Thus, the predictive power obtained in leave-one-out cross-validation runs did not exceed 0.55. Also r^2 values for an external prediction set were rather low, but could be improved when increasing the structural diversity of the training set by the incorporation of additional compounds. The author thus concluded that "a better predictive model should be derived, once a larger data set of high quality becomes available".

A complete different set of descriptors was used recently by Rose and Hall, who published a model based on the electrotopological state indices. In the E state representation, the central feature is the intrinsic state term I_i, which encodes in an integrated fashion both electronic and topological attributes. Thus, E states somehow represent the electronic distribution across a molecule which forms the basis for both steric and electronic properties of the molecule. For a set of 102 structurally diverse compounds, use of the hydrogen E state index for hydrogen bond donors [$HS^T(HBd)$], the hydrogen E state index for aromatic CHs [$HS^T(arom)$] and the second-order difference valence molecular connectivity index ($d^2\chi^v$) gave a model with reasonable predictive power [$q^2 = 0.62$; Eq. (18.10)].

$$logBB = -0.202\,HS^T(HBd) + 0.00627\,[HS^T(arom)]^2 - 0.105\,(d^2\chi^v)^2 - 0.425$$

(18.10)

$$r^2 = 0.66,\ s = 0.45,\ F = 62.4,\ n = 102,\ q^2 = 0.62$$

Detailed analysis of the model indicates that molecules with large $HS^T(arom)$ values (presence of aromatic rings), small $HS^T(HBd)$ values (few and weak H-bond donors) and low $d^2\chi^v$ values (less branched molecules with few electronegative atoms) are able to penetrate the BBB. Despite cross-validation and the use of an external test set selected randomly from the data set available, an external prediction of the CNS$^\pm$ data set of Crivori et al. was also performed. The model correctly classified 27 out of 28 molecules. Furthermore, logBB values in a large database of drug-like molecules (20039) were computed to demonstrate the general applicability of the approach to the drug-like chemical space.

In their work, Pan et al. argued that a common mechanism of action, which is one of the most basic assumptions of QSAR analysis, may not be satisfied for ADMET data sets [21]. Thus, structurally diverse compounds might interact in different modes with the membrane. To overcome this problem, the authors divided the whole data set of 150 compounds into subsets based on 4D molecular similarity (4D-MS) measures. Subsequently, predictive QSAR models were constructed for each cluster subset using the already described MI-QSAR descriptors derived from membrane interaction simulations. 4D-MS developed by Hopfinger and co-workers also takes the thermodynamic distribution of different conformers of a given molecule into account [22]. Thus, a conformation ensemble profile of each member of a set of molecules is generated using molecular dynamics simulations. This is followed by construction of the main distance-dependent matrix (MDDM) for each pair of interaction pharmacophore elements (IPE) of each molecule. Subsequent PCA transforms the MDDM into a set of eigenvalues which are normalized and sorted. This set of normalized eigenvalues is used as a fingerprint for each molecule and serves as basis for the clustering of the compounds into structurally related subsets. In the current study, the set of 150 compounds was divided into three subsets. The optimized QSAR model for the complete data set is similar to the model reported by Clark and includes ClogP and TPSA as descriptors (Eq. 18.11).

$$\log BB = 0.20\,\text{ClogP} - 0.01\,\text{TPSA} + 0.064 \tag{18.11}$$

$n = 150$, $r^2 = 0.69$, $q^2 = 0.60$

Subset 1 includes 37 compounds which are mainly simple, nondrug-like molecules with less than two heteroatoms (excluding halogens). Dividing the subset into a training and test set and performing a multiple linear regression analysis gave Eq. (18.12), in which Echg represents the electrostatic interaction energy for the solute-membrane complex.

$$\log BB = 0.26\,\text{ClogP} + 0.00077\,\text{Echg} + 0.26 \tag{18.12}$$

$n = 24$, $r^2 = 0.85$, $q^2 = 0.83$

For subset 2, which contains 88 compounds (63 training, 25 test), in addition to PSA and ClogP, the two topological descriptors S_sF and Kappa3 also showed statistically significant contributions Eq. (18.13). S_sF is the electrotopological state index for a fluorine atom with one single bond and Kappa3 is generally considered an index which reflects the molecular shape of a molecule.

$$\log BB = 0.25\,\text{ClogP} - 0.01\,\text{PSA} - 0.025\,\text{S\_sF} - 0.11\,\text{Kappa3} + 0.66 \qquad (18.13)$$

$n = 63$, $r^2 = 0.69$, $q^2 = 0.66$

Interestingly, in subset 3, which contains 25 compounds (17 training, 8 test), ClogP is not significant in explaining logBB values (Eq. 18.14). Instead, the two descriptors Ecoh and ΔE_{TT}(vdw) are found in its place. Ecoh is the cohesive energy of the solute, which reflects how well a compound self-aggregates. ΔE_{TT}(vdw) is an explicit MI-QSAR descriptor of the van der Waals energy difference between the solute-membrane complex and the free states of the solute molecule and the membrane monolayer.

$$\log BB = 0.069\,E\text{coh} - 0.095\,\Delta E_{TT}(\text{vdw}) - 0.031\,\text{PSA} - 0.58 \qquad (18.14)$$

$n = 17$, $r^2 = 0.80$, $q^2 = 0.72$

To better evaluate the advantage of dividing the whole data set into three subsets, a QSAR model of the combined training sets was also performed. The optimum model for this overall training set is shown in Eq. (18.15):

$$\log BB = 0.19\,\text{ClogP} - 0.011\,\text{TPSA} + 0.05 \qquad (18.15)$$

$n = 104$, $r^2 = 0.69$, $q^2 = 0.64$

As exemplified by the statistical parameters, each of the three models for the subsets showed a higher predictive power (q^2) than the overall model. A more detailed analysis showed that the general model captures almost the same information as submodels 1 and 2, whereas new, additional information is provided by model 3. Overall, it appears to be advantageous to build individual QSARs for subsets of large data sets, because "minority" descriptors which are only descriptive for a small subset of compounds do not survive in a general model built for the whole data set. This might be an excellent approach to deal with the different mechanisms influencing BBB permeation, such as passive diffusion and active influx/efflux.

Last but not least the recently introduced VSA descriptors have to be mentioned. These descriptors are based on the atomic contributions of logP (10), molar refractivity (8) and partial charge (14) to the van der Waals surface area. In total this yields 32 descriptors which were used to build models for boiling point, vapor pressure, free energy of solvation in water, solubility in water, thrombin/trypsin/factor Xa activity and BBB permeability [23]. For the latter, the data set of Luco was used. Principal component regression analysis led to a final model based on 15 descriptors, which showed a remarkably high r^2 value (0.83).

18.5.2
Classification Systems

In the early stages of drug discovery, there is no need for highly sophisticated prediction tools for logBB values. These are definitely advantageous in the lead optimization process, where the compound of interest has already a given pharmacological activity and BBB permeation. In the discovery of new lead compounds, high-throughput in silico tools are favorable which give early alerts to the medicinal chemist, thus guiding both the design of combinatorial libraries and the acquisition of compounds for the HTS library. For this purpose, simple descriptors such as atom counts or easy-to-calculate physicochemical parameters seem to be preferred over methods which require the calculation of interaction fields or quantum chemical parameters. Undoubtedly the pioneering work of Lipinski paved the way for rule-based methods for assessing drug-likeness, lead-likeness or even hit-likeness. Thus, in analogy to the classic rule of five, several rule-based methods for BBB permeation have been published. Norinder and Haeberlein proposed a very simple system based on only two rules:

1. If the number of N + O atoms in a molecule is five or less, then it has a high chance of entering the brain.
2. If ClogP–(N + O) > 0, then logBB is positive.

These rules are based on the fact that there is a strong correlation between the number of nitrogens and oxygens and the respective PSA value. The polar surface area has already been recognized as a versatile descriptor by both van de Waterbeemd and Kelder, using an upper limit for PSA to assess the likelihood of brain uptake. Analyzing 125 marketed drugs, van de Waterbeemd proposed upper limits for PSA (90) and molecular weight (450). Extending the training set to compounds being at least in clinical phase II studies, but not on the market yet, Kelder et al. found a slightly lower limit for PSA (60–70).

For classification purposes, data sets used are very often clustered into CNS active and nonactive drugs. This leads to a remarkable increase of the number of compounds available for training and testing. However, one has to bear in mind that it is not necessarily true that nonCNS drugs are not entering the brain. While CNS^+ automatically includes BBB^+, CNS^- might also be BBB^+. Nevertheless, several authors followed this approach and several interesting models have been published, mainly using artificial neural networks (ANNs) for classification. ANNs are robust to noise and, once trained, allow fast and accurate predictions. Thus, both Sadowski and Kubinyi [24] and Ajay et al. [25] presented networks which are able to distinguish between drugs and nondrugs. Molecules are encoded using Ghose-Crippen descriptors and the networks are used for in silico screening of combinatorial libraries prior to synthesis and for analysis of commercially available libraries. An identical approach was applied by Ajay et al. to distinguish between CNS and nonCNS drugs [26]. The authors used a Bayesian neural network, which is known to be robust to noise introduced by misclassifications. This is of utmost importance, because the assignment whether a given compound is regarded as CNS$^+$ or CNS$^-$ was made solely on the basis of the therapeutic use of the compound. As outlined above, this approach may lead to a lot of misclassifications in the CNS$^-$ group. Additionally, CNS$^-$ does not necessarily mean BBB$^-$, so this network should not be used for predicting BBB permeation behavior. However, the network correctly predicted 80% of a test set comprising 13 000 CNS$^+$ and 53 000 CNS$^-$ compounds from the databases used for verification.

Recently, Engkvist et al. presented an in silico ADMET prediction tool based on substructural analysis [27]. Substructural analysis has a long history and is based on the rather simple principle that the pharmacokinetic properties of a molecule depend on its chemical structure, i.e. atom types and topological information. Thus, it should be possible to assign a molecule with a specific fragment composition to a certain cluster (i.e. CNS$^+$) by analyzing the distribution of molecular fragments within a large data set with known CNS activities. For training, the authors divided the World Drug Index (WDI) into three groups: (1) drugs that must pass the BBB for their action (psychosedatives, psychostimulants, antidepressants), (2) drugs that might pass the BBB (antiemetics, antiparkinsonians) and (3) drugs that most likely do not pass the BBB (cytostatics, antibiotics, antihypertensives). Class (1) contained 3678 molecules and was designated as CNS$^+$, class (2) was omitted and, out of the 50 025 molecules from class (3), 5000 were randomly selected to serve as CNS$^-$ compounds. Models obtained with the SUBSTRUCT software package were able to separate CNS active and nonactive compounds with approximately 80% accuracy. Analysis of the model showed that protonated nitrogens, aromatic rings, chlorine and fluorine are more frequent among CNS active compounds, while the opposite is true for oxygen-containing compounds. In comparison to an ANN, the ANN was faster (2 s for 1000 molecules) than SUBSTRUCT (18 s) in classifying unknown compounds, but did not allow to identify the crucial features for separating CNS$^+$ and CNS$^-$ compounds.

Adenot and Lahana used discriminant analysis (DA) and PLS-DA to develop a model which is able to filter CNS drugs from large virtual libraries, taking into account simultaneously both passive diffusion and efflux transport components [28]. After carefully analyzing the WDI, they ended up with a heterogenous CNS library of approximately 1700 compounds, including 1336 BBB$^+$ drugs, 259 BBB$^-$ drugs and 91 P-gp substrates. Out of a large set of descriptors combining molecular properties, surface areas, electronic parameters and topology, DA identified a small set of simple descriptors as being sufficient to evaluate BBB permeation in most cases. Thus, the number of heteroatoms is sufficient to predict BBB permeation with a high rate of well classified compounds (92%). All compounds were characterized by both a passive diffusion component and a P-gp efflux component. Combining these two components allowed mapping of the CNS compounds. This mapping clearly delineates three distinct areas, i.e. P-gp$^+$ drugs, BBB$^+$ drugs and BBB$^-$ drugs. Interestingly, several CNS drugs, like amitryptiline, chlorpromazine, morphine and disulfiram, are also P-gp substrates in vitro, but have been predicted with low P-gp activity. This is somewhat contradictory and indicates the complexity of the in vivo system. Additionally it has to be stressed that, in the case of P-gp substrates, properties are difficult to measure. Using inside-out membrane vesicles from P-gp-expressing cells Schmid et al. could demonstrate that some propafenone-type P-gp inhibitors stimulate the ATPase activity of P-gp [29]. This indicates that inhibitors might be substrates which block the pump via a rapid rediffusion process. In this case, no net transport is observed and the pump is kept busy, which gives the macroscopic picture of being blocked.

With the armory of supervised machine learning approaches, support vector machines (SVMs) have been introduced as a powerful, theoretically well founded algorithm, capable of dealing with large, high-dimensional, nonlinear classification problems [30]. A particularly important feature of SVMs is that they explicitly rely on statistical learning theory [31] and thus avoid overfitting. The key concept is the structural risk minimization principle proposed by Vapnik and Chervonenkis [32]. SVMs have consistently achieved a performance competitive with ANNs, especially in the field of ADME property classification. They have been used for compound classification in several stages of the drug discovery process, including the drug/nondrug problem [33] and prediction of BBB permeation. For the latter, the data set implemented in VolSurf was used, which is the result of in vivo studies and contains 337 BBB^+ and 139 BBB^- compounds [34]. Applying the 72 VolSurf descriptors, a SVM based on a radial basis function outperformed ANNs as well as a C5.0 decision tree and a nearest-neighbor classifier, showing an overall accuracy of 0.87 [35].

18.6
Field-Based Methods

Field-based methods are gaining an increasing interest in the in silico modeling of complex phenomena. In the VolSurf approach recently introduced by the group of Cruciani, molecular interaction fields computed by GRID [36] are used for calculation of a set of descriptors. Thus, compounds are placed in a 3D grid with a given distance between grid points and so-called probes are placed on the grid points. Then the steric, electrostatic, hydrophobic, H-bond acceptor and H-bond donor interaction energy between the probe and the compound is calculated. The probes are designed to mimic protein-ligand interaction forces and include, e.g., H^+, carbonyl, DRY (methane), aromatic ring and tyrosine-OH. This interaction energy matrix is subsequently used to extract a set of 72 descriptors. These so-called VolSurf descriptors refer to molecular size and shape, the size and shape of hydrophobic and hydrophilic regions, hydrogen bonding, amphiphilic moments and critical packing [37]. These descriptors have been shown to be very useful for the modeling and prediction of pharmacokinetic properties [38]. For the BBB model, the authors used the water, DRY and carbonyl oxygen probes for calculating the interaction energies of 110 compounds. Principal component analysis led to a model which was able to separate BBB^+

from BBB⁻ compounds quite well and showed correct classifications of 90% of the BBB⁺ and 65% of the BBB⁻ compounds in an external test set of 120 compounds. In a further refinement of the model, the combined data set (229 compounds) was used and a categorial score of +1 was given to BBB⁺ compounds and –1 to BBB⁻ and BBB± compounds. Subsequent PLS analysis gave a model which distinguishes well between BBB⁺ and BBB⁻ compounds [37]. The coefficient plot revealed that descriptors for hydrophilic regions, capacity factors, H-bonding and polar water accessible surface area are inversely correlated with BBB permeation, whereas those encoding hydrophobic interactions are directly correlated with BBB permeation.

In the course of our studies, we projected both a set of benzodiazepines and a series of polyamines on the pre-defined BBB model. The benzodiazepines are all classified as BBB⁺, which is quite obviously in accordance to their pharmacological profile as CNS drugs. However, the plot of the polyamines shows that they cover a new chemical space (Fig. 18.4). The same was in part true for a set of glycine-antagonists we are currently testing on our in vitro BBB model. Thus, the results have to be taken with caution and an extension of the training set with focus on a higher structural diversity is highly recommended. As mentioned already several times, 100–200 molecules are by far too few to allow in silico models with general predictivity.

In a recent paper, the VolSurf approach was extended to yield a quantitative model based on logBB values of 83 compounds [39]. A four-component PLS model was obtained which, after removal of four outlyers, showed a q^2 value of 0.65. Analyzing the contribution of each descriptor the following conclusions were drawn:

1. Descriptors of polarity (hydrophilic regions, capacity factors) are inversely correlated with logBB values.
2. Descriptors for hydrophobic interactions are directly correlated to BBB permeation.
3. Descriptors for size and shape have no pronounced effect on brain uptake.

Although the VolSurf approach is a very powerful tool not only for predicting BBB permeation behavior, but also for modeling ADME parameters, several drawbacks have to be considered. First, VolSurf is a 3D-QSAR method, which implicates that it is conformation-dependent. Although it has been shown that the influence of different conformations within, e.g., homologous series of molecules is relatively low, we observed deviations in q^2 values in the range of 0.1 when using different methods for the generation of 3D structures from 2D-sdf files. Second, it is relatively time-consuming and calculation of the descriptors for a set of 10 000 molecules needs about 2 days on a Pentium 2 MHz Linux PC. Third, the descriptors derived are rather abstract and information on how an optimal molecule should look are difficult to retrieve. However, VolSurf represents an interesting approach to derive predictive models for complex types of biological phenomena and models for P-gp, cytochromes and the human ether-a-go-go-related gene potassium channel (hERG) are on the way.

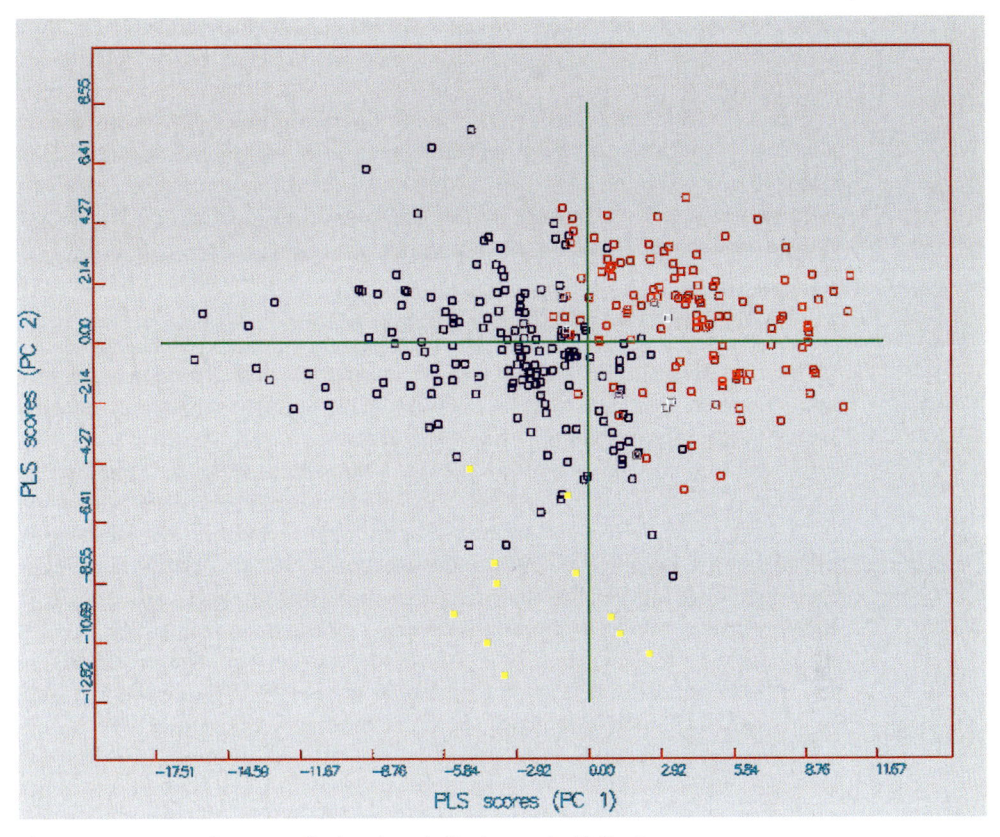

Fig. 18.4 Projection of a series of polyamines (yellow) onto the VolSurf model provided by Tripos (www.tripos.com). Red: BBB$^+$, blue: BBB$^-$.

18.7
Active Transport

In vivo, transcellular transport of compounds into the brain is mediated either via passive diffusion or by carrier transport and receptor transcytosis. With the increasing knowledge on the transport processes involved in brain uptake, a series of efflux transport proteins have also been identified, such as the multidrug transporter P-glycoprotein (P-gp) and the organic anion transport protein (OATP) family of transporters. These proteins function as gate-keepers, thus preventing the entry of toxic compounds into the brain [40]. Most of them show a broad substrate specificity coupled with a high efficiency. Undoubtedly, active transport is an important functional characteristic and in silico prediction methods have to take this into account. However, due to its original discovery as a main contributor to multiple drug resistance in tumors, research on P-gp focused on the development of inhibitors rather than the prediction of substrate properties [41–43]. Nowadays, the role of P-gp in ADME and especially brain uptake [44] is well understood and thus P-gp is increasingly considered as an antitarget [45–47].

Within the past few years, several models have been published for the prediction of P-gp substrate properties. Didziapetris et al. developed a classification system based on simple descriptors. Based on a set of 220 compounds with data from polarized transport across MDR1 transfected cell monolayers, the authors derived the "rule of fours" for a crude estimation of P-gp substrates [48]:

1. substrates: (N + O) ≥8, MW >400, acid pK_a >4
2. nonsubstrates: (N + O) ≤4, MW <400, base pK_a <8

These rules are based on the compound size, H-bond accepting properties and ionization; and they support the view that P-gp functioning can be compared to a complex system with fuzzy specificity.

Gombar and colleagues built a QSAR computational model that predicts the outcome of their inhouse monolayer P-gp efflux assay. Based on a training set of 95 compounds, a two-group linear discriminant model has been developed, which shows a sensitivity of 100% and a specificity of 90.6%. The 27 descriptors used mainly account for the ability to partition into membranes, molecular bulk and the counts and electrotopological values of certain hydrides. Within this work, several simple relationships have also been derived. The most striking one is the relationship between the molecular E state (MolES) and the P-gp substrate classifier (0 or 1). Among the 95 training set molecules, those with MolES >110 were predominantly P-gp substrates (95.0%) and those with MolES <49 were nonsubstrates (84.6%). This "Gombar-Polli rule" may be a rapid way to initially screen large combinatorial and virtual compound libraries [49].

The challenging task of predicting P-gp substrates also served as a model case to prove the accuracy of support vector machines. In total, 159 descriptors comprising 18 descriptors in the class of simple molecular properties, 28 descriptors for molecular connectivity and shape, 84 electrotopological state descriptors, 13 descriptors in the class of quantum chemical properties and 16 descriptors for geometrical properties were used as the input vector [50]. The SVM model obtained gave a prediction accuracy of 81% for P-gp substrates and 79% for nonsubstrates. This is slightly better than the performance obtained with k-nearest neighbor, ANNs and a C4.5 decision tree.

Extending their CATALYST pharmacophore model for P-gp inhibitors, Ekins et al. generated a pharmacophore model for P-gp substrates. Using a set of 16 inhibitors of verapamil-binding to P-gp, the authors generated a pharmacophore that consisted of one hydrogen bond acceptor, one ring aromatic feature and two hydrophobes [51]. This model correctly predicted the rank order of the four data sets used for the inhibitor models and also fit the three substrate probes verapamil, vinblastine and digoxin. This indicates the presence of partially overlapping binding sites within P-gp.

Very recently, Pleban et al. used a combined photoaffinity labeling-protein homology modeling approach to generate a 3D model of P-gp. Briefly, a series

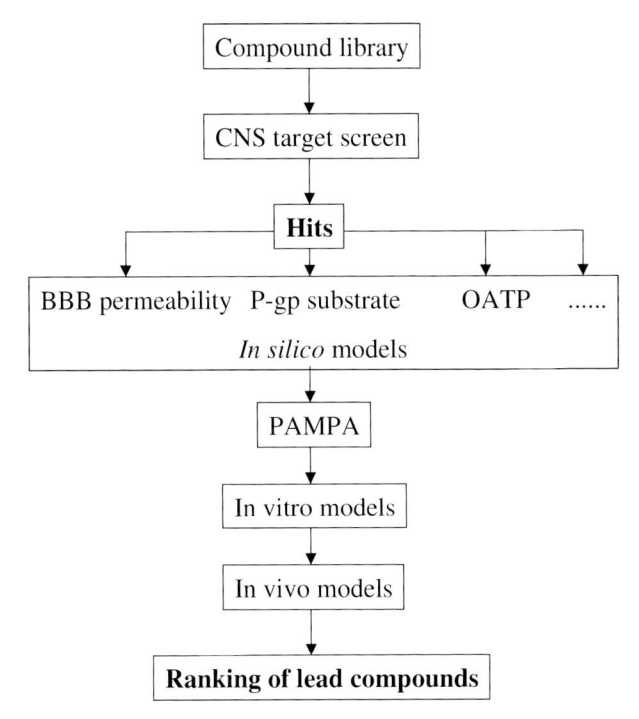

```
        Compound library
              │
              ▼
        CNS target screen
              │
              ▼
            Hits
    ┌─────────┼─────────┐
    ▼         ▼         ▼
BBB permeability  P-gp substrate   OATP   ......
              In silico models
              │
              ▼
            PAMPA
              │
              ▼
         In vitro models
              │
              ▼
         In vivo models
              │
              ▼
       Ranking of lead compounds
```

Fig. 18.5 Proposed workflow for combined BBB permeation models. (Modified from Ref. [54]).

of propafenone-type benzophenones were synthesized and used as photoaffinity probes to map the binding regions of P-gp. After UV irradiation in the presence of the benzophenone ligands, the protein was isolated and digested. Subsequent MALDI-TOF mass spectrometry revealed a series of fragments which were covalently labeled by the photo probes [52]. Generation of a protein homology model on the basis of the recently published X-ray structure of the ABC transporter MsbA from *Vibrio cholerae* and projection of the regions with the highest labeling frequency showed two distinct regions involved in ligand binding. Labeling is predicted by the model to predominantly occur at the two transmembrane domain/transmembrane domain interfaces formed between the amino- and carboxy-terminal halves of P-gp. These interfaces are formed by transmembrane (TM) helices 3 and 11 on the one hand and TM segments 5 and 8 on the other hand (Fig. 18.3) [53]. After subsequent molecular dynamic simulations, these model may serve as a versatile tool for docking experiments and target-based in silico screening of compound libraries.

18.8
Conclusions and Future Directions

Although great progress has been made in the development of in vivo and in vitro models for BBB permeation, their throughput is still rather limited. In silico methods can help to establish predictive rules or equations for high-throughput permeability prediction as a first screen in the drug discovery process. However, with the increasing knowledge of the numerous active transport processes involved, it is becoming evident that models based on passive diffusion alone are not suitable to cover the whole drug-like chemical space. First, hybrid (in vitro/in silico) models capable of predicting both passive and transporter-mediated contributions appeared in the literature and several models for predicting P-gp substrate properties have been described. As outlined in a recent technological report, a combination of techniques is recommended to cover the whole range of drug entry and efflux mechanisms. Combining serial and parallel screening processes may guide the selection of the right hits and facilitate the ranking of lead compounds (Fig. 18.5) [54]. The outcome of an expert system which combines different models undoubtedly will be greater than the sum of its parts.

Acknowledgment

We gratefully acknowledge financial support by the Austrian Science Fund, grant P14582-N03.

References

1 C.A. Lipinski, F. Lombardo, B.W. Dominy, P.J. Feeney **1997**, *Adv. Drug Delivery Rev.* 23, 3-25.
2 J. Sadowski, H. Kubinyi **1998**, *J. Med. Chem.* 41, 3325–3329.
3 A. Ajay, W.P. Walters, M.A. Murcko **1998**, *J. Med. Chem.* 41, 3314–3324.
4 M. Wagener, J.V. van Geerestein **2000**, *J. Chem. Inform. Comput. Sci.* 40, 280–292.
5 E. Byvatov, U. Fechner, J. Sadowski, G. Schneider **2003**, *J. Chem. Inform. Comput. Sci.* 43, 1882–1889.
6 J.F. Lowrie, R.K. Delisle, D.W. Hobbs, D.J. Diller **2004**, *Combin. Chem. High Throughput Screen.* 7, 495–510.
7 R.W. de Simone, K.S. Currie, S.A. Mitchell, J.W. Darrow, D.A. Pippin **2004**, *Combin. Chem. High Throughput Screen.* 7, 473–493.
8 S. Ekins **2004**, *Drug Discov. Today* 9, 276–285.
9 G.F. Ecker, C.R. Noe **2004**, *Curr. Med. Chem.* 11, 1617–1628.
10 D. Pan, M. Iyer, J. Liu, Y. Li, A.J. Hopfinger **2004**, *J. Chem. Inform. Comput. Sci.* 44, 2083.
11 R. Lauer, W. Neuhaus, J. Kainz, G.F. Ecker, C.R. Noe **2005**, submitted.

12 M. C. Hutter **2003**, *J. Comput. Aided Mol. Des.* 17, 415–433.
13 M. Iyer, R. Mishra, Y. Han, A. J. Hopfinger **2002**, *Pharm. Res.* 19, 1611–1621.
14 A. S. Kulkarni, A. J. Hopfinger **1999**, *Pharm. Res.* 16, 1244–1252.
15 A. Kulkarni, Y. Han, A. J. Hopfinger **2002**, *J. Chem. Inform. Comput. Sci.* 42, 331–342.
16 M. Lobell, L. Molnar, G. M. Keserue **2003**, *J. Pharm. Sci.* 92, 360–370.
17 H. Fischer, R. Gottschlich, A. Seelig **1998**, *J. Membr. Biol.* 165, 201–211.
18 T. Hou, X. Xu **2002**, *J. Mod. Model.* 8, 337–349.
19 D. Rogers, A. J. Hopfinger **1994**, *J. Chem. Inf. Comp. Sci.* 34, 854–866.
20 H. Kubinyi **1994**, *Quant. Struct. Act. Rel.* 13, 285–294.
21 D. Pan, M. Iyer, J. Liu, Y. Li, A. J. Hopfinger **2004**, *J. Chem. Inform. Comput. Sci.* 44, 2083–2098.
22 J. S. Duca, A. J. Hopfinger **2001**, *J. Chem. Inform. Comput. Sci.* 41, 1367–1387.
23 P. Labute **2000**, *J. Mol. Graph. Model.* 18, 464.
24 J. Sadowski, H. Kubinyi **1998**, *J. Med. Chem.* 41, 3325.
25 A. Ajay, W. P. Walters, M. A. Murcko **1998**, *J. Med. Chem.* 41, 3314.
26 A. Ajay, G. W. Bemis, M. A. Murcko **1999**, *J. Med. Chem.* 42, 4942.
27 O. Engkvist, P. Wrede, U. Rester **2003**, *J. Chem. Inform. Comput. Sci.* 43, 155.
28 M. Adenot, R. Lahana **2004**, *J. Chem. Inform. Comput. Sci.* 44, 239.
29 D. Schmid, G. Ecker, S. Kopp, M. Hitzler, P. Chiba **1999**, *Biochem. Pharmacol.* 58, 1448.
30 V. V. Zernov, K. V. Balakin, A. A. Ivaschenko, N. P. Savchuk, I. V. Pletnev **2003**, *J. Chem. Inf. Comput. Sci.* 43, 2048.
31 V. Vapnik **1998**, *Statistical Learning Theory*, Wiley, New York.
32 V. Vapnik, A. Chervonenkis **1974**, *Autom. Remote Control* 8, 9.
33 E. Byvatov, U. Fechner, J. Sadowski, G. Schneider **2003**, *J. Chem. Inform. Comput. Sci.* 43, 1882.
34 P. Crivori, G. Cruciani, P. A. Carrupt, B. Testa **2000**, *J. Med. Chem.* 43, 2204.
35 M. W. B. Trotter, S. B. Holden **2003**, *QSAR Comb. Sci.* 22, 533.
36 P. J. Goodford **1985**, *J. Med. Chem.* 28, 849.
37 G. Cruciani, P. Crivori, P. A. Carrupt, B. Testa **2000**, *J. Mol. Struct.* 503, 17.
38 G. Cruciani, M. Pastor, W. Guba **2000**, *Eur. J. Pharm. Sci.* 11[Suppl. 2], S29.
39 F. Ooms, P. Weber, P. A. Carrupt, B. Testa **2002**, *Biochim. Biophys. Acta* 1587, 118.
40 A. H. Schinkel **1999**, *Adv. Drug Deliv. Rev.* 36, 179.
41 M. Wiese, I. K. Pajeva **2001**, *Curr. Med. Chem.* 8, 685.
42 C. Avenado, J. C. Menendez **2002**, *Curr. Med. Chem.* 9, 159.
43 K. Pleban, G. F. Ecker **2005**, *Mini Rev. Med. Chem.* 5, 153.
44 T. Terasaki, K. Hosoya **1999**, *Adv. Drug Deliv. Rev.* 36, 195.
45 S. Ekins **2004**, *Drug Discov. Today* 9, 276.
46 M. Recanatini, G. Bottegoni, A. Cavalli **2004**, *Drug Discov. Today Technol.* 1, 209.
47 G. F. Ecker **2005**, *Chem. Today* (in press).
48 R. Didziapetris, P. Japertas, A. Avdeef, A. Petrauskas **2003**, *J. Drug Target.* 11, 391.

49 V. K. Gombar, J. W. Polli, J. E. Humphreys, S. A. Wring, C. S. Serabijt-Singh **2004**, *J. Pharm. Sci.* 93, 957.
50 Y. Xue, C. W. Yap, L. Z. Sun, Z. W. Cao, J. F. Wang, Y. Z. Chen **2004**, *J. Chem. Inform. Comput. Sci.* 44, 1497.
51 S. Ekins, R. B. Kim, B. F. Leake, A. H. Dantzig, E. G. Schuetz, L.-B. Lan, K. Yasuda, R. L. Shepard, M. A. Winter, J. D. Schuetz, J. H. Wikel, S. A. Wrighton **2002**, *Mol. Pharmacol.* 61, 974.
52 G. Ecker, E. Csaszar, S. Kopp, B. Plagens, W. Holzer, W. Ernst, P. Chiba **2002**, *Mol. Pharmacol.* 61, 637.
53 K. Pleban, S. Kopp, E. Csaszar, M. Peer, T. Hrebicek, A. Rizzi, G. F. Ecker, P. Chiba **2005**, *Mol. Pharmacol.* 67, 365.
54 N. J. Abbott **2004**, *Drug Discov. Today Technol.* 1, 407.